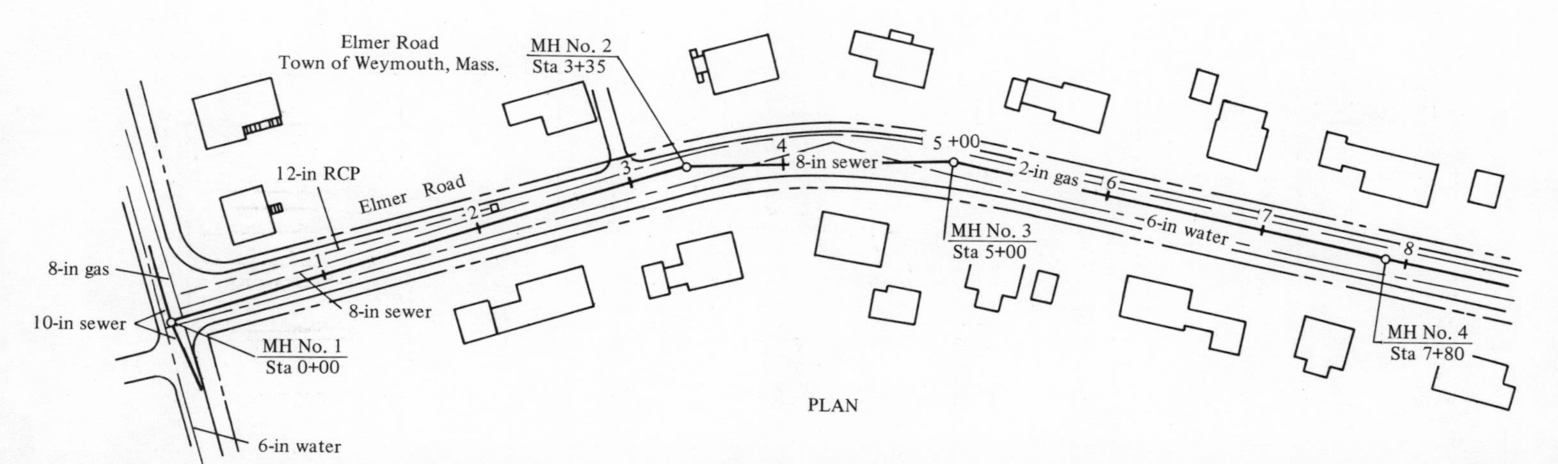

Elmer Road
Town of Weymouth, Mass.
MH No. 2
Sta 3+35
12-in RCP
Elmer Road
8-in gas
10-in sewer
8-in sewer
MH No. 1
Sta 0+00
6-in water
8-in sewer
5 +00
2-in gas
6-in water
MH No. 3
Sta 5+00
MH No. 4
Sta 7+80
PLAN

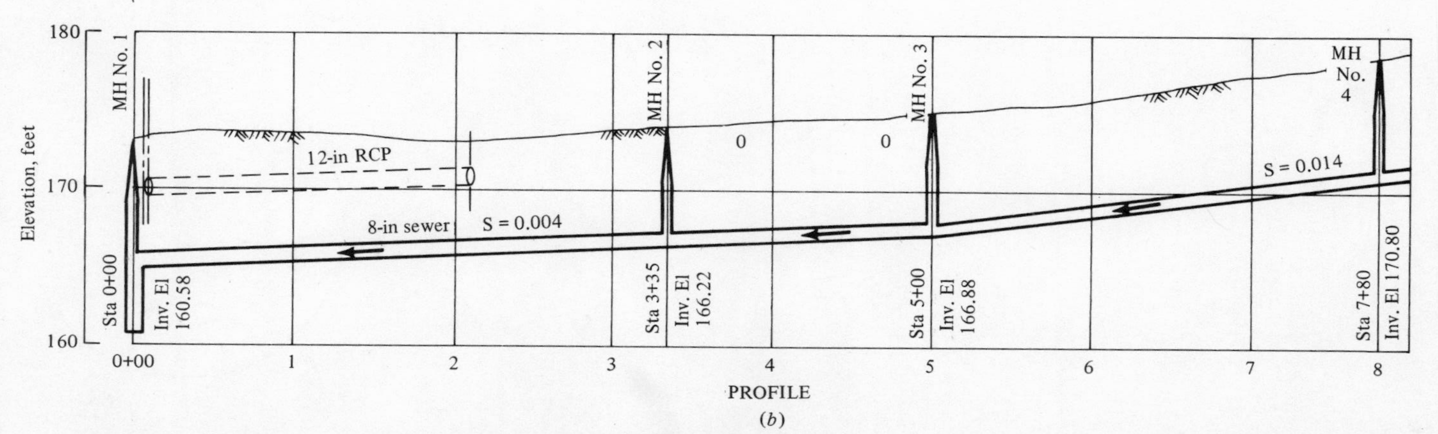

Elevation, feet
180
170
160
MH No. 1
MH No. 2
MH No. 3
MH No. 4
12-in RCP
8-in sewer
S = 0.004
S = 0.014
Sta 0+00
Inv. El 160.58
Sta 3+35
Inv. El 166.22
Sta 5+00
Inv. El 166.88
Sta 7+80
Inv. El 170.80
0+00
1
2
3
4
5
6
7
8

PROFILE
(*b*)

WASTEWATER ENGINEERING: COLLECTION AND PUMPING OF WASTEWATER

McGraw-Hill Series in Water Resources and Environmental Engineering

Ven Te Chow, Rolf Eliassen, Paul H. King, and Ray K. Linsley
Consulting Editors

Bailey and Ollis: *Biochemical Engineering Fundamentals*
Bockrath: *Environmental Law for Engineers, Scientists, and Managers*
Bouwer: *Groundwater Hydrology*
Canter: *Environmental Impact Assessment*
Chanlett: *Environmental Protection*
Gaudy and Gaudy: *Microbiology for Environmental Scientists and Engineers*
Graf: *Hydraulics of Sediment Transport*
Haimes: *Hierarchical Analysis of Water Resources Systems: Modelling and Optimization of Large-Scale Systems*
Hall and Dracup: *Water Resources Systems Engineering*
James and Lee: *Economics of Water Resources Planning*
Linsley and Franzini: *Water Resources Engineering*
Linsley, Kohler, and Paulhus: *Hydrology for Engineers*
Metcalf & Eddy, Inc.: *Wastewater Engineering: Collection and Pumping of Wastewater*
Metcalf & Eddy, Inc.: *Wastewater Engineering: Treatment, Disposal, Reuse*
Nemerow: *Scientific Stream Pollution Analysis*
Rich: *Environmental Systems Engineering*
Rich: *Low-Maintenance, Mechanically-Simple Wastewater Treatment Systems*
Sawyer and McCarty: *Chemistry for Environmental Engineering*
Schroeder: *Water and Wastewater Treatment*
Steel and McGhee: *Water Supply and Sewerage*
Tchobanoglous, Theisen, and Eliassen: *Solid Wastes: Engineering Principles and Management Issues*
Walton: *Groundwater Resources Evaluation*

WASTEWATER ENGINEERING: COLLECTION AND PUMPING OF WASTEWATER

METCALF & EDDY, INC.

Written and edited by

GEORGE TCHOBANOGLOUS

Professor of Civil Engineering
University of California, Davis

McGraw-Hill Book Company

New York St. Louis San Francisco Auckland Bogotá Hamburg
Johannesburg London Madrid Mexico Montreal New Delhi
Panama Paris São Paulo Singapore Sydney Tokyo Toronto

This book was set in Times Roman by Holmes Composition Service.
The editors were Julienne V. Brown and J. W. Maisel;
the production supervisor was Charles Hess.
The drawings were done by J & R Services, Inc.
R. R. Donnelley & Sons Company was printer and binder.

WASTEWATER ENGINEERING:
COLLECTION AND PUMPING OF WASTEWATER

1234567890 DODO 8987654321

Library of Congress Cataloging in Publication Data

Metcalf & Eddy.
Wastewater engineering.

(McGraw-Hill series in water resources and environmental engineering)
Includes bibliographies and index.
1. Sewerage. 2. Pumping stations. I. Tchobanoglous, George. II. Title.
TD673.M47 628′.2 80-20074
ISBN 0-07-041680-X

CONTENTS

Appendixes

PREFACE

This text is a natural outgrowth of the first edition of the widely accepted *Wastewater Engineering: Collection, Treatment, Disposal*. In the second edition, *Wastewater Engineering: Treatment, Disposal, Reuse,* chapters dealing with the collection and pumping of wastewater were deleted to meet the expanded objectives established for the revised edition. The material—in a completely revised and expanded form—is contained in this text, *Collection and Pumping of Wastewater*.

New material presented reflects changes brought about by the passage of the Federal Water Pollution Control Act Amendments of 1972 (Public Law 92-500) and subsequent legislation that has had a major impact on wastewater engineering. In addition, separate chapters have been prepared on the subjects of (1) infiltration/inflow and associated planning requirements, and (2) the effects of biological transformations that occur in collection systems.

As in the second edition of *Wastewater Engineering,* this text was prepared (1) to present the latest technical developments in the field; (2) to make the material covered available in a usable form for students, teachers, practicing engineers, and other users; and (3) to provide leadership in the wider adoption and use of the metric system in the design and analysis of collection and pumping facilities.

Because most of the world is now using some form of metric units, and because the United States is gradually adopting them, the metric units, along with conversion to U.S. customary units, are used in this text. Inasmuch as both sets of units will be in use for some time, conversion tables from metrics to U.S. customary and U.S. customary to metrics are included in Appendix A. To increase the usefulness of the text, conversions from metric units to U.S. customary units are included in figure legends and in the footnotes to all tables.

To make this textbook useful as a teaching and reference text, detailed example problems have been prepared with units carried through all the computational steps to aid the reader's understanding of the principles involved.

Where appropriate, comments are included at the end of an example problem to elucidate basic concepts and highlight additional applications. Approximately 120 discussion topics and problems have been prepared to test the reader's understanding of the material presented. In addition, more than 70 tables containing design data and information are included. To illustrate basic concepts and physical applications more clearly, there are approximately 155 drawings and photographs.

It is hoped that this book will serve the needs of students who are interested in applying their knowledge of fluid mechanics to the problems of collection and pumping of wastewater. This phase of applied hydraulics is the key to success in achieving economy and reliability of design, construction, and operation of entire systems for municipalities and industries. It is also hoped that consulting engineers, public works engineers, and industrial engineers will find the material in this book of assistance in achieving the goals noted above, both in this country and abroad. The problems of the collection and pumping of wastewater are indeed international in scope.

Rolf Eliassen
Chairman of the Board
Metcalf & Eddy, Inc.

George Tchobanoglous
Professor of Civil Engineering
University of California, Davis

ACKNOWLEDGMENTS

Dr. George Tchobanoglous, who has served as a consultant on engineering projects to Metcalf & Eddy, Inc. for over 10 years, was assigned responsibility for overall concept and preparation of this book. He wrote many of the chapters, provided the many detailed examples as well as the problems at the end of each chapter, and coordinated the efforts of the Metcalf & Eddy staff members who prepared other chapters. The clarity and fluid style of his writing, always with the student in mind, is evident throughout the book. We at Metcalf & Eddy are grateful for the engineering skill and high standards of technical excellence that Dr. Tchobanoglous has brought to this manuscript.

The book is a project of the Metcalf & Eddy office in Palo Alto, California, under the administrative direction of Franklin L. Burton, Vice President, who reviewed the entire manuscript and made valuable technical editorial comments. Karen A. Edlefsen served as technical editor and overall coordinator; her editing skills and concern for logic are reflected throughout the text, especially in its readability. Arthur L. Holland coordinated the standardization and preparation of the figures, most of which were originally drawn by Disodada C. Cantimbuhan and Lloyd R. Pound. Kandi B. Masters and Mary G. Spont typed the manuscript.

Personnel from the Boston office of Metcalf & Eddy reviewed the manuscript and prepared drafts of several chapters. Francis C. Tyler served as general coordinator, reviewing all the chapters and providing current design data and figures. His concern for detail contributed significantly to the potential usefulness of this text. James A. Ryan, Jr., prepared the original draft of Chapter 6. Frank M. Gunby, Jr., Allen J. Burdoin, Francis C. Tyler, and Abu M. Z. Alam were responsible for the preparation of the original drafts of Chapters 8 and 9. John G. Chalas, Allen J. Burdoin, Lyle E. Branagan, Eugene S. Grafton, and Jakobs P. Vittands reviewed various chapters.

A number of other individuals reviewed various sections and contributed to the preparation of the text. Jeffrey R. Hauser reviewed all the chapters and helped prepare Chapters 2 and 7. Mark R. Matsumoto and George D. Warren, II, reviewed portions of the manuscript. Rosemary Tchobanoglous typed the rough draft and provided moral support.

We acknowledge the continued leadership of Peter J. Gianacakes, President of Metcalf & Eddy, Inc., in making this text a reality by the commitment of the resources of the firm to the accomplishment of this important task.

Rolf Eliassen
Chairman of the Board
Metcalf & Eddy, Inc.

CHAPTER

ONE

WASTEWATER COLLECTION AND PUMPING: AN OVERVIEW

Every community produces both liquid and solid wastes. The liquid portion—wastewater—is essentially the water supply of the community after it has been fouled by a variety of uses. From the standpoint of sources of generation, wastewater may be defined as a combination of the liquid or water-carried wastes removed from residences, institutions, and commercial and industrial establishments, together with such groundwater, surface water, and storm water as may be present.

If untreated wastewater is allowed to accumulate, the decomposition of the organic materials it contains can lead to the production of large quantities of malodorous gases. In addition, untreated wastewater usually contains numerous pathogenic or disease-causing microorganisms that dwell in the human intestinal tract or that may be present in certain industrial wastes. It also contains nutrients which can stimulate the growth of aquatic plants, and may contain toxic compounds. For these reasons, the immediate and nuisance-free removal of wastewater from its sources of generation, followed by treatment and disposal, is not only desirable but also necessary in an industrial society. In the United States, it is now mandated by numerous federal and state laws [3]. The development of sanitary water supplies and the collection, treatment, and disposal of domestic wastewater are among the most important factors responsible for the general level of good health enjoyed by the population of the United States.

The planning and design of wastewater collection and pumping facilities is thoroughly discussed in this book. Typically, it involves the determination of wastewater flowrates (Chap. 3), the hydraulic design of sewers (Chap. 4), and the selection of appropriate sewer appurtenances (Chap. 5). The rehabilitation of existing collection systems and the design of new systems also involve analyzing the infiltration and inflow of extraneous flows into sewers and the means to limit their occurrence (Chap. 6), as well as analyzing the odors and corrosion that develop in sewers, and the implications of these factors in design (Chap. 7). The final step involves the selection of pumps and the design of pumping stations to transport wastewater to treatment facilities (Chaps. 8 and 9, respectively). Because a knowledge of hydraulics is fundamental in planning and designing wastewater collection and pumping facilities, the basic hydraulic principles and important design equations are briefly discussed in Chap. 2.

To provide an overview of the collection and pumping of wastewater, historical developments in these areas of wastewater engineering, current data on collection systems and pumping stations, and recent trends and developments are briefly reviewed in this chapter. The relationship of the subjects covered in this book to the overall field of wastewater engineering and the role of the engineer are also described.

1-1 HISTORICAL DEVELOPMENTS

Of the many early sewers that have been described in the literature, most is known about the great underground drains of ancient Rome. On the basis of early writings, it is known that direct connections from the houses to these channels and conduits were not used extensively because the requirements of public health were little recognized and because compulsory sanitation would have been considered an invasion of individual rights. Following Roman practice, early sewers, both in Europe and the United States, were constructed originally to collect storm water. All human excreta were excluded from the sewers of London until 1815, from those of Boston until 1833, and from those of Paris until 1880 [1].

It is astonishing that, although many sewers were built following the days of the Roman empire, little if any progress in the design and construction of wastewater collection systems occurred until the 1840s. The renaissance began in Hamburg, Germany, in 1842 after a severe fire destroyed part of the city. For the first time, a complete new wastewater conveyance system was designed according to the modern theories of the day, which took into account topographic conditions and recognized community needs [1]. This was a spectacular advance when it is considered that the fundamental principles on which the design was based are in use today but were not generally applied before the twentieth century.

London Sewers

In London, as late as 1845, no survey of the metropolis was adequate as a basis for planning sewerage systems. The sewers in adjoining parishes were at different elevations, making junctions with them impracticable. Some of the smaller sewers were higher than the cesspools they were supposed to drain, while other large sewers had been constructed in such a way that wastewater would have had to flow uphill. Some of the large sewers were also made to discharge into small sewers.

Following the great epidemic of Asiatic cholera in 1832, cholera again erupted in London in 1848 and claimed over 25,000 victims during the next six years. Although the connection between a contaminated water supply and the rapid spread of the disease was clearly shown, the filthy living conditions in most houses, which resulted from the lack of domestic sewers, were a great hindrance in combating the epidemic. It was not until 1855 that Parliament provided for the Metropolitan Board of Works, which soon after undertook the development of an adequate wastewater collection system.

American Sewers

Little is known about the early wastewater collection systems in the United States. Often they were constructed by individuals or by inhabitants of small districts at their own expense and with little or no public supervision. There was a tendency in this country, as elsewhere, to make sewers larger than necessary. One of the oldest sewers in Brooklyn that drained less than 20 acres had a grade of 1 in 36, and was 1.25 m (4 ft) deep and 1.5 m (5 ft) wide. In some cases, the sewers were very large, not only at their outlets but also all the way to their starting points. It was impossible to obtain adequate velocity in such sewers unless they were laid on steep grades and, consequently, some of them became offensive as the accumulated wastewater solids decomposed. There were even some instances in which the slopes were laid in the wrong direction.

Although, as noted previously, the fundamental principles governing the flow of wastewater were known from the early 1840s, their application to the design of sewers has been evolutionary rather than marked by clear, progressive steps. Many of the same equations are used today, but their fundamental basis and limits of applicability are now better understood.

Early Pumping Stations

In 1910, less than 200 wastewater pumping stations existed in the United States. Although the types of pumps employed at that time were essentially the same as those used today, the motive power of the pumps was varied. Before the widespread distribution of electricity, the motive power was supplied by steam, town gas, gasoline, and hot-air engines. The use of electricity as the motive power for pumps became common in the early 1900s.

Even after the use of electricity became common, engineers were sometimes reluctant to use pumping stations because power outages were routine. Problems occurred in many communities where standby power was either not available or not included as part of the pumping station installation. The practice of avoiding the use of pumping stations led to the construction of some very deep and costly large sewers (interceptors). Today, as discussed in the following section, pumping stations are a common feature of any wastewater collection system.

1-2 CURRENT STATUS

The current status of collection systems and pumping stations is briefly described in this section. Additional details may be found in Ref. 2 and in current U.S. Environmental Protection Agency (EPA) publications.

Collection Systems

In an analysis made by the Federal Water Quality Administration in 1968 [4], it was determined that approximately 140 million people were served by public sewers in the United States. The distribution of this population by community size is shown in Table 1-1. Even though many communities have a population of less than 10,000, only a small fraction of the nation's population in these communities is served by sewers. Conversely, over 40 percent of the nation's population in sewered areas live in the relatively few communities with a population of more than 100,000.

Collection systems used for wastewater are of two basic types—separate or combined. Separate systems are designed for the exclusive transport of

Table 1-1 Distribution of United States population served by sewers [a]

Size of community, persons	Estimated 1968 population in sewered areas		Percent of communities with combined sewers
	Persons, millions	Percent of total	
Under 500	0.72	0.5	5
500– 1,000	1.91	1.4	10
1,000– 5,000	13.49	9.6	15
5,000– 10,000	11.14	8.0	20
10,000– 25,000	19.31	13.8	22
25,000– 50,000	15.90	11.3	30
50,000–100,000	16.19	11.5	36
100,000–250,000	13.66	9.7	49
250,000–500,000	16.01	11.4	42
Over 500,000	31.90	22.8	73
Total	140.23	100.0	

[a] Adapted from Ref. 2.

sanitary wastewater. (The terms *separate* and *sanitary* are often used interchangeably in references to sewers and collection systems.) Combined systems are designed for the transport of both sanitary wastewater and storm water. The existence of even a relatively small percentage of combined sewers within a municipal wastewater collection system is usually enough to classify the system as combined. Most state regulations now permit the construction of separate sewers only.

The types of sewer systems in use in 1968 are also shown in Table 1-1. Combined sewer systems are located primarily in large cities. Over 45 percent of all communities with populations of over 100,000 have combined sewer systems or both combined and separate systems. However, in terms of national totals, less than 15 percent of all communities have combined sewers, and approximately 25 percent of the total population in sewered areas is served by combined sewers [2].

The number of people served by collection systems in 1972–1973, along with information on the total length of sewers and the number of pumping stations in use, is listed in Table 1-2. As shown, there is about 4.53 m (14.86 ft) of sewer for each person served. Detailed information on the distribution of sizes of sewer pipes used is reported in Table 4-3 in Chap. 4. As shown in Table 4-3, about 75 percent of all sewers are 200 mm (8 in) or less in diameter. Data on the materials most commonly used to manufacture sewer pipe are given in Table 1-3.

Pumping Stations

As shown in Table 1-2, there were 36,900 wastewater pumping stations in the United States in 1972–1973. (The number in 1910 was less than 200.) On the

Table 1-2 Wastewater collection systems in the United States[a]

	Topography class[b]			
Item	1	2	3	Total
Population served by sewers, millions	87.1	37.8	28.1	153.0
Sewer length, km	394,300	171,100	127,500	692,900
Number of pumping stations	23,800	8,400	4,700	36,900

[a] Adapted from Ref. 2, data for 1972–1973.

[b] Class 1 (flat)—the requirement for a minimum velocity of flow of 0.6 m/s is a design constraint.

Class 2 (moderate)—a velocity of 0.75 m/s is attainable.

Class 3 (steep)—a velocity of 0.9 m/s is attainable.

Note: km × 0.6214 = mi

m/s × 3.2808 = ft/s

Table 1-3 Percentage distribution of sewer pipe materials by pipe size[a]

Type of pipe material,[b] percent	Size range[c]		
	$D \leq 300$ mm	$D > 300$ to < 600 mm	$D \geq 600$ mm
Vitrified clay (VC)	72	55	1
Asbestos cement (AC)	23	10	8
Reinforced concrete (RC)	—	25	87
Cast iron (CI)	5	10	4
	100	100	100

[a]Adapted from Ref. 2; the percentage distributions are based on 1968 data.

[b]For new sewers ($D \leq 300$ mm), more polyvinyl chloride (PVC) pipe instead of VC and AC pipe is now being chosen. For larger sizes, other plastic materials may take over a significant portion of the market.

[c]D = pipe diameter, mm.

Note: mm × 0.03937 = in

basis of the data reported in Table 1-2, one pumping station is required for every 18.8 km (11.7 mi) of sewer. Eighty percent of the existing pumping stations are factory-assembled or packaged installations and 20 percent are conventional built in-place. Of the factory-assembled units, 10 percent are ejector stations, and 70 percent are standard pumping types. The percentage of factory-assembled installations is high because most municipal pumping stations serve relatively small populations, such as those in subdivisions. Design features of typical pumping stations are reported in Table 1-4.

1-3 RECENT TRENDS AND DEVELOPMENTS

The results of continuing research in wastewater engineering and the application of developments in other fields to wastewater engineering are responsible for many recent changes in the design and analysis of wastewater collection systems. Important changes include the use of advanced techniques in designing both sanitary and storm water sewers, improvements in construction materials, and recognition of the need to manage wastewater and storm water collection systems.

Design

In the past, one of the most time-consuming aspects of sewer design was the preparation of maps and ground surface profiles. Today, modern photogram-

Table 1-4 Design features of typical pumping stations in wastewater collection systems[a]

	Type of pump station		
Design feature	Factory-assembled ejector	Factory-assembled	Conventional (built in-place)
Flow, L/s			
Design peak	3.8	18.9	95
Design average	2.2	8.5	45
Discharge head, m	9.1	13.7	18.3
Number of installed pumps including standby	2	2	3

[a]Adapted from Ref. 2.
Note: L/s × 15.8508 = gal/min
m × 3.2808 = ft

metric techniques have eliminated most of the tedium once involved in this work. Several firms and government agencies have developed computer programs that can be used to design various parts of wastewater collection systems. In the future, it is anticipated that complete computer design facilities will be developed so that the design of selected sewer systems can be automated entirely. However, conventional methods will probably be used to design some systems because of their small size or inherent constraints.

Materials

Recent developments in both the manufacture and use of natural and synthetic materials have significantly affected the design and construction of wastewater collection systems. Along with the use of plastics and various other materials for the construction of sewer pipes, the use of protective coatings and epoxy linings is also increasing. Precast concrete manholes and improved pipe joints made with gaskets of elastomeric materials are now used almost without exception. As a result, it is now possible, with proper construction and inspection, to build sewers from which most of the potential groundwater infiltration is excluded.

Alternative Collection Systems

Several different types of wastewater collection systems have been developed recently as alternatives to conventional sewers. The systems receiving the most attention are those in which mechanically applied low pressure is the motive force for the movement of wastewater through small-diameter sewers. Individually they are known as pressure and vacuum sewers. Both systems are commercially available and have been applied in various configurations. In one pressure sewer system, small-diameter [32 mm (1.25 in)] sewers are used in conjunction with grinder pumps at each point of waste discharge. The grinder pumps eliminate the problems associated with the discharge of solids to small-diameter sewers. These systems are expected to be especially suitable for small communities.

Management of Wastewater Collection Systems

During the past few years, the need for the proper management and control of the wastewater sewers has become better understood and accepted more widely. The requirement that an infiltration/inflow analysis (see description in Chap. 6) be performed before federal funds can be obtained for the construction of facilities has served to further highlight the importance of this concept.

In the future, the design of wastewater collection systems is expected to be coordinated to a much greater extent with the design of treatment facilities, especially as the transformations that occur in wastewater during transport become more clearly understood. Many of these transformations are discussed in Chap. 7.

Storm-Water Management

Increased attention is being given to the means of controlling or mitigating the adverse effects of discharging untreated and unregulated storm water overflows. Computers, coupled to remote rainfall and receiving-water sensing devices operating on a real-time basis, are now being used to control the various overflow, pumping, and storage facilities and to minimize the effects of discharging storm water to the environment. The increasing use of computers and the construction of more facilities designed specifically for the treatment of storm water overflows are expected to continue in the future.

1-4 WASTEWATER ENGINEERING AND THE ROLE OF THE ENGINEER

Wastewater engineering is that branch of environmental engineering in which the basic principles of science and engineering are applied to the problems of water pollution control. The ultimate goal—wastewater management—is the protection of the environment in a manner commensurate with economic, social, and political concerns.

Practicing wastewater engineers are involved in the conception, planning, evaluation, design, construction, and operation and maintenance of the systems that are needed to meet wastewater management objectives. The major elements of wastewater systems, the associated engineering tasks, and the related chapters in this book are identified in Table 1-5.

The primary focus of this book (Chaps. 4 through 9) concerns the third and fourth elements listed in Table 1-5: collection and transmission, and pumping.

Table 1-5 Major elements of wastewater management systems and associated engineering tasks[a]

Element	Engineering task	See chap.
Source of generation	Estimation of the quantities of wastewater, evaluation of techniques for wastewater reduction, and determination of wastewater characteristics.	2, 3[b]
Source control	Design of on-site systems to provide partial treatment of the wastewater before it is discharged to collection systems (principally involves industrial dischargers).	3
Collection and transmission	Design of sewers used to collect wastewater from the various sources of generation and to transport it to treatment facilities or to other locations for processing.	2, 4–7
Pumping	Design of pumping stations and force mains to lift and transport wastewater.	2, 8, 9
Treatment (wastewater and sludge)	Selection, analysis, and design of treatment operations and processes to meet specified treatment objectives related to the removal of wastewater contaminants of concern.	[b]
Disposal and reuse	Design of facilities used for the disposal and reuse of treated effluent in the aquatic and land environment, and the disposal and reuse of sludge on land.	[b]

[a] Adapted from Ref. 3.

[b] The subjects of wastewater characteristics, treatment, disposal, and reuse are covered in Ref. 3, the companion text to this book.

These areas of wastewater engineering, like the others, have been and continue to be in a dynamic period of development. Old ideas are being reevaluated, and new concepts are being formulated. To play an active role in the development of this field, the engineer must know the fundamentals on which it is based. The delineation of these fundamentals is the main purpose of this book.

REFERENCES

1. Metcalf. L., and H. P. Eddy: *American Sewerage Practice*, vol. 1, 2d ed., McGraw-Hill, New York, 1928.
2. Metcalf & Eddy, Inc.: *Report to National Commission on Water Quality on Assessment of Technologies and Costs for Publicly Owned Treatment Works,* vol. 1, prepared under Public Law 95-500, Boston, 1975.
3. Metcalf & Eddy, Inc.: *Wastewater Engineering: Treatment, Disposal, Reuse,* 2d ed., McGraw-Hill, New York, 1979.
4. *Municipal Waste Facilities in the United States: Statistical Summary, 1968 Inventory,* U.S. Department of the Interior, Federal Water Quality Administration, Publication CWT-6, Washington, D.C., 1970.

CHAPTER

TWO

REVIEW OF APPLIED HYDRAULICS

The principal factors that affect the flow of wastewater in sewers are:

1. The slope
2. The cross-sectional area and shape of the conduit
3. The roughness of the interior pipe surface
4. The conditions of flow, such as whether the conduit is flowing full or partly full and whether the flow is steady or varied
5. The presence or absence of obstructions, bends, and other flow disturbances
6. The character, specific gravity, and viscosity of the liquid

The purpose of this chapter is to review the applied hydraulics used in the analysis and design of collection, pumping, and wastewater-management facilities. Applications of the principles discussed here to the measurement of flow are discussed in Chap. 3. Sewer design is discussed in Chap. 4. The design of overflow and diversion structures is discussed in Chap. 5. The details of pump hydraulics and pumping station design are discussed in Chaps. 8 and 9, respectively. It must be stressed that the material presented in this chapter is meant to serve only as a review. Standard texts in fluid mechanics should be consulted for additional details on the topics discussed.

2-1 FUNDAMENTALS OF PIPE AND OPEN-CHANNEL FLOW

The analysis of flow in both pipes (closed conduits) and open channels is based on an adaptation of three basic equations of fluid mechanics: the equation of continuity, the energy equation, and the momentum equation. Before discussing these as well as other concepts some of the terms commonly used in the field of hydraulics must be defined.

Definition of Terms

The following terms are basic to an understanding of both pipe and open-channel flow. A more complete listing of hydraulic terms may be found in Vennard [14].

Laminar flow. Laminar flow involves the mixing of fluid particles on a molecular scale so that, when in motion, the fluid particles move in parallel paths under the action of viscosity. In practice, laminar flow occurs when the Reynolds number is less than 1500 to 2000.

Turbulent flow. Turbulent flow involves the mixing of fluid particles on a molar scale so that, when in motion, an exchange of momentum occurs between adjacent fluid particles which results in the rapid and continuous mixing of the fluid. In practice, turbulent flow occurs at Reynolds numbers greater than 6000 to 10,000.

Pipe and open-channel flow. The flow of liquid in a conduit may be classified as *pipe* (or closed-conduit) flow or *open-channel* flow depending on whether the free-liquid surface is subject to atmospheric pressure. For example, when a sewer is flowing full or under pressure, the flow is referred to as pipe flow. If the flow is in a partially filled sewer or in an open channel, the flow is referred to as open-channel flow. Pipe flow and open-channel flow are compared schematically in Fig. 2-1.

Head loss. Head loss is the loss of energy that occurs when liquids flow in pipes and open channels. The energy required to overcome the effects of friction in

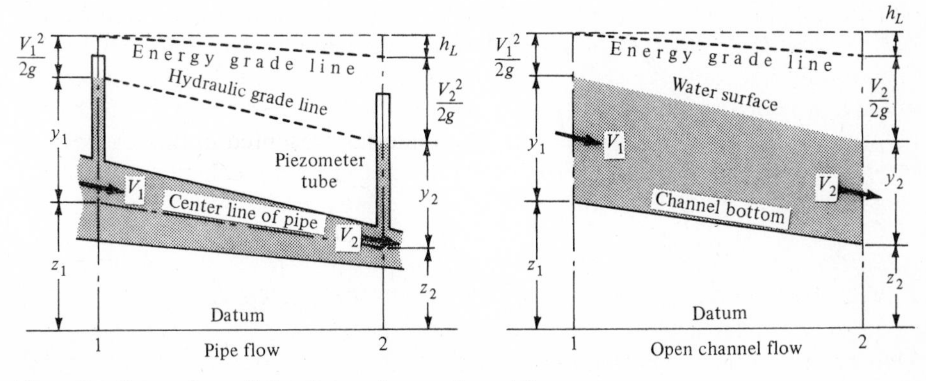

Figure 2-1 Comparison of pipe flow and open-channel flow.

turbulent flow is the head loss. The loss of energy caused by the turbulence induced by the appurtenances used with pipelines and open channels is also head loss. Head loss usually is denoted by the symbol h_L.

Hydraulic grade line. The hydraulic grade line, also shown in Fig. 2-1, is a line connecting the points to which the liquid would rise at various places along any pipe or open channel if piezometer tubes were inserted in the liquid. It is a measure of the pressure head available at these various points. When water flows in an open channel, the hydraulic grade line coincides with the profile of the water surface.

Energy grade line. The total energy of flow in any section with reference to some datum is the sum of the elevation head z, the pressure head y, and the velocity head $V^2/2g$. The energy from section to section is usually represented by a line called the *energy grade line* or *energy gradient* (Fig. 2-1). In the absence of frictional losses, the energy grade line would remain horizontal, although the relative distribution of energy could vary between the elevation, pressure, and velocity heads. However, in all real systems, losses of energy will occur because of resistance to flow, and the resulting energy grade line will be sloped.

Specific energy. The specific energy E, sometimes called the *specific head,* is the sum of the pressure head y and the velocity head $V^2/2g$. The specific-energy concept is especially useful in analyzing flow in open channels.

The relationship between specific energy and depth of flow for a constant rate of flow is illustrated in Fig. 2-2*a*, which usually is called a specific-energy diagram. The relationship between the depth of flow and discharge per unit width of channel for constant specific energy is called a *q curve* and is illustrated in Fig. 2-2*b*.

Steady flow. Steady flow occurs when the discharge or rate of flow at any cross section is constant.

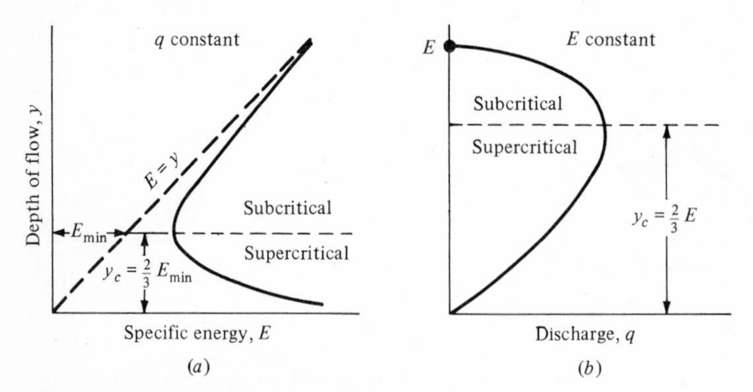

Figure 2-2 (*a*) Specific-energy diagram; (*b*) *q* curve [14].

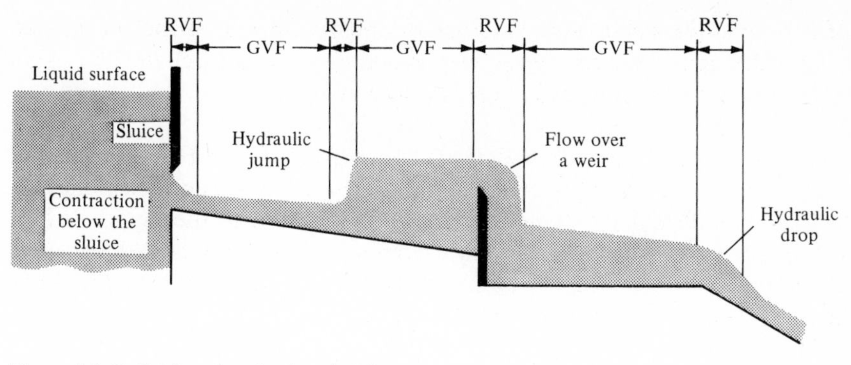

Figure 2-3 Definition sketch of varied flow.

Uniform and nonuniform flow. Uniform flow occurs when the depth, cross-sectional area, and other elements of flow are substantially constant from section to section. Nonuniform flow occurs when the slope, cross-sectional area, and velocity are changing from section to section. An example of steady nonuniform flow is the flow through a Venturi section used for flow measurements (see Chap. 3, Sec. 3-4, "Venturi meter").

Varied flow. The flow in a channel is considered varied if the depth of flow changes along the length of the channel. In general, the flow may be gradually varied (GVF) or rapidly varied (RVF), as shown in Fig. 2-3. Rapidly varied flow occurs when the depth of flow changes abruptly.

Equation of Continuity

The equation of continuity expresses the conservation of mass from section to section in a streamtube control volume, as shown in Fig. 2-4. According to the

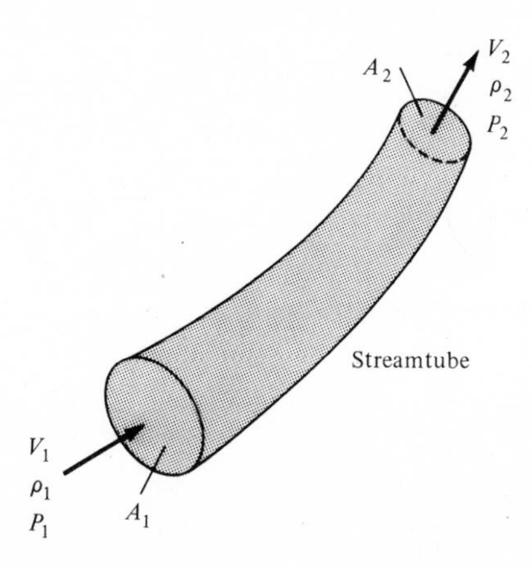

Figure 2-4 Flow through a streamtube control volume.

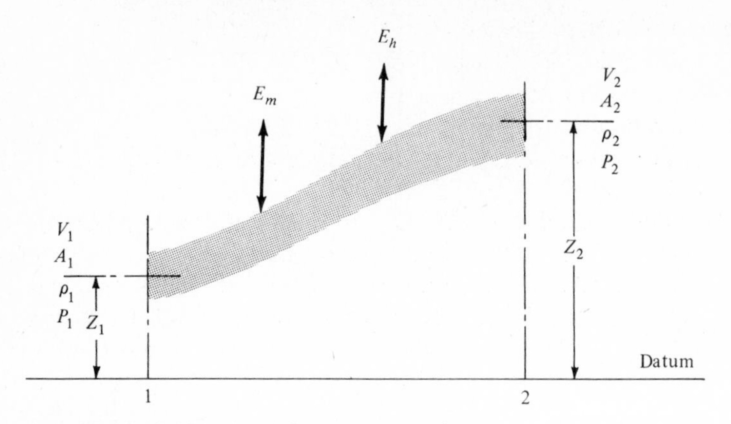

Figure 2-5 Definition sketch for energy equation.

principle of the conservation of mass, mass can be neither created nor destroyed between sections A_1 and A_2. Thus, the equation of continuity becomes

$$\rho_1 A_1 V_1 = \rho_2 A_2 V_2 = \rho_1 Q_1 = \rho_2 Q_2 \tag{2-1}$$

where ρ = fluid density, kg/m³ (slug/ft³)
A = cross-sectional area, m² (ft²)
V = velocity, m/s (ft/s)
Q = flowrate, m³/s (ft³/s)

If the fluid is incompressible, then $\rho_1 = \rho_2$, and

$$A_1 V_1 = A_2 V_2 = Q_1 = Q_2 \tag{2-2}$$

Energy Equation

A flowing fluid may have four types of energy: pressure energy E_p, velocity energy E_v, potential energy E_q, and thermal or internal energy E_i. If E_m represents the mechanical energy transferred to (+) or from (−) the fluid (e.g., in a pump, fan, or turbine), and E_h represents the heat energy transferred to (+) or from (−) the fluid (e.g., in a heat exchanger), then application of the law of conservation of energy between points 1 and 2 in Fig. 2-5 yields the following equation:

$$(E_p + E_v + E_q + E_i)_1 \pm E_m \pm E_h = (E_p + E_v + E_q + E_i)_2 + \text{losses} \tag{2-3}$$

The losses in Eq. 2-3 represent energy that has been transformed into nonrecoverable forms by irreversibilities caused by friction (e.g., energy dissipated as heat or noise).

The general expression for an incompressible liquid may be rewritten as

$$\frac{p_1}{\gamma} + \alpha_1 \frac{V_1^2}{2g} + z_1 \pm E_m \pm E_h = \frac{p_2}{\gamma} + \alpha_2 \frac{V_2^2}{2g} + z_2 + h_L \tag{2-4}$$

where p_1, p_2 = pressure, kN/m^2 (lb_f/in^2)
γ = specific weight of water, kN/m^3 (lb/ft^3)
α_1, α_2 = kinetic-energy correction factors
g = acceleration due to gravity, 9.81 m/s^2 (32.2 ft/s^2)
z_1, z_2 = height of streamtube above any assumed datum plane, m (ft)
h_L = head loss, m (ft)

For laminar flow in pipes, the value of α is 2.0. For turbulent flow in pipes, the value of α ranges between 1.01 and 1.10. Turbulent flow is by far the most common and, in practice, α is usually taken to be equal to unity. The head-loss term h_L represents the losses and the change in internal energy E_i. If the fluid in question is ideal (frictionless) and no mechanical or heat energy is transferred, Eq. 2-4 reduces to

$$\frac{p_1}{\gamma} + \frac{V_1^2}{2g} + z_1 = \frac{p_2}{\gamma} + \frac{V_2^2}{2g} + z_2 \tag{2-5}$$

which is the familiar form of the Bernoulli equation for incompressible flow.

Application of the energy equation or Bernoulli's equation to flow in a pipeline supplied from a reservoir is shown in Fig. 2-6. The energy equation written between points 1 and 2 would be

$$H = \frac{p_1}{\gamma} + \frac{V_1^2}{2g} + z_1 + h_{en} = \frac{p_2}{\gamma} + \frac{V_2^2}{2g} + z_2 + h_{en} + h_{f_{1-2}} \tag{2-6}$$

where H = total head, m (ft)
h_{en} = head loss at pipe entrance, m (ft)
$h_{f_{1-2}}$ = head loss in pipe due to friction between points 1 and 2, m (ft)

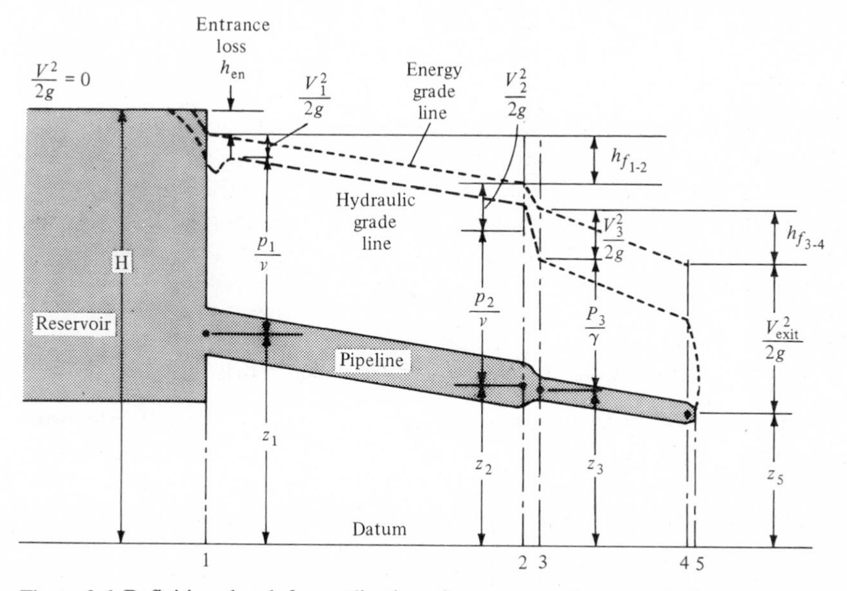

Figure 2-6 Definition sketch for application of energy equation to a pipeline.

Pumps offer another example of the application of the energy equation, as shown in Fig. 2-7. In this case, the energy equation written between points 1 and 2 is

$$\frac{p_1}{\gamma} + \frac{V_1^2}{2g} + z_1 + E_p = \frac{p_2}{\gamma} + \frac{V_2^2}{2g} + z_2 \tag{2-7}$$

The head-loss term h_L is embedded in all applications of the energy equation to fluid flow. In Eq. 2-7, E_p represents the net energy supplied by the pump, allowing for head losses within the pump. Several equations can be used to determine h_L as a function of geometric considerations, fluid characteristics, and flowrate (for both open-channel and closed-conduit flow). Some of those used most frequently are discussed in Sec. 2-2.

The head-loss term h_L includes the frictional head loss h_f and other head losses that occur at discontinuities in flow geometry (e.g., contractions, expansions, bends). These are called *minor losses* (see Sec. 2-5).

Momentum Equation

Unlike the continuity and energy equations, which are scalar relationships, the momentum equation is a vector relationship, that is, both the magnitude and direction of the forces and velocities are important. The law of the conservation of momentum may be stated as follows: The time rate of change in momentum (defined as the mass rate of flow ρAV multiplied by the velocity V) along the path of flow will result in a force called the *impulse force*.

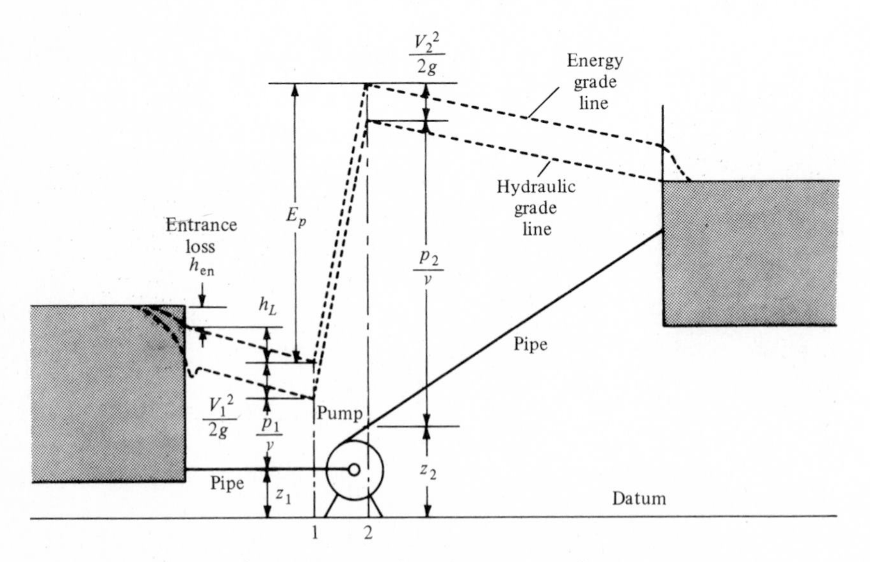

Figure 2-7 Definition sketch for application of energy equation to a pump.

The application of the momentum equation to the analysis of flow in the section of pipe shown in Fig. 2-4 is as follows: The momentum carried across area A_1 during the time dt is equal to

$$(\text{Mass transferred across } A_1)V_1 = (\rho_1 A_1 V_1\, dt)V_1 \tag{2-8}$$

and is in the direction of V_1. In the same manner, the momentum carried across area A_2 during time dt is equal to

$$(\rho_2 A_2 V_2\, dt)V_2 \tag{2-9}$$

The net force on the fluid caused by the change of momentum between sections 1 and 2 is

$$F = \frac{(\rho_2 A_2 V_2\, dt)V_2 - (\rho_1 A_1 V_1\, dt)V_1}{\text{dt}} \tag{2-10}$$

$$F = M(V_2 - V_1) \tag{2-11}$$

where M = mass = $\rho_2 A_2 V_2 = \rho_1 A_1 V_1$

The quantity $V_2 - V_1$ in Eq. 2-11 represents the vector change in the velocity.

The momentum equation is often used to analyze the forces that develop in a pipe bend. As shown in Fig. 2-8, the forces F_x and F_y are applied to maintain equilibrium with the forces caused by the change in momentum as the water flows through the bend. Application of the momentum equation yields

In the x direction,

$$p_1 A_1 - p_2 A_2 \cos\theta - F_x = \rho Q(V_2 \cos\theta - V_1) \tag{2-12}$$

In the y direction,

$$p_2 A_2 \sin\theta - W - F_y = -\rho Q V_2 \sin\theta \tag{2-13}$$

The forces F_x and F_y needed to maintain equilibrium are applied through the pipe wall by supporting structures, hangers, tie rods, thrust blocks, or the like.

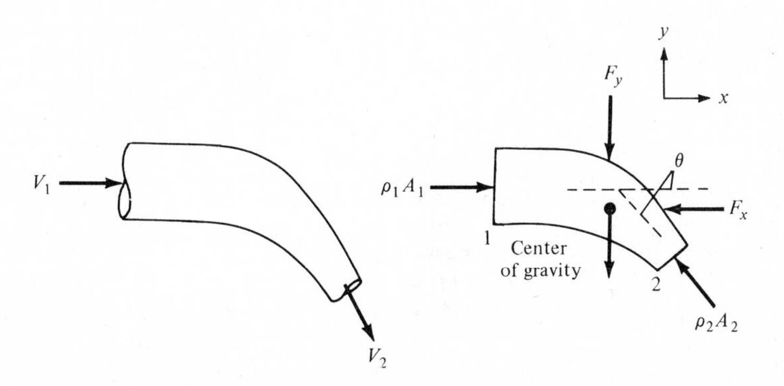

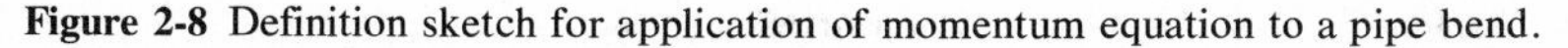

Figure 2-8 Definition sketch for application of momentum equation to a pipe bend.

The principles of momentum and continuity are also used in the equation for a hydraulic jump in rectangular open channels. If a unit width of channel is considered, application of the momentum equation to the flow shown in Fig. 2-9 results in

$$\Sigma F_x = F_1 - F_2 = \frac{\gamma y_1^2}{2} - \frac{\gamma y_2^2}{2} = \rho y_2 V_2^2 - \rho y_1 V_1^2 \qquad (2\text{-}14)$$

From the equation of continuity,

$$V_1 y_1(\text{unit width}) = V_2 y_2(\text{unit width}) \qquad (2\text{-}15)$$

Using Eq. 2-15 to eliminate V_2 in Eq. 2-14 leads to

$$y_2 = \frac{-y_1}{2} + \sqrt{\left(\frac{y_1}{2}\right)^2 + \frac{2y_1V_1^2}{g}} \qquad (2\text{-}16)$$

To calculate y_1 with known downstream conditions, interchange the subscripts 1 and 2 in Eq. 2-16.

The depths before and after the jump (y_1 and y_2, respectively) are referred to as conjugate depths.

2-2 FLOW EQUATIONS

To design facilities for the transmission of fluids—whether the flow is in closed conduits or open channels—one must know (1) the relationship between the head loss or the slope of the energy grade line and the flowrate, (2) the fluid characteristics, and (3) the conduit or channel roughness and configuration. Several equations relating these factors are discussed in this section. Because it is assumed that the reader is familiar with the fundamentals of fluid flow, lengthy derivations are not included, and the equations are presented without discussing all of the limitations involved in their use.

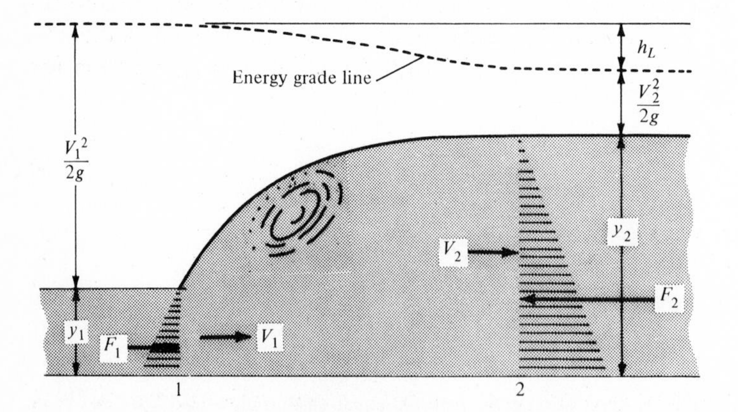

Figure 2-9 Definition sketch for application of momentum equation to a hydraulic jump.

Flow in Closed Conduits

Equations for the flow of fluids in closed conduits may be derived from theoretical considerations or empirically. The Poiseuille equation for laminar flow and the Darcy-Weisbach universal equation are examples of equations derived from theoretical considerations. The Manning equation and the Hazen-Williams equation, which are used in the design of sewers and force mains, are examples of equations derived empirically.

Poiseuille equation. In laminar flow, the forces of viscosity predominate in relation to other forces, such as inertia. An example of laminar flow is the pumping of sludge at low velocities at a wastewater treatment plant. Under laminar conditions, the Poiseuille equation for head loss h_f may be expressed as

$$h_f = \frac{32\mu LV}{\rho g D^2} = \frac{32\nu LV}{gD^2} \tag{2-17}$$

where h_f = head loss, m (ft)
μ = dynamic viscosity of fluid, $N \cdot s/m^2$ ($lb_f \cdot s/ft^2$)
L = length of pipeline, m (ft)
V = velocity, m/s (ft/s)
ρ = fluid density, kg/m^3 ($slug/ft^3$)
g = acceleration due to gravity, 9.81 m/s^2 (32.2 ft/s^2)
D = diameter of pipeline, m (ft)
ν = kinematic viscosity of fluid, m^2/s (ft^2/s)

The corresponding expression for the flowrate Q is

$$Q = \frac{\pi D^4 g h_f}{128\nu L} \tag{2-18}$$

where Q = flowrate, m^3/s (ft^3/s).

Darcy-Weisbach equation. In about 1850, Darcy, Weisbach, and others deduced a formula for pipe friction from the results of experiments conducted on various pipes. The formula now known as the Darcy-Weisbach equation for circular pipes is

$$h_f = f\frac{LV^2}{D2g} \tag{2-19}$$

In terms of the flowrate Q, the equation becomes

$$h_f = \frac{8fLQ^2}{\pi^2 g D^5} \tag{2-20}$$

where h_f = head loss, m (ft)
f = coefficient of friction (in many parts of the world the symbol λ is used to denote the coefficient of friction)

L = length of pipe, m (ft)
V = mean velocity, m/s (ft/s)
D = diameter of pipe, m (ft)
g = acceleration due to gravity, 9.81 m/s^2 (32.2 ft/s^2)
Q = flowrate, m^3/s (ft^3/s)

The value of f has been found to vary with the Reynolds number N_R, pipe roughness, pipe size, and other factors. The relationships among these factors are presented graphically in Figs. 2-10 and 2-11, commonly known as Moody diagrams. The effect of size and roughness is expressed by the relative roughness, which is the ratio of absolute roughness of the pipe ϵ to the pipe diameter D, both expressed in the same unit of length. The Reynolds number is

$$N_R = \frac{VD\rho}{\mu} = \frac{VD}{\nu} \tag{2-21}$$

where N_R = Reynolds number, unitless
V = velocity, m/s (ft/s)
D = diameter of pipe, m (ft)
ρ = fluid density, kg/m^3 (slug/ft^3)
μ = dynamic viscosity of fluid, N·s/m^2 (lb$_f$·s/ft^2)
ν = kinematic viscosity of fluid, m^2/s (ft^2/s)

If the value of ϵ is known or can be estimated, the appropriate friction factor f for wholly turbulent flow may be obtained from Fig. 2-10 or Fig. 2-11, or computed from the following equation:

$$\frac{1}{\sqrt{f}} = 2 \log \frac{D}{2\epsilon} + 1.74 \tag{2-22}$$

When flow conditions fall within the transition zone, values of f are obtained from Fig. 2-10, based on a computed Reynolds number and the relative roughness. If the flow is laminar, the roughness is not a consideration, and it can be shown from theoretical considerations that

$$f = \frac{64}{N_R} \tag{2-23}$$

Equation 2-22 is often considered the general equation for determining the friction factor for rough pipes and is sometimes referred to as the *rough-pipe law* or the *quadratic law* [15].

Several equations have also been developed for use in the transition zone. A more complete discussion of this topic may be found in Ref. 15.

Figure 2-10 Moody diagram for friction factor in pipes vs. Reynolds number and relative roughness [13].

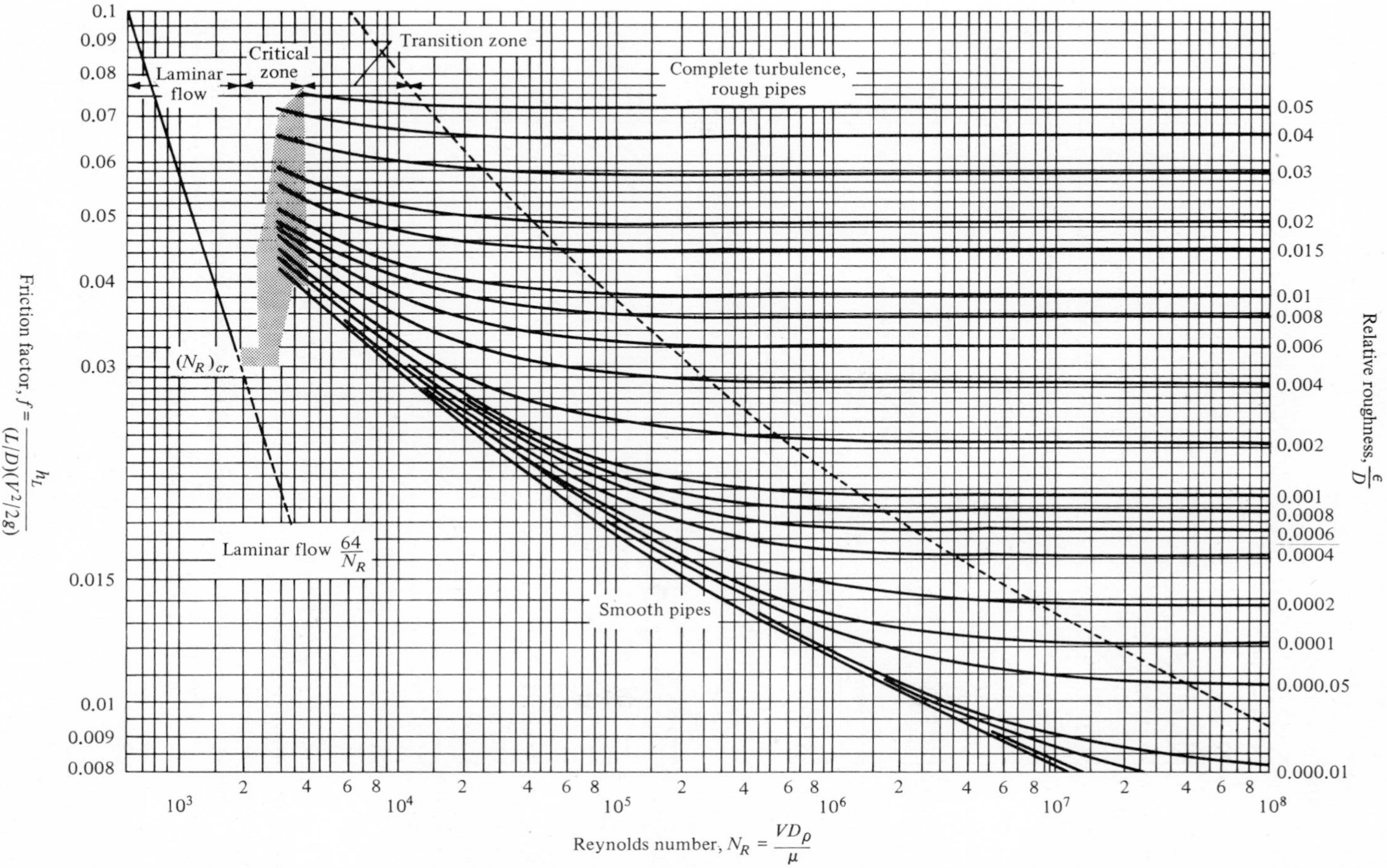

Transition zone
Critical zone
Laminar flow
Complete turbulence, rough pipes
$(N_R)_{cr}$
Laminar flow $\frac{64}{N_R}$
Smooth pipes
Friction factor, $f = \frac{h_L}{(L/D)(V^2/2g)}$
0.1
0.09
0.08
0.07
0.06
0.05
0.04
0.03
0.015
0.01
0.009
0.008
Relative roughness, $\frac{\epsilon}{D}$
0.05
0.04
0.03
0.02
0.015
0.01
0.008
0.006
0.004
0.002
0.001
0.0008
0.0006
0.0004
0.0002
0.0001
0.000.05
0.000.01
10^3
10^4
10^5
10^6
10^7
10^8
Reynolds number, $N_R = \frac{VD\rho}{\mu}$

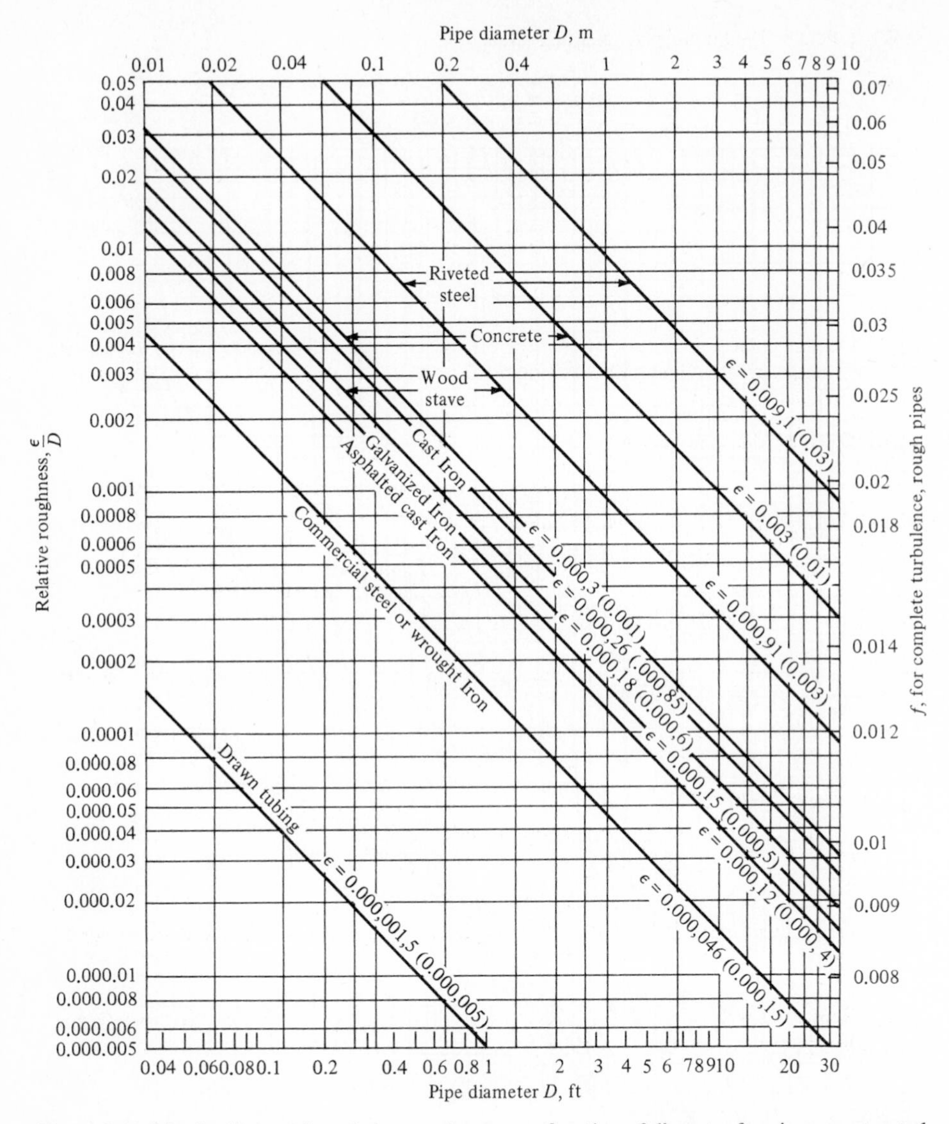

Figure 2-11 Moody diagram for relative roughness as a function of diameter for pipes constructed of various materials. Note: m × 2.2808 = ft.

Example 2-1 Use of the Darcy-Weisbach equation and the Moody diagram Determine the flowrate in a 500-m (1640-ft) section of a 1-m (3.28-ft) -diameter commercial-steel pipe when there is a 2-m (6.6-ft) drop in the energy grade line over the section.

SOLUTION

1. Select a value for the friction factor f. For a first approximation, choose an f value from Fig. 2-11 assuming the flow to be wholly turbulent.

$$f \cong 0.0105$$

2. Calculate the flowrate by using Eq. 2-20.

$$h_f = \frac{8fLQ^2}{\pi^2 g D^5}$$

$$Q = \left(\frac{h_L \pi^2 g D^5}{8fL}\right)^{1/2}$$

$$Q = \left[\frac{(2\text{ m})(\pi^2)(9.81\text{ m/s}^2)(1\text{ m})^5}{8(0.0105)(500\text{ m})}\right]^{1/2}$$

$$Q = 2.15\text{ m}^3/\text{s}$$

At this point, it is necessary to check the friction factor f that was chosen by assuming wholly turbulent flow. If it is determined from Fig. 2-10 that the f value is incorrect, the calculations must be repeated with a revised value of f.

3. Calculate the fluid velocity.

$$V = \frac{Q}{A}$$

$$V = \frac{2.15\text{ m}^3/\text{s}}{(\pi/4)(1\text{ m})^2}$$

$$V = 2.74\text{ m/s}$$

4. Calculate the Reynolds number. Assume that the temperature and the kinematic viscosity are about 15°C and 1.14×10^{-6} m²/s, respectively.

$$N_R = \frac{VD}{\nu}$$

$$N_R = \frac{(2.74\text{ m/s})(1\text{ m})}{(1.14 \times 10^{-6}\text{ m}^2/\text{s})}$$

$$N_R = 2.4 \times 10^6$$

5. Obtain a revised f value from Fig. 2-10 based on the Reynolds number computed in step 4.

$$f \cong 0.0115$$

6. Repeat steps 2 to 4 with the new f value. The resulting values for the flowrate and Reynolds number are:

$$Q = 2.05\text{ m}^3/\text{s }(72.4\text{ ft}^3/\text{s})$$

$$N_R = 2.3 \times 10^6$$

7. Check Fig. 2-10 to determine if the f value must be revised from the value used in the previous iteration. Since there is essentially no change in the f value, the flowrate calculated in step 6 is correct.

Manning equation. On the basis of work conducted during the latter part of the nineteenth century, Robert Manning published his now well-known equation for flow in open channels. Although this equation was originally intended for the design of open channels, it is now used for both open channels and closed conduits. The Manning equation is

$$V = \frac{1}{n} R^{2/3} S^{1/2} \quad \text{(SI units)} \tag{2-24}$$

$$V = \frac{1.486}{n} R^{2/3} S^{1/2} \quad \text{(U.S. customary units)} \tag{2-24a}$$

where V = velocity, m/s (ft/s)
n = coefficient of roughness
R = hydraulic radius, m (ft)
S = slope of energy grade line, m/m (ft/ft)

The hydraulic radius R is defined as

$$R = \frac{\text{cross-sectional area of flow, m}^2\text{ (ft}^2\text{)}}{\text{wetted perimeter, m (ft)}} \tag{2-25}$$

For a pipe flowing full, the hydraulic radius is

$$R = \frac{(\pi/4)(D^2)}{\pi D} = \frac{D}{4} \tag{2-26}$$

Substituting for R, the Manning equation for pipes flowing full becomes

$$V = \frac{0.397}{\text{n}} D^{2/3} S^{1/2} \quad \text{(SI units)} \tag{2-27}$$

$$V = \frac{0.590}{\text{n}} D^{2/3} S^{1/2} \quad \text{(U.S. customary units)} \tag{2-27a}$$

or, in terms of flowrate,

$$Q = \frac{0.312}{\text{n}} D^{8/3} S^{1/2} \quad \text{(SI units)} \tag{2-28}$$

$$Q = \frac{0.463}{n} D^{8/3} S^{1/2} \quad \text{(U.S. customary units)} \tag{2-28a}$$

where Q = flowrate, m^3/s (ft^3/s).

Typical n values for various types of pipes and open channels are presented in Table 2-1.

Hazen-Williams equation. Of the numerous exponential types of equations that describe the flow of water in pipes, the Hazen-Williams equation, which was developed in 1902, has been the one most commonly used for water pipes and wastewater force mains (see Sec. 9-4). The Hazen-Williams equation is

$$V = 0.849\, CR^{0.63}S^{0.54} \quad \text{(SI units)} \tag{2-29}$$

$$V = 1.318\, CR^{0.63}S^{0.54} \quad \text{(U.S. customary units)} \tag{2-29a}$$

where V = velocity, m/s (ft/s)
C = coefficient of roughness (C decreases with roughness)

Table 2-1 Values of n to be used with the Manning equation [2]

Surface	Best	Good	Fair	Bad
Uncoated cast-iron pipe	0.012	0.013	0.014	0.015
Coated cast-iron pipe	0.011	0.012[a]	0.013[a]	
Commercial wrought-iron pipe, black	0.012	0.013	0.014	0.015
Commercial wrought-iron pipe, galvanized	0.013	0.014	0.015	0.017
Smooth brass and glass pipe	0.009	0.010	0.011	0.013
Smooth lockbar and welded "OD" pipe	0.010	0.011[a]	0.013[a]	
Riveted and spiral steel pipe	0.013	0.015[a]	0.017[a]	
Vitrified sewer pipe	{0.010 / 0.011}	0.013[a]	0.015	0.017
Common clay drainage tile	0.011	0.012[a]	0.014[a]	0.017
Glazed brickwork	0.011	0.012	0.013[a]	0.015
Brick in cement mortar; brick sewers	0.012	0.013	0.015[a]	0.017
Neat cement surfaces	0.010	0.011	0.012	0.013
Cement mortar surfaces	0.011	0.012	0.013[a]	0.015
Concrete pipe	0.012	0.013	0.015[a]	0.016
Wood stave pipe	0.010	0.011	0.012	0.013
Plank flumes				
Planed	0.010	0.012[a]	0.013	0.014
Unplaned	0.011	0.013[a]	0.014	0.015
With battens	0.012	0.015[a]	0.016	
Concrete-lined channels	0.012	0.014[a]	0.016[a]	0.018
Cement-rubble surface	0.017	0.020	0.025	0.030
Dry-rubble surface	0.025	0.030	0.033	0.035
Dressed-ashlar surface	0.013	0.014	0.015	0.017
Semicircular metal flumes, smooth	0.011	0.012	0.013	0.015
Semicircular metal flumes, corrugated	0.0225	0.025	0.0275	0.030
Canals and ditches				
Earth, straight and uniform	0.017	0.020	0.0225[a]	0.025
Rock cuts, smooth and uniform	0.025	0.030	0.033[a]	0.035
Rock cuts, jagged and irregular	0.035	0.040	0.045	
Winding sluggish canals	0.0225	0.025[a]	0.0275	0.030
Dredged-earth channels	0.025	0.0275[a]	0.030	0.033
Canals with rough stony beds, weeds on earth banks	0.025	0.030	0.035[a]	0.040
Earth bottom, rubble sides	0.028	0.030[a]	0.033[a]	0.035
Natural-stream channels				
1. Clean, straight bank, full stage, no rifts or deep pools	0.025	0.0275	0.030	0.033
2. Same as (1), but some weeds and stones	0.030	0.033	0.035	0.040
3. Winding, some pools and shoals, clean	0.033	0.035	0.040	0.045
4. Same as (3), lower stages, more ineffective slope and sections	0.040	0.045	0.050	0.055
5. Same as (3), some weeds and stones	0.035	0.040	0.045	0.050
6. Same as (4), stony sections	0.045	0.050	0.055	0.060
7. Sluggish river reaches, rather weedy or with very deep pools	0.050	0.060	0.070	0.080
8. Very weedy reaches	0.075	0.100	0.125	0.150

[a]Values commonly used in designing.

R = hydraulic radius, m (ft)
S = slope of energy grade line, m/m (ft/ft)

This equation was developed originally in U.S. customary units and was written

$$V = CR^{0.63}S^{0.54}(0.001)^{-0.04}$$

Hazen and Williams stated that "the last term . . . (was) introduced to equalize the value of C with the value in . . . other exponential formulas . . . at a slope of 0.001 instead of at a slope of 1" [16]. The term $(0.001)^{-0.04}$, which appears as the term 1.318 in Eq. 2-29*a*, is combined with unit conversion factors to yield the constant 0.849 in Eq. 2-29.

If $D/4$ is substituted for the hydraulic radius R, the Hazen-Williams equation written in terms of the flowrate Q is

$$Q = 0.278\ CD^{2.63}S^{0.54} \quad \text{(SI units)} \tag{2-30}$$

$$Q = 0.432\ CD^{2.63}S^{0.54} \quad \text{(U.S. customary units)} \tag{2-30a}$$

where Q = flowrate, m^3/s (ft^3/s).

Typical C values are shown in Table 2-2.

Example 2-2 Use of the Hazen-Williams equation Determine the head loss in a 1000-m (3280-ft) pipeline with a diameter of 500 mm (20 in) that is discharging 0.25 m^3/s (8.8 ft^3/s). Assume that the Hazen-Williams coefficient for the pipe equals 130.

SOLUTION

1. Compute the head loss with the Hazen-Williams equation. Rearrange Eq. 2-30 to allow calculation of head loss.

$$Q = 0.278\ CD^{2.63}\left(\frac{h_f}{L}\right)^{0.54}$$

$$\frac{h_f}{L} = \left(\frac{Q}{0.278CD^{2.63}}\right)^{1/0.54}$$

$$h_f = \frac{10.7Q^{1.85}L}{C^{1.85}D^{4.87}}$$

Table 2-2 Hazen-Williams coefficients [14]

Type of pipe	C
Pipes extremely straight and smooth	140
Pipes very smooth	130
Smooth wood, smooth masonry	120
New riveted steel, vitrified clay	110
Old cast iron, ordinary brick	100
Old riveted steel	95
Old iron in bad condition	60–80

2. Substitute known values and compute the head loss h_f.

$$h_f = \frac{10.7(0.25 \text{ m}^3/\text{s})^{1.85} (1000 \text{ m})}{(130)^{1.85} (0.5 \text{ m})^{4.87}}$$

$$h_f = 2.96 \text{ m } (9.71 \text{ ft})$$

Comparison of the Darcy-Weisbach, Manning, and Hazen-Williams equations. Because the Darcy-Weisbach, Manning, and Hazen-Williams equations are all used frequently in practice, it is important to know their similarities and differences. They can be compared more easily if each is solved for the slope of the energy grade line:

Darcy-Weisbach:

$$S = \frac{h_f}{L} = \frac{8fQ^2}{\pi^2 g D^5} \tag{2-31}$$

Manning:

$$S = 10.3n^2 \frac{Q^2}{D^{16/3}} \quad \text{(SI units)} \tag{2-32}$$

$$S = 4.66n^2 \frac{Q^2}{D^{16/3}} \quad \text{(U.S. customary units)} \tag{2-32a}$$

Hazen-Williams:

$$S = \frac{10.7Q^{1.85}}{C^{1.85}D^{4.87}} \quad \text{(SI units)} \tag{2-33}$$

$$S = \frac{4.73Q^{1.85}}{C^{1.85}D^{4.87}} \quad \text{(U.S. customary units)} \tag{2-33a}$$

It can be seen that all three expressions are approximately of the form

$$S = \frac{KQ^2}{D^5} \tag{2-34}$$

where K = constant dependent on pipe roughness.

If these expressions are equated and simplified, the following relationship between f, n, and C is obtained:

$$0.0827f = \frac{10.3n^2}{D^{1/3}} = \frac{10.7D^{0.13}}{C^{1.85}D^{0.15}} \quad \text{(SI units)} \tag{2-35}$$

$$0.0252f = \frac{4.66n^2}{D^{1/3}} = \frac{4.73D^{0.13}}{C^{1.85}Q^{0.15}} \quad \text{(U.S. customary units)} \tag{2-35a}$$

When one of the coefficients is known, the other two can be calculated with Eq. 2-35. The resulting values will lead to identical slopes of the energy grade line as calculated with the three equations. (If the Hazen-Williams C value is sought, the flowrate must be known or estimated.) With Eq. 2-35, it can also be shown that the choice of the Hazen-Williams C and the Manning n that is appropriate for one diameter of pipe may not be appropriate for other diameters, even though the pipe material remains the same. Furthermore, for the Hazen-Williams equation to be consistent with the other two equations, the C value will vary slightly with flowrate. This variation is quite small, however, and usually can be neglected.

The Darcy-Weisbach f will vary with pipe size and, unless the flow is wholly turbulent, with the flowrate. Most of the flows encountered in designing wastewater facilities fall within the transition zone between turbulent and laminar flow.

General comments on pipe-flow equations. Before discussing the equations used for open-channel flow, a few comments about the preceding pipe-flow equations are warranted. The results obtained from these equations in practice are generally not precise. They are affected by the values of the friction factor, the choice of which is not precise in the first place.

In addition, when the equations are applied to wastewater flows, there are complications because the fluid characteristics differ; wastewater, for example, contains solid material and clean water does not. Joints and various other discontinuities in pipelines disrupt the flow patterns. As the pipelines age, their characteristics are changed by corrosion and the deposition of solids. If the analyst recognizes these and other complications, the results obtained with these equations are quite satisfactory.

Flow in Open Channels

In the late nineteenth and early twentieth centuries, a number of empirical equations were developed for determining flows in open channels, including those proposed by Chezy, Ganguillet and Kutter, Manning, and Scobey. Of these, only the Manning equation is now in general use.

Because of its simplicity and because considerable experimental data are available for estimating values of the friction factor, the Manning equation (as presented previously and repeated here for convenience) is now the one most commonly used in the design of sewers. The Manning equation is

$$V = \frac{1}{n} R^{2/3} S^{1/2} \qquad \text{(SI units)} \qquad (2\text{-}24)$$

$$V = \frac{1.486}{n} R^{2/3} S^{1/2} \qquad \text{(U.S. customary units)} \qquad (2\text{-}24a)$$

The S in the Manning equation is the slope of the energy grade line, not the slope of the channel bottom. However, in uniform flow, these slopes are equivalent and the slope of the channel bottom can be used.

Typical n values for various types of open channels were listed in Table 2-1. In general, n values varying from 0.013 to 0.015 are used in sewer design. Furthermore, these n values usually are assumed to be valid for all depths of flow. However, in experiments conducted to determine the effect of variation in depth of flow on the friction factor n, it has been shown conclusively that n values are greater in partially filled sewers than in sewers flowing full [10]. The variation in n with depth of flow in a circular sewer is shown in Fig. 2-16 (see Sec. 2-4). Thus, when designing sewers that will usually flow partly full, the n value selected should be adjusted accordingly. The application of the Manning equation in developing sewer-design charts is illustrated in Sec. 2-4.

Example 2-3 Use of the Manning equation Determine the flowrate in a rectangular concrete channel with a width of 3 m (9.8 ft) and a slope of 0.001 m/m (ft/ft) when the depth of flow is 1.5 m (4.9 ft). Assume that $n = 0.014$.

SOLUTION

1. Determine the hydraulic radius R.

$$R = \frac{A}{P}$$

$$R = \frac{(3\text{ m})\,(1.5\text{ m})}{3\text{ m} + 2(1.5\text{ m})}$$

$$R = 0.75\text{ m (2.46 ft)}$$

2. Calculate flowrate Q with Eq. 2-24 by multiplying by the area.

$$Q = \frac{1}{n}AR^{2/3}S^{1/2}$$

$$Q = \frac{1}{0.014}(3\text{ m} \times 1.5\text{ m})\,(0.75\text{ m})^{2/3}\,(0.001)^{1/2}$$

$$Q = 8.39\text{ m}^3/\text{s } (296\text{ ft}^3/\text{s})$$

2-3 PIPE SIZES

In finding the correspondence between pipe sizes expressed in metric units and those expressed in U.S. customary units, a problem has developed with respect to pipe sizes. As shown in Table 2-3, existing pipe sizes in the United States are reported in both inches and inches converted to millimeters. Because the United States pipe industry has a high financial investment in existing manufacturing equipment, the switchover to metric sizes will probably occur gradually over a period of time, as the existing manufacturing equipment is replaced. Therefore, to ease the conversion process, it has been proposed that the designation of pipe sizes be changed but not the actual sizes of the pipes.

Table 2-3 Sewer pipe sizes in metric and U.S. customary units

Metric sizes, mm	Existing U.S. pipe sizes	
	in	mm
100	4	101.6
125	5	127.0
150	6	152.4
200	8	203.2
250	10	254.0
300	12	304.8
350	14	355.6
375	15	381.0
400	16	406.4
450	18	457.2
500	20	508.0
525	21	533.4
600	24	609.6
675	27	685.8
750	30	762.0
900	36	914.4
1,050	42	1,066

Note: mm $\times$ 0.03937 = in

For example, an 8-in pipe, which is actually 203.2 mm in diameter, will be called a 200-mm pipe. In effect, if this system of designation is adopted, then the capacities and velocities computed will be about 4 and 2 percent smaller, respectively, than the actual values. In computations for pipes manufactured in metric sizes, however, capacities and velocities computed will be the actual values. In this book, metric pipe sizes are used for all computations.

2-4 DESIGN CHARTS AND TABLES

To aid in the solution of the flow equations discussed previously in this chapter, various design charts and tables have been developed. Several that are applicable to sewers when they are flowing full are presented in this section. They are followed by a discussion of Manning's equation, which is applicable to sewers flowing partly full. Finally, charts for the hydraulic elements of noncircular sewers are presented.

Sewers Flowing Full

As mentioned previously, the Manning equation is the one most common in sewer design. The most direct method of using this equation to solve flow problems in sewers and pipes in general is by preparing nomographs, such as those shown in Figs. 2-12 through 2-15, which can be prepared for any design conditions. Values of 0.013 and 0.015 for Manning's n were selected for these diagrams because they are the most frequently used values in sewer design.

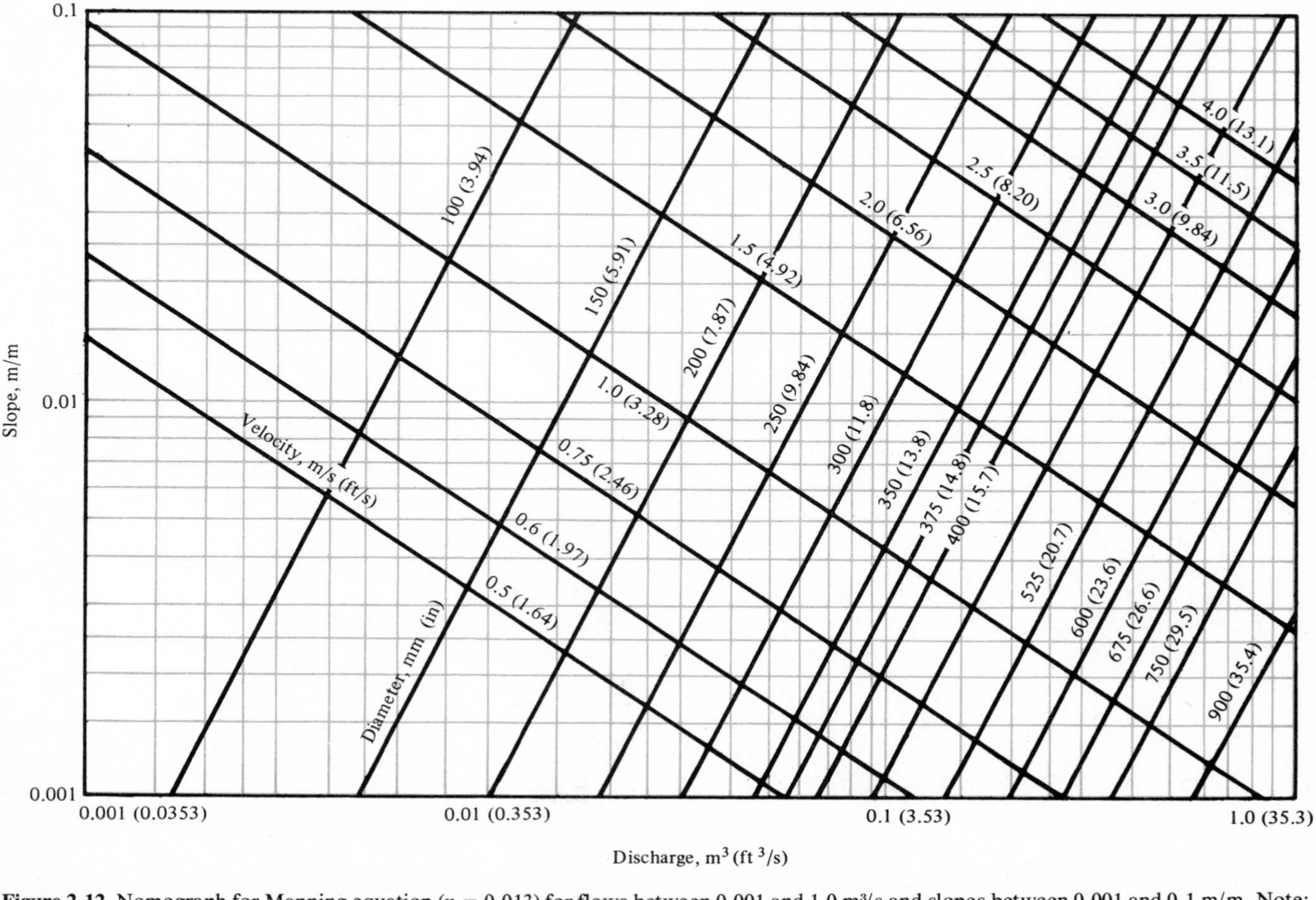

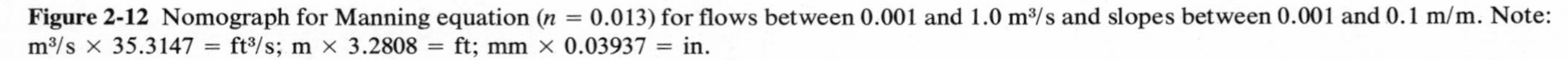

Figure 2-12 Nomograph for Manning equation ($n = 0.013$) for flows between 0.001 and 1.0 m^3/s and slopes between 0.001 and 0.1 m/m. Note: $m^3/s \times 35.3147 = ft^3/s$; $m \times 3.2808 = ft$; $mm \times 0.03937 = in$.

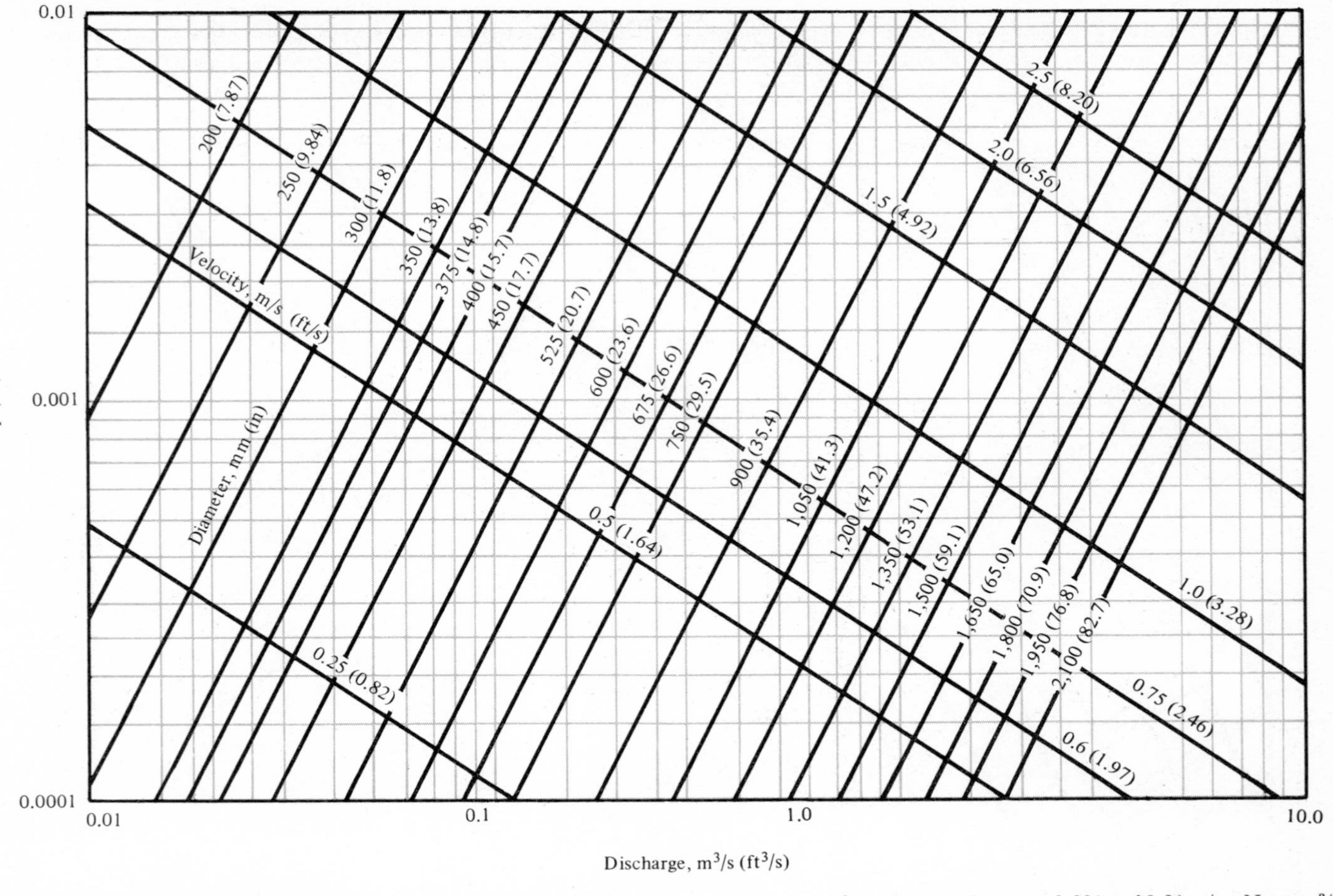

Figure 2-13 Nomograph for Manning equation ($n = 0.013$) for flows between 0.01 and 10.0 m^3/s and slopes between 0.001 and 0.01 m/m. Note: $m^3/s \times 35.3147 = ft^3/s$; m $\times$ 3.2808 = ft; mm $\times$ 0.03937 = in.

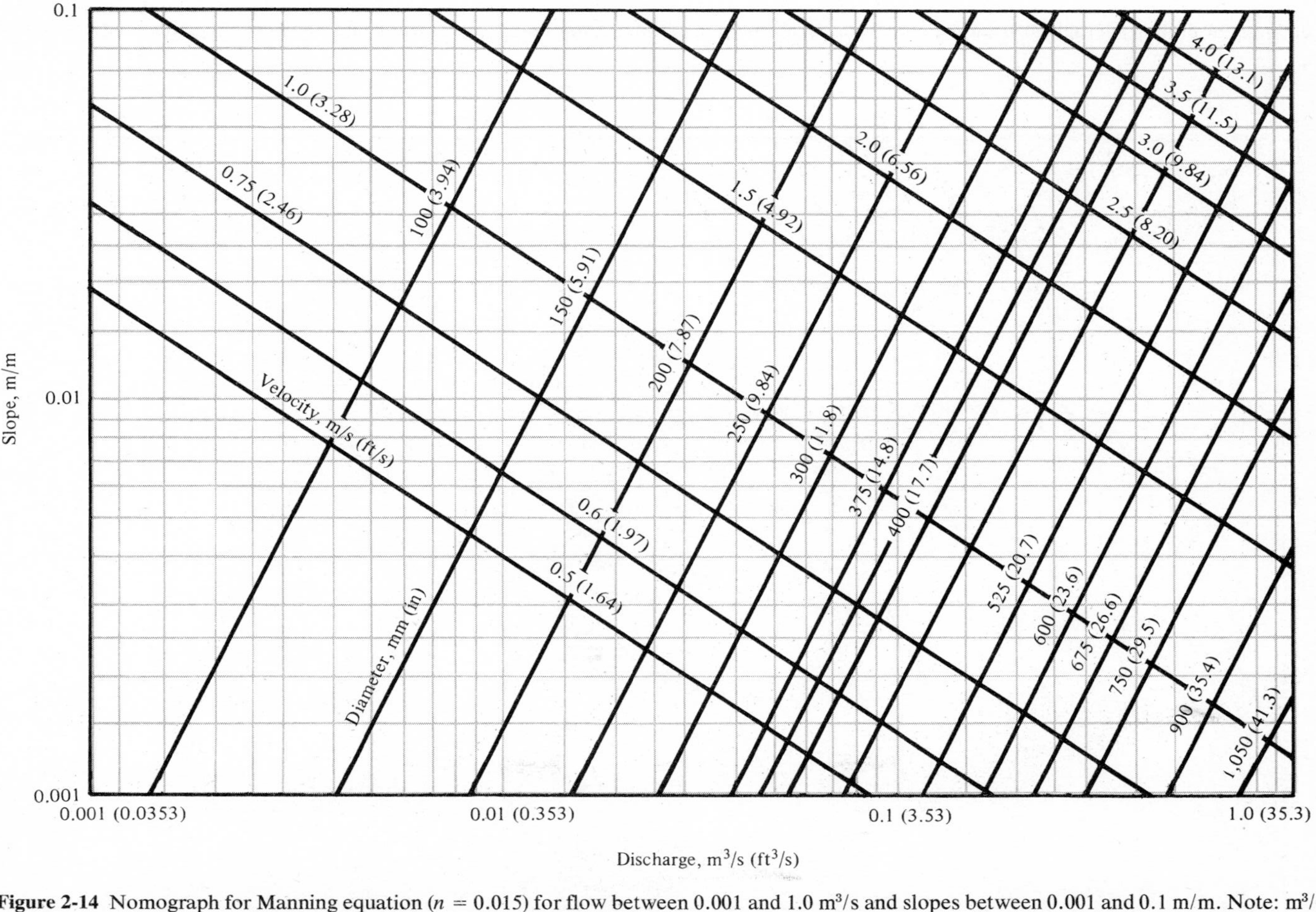

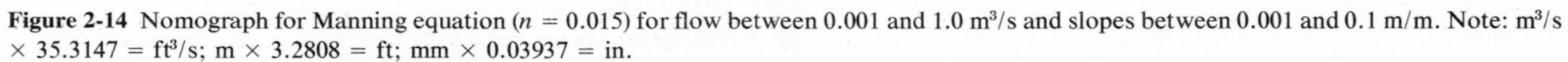

Figure 2-14 Nomograph for Manning equation ($n = 0.015$) for flow between 0.001 and 1.0 m^3/s and slopes between 0.001 and 0.1 m/m. Note: $m^3/s \times 35.3147 = ft^3/s$; $m \times 3.2808 = ft$; $mm \times 0.03937 = in$.

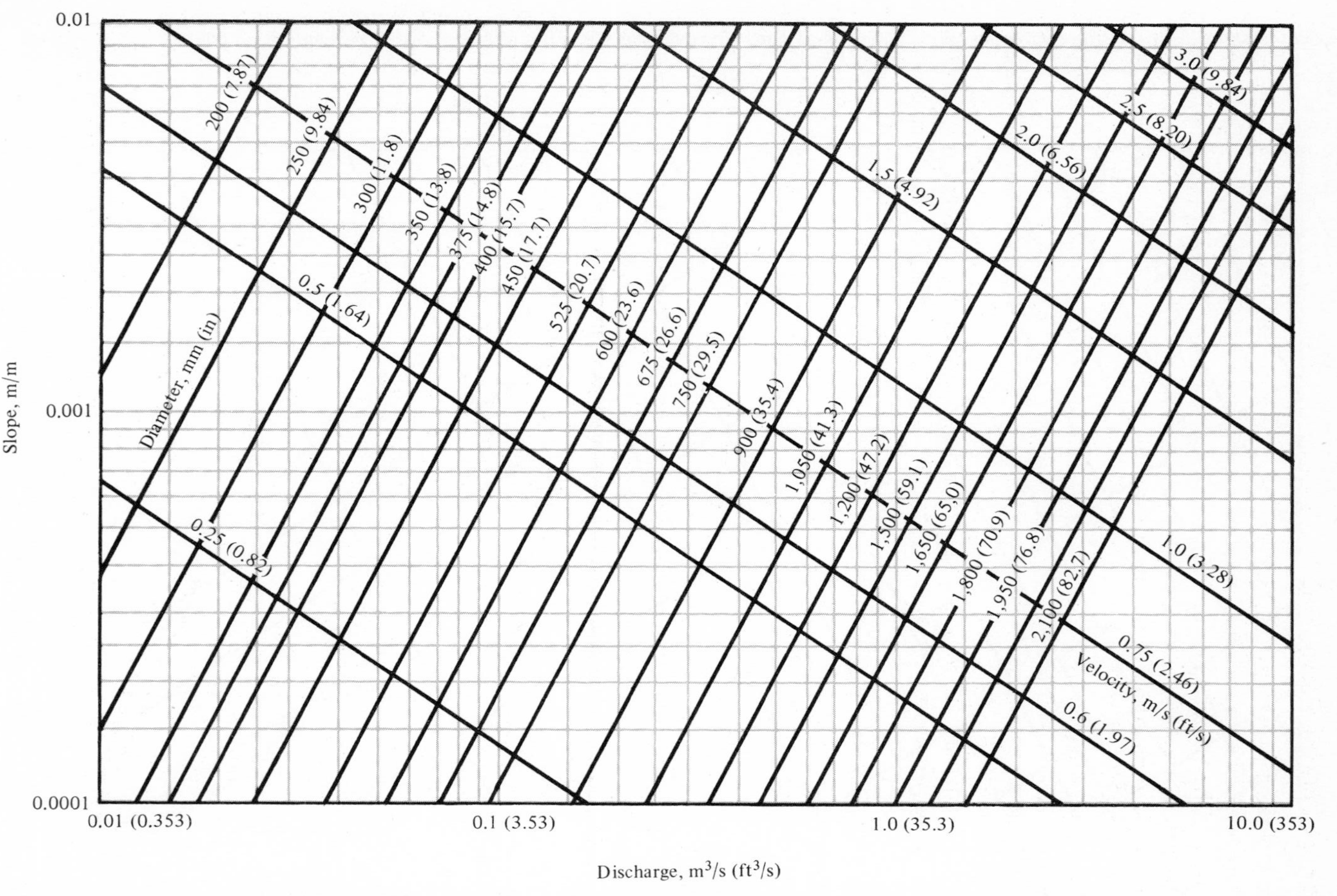

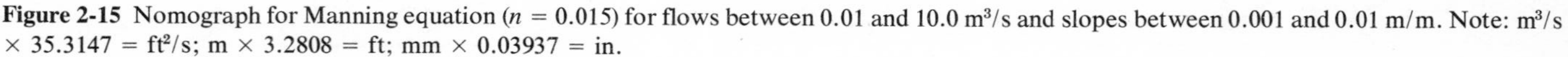

Figure 2-15 Nomograph for Manning equation ($n = 0.015$) for flows between 0.01 and 10.0 m^3/s and slopes between 0.001 and 0.01 m/m. Note: $m^3/s \times 35.3147 = ft^3/s$; m × 3.2808 = ft; mm × 0.03937 = in.

Sewers Flowing Partly Full

In general, sewers are designed to flow full only under maximum conditions. Therefore, in many of the problems arising in sewer design, it is necessary to estimate the velocity and discharge when a sewer is partly filled. The relations between hydraulic elements for flow at full depth and at other depths in circular sewers, computed according to the Manning equation, are shown in Fig. 2-16. The hydraulic elements for a circular sewer, as shown in Fig. 2-16, are the hydraulic radius R, the cross-sectional area of the flowing stream A, the average velocity V, and the rate of discharge Q.

If the variation of n with depth is to be neglected, calculations involving flow in partly filled sewers can easily be handled by using the data in Tables 2-4 and 2-5 adapted from Brater and King's *Handbook of Hydraulics* [2]. The following two examples illustrate the use of Fig. 2-16 and Tables 2-4 and 2-5.

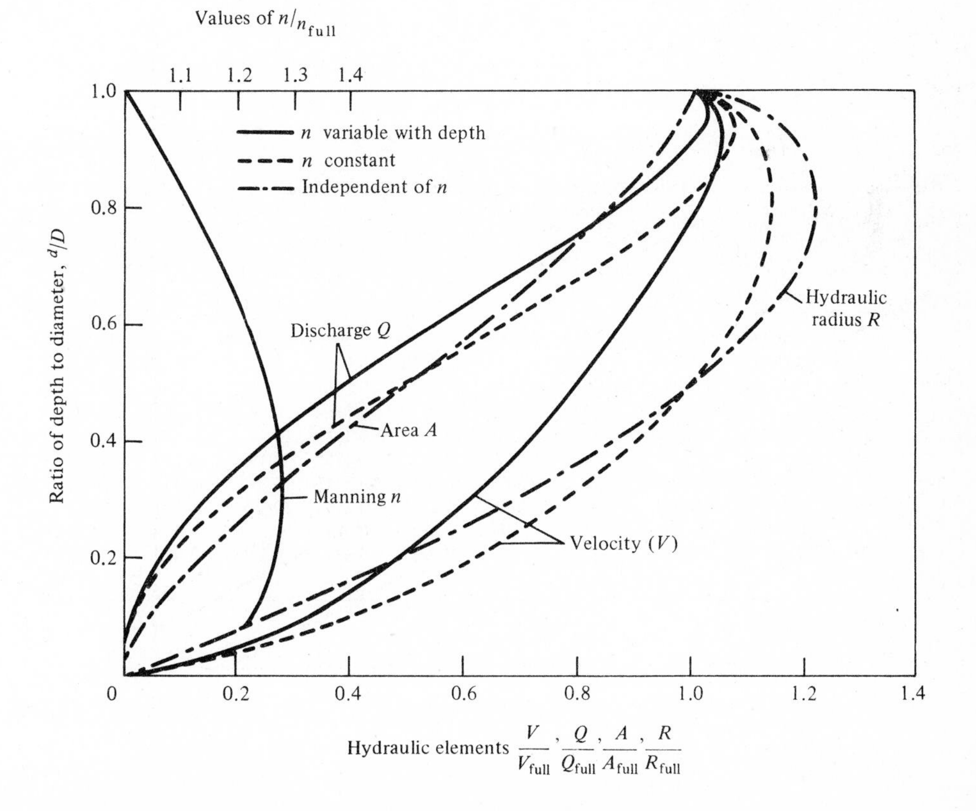

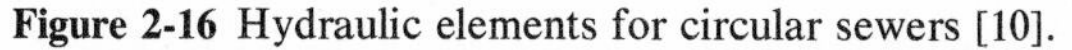

Figure 2-16 Hydraulic elements for circular sewers [10].

Table 2-4 Values of K for circular channels in terms of depth of flow in the equation[a] $Q = (K/n)d^{8/3}S^{1/2}$

$\frac{d^a}{D}$	0.00	0.01	0.02	0.03	0.04	0.05	0.06	0.07	0.08	0.09
0.0	....	10.11	7.11	5.77	4.96	4.41	4.00	3.68	3.42	3.20
0.1	3.02	2.85	2.72	2.60	2.49	2.39	2.29	2.21	2.13	2.06
0.2	2.00	1.93	1.88	1.82	1.77	1.72	1.68	1.63	1.59	1.55
0.3	1.51	1.48	1.44	1.41	1.38	1.35	1.32	1.29	1.26	1.23
0.4	1.21	1.18	1.16	1.14	1.11	1.09	1.07	1.05	1.03	1.01
0.5	0.990	0.971	0.952	0.934	0.916	0.899	0.882	0.805	0.849	0.833
0.6	0.818	0.802	0.787	0.772	0.758	0.744	0.730	0.716	0.702	0.690
0.7	0.676	0.662	0.650	0.637	0.624	0.612	0.600	0.588	0.576	0.564
0.8	0.552	0.541	0.529	0.518	0.507	0.495	0.484	0.473	0.462	0.451
0.9	0.440	0.429	0.418	0.407	0.395	0.384	0.372	0.360	0.348	0.334
1.0	0.312									

[a]Adapted from Ref. 2,

where Q = flowrate, m^3/s

n = Manning coefficient of friction

d = depth of flow, m

S = slope of energy grade line, m/m.

[b]D = diameter of conduit, m.

Note: $m^3/s \times 35.3147 = ft^3/s$

m × 3.2808 = ft

Table 2-5 Values of K' for circular channels in terms of diameter in the equation[a] $Q = (K'/n)D^{8/3}S^{1/2}$

$\frac{d^b}{D}$	0.00	0.01	0.02	0.03	0.04	0.05	0.06	0.07	0.08	0.09
0.0	....	0.000047	0.00021	0.00050	0.00093	0.00150	0.00221	0.00306	0.00407	0.00521
0.1	0.00651	0.00795	0.00953	0.0113	0.0131	0.0152	0.0173	0.0196	0.0220	0.0246
0.2	0.0273	0.0301	0.0331	0.0362	0.0394	0.0427	0.0461	0.0497	0.0534	0.0572
0.3	0.0610	0.0650	0.0691	0.0733	0.0776	0.0820	0.0864	0.0910	0.0956	0.1003
0.4	0.1050	0.1099	0.1148	0.1197	0.1248	0.1298	0.1349	0.1401	0.1453	0.1506
0.5	0.156	0.161	0.166	0.172	0.177	0.183	0.188	0.193	0.199	0.204
0.6	0.209	0.215	0.220	0.225	0.231	0.236	0.241	0.246	0.251	0.256
0.7	0.261	0.266	0.271	0.275	0.280	0.284	0.289	0.293	0.297	0.301
0.8	0.305	0.308	0.312	0.315	0.318	0.321	0.324	0.326	0.329	0.331
0.9	0.332	0.334	0.335	0.335	0.335	0.335	0.334	0.332	0.329	0.325
1.0	0.312									

[a]Adapted from Ref. 2,

where Q = flowrate, m^3/s

n = Manning coefficient of friction

D = diameter of conduit

S = slope of energy grade line, m/m.

[b]d = depth of flow

Note: $m^3/s \times 35.3147 = ft^3/s$

m × 3.2808 = ft

Example 2-4 Determination of depth and velocity in a sewer flowing partly full Determine the depth of flow and velocity in a sewer with a diameter of 300 mm (12 in) laid on a slope of 0.005 m/m (ft/ft) with an n value of 0.015 when discharging 0.01 m^3/s (0.35 ft^3/s).

SOLUTION

1. Compute the value of K' in the equation

$$Q = (K'/n)\, D^{8/3}S^{1/2}$$

$$K' = \frac{nQ}{D^{8/3}S^{1/2}} = \frac{0.015 \times 0.01 \text{ m}^3/\text{s}}{(0.3 \text{ m})^{8/3}\,(0.005 \text{ m/m})^{1/2}} = 0.0526$$

2. To determine depth of flow, locate in Table 2-5 the computed value of K' and find the corresponding value of d/D. The value from Table 2-5 is approximately 0.28.

$$\text{Depth of flow } d = 300 \text{ mm} \times 0.28 = 84 \text{ mm (3.31 in)}$$

3. Compute the velocity. From Fig. 2-16, for a value of d/D of 0.28 (84/300), the area of flow is equal to 0.22 times the area of the pipe, which is equal to 0.22(0.0707 m^2) = 0.0156 m^2 (0.168 ft). The velocity is equal to:

$$\text{Velocity} = \frac{0.01 \text{ m}^3/\text{s}}{0.0156 \text{ m}^2} = 0.641 \text{ m/s (2.11 ft/s)}$$

Example 2-5 Determination of the diameter of a sewer flowing partly full Determine the diameter of a sewer required to handle a flow of 0.15 m^3/s (5.3 ft^3/s) when flowing 65 percent full. The sewer is to be laid at a slope of 0.001 m/m (ft/ft) and the n value is assumed to be 0.013.

SOLUTION

1. Determine K' in the equation

$$Q = (K'/n)\, D^{8/3}S^{1/2}$$

Entering Table 2-5 with a d/D value of 0.65, the corresponding value of K' is 0.236.

2. Compute diameter D.

$$D = \left(\frac{Qn}{K'S^{1/2}}\right)^{3/8} = \left[\frac{0.15 \text{ m}^3/\text{s} \times 0.013}{0.236 \times (0.001 \text{ m/m})^{1/2}}\right]^{3/8}$$

$$= 0.605 \text{ m (1.98 ft)}$$

Use a 600-mm (24-in) pipe.

Hydraulic Elements of Noncircular Sewer Sections

Almost all the sewers now constructed in the United States are circular cross sections. In the past, however, a wide variety of noncircular sewer sections were used, including egg-shaped, semielliptical, horseshoe, basket handle, oval, catenary, gothic, parabolic, and elliptical. The first four of these were the more popular shapes. Typical examples of these sections, along with data on their shapes and hydraulic elements, are shown in Figs. 2-17 and 2-18 and in

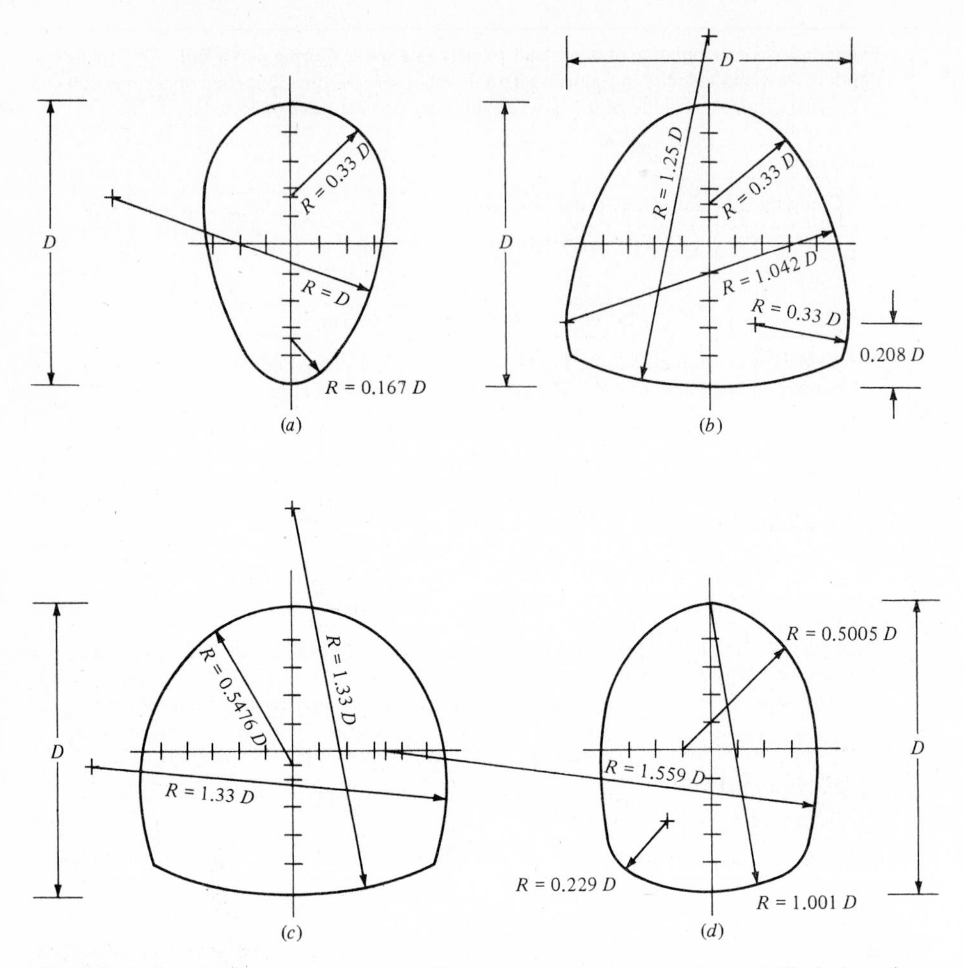

Figure 2-17 Noncircular sewer sections [12]. (*a*) Egg-shaped section; (*b*) semielliptical section; (*c*) horseshoe section; (*d*) basket handle section.

Table 2-6. Detailed data on these and other sections may be found in Refs. 1 and 11.

2-5 MINOR LOSSES

Head losses in sewerage systems other than the normal frictional losses that typically occur where the magnitude or direction of the fluid velocity changes are called *minor losses*. These minor losses can usually be expressed as functions of the squares of representative fluid velocities. Minor losses in closed conduits are somewhat different from those in open channels.

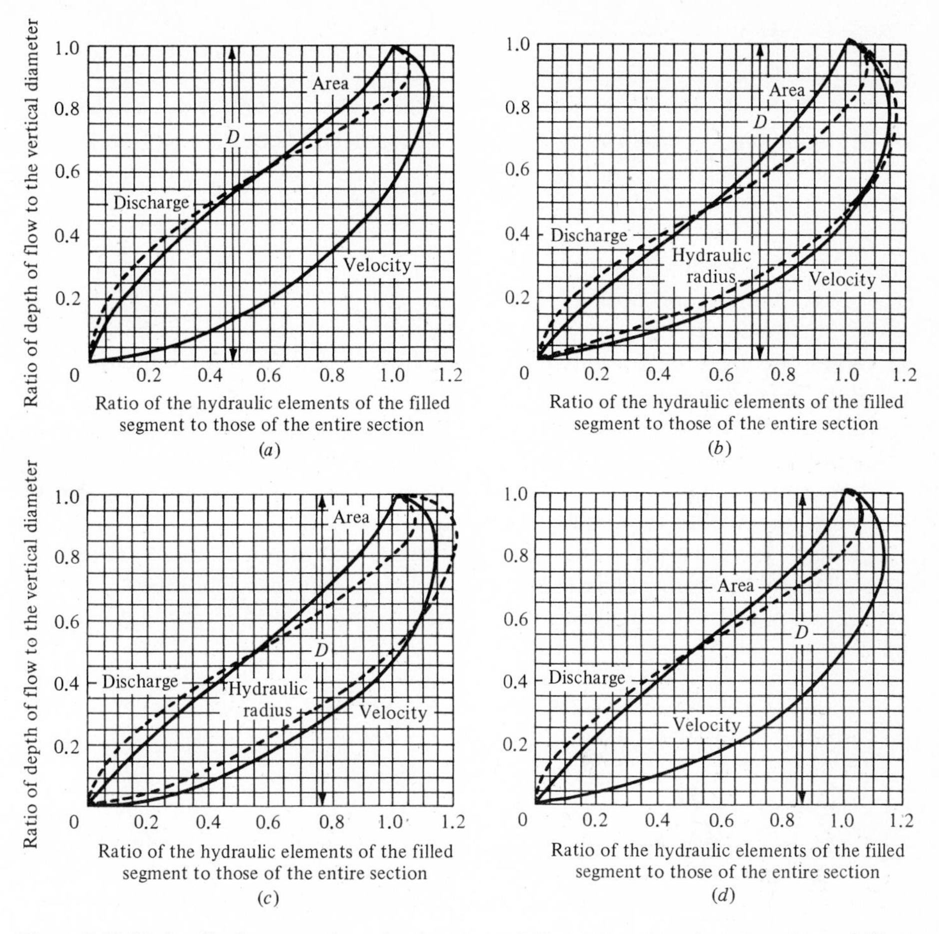

Figure 2-18 Hydraulic elements of noncircular sewers [12]. (*a*) Egg-shaped section; (*b*) semielliptical section; (*c*) horseshoe section; (*d*) basket handle section.

Table 2-6 Hydraulic elements of various sewer sections [1, 11] (see Figs. 2-17 and 2-18)

	Hydraulic elements of full sections[a]		
Type of sewer	Area	Wetted perimeter	Hydraulic radius
Egg-shaped	$0.510D^2$	$2.643D$	$0.193D$
Semielliptical	$0.783D^2$	$3.258D$	$0.240D$
Horseshoe	$0.913D^2$	$3.466D$	$0.263D$
Basket handle	$0.786D^2$	$3.193D$	$0.246D$

[a] D = diameter of conduit, m.
Note: m × 3.2808 = ft

Minor Losses in Closed-Conduit Flow

Most minor losses in a closed-conduit system can be expressed as a multiple of the velocity head immediately upstream, within, or immediately downstream of the fitting or pipe configuration causing the loss. The general equation is

$$h_m = K\frac{V^2}{2g} \qquad (2\text{-}36)$$

where h_m = minor loss, m (ft)
K = minor loss coefficient
$V^2/2g$ = velocity head, m (ft)

When the velocity head changes as the flow proceeds through the fitting in question, it is important for the analyst to know on which velocity head the minor-loss coefficient is based. In most tables of minor-loss coefficients, this information is given.

For some fittings, such as expansions and contractions, the equation for the minor loss will differ slightly from Eq. 2-36.

An extensive listing of the types of fittings and appurtenances frequently encountered in closed-conduit systems and the associated minor-loss coefficients (and minor-loss equations, if different from Eq. 2-36) is given in Appendix C.

Example 2-6 Calculation of flowrate with minor losses involved Determine the flowrate from reservoir A to reservoir B for the system shown in Fig. 2-19. Assume that the friction factors f are constant at the values given.

SOLUTION

1. Develop an expression that can be used to compute the frictional losses in the straight sections of pipe as a function of the velocity in the 200-mm pipe.
   a The velocity in the 500-mm pipe can be found from the continuity equation as follows:

$$V_{500} = V_{200}\frac{A_{200}}{A_{500}}$$

$$V_{500} = V_{200}\frac{(\pi/4)(0.2\text{ m})^2}{(\pi/4)(0.5\text{ m})^2}$$

$$V_{500} = 0.16V_{200}$$

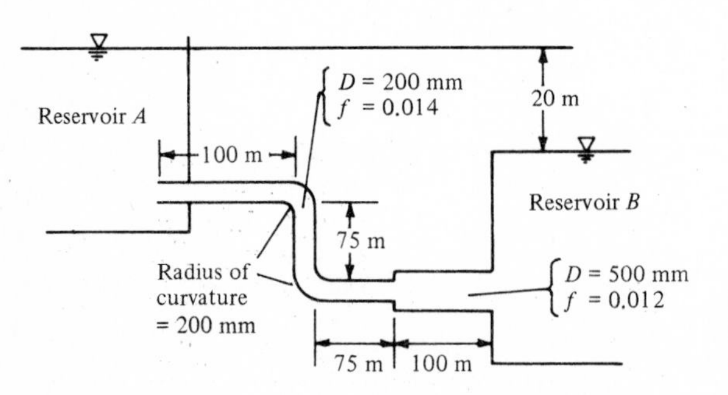

Figure 2-19 Definition sketch for Example 2-16. Note: m × 3.2808 = ft; mm × 0.03937 = in.

b The total frictional loss in the two straight sections of pipe is

$$h_f = 0.014 \frac{250 \text{ m}}{0.20 \text{ m}} \frac{V_{200}^2}{(2)9.81 \text{ m/s}^2} + 0.012 \frac{100 \text{ m}}{0.50 \text{ m}} \frac{(0.16V_{200})^2}{(2)9.81 \text{ m/s}^2}$$

$$h_f = (0.895 \text{ s}^2/\text{m})V_{200}^2$$

2. Develop relationships for the minor losses as a function of V_{200}.

a The entrance loss at reservoir *A* as a function of V_{200} is

$$h_{en} = 1.0 \frac{V_{200}^2}{2g}$$

$$h_{en} = (0.051 \text{ s}^2/\text{m})V_{200}^2$$

b The loss in the bends as a function of V_{200} is

$$h_b = 2(0.30) \frac{V_{200}^2}{2g}$$

$$h_b = (0.031 \text{ s}^2/\text{m})V_{200}^2$$

c The loss at the sudden expansion as a function of V_{200} is

$$h_{se} = \frac{(V_{200} - V_{500})^2}{2g}$$

$$h_{se} = \frac{(V_{200} - 0.16V_{200})^2}{(2)9.81 \text{ m/s}^2}$$

$$h_{se} = (0.036 \text{ s}^2/\text{m})V_{200}^2$$

d The exit loss at reservoir *B* as a function of V_{200} is

$$h_{ex} = 1.0 \frac{(0.16V_{200})^2}{2g}$$

$$h_{ex} = (0.001 \text{ s}^2/\text{m})V_{200}^2$$

3. Determine the velocity in the 200-mm pipe. The velocity is found by equating the sum of all the head losses (expressed in terms of the velocity in the 200-mm pipe) to the head available for driving the flow which, in this case, is the difference in water-surface elevations between the two reservoirs, and solving for V_{200}.

$$H = h_f + h_{en} + h_b + h_{se} + h_{ex}$$

$$20 \text{ m} = V_{200}^2(0.895 + 0.051 + 0.031 + 0.036 + 0.001) \text{ s}^2/\text{m}$$

$$V_{200}^2 = \frac{20 \text{ m}}{1.014 \text{ s}^2/\text{m}}$$

$$V_{200} = 4.44 \text{ m/s } (14.57 \text{ ft/s})$$

4. Determine the flowrate from reservoir *A* to reservoir *B*.

$$Q = V_{200}A_{200}$$

$$Q = (4.44 \text{ m/s})\left[\frac{\pi}{4}(0.2 \text{ m})^2\right]$$

$$Q = 0.139 \text{ m}^3/\text{s } (4.91 \text{ ft}^3/\text{s})$$

Minor Losses in Open-Channel Flow

Minor losses in open-channel flow frequently include those at channel entrances, contractions and expansions, bends, curves, transitions, and channel exits. Although there are no universally accepted equations for determining the magnitude of these minor losses, some of the more common methods and equations are presented in this section. For additional information, Refs. 2, 5, 7, 8, and 10 are recommended.

Entrance losses. The conditions that govern head losses at the entrance to an open channel or conduit leading from a reservoir are similar to those for a pipe flowing under pressure. However, experimental data are lacking. It is probably safe to estimate the loss on the same basis as in a closed conduit.

Contractions and expansions. When all depths of flow are greater than the critical depth, the loss of head at contractions and expansions can be related directly to the difference in velocity heads before and after the change in cross section. (Critical depth is the depth of flow that occurs at minimum specific energy and is discussed in more detail in Sec. 2-6.) The following equations can be used [2]:

For contractions

$$h_c = K_c\left(\frac{V_2^2}{2g} - \frac{V_1^2}{2g}\right) \tag{2-37}$$

For expansions

$$h_e = K_e\left(\frac{V_1^2}{2g} - \frac{V_2^2}{2g}\right) \tag{2-38}$$

where h_c = head loss at contraction, m (ft)
h_e = head loss at expansion, m (ft)
K_c = constant for contraction
K_e = constant for expansion
V_1 = fluid velocity approaching the change in cross section, m/s (ft/s)
V_2 = fluid velocity after the change in cross section, m/s (ft/s)
g = acceleration due to gravity, 9.81 m/s^2 (32.2 ft/s^2)

Typical values for K_c and K_e are given in Table 2-7.

Bends. Information on the head loss in bends for open-channel flow in sewers is almost nonexistent. However, it seems reasonable to use the same analysis that was used for bends in closed conduits. When channels or conduits join at a 90° angle without a curved section, as in some treatment-plant channels, for example, it usually is appropriate (in conservative design) to assume a loss of the entire entering velocity head.

Table 2-7 Typical K_c and K_e values for determining head loss at contractions and expansions in open channel flow[a] [2]

Form of transition	K_c	K_e
Sudden change in area, sharp corners	0.5	1.0
"Well designed"[b]		
Best	0.05	0.10
Design value	0.10	0.20

[a] At depths greater than critical depth.
[b] All surfaces connected by tangent curves and flow lines at angles less than 12½° with the axis of channel.

Curves. Even a slight change in direction can disturb flow, which increases the frictional resistance. The length in which such disturbance occurs is of greater significance than the sharpness of the deflection. The total head loss through a bend is equal to the sum of the frictional loss that would occur in an equal length of straight pipe and the eddy losses caused by the change in direction. A curve of short radius and correspondingly short length may cause a smaller loss of head than a curve of long radius with the same change of direction. There are cases, however, in which the opposite effect is probable.

From the fragmentary data available, it appears that the effect of curvature has generally been equivalent to an increase in the value of n by an amount varying from 0.003 to 0.005 in the sections containing significant curvature.

Transitions. Transitions in sewerage systems are required between sewers of different sizes, between sewers laid on different slopes, and at sewer junctions. In general, the function of the transition is to change either the shape or the cross section of the flow. The hydraulic head losses associated with transitions are friction losses and conversion losses. As a first approximation in the design of transition structures, frictional losses can usually be neglected. Conversion losses, such as momentum losses that depend on the geometry of the structure, must be determined or estimated individually for each structure. For this reason, there is no fixed approach to the design of transition structures. When designing a transition structure, it is recommended that the reader refer to Refs. 3 to 7 and 9.

Exit losses. When an open channel discharges into a tank or reservoir, it is difficult to recover any of the velocity head associated with the flow in the channel. As the flowing water enters the body of still water, the velocity head is dissipated through turbulent mixing. Thus, the exit loss is taken to be $1.0V^2/2g$.

2-6 NONUNIFORM OPEN-CHANNEL FLOW

Conditions of steady nonuniform flow exist when a constant quantity of water flows through variable cross sections, slopes, and velocities. The surface of the water, therefore, is not parallel to the invert of the conduit. This condition always exists at points of changing equilibrium, such as at and near changes of grade and in cross sections, and above obstructions of free outlets. Typical examples of nonuniform flow are shown in Figs. 2-20 through 2-22.

General Equation for Nonuniform Flow

To analyze nonuniform flow, it is necessary to consider a reach of channel short enough so that flow conditions do not change greatly over its length. The total drop in the energy grade line over this reach of channel can be approximated from the slopes of the energy grade line at upstream and downstream

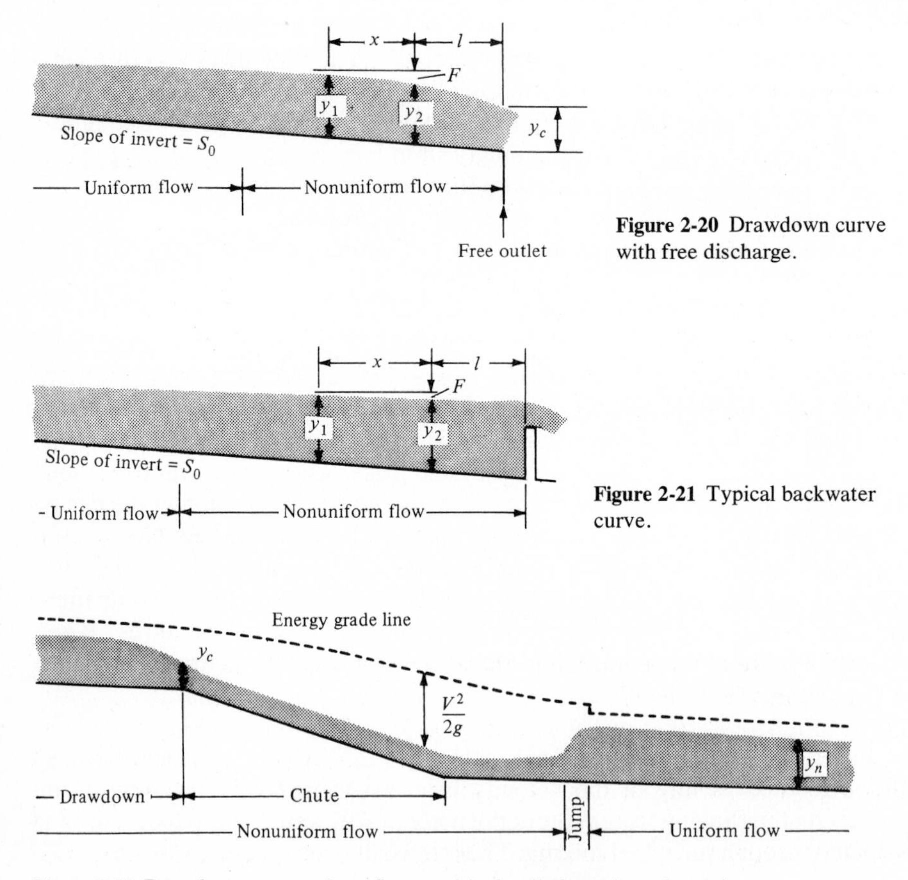

Figure 2-20 Drawdown curve with free discharge.

Figure 2-21 Typical backwater curve.

Figure 2-22 Drawdown curve, chute flow, and hydraulic jump in a channel.

ends of the reach. The slope at each end is determined by rearranging the Manning equation (Eq. 2-24) as follows:

$$S = \left(\frac{Vn}{R^{2/3}}\right)^2 \qquad \text{(SI units)} \tag{2-39}$$

$$S = \left(\frac{Vn}{1.486\, R^{2/3}}\right)^2 \qquad \text{(U.S. customary units)} \tag{2-39a}$$

or, in terms of flowrate Q,

$$S = \left(\frac{Qn}{AR^{2/3}}\right)^2 \qquad \text{(SI units)} \tag{2-40}$$

$$S = \left(\frac{Qn}{1.486\, AR^{2/3}}\right)^2 \qquad \text{(U.S. customary units)} \tag{2-40a}$$

If subscript 1 is used to denote upstream conditions and subscript 2 is used to denote downstream conditions, the effective slope of the energy grade line over the reach of channel is

$$S_a = \frac{S_1 + S_2}{2} \tag{2-41}$$

where S_1 and S_2 are calculated with Eq. 2-39 or Eq. 2-40.

The total fall in the water surface F is equal to the sum of the frictional loss and the difference in the velocity heads

$$F = xS_a + \frac{V_1^2}{2g} - \frac{V_2^2}{2g} \tag{2-42}$$

Any equation for frictional loss may be used to evaluate the term xS_a.

If y_1 and y_2 represent the depths of water at the two ends of the reach, and S_0 represents the inclination of the bottom, then

$$F = xS_0 + (y_1 - y_2) \tag{2-43}$$

Equating the values of F in Eqs. 2-43 and 2-44 yields

$$x = \frac{\left(y_1 + \dfrac{V_1^2}{2g}\right) - \left(y_2 + \dfrac{V_2^2}{2g}\right)}{S_a - S_0} \tag{2-44}$$

With the use of this expression, the distance x between any two sections of the stream, in which the change in depth is $y_1 - y_2$, can be computed approximately. The sign of x is always positive (+) in the direction of computation. The foregoing expressions are general and may be applied to any case of steady nonuniform flow, within the limits of the accuracy of the assumptions.

Control sections. All water-surface calculations must begin at a section in which the depth of flow is known. Often this section is a control section in which the depth is uniquely determined by the rate of discharge and the geometry of the section. Examples of controls include weirs; sudden expansions into large bodies of water such as reservoirs or oceans; falls; and abrupt changes in channel bottom slope from mild to steep; and in long reaches of channel in which steady uniform flow can take place.

Critical depth. Control sections occur where there is a free discharge over a fall, illustrated in Fig. 2-20, or where there is a decided increase in inclination of the channel, illustrated in Fig. 2-22. In such sections the depth of flow at the outlet or at the break in grade will be definitely fixed by the rate of discharge for any given conduit. This depth is called the *critical depth,* designated as y_c, and is the depth for which the specific energy (that is, $y + V^2/2g$) is a minimum (see Figs. 2-2*a* and 2-2*b*).

The expression for the specific energy in a rectangular channel may be written in terms of the discharge q per unit width as

$$E = y + \frac{q^2}{2gy^2} \tag{2-45}$$

Differentiating Eq. 2-45 with respect to depth, and setting the result to zero, yields

$$\frac{dE}{dy} = 1 - \frac{q^2}{gy_c^3} = 0 \tag{2-46}$$

or

$$y_c = \left(\frac{q^2}{g}\right)^{1/3} \tag{2-47}$$

It can also be shown that, for any channel shape, the following relationships apply when the flow is at critical depth:

$$\frac{Q^2}{g} = \frac{A^3}{T} \tag{2-48}$$

$$\frac{V^2}{2g} = \frac{A}{2T} \tag{2-49}$$

where g = acceleration due to gravity, 9.81 m/s² (32.2 ft/s²)
A = area at critical depth, m² (ft²)
T = top width of water surface, m (ft)
V = velocity at critical depth, m/s (ft/s)

Equations 2-48 and 2-49 are solved by successive trials assuming a new value of y_c for each trial. The derivation and the application of these equations to the development of a rating curve for a Palmer-Bowlus flume are discussed in Chap. 3, Sec. 3-4.

Within the specified limits, the following equation may be used for computing critical depth in a circular section flowing partly full:

$$y_c = 0.483\left(\frac{Q}{D}\right)^{2/3} + 0.083D \qquad \text{(SI units)} \tag{2-50}$$

$$y_c = 0.325\left(\frac{Q}{D}\right)^{2/3} + 0.083D \qquad \text{(U.S. customary units)} \tag{2-50a}$$

where y_c = critical depth, m (ft)
Q = discharge, m^3/s (ft^3/s)
D = pipe diameter, m (ft)
and $0.3 < y_c/D < 0.9$.

Drawdown. The transition from a condition of uniform flow in the conduit to the discharge at a free outlet or the drop at a chute is accomplished by a gradual decrease in depth of flow with the stream surface similar to the surface curve above a weir. If the conduit is on a relatively flat slope and the normal velocity is low, the difference between the normal depth and the critical depth will be substantial, the drawdown will be consequential, and the drawdown curve will extend upstream for a considerable distance.

When these three events occur, it may sometimes be possible to reduce construction costs (1) by reducing the size of the conduit and eliminating the drawdown, or (2) by lowering the roof while leaving the width unchanged. On the other hand, if the conduit is on such a steep slope that the velocity is high and the critical depth is only slightly lower than the normal depth, no substantial reduction in the size of the conduit can be made, and the drawdown will not be significant. In those cases, too, where there is a possibility that the conduit may flow full under pressure at times, reduction in the section for drawdown would be undesirable.

The length of the drawdown from the control section (the point of critical depth) can be computed approximately by successive steps with the general equation for nonuniform flow. The computational procedure is illustrated in the following example.

Example 2-7 Determination of length of drawdown curve in rectangular section Determine the length of the drawdown curve for a rectangular channel 3 m (9.84 ft) wide, with a slope of 0.001 m/m (ft/ft), with $n = 0.015$, and $Q = 7\ m^3/s$ (247 ft^3/s), that discharges freely, as shown in Fig. 2-20.

SOLUTION

1. Determine the normal depth of flow y_n, with the use of the Manning equation (Eq. 2-24).

$$Q = AV = \frac{A}{n} R^{2/3} S^{1/2}$$

where A = cross-sectional area of flow

$$= (3\ \text{m})y_n$$

$$R = \frac{(3\ \text{m})y_n}{2y_n + 3\ \text{m}}$$

Thus,

$$7\ \text{m}^3/\text{s} = \frac{(3\ \text{m})y_n}{0.015}\left[\frac{(3\ \text{m})y_n}{2y_n + 3\ \text{m}}\right]^{2/3}(0.001)^{1/2}$$

$$0.532 = \frac{y_n^{5/3}}{(2y_n + 3)^{2/3}}$$

and, by trial and error,

$$y_n = 1.38\ \text{m}\ (4.53\ \text{ft})$$

This is the depth of flow when the depth is no longer affected by the drawdown.

2. Determine the depth of flow at the control section.

$$y_c = \left(\frac{q^2}{g}\right)^{1/3} = \left\{\frac{[(7\ \text{m}^3/\text{s})/3\ \text{m}]^2}{9.81\ \text{m/s}^2}\right\}^{1/3}$$

$$= 0.82\ \text{m}\ (2.69\ \text{ft})$$

3. Determine the length of the water surface between the depths of 1.35 m (see comment) and 0.82 m. The iterative computations are summarized in Table 2-8.

Comment Theoretically the length of the drawdown (or backwater) curve to the normal depth is infinite. The computed length will vary with the size of the increment in depth used in each computation step (see col. 1, Table 2-8). In most practical situations, the increment of depth used for each computation step is on the order of 10 percent of the depth of the control section.

Backwater. The surface curve of a stream of water when backed up by a dam or other obstruction is called the backwater curve (see Fig. 2-21). It may be necessary to determine the amount by which the depth is increased at specified points or the distance upstream to which the effect of backwater can be detected. The computational procedure is illustrated in Example 2-8.

Example 2-8 Determination of length of backwater curve in circular sewer section Determine the length of the backwater curve for a 2-m (6.56-ft) diameter reinforced-concrete sewer that discharges 2.5 m^3/s (88 ft^3/s) to a wet well. The level of water in the wet well is coincident with the top of the inside of the sewer. The slope is 0.001 m/m (ft/ft) and n equals 0.013. Determine the length of the curve to the point where the depth of flow is 1.4 m (4.59 ft), assuming the value of n to be constant. Evaluate the effect of using a constant n value by computing the distance, assuming n to vary with depth.

SOLUTION

1. Determine the length of backwater curves. Computations similar to those summarized in Table 2-8 are required. They are summarized in Table 2-9 for constant n and in Table 2-10

Table 2-8 Computation of drawdown curve for Example 2-7[a]

y,	A,	P,	R,	V,	$\frac{V^2}{2g}$,	$y + \frac{V^2}{2g}$,	S,	S_1,	$(S_a - S_0)$,[b]	$\Delta\left(y + \frac{V^2}{2g}\right)$	Δx,	Σx,
m	m²	m	m	m/s	m	m	m/m	m/m	m/m	m	m	m
0.82	2.46	4.64	0.53	2.85	0.41	1.23	0.00426					
								0.00377	0.00277	0.01	3.6	3.6
0.90	2.70	4.80	0.56	2.59	0.34	1.24	0.00327					
								0.00284	0.00184	0.04	21.7	25.3
1.00	3.00	5.00	0.60	2.33	0.28	1.28	0.00241					
								0.00214	0.00114	0.05	43.9	69.2
1.10	3.30	5.20	0.63	2.12	0.23	1.33	0.00187					
								0.00166	0.00066	0.06	90.9	160.1
1.20	3.60	5.40	0.67	1.94	0.19	1.39	0.00144					
								0.00130	0.00030	0.07	233.3	393.4
1.30	3.90	5.60	0.70	1.79	0.16	1.46	0.00116					
								0.00111	0.00011	0.04	363.6	757.0
1.35	4.05	5.70	0.71	1.73	0.15	1.50	0.00106					

Length of drawdown[c] ≃ 757.0 m

[a]Channel width = 3 m, $S = 0.001$, $n = 0.015$, $Q = 7.0\ m^3/s$.
[b]S_0 = invert slope.
[c]See comment at end of Example 2-7.
Note: m × 3.2808 = ft
$m^2 \times 10.7639 = ft^2$
$m^3/s \times 35.3147 = ft^3/s$

Table 2-9 Computation of backwater curve with constant n for Example 2-8[a]

d, m	$\frac{d}{D}$	$(K')^b$	$\left(\frac{A}{A_{full}}\right)^c$	A, m²	V, m/s	$\frac{V^2}{2g}$, m	$d + \frac{V^2}{2g}$, m	$\frac{K'}{N}D^{8/3}$	$S^{1/2}$, m/m	S, m/m	S_a, m/m	$(S_a - S_0)^d$, m/m	$\Delta\left(d + \frac{V^2}{2g}\right)$, m	Δx, m	Σx, m
2.0	1.00	0.312	1.00	3.14	0.80	0.033	2.033	152	0.0164	0.000269	0.000253	−0.000747	−0.197	264	264
1.8	0.90	0.332	0.95	2.98	0.84	0.036	1.836	162	0.0154	0.000237	0.000260	−0.000740	−0.192	259	523
1.6	0.80	0.305	0.86	2.70	0.93	0.044	1.644	149	0.0168	0.000282	0.000335	−0.000665	−0.187	281	804
1.4	0.70	0.261	0.75	2.36	1.06	0.057	1.457	127	0.0197	0.000388					

Length of backwater[e] ≃ 804 m

[a]Sewer diameter = 2.0 m, S_0 = 0.001, n = 0.013, Q = 2.5 m³/s.
[b]From Table 2-5.
[c]From Fig. 2-16.
[d]S_0 = invert slope.
[e]See comment at end of Example 2-7.

Note: m × 3.2808 = ft
m² × 10.7639 = ft²
m³/s × 35.3147 = ft³/s

Table 2-10 Computation of backwater curve with variable n for Example 2-8[a]

d, m	$\frac{d}{D}$	$(K')^b$	$\left(\frac{A}{A_{full}}\right)^c$	A, m^2	$\left(\frac{n}{n_{full}}\right)^c$	N	V, m/s	$\frac{V^2}{2g}$, m	$d+\frac{V^2}{2g}$, m	$\frac{K'}{n}D^{8/3}$	$S^{1/2}$, m/m	S, m/m	S_a, m/m	$(S_a - S_0)^d$, m/m	$\Delta\left(d+\frac{V^2}{2g}\right)$, m	Δx, m	Σx, m
2.0	1.00	0.312	1.00	3.14	1.0	0.0130	0.80	0.033	2.033	152	0.0164	0.000269					
1.8	0.90	0.332	0.95	2.98	1.07	0.0139	0.84	0.036	1.836	152	0.0164	0.000269	0.000269	−0.000731	−0.197	269	269
1.6	0.80	0.305	0.86	2.70	1.13	0.0147	0.93	0.044	1.644	132	0.0189	0.000357	0.000313	−0.000687	−0.192	279	548
1.4	0.70	0.261	0.75	2.36	1.17	0.0152	1.06	0.057	1.457	109	0.0229	0.000524	0.000441	−0.000559	−0.187	335	883

Length of backwater[e] ≃ 883 m

[a]Sewer diameter = 2.0 m, $S_0 = 0.001$, $n_{full} = 0.013$, $Q = 2.5\ m^3/s$.
[b]From Table 2-5.
[c]From Fig. 2-16.
[d]S_0 = invert slope.
[e]See comment at end of Example 2-7.

Note: m × 3.2808 = ft
$m^2 \times 10.7639 = ft^2$
$m^3/s \times 35.3147 = ft^3/s$

for n varying with depth. Although the tables are self-explanatory, for the most part, it may be helpful to comment on how the slope of the energy grade line was determined. The general equation used to determine the slope of the energy grade line is

$$S^{1/2} = \frac{Qn}{K'D^{8/3}}$$

where $Q = 2.5\ \text{m}^3/\text{s}$, $D = 2.0$ m, and values for K' were selected from Table 2-5. For the case with n constant (Table 2-9), the n value given in the problem statement (0.013) was used. For the case with n varying with depth (Table 2-10), the value of n was adjusted by using the ratio of n/n_f taken from Fig. 2-16.

2. Comparison of lengths. Comparing the lengths given in Tables 2-9 and 2-10, it can be seen that the overall length of the backwater curve was increased by about 9 percent, assuming n varies with depth.

Chute. A chute is a channel with so steep a grade that uniform flow can take place at a depth less than the critical depth (see Fig. 2-22). The computation of flow in a chute, insofar as such flow is uniform, is accomplished by the use of the ordinary equations, such as Manning's, although the applicability of these equations at very high velocities is doubtful. However, chutes are usually short, and the portion of their length in which uniform flow conditions exist is often insignificant. Nonuniform flow occurs at the upper end of the chute, and a hydraulic jump may occur at the lower end if conditions beyond the chute are such as to produce it.

Hydraulic jump. It is not uncommon in practice to have water flowing at a supercritical velocity enter a channel of mild slope or a stilling basin in which the depth is significantly greater. Two examples of this condition are (1) at the junction of a chute and a channel of mild slope, and (2) where water flowing under a gate at a supercritical velocity is discharged into a channel of mild slope. When such conditions occur, a hydraulic jump could form (see Fig. 2-22). However, for the jump to form, the depths just before and just after the jump must be conjugate depths. That is, they must satisfy the momentum relationships discussed in Sec. 2-1 (Eq. 2-16 for rectangular channels).

To determine if a jump will form, it is necessary to know the depth of flow immediately downstream from the point where the jump is expected to form. If the channel is sufficiently long, the depth of flow will be the normal depth. However, if there is a change in channel slope or cross section, or in any other condition that produces nonuniform flow at a relatively short distance downstream, then the water-surface profile must be calculated by working upstream from a control section or any other point where the depth is known.

Based on the depth just downstream from the expected jump location, the conjugate depth is calculated with Eq. 2-16. The conjugate depth is then compared with the actual depth just upstream from the expected jump location. If the two are equivalent, a jump will, in fact, form at the expected location. If the conjugate depth is greater than the depth immediately upstream from the expected jump location, the jump will form farther downstream at a point where the water surface has risen so that conjugate depths are obtained.

If the opposite is true (upstream depth greater than conjugate depth), the jump will form farther upstream in the channel. If this condition occurs at the upstream end of the mild-slope channel, a jump will not form in this section of channel. If it occurs at the outlet of a gate structure, water will be backed up to the gate, and the gate opening will be submerged. If it is produced at the end of a chute, the jump will form in the chute at a point where the depth of water backed up in the chute equals the conjugate depth for the depth of flow just upstream. Determination of the location of a hydraulic jump is illustrated in Example 2-9.

Example 2-9 Analysis of hydraulic jump in a rectangular section A rectangular section 3 m (9.84 ft) wide with $n = 0.015$ is carrying a flow of 7 m³/s (247 ft³/s). If the inclination of the upper section (chute) is 0.05 m/m (ft/ft), and the inclination of the lower section is 0.001 m/m (ft/ft), determine the length below the end of the chute where the hydraulic jump will occur. Assume that the flow has reached normal depth in the chute and that normal depth will be obtained downstream from the hydraulic jump.

SOLUTION

1. Determine the normal depth in the lower section and in the chute. The normal depth in the lower section was calculated in Example 2-7 and is 1.38 m. The normal depth in the chute is calculated with the same procedure:

$$7 \text{ m}^3/\text{s} = \frac{(3 \text{ m})y_n}{0.015}\left(\frac{(3 \text{ m})y_n}{2y_n + 3 \text{ m}}\right)^{2/3}(0.05)^{1/2}$$

$$0.075 = \frac{y_n^{5/3}}{(2y_n + 3)^{2/3}}$$

$$y_n = 0.36 \text{ m}$$

This depth is less than the critical depth of 0.82 m calculated in Example 2-7.

2. Compute the velocity in the lower section when flowing at normal depth.

$$V_2 = Q/A_2 = \frac{7 \text{ m}^3/\text{s}}{3 \text{ m} \times 1.38 \text{ m}} = 1.69 \text{ m/s}$$

3. Assuming that the depth of the jump will be normal depth in the lower section, compute the depth before the jump. The converse of Eq. 2-16 is used.

$$y_1 = -\frac{y_2}{2} + \sqrt{\left(\frac{y_2}{2}\right)^2 + \frac{2y_2V_2^2}{g}}$$

$$y_1 = -\frac{1.38 \text{ m}}{2} + \sqrt{\left(\frac{1.38 \text{ m}}{2}\right)^2 + \frac{2(1.38 \text{ m})(1.69 \text{ m/s})^2}{9.81 \text{ m/s}^2}}$$

$$y_1 = 0.44 \text{ m}$$

4. Determine the distance below the end of the chute to the point where the hydraulic jump will form. When the flow reaches the bottom of the chute, the slope retards the flow and

the depth increases. The flow must continue at a depth below critical depth in the lower section until the retardation has increased the depth of flow from 0.36 to 0.44 m, at which point the jump will form. The required distance for this gradual increase in depth to occur is determined with the use of the computational procedure summarized in Table 2-11. As shown, the jump will be about 17.8 m (58.4 ft) below the end of the chute.

Comment The existence and location of the hydraulic jump depend on the slope, shape, and length of the lower channel. A jump will not form if the lower channel is too steeply inclined or, if flat enough, too short. If for some reason the water depth in the lower channel is deeper than the upper conjugate depth, the jump may form on the steeper incoming channel, or it may be drowned out entirely.

Banking on Curves

In some cases, the banking or superelevation of the water surface along the outer wall of a curved channel may be very important. It is especially important when setting bank or side-wall elevations for stream channels or open flumes that should not overflow, or when designing flat-topped conduits in which the water should not touch the roof of the conduit.

The excess in elevation of the water at the outer bank may be computed approximately by the equation

$$E = \frac{V^2 b}{gr} \tag{2-51}$$

where E = difference in elevation of water surface at the two banks, m (ft)
V = average velocity in cross section, m/s (ft/s)
b = breadth of channel or stream, m (ft)
g = acceleration due to gravity, 9.81 m/s^2 (32.2 ft/s^2)
r = radius of curvature on centerline of channel, m (ft)

In some cases, the actual difference in elevation has been slightly greater than would be given by this equation.

DISCUSSION TOPICS AND PROBLEMS

2-1 In the pipe shown in Fig. 2-23, a tube has been inserted through the pipe wall and is pointing directly upstream into the flowing water. A second tube has been inserted into the pipe wall, but does not extend into the flowing water. Determine the flowrate in the pipe when the height of the water column in each of the two tubes is as shown in Fig. 2-23. Assume that the pressure distribution in the vertical direction within the pipe is the same if the fluid were still and that the velocity over the pipe cross section is uniform. Neglect all losses.

2-2 Determine the force exerted by the nozzle on the pipe shown in Fig. 2-24 when the flowrate is 0.01 m^3/s. Neglect all losses.

2-3 Determine the magnitude and direction of the force needed to counteract the force resulting from the change in momentum in a horizontal 90° bend in a 200-mm force main. The rate of flow through the force main is 0.1 m^3/s.

Table 2-11 Computation of hydraulic jump for Example 2-9

y, m	A, m^2	V, m/s	P, m	R, m	$\frac{V^2}{2g}$, m	$y + \frac{V^2}{2g}$, m	$S^{1/2}$ m/m	S, m/m	s_a, m/m	$(S_a - S_0)$ m/m	$\Delta\left(y + \frac{V^2}{2g}\right)$, m	Δx, m	Σx, m
0.36	1.08	6.48	3.72	0.29	2.14	2.50	0.2219	0.0492	0.0457	0.0447	0.20	4.47	4.47
0.38	1.14	6.14	3.76	0.30	1.92	2.30	0.2055	0.0422	0.0386	0.0376	0.17	4.52	8.99
0.40	1.20	5.83	3.80	0.32	1.73	2.13	0.1870	0.0350	0.0328	0.0318	0.13	4.09	13.08
0.42	1.26	5.56	3.84	0.33	1.58	2.00	0.1745	0.0305	0.0286	0.0276	0.13	4.71	17.79
0.44	1.32	5.30	3.88	0.34	1.43	1.87	0.1633	0.0267					

Total distance below chute ≃ 17.79 m

Note: m × 3.2808 = ft
m^2 × 10.7639 = ft^2
m^3/s × 35.3147 = ft^3/s

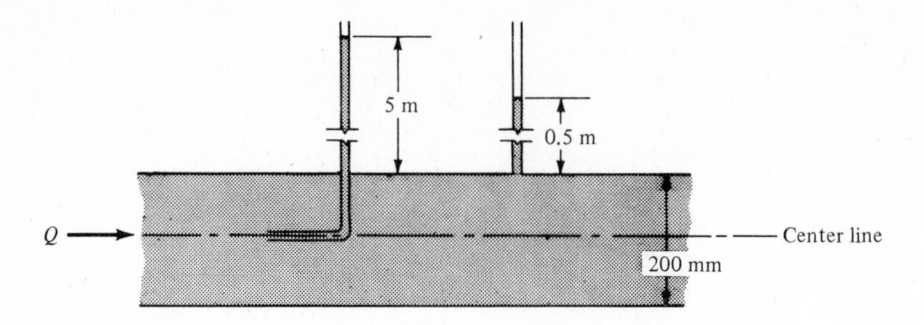

Figure 2-23 Definition sketch for Prob. 2-1. Note: m × 3.2808 = ft; mm × 0.03937 = in.

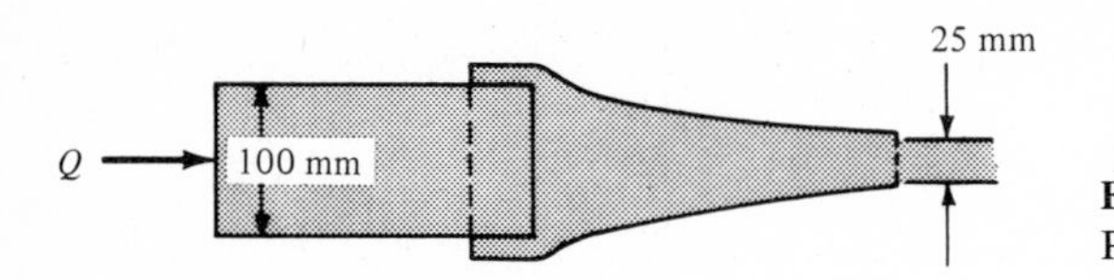

Figure 2-24 Definition sketch for Prob. 2-2. Note: mm × 0.03937 = in.

2-4 Pressure taps have been installed 150 m apart at the same elevation on a 250-mm-diameter force main used for the pumping of effluent from a wastewater treatment plant. The pressure taps are connected to a differential mercury manometer by small tubing. If the flowrate is 0.2 m^3/s and the deflection of the mercury manometer is 1.0 m, determine the friction factor in the force main. (*Note:* The specific gravity of mercury is equal to 13.57 g/cm^3.)

2-5 It has been determined that a Hazen-Williams coefficient of roughness C value of 130 is applicable to a 1.0-m-diameter pipe that is carrying a flow of 3.0 m^3/s. Determine the corresponding values of the Manning n and the Darcy-Weisbach f that would also be applicable.

2-6 A cast-iron force main, 200 mm in diameter and 8 m long, is used to transport wastewater from a pumping station up a hill to a nearby trunk sewer. The Hazen-Williams coefficient of roughness C is equal to 130. The elevation of the force main at the pumping station is 15 m below the elevation at the trunk sewer. Determine the pressure in the force main at the pumping station when the flowrate is 0.08 m^3/s. Assume that the pressure at the trunk sewer is atmospheric.

2-7 A 1.5-m circular sewer is on a slope of 0.0025 m/m. If n is 0.013 at all depths of flow, find:
(*a*) Q and V when the sewer is flowing full.
(*b*) Q and V when wastewater is flowing at a depth of 0.4 m.
(*c*) Q and V when the sewer is carrying 0.6 of its capacity.
(*d*) V and depth of flow when Q is 1.5 m^3/s.

2-8 Solve Prob. 2-7 by assuming that n is variable. The value of n is 0.013 when the sewer is flowing full.

2-9 A 2.0-m circular sewer has a slope of 0.0065 m/m. If n is 0.0015 at all depths of flow, find:
(*a*) Q and V when the sewer is flowing full.
(*b*) Q and V when wastewater is flowing at a depth of 0.5 m.
(*c*) Q and V when the sewer is carrying 0.7 of its capacity.
(*d*) V and the depth of flow when Q is 2.0 m^3/s.

2-10 Solve Prob. 2-9 by assuming the n value is variable and is equal to 0.015 when the sewer is flowing full.

2-11 As part of the feasibility study for a wastewater-treatment plant to be located on the coast, it is necessary to determine if effluent pumping facilities are needed. The plant effluent is to be discharged through an ocean outfall to a point 3.0 km from the treatment plant at a depth of 125 m during high-tide conditions. The outfall is to have a diameter of 2.0 m with n equal to 0.014 and must carry a peak flowrate of 5.5 m^3/s. The hydraulic grade line of the plant effluent is to be at an elevation 3.20 m above sea level at high tide without pumping. Will effluent pumping be necessary under these conditions, or can a sufficient flowrate be obtained by gravity flow? If pumping is needed under these conditions, to what diameter must the outfall be increased for gravity flow to be sufficient? Neglect all minor losses including losses at the diffuser. Assume that the density of the effluent and seawater are the same.

2-12 The following equations were used to develop Fig. 2-16:

$$\frac{A}{A_{\text{full}}} = \frac{1}{\pi}\cos^{-1}\phi - \frac{1}{\pi}\phi\sqrt{1-\phi^2}$$

$$\frac{P}{P_{\text{full}}} = \frac{1}{\pi}\cos^{-1}\phi$$

$$\frac{R}{R_{\text{full}}} = \frac{A}{A_{\text{full}}}\left(\frac{P}{P_{\text{full}}}\right)^{-1}$$

$$\frac{V}{V_{\text{full}}} = \left(\frac{R}{R_{\text{full}}}\right)^{2/3}$$

$$\frac{Q}{Q_{\text{full}}} = \frac{A}{A_{\text{full}}}\frac{V}{V_{\text{full}}}$$

where A = cross-sectional area of flow, m^2
A_{full} = cross-sectional area of flow for full pipe, m^2
$\phi = 1\text{–}2\,(d/D)$
d = depth of flow, m
D = diameter of pipe, m
P = wetted perimeter, m
P_{full} = wetted perimeter for full pipe, m
R = hydraulic radius, m
R_{full} = hydraulic radius for full pipe, m
V = velocity, m/s
V_{full} = velocity for full pipe, m/s
Q = discharge, m^3/s
Q_{full} = discharge for full pipe, m^3/s
and, the angle $\cos^{-1}\phi$ is in radians.

Derive the equations for A/A_{full} and P/P_{full}. Perform separate derivations for the case of a pipe flowing less than half full and for the case of a pipe flowing greater than half full. Show that both derivations lead to the equations listed previously. Then derive the relationships for R/R_{full}, V/V_{full}, and Q/Q_{full}. Refer to the sketches shown in Fig. 2-25 for useful geometric relationships.

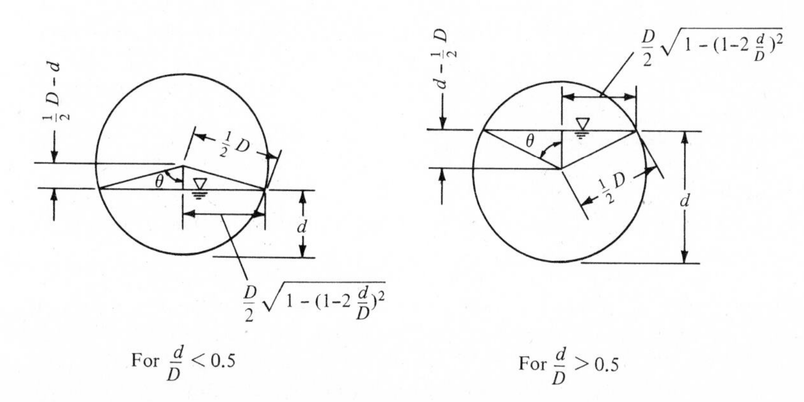

Figure 2-25 Definition sketch for Prob. 2-12.

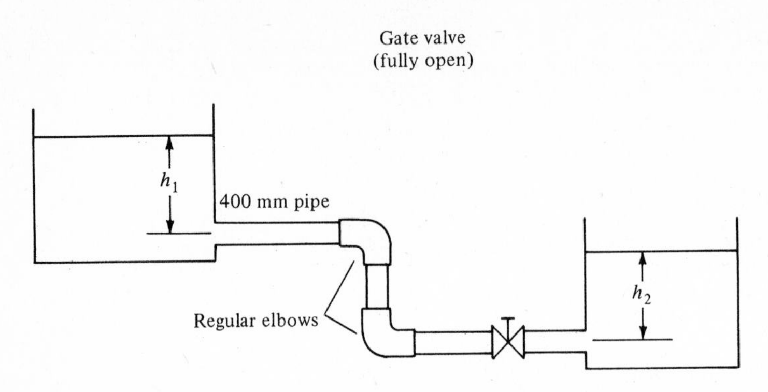

Figure 2-26 Definition sketch for Prob. 2-13. Note: mm × 0.03937 = in.

2-13 Express the total loss of head due to the minor losses in the system shown in Fig. 2-26 as a function of flowrate Q.

2-14 Determine the drop in pressure across a wide-open gate valve in a 300-mm pipe when the flowrate is 0.15 m^3/s.

2-15 In the system shown in Fig. 2-27, a pipe extends into the interior of a tank. Determine the flowrate at the instant when the level in the tank is as shown. Both pipes are cast iron with a diameter of 150 mm. Include all losses. Assume that the minor loss in the elbow is equal to $0.3V^2/2g$.

2-16 A sluice gate is used to control the flowrate in a rectangular concrete channel (see Fig. 2-28). The channel is 4 m wide and has a bottom slope of 0.0025 m/m; the n value is 0.013. For a certain upstream condition, the depth of water just downstream from the sluice gate is 1.5 m and the flowrate through the gate is 50 m^3/s. Assuming the downstream conditions are such that water will flow at normal depth at some distance down the channel, prove that a hydraulic jump will form and find the distance L downstream from the sluice gate to the point where the jump begins to form (see Fig. 2-28). Also, compare the critical depth with the depths just upstream and downstream from the jump.

2-17 In a rectangular channel, there is a transition in which the width of the channel decreases gradually from 4 to 3 m. The flowrate in the channel is 20 m^3/s and the depth in the wide section is 3 m. Draw a q curve for the system and determine the depth in the narrow section. Neglect all losses. (The flow is subcritical in both sections of the channel.)

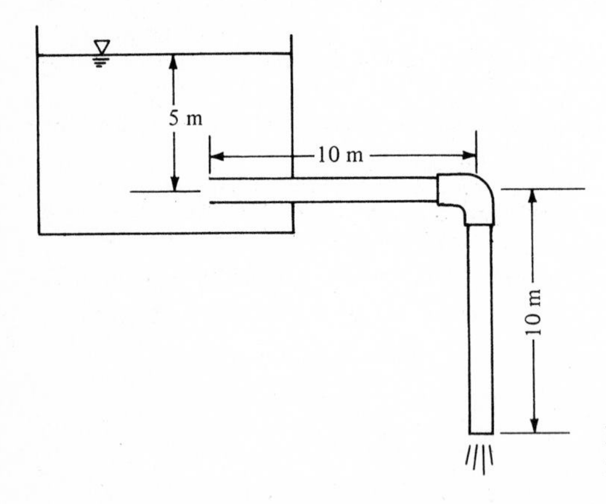

Figure 2-27 Definition sketch for Prob. 2-15. Note: m × 3.2808 = ft.

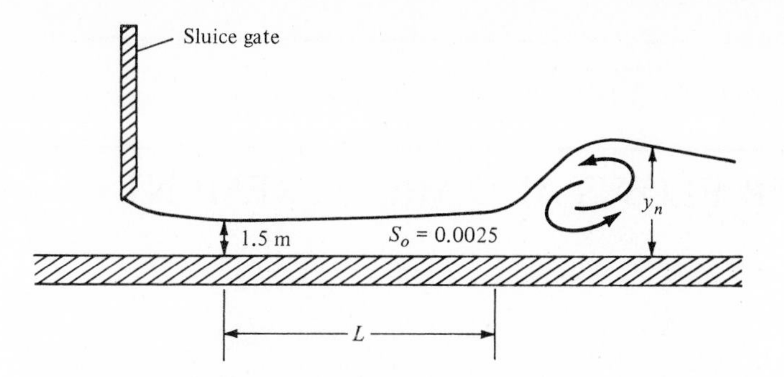

Figure 2-28 Definition sketch for Prob. 2-16. Note: m × 3.2808 = ft.

2-18 A 1.5-m-diameter trunk sewer with $n = 0.013$ and $S = 0.001$ m/m flows into a wet well at a pumping station. If the flowrate in the sewer is 0.3 m^3/s and the water surface in the wet well is 1.2 m above the invert of the incoming sewer, determine the volume of wastewater stored in the sewer above the normal depth of flow.

REFERENCES

1. Babbitt, H. E., and E. R. Baumann: *Sewerage and Sewage Treatment,* 8th ed., Wiley, New York, 1958.
2. Brater, E. F., and H. W. King: *Handbook of Hydraulics,* 6th ed., McGraw-Hill, New York, 1976.
3. Camp, T. R.: Hydraulics of Sewers, *Public Works,* vol. 83, no. 6, 1952.
4. Camp, T. R.: Hydraulics of Sewer Transitions, *J. Boston Soc. Civ. Eng.,* vol. 19, no. 6, 1932.
5. Chow, V. T.: *Open-Channel Hydraulics,* McGraw-Hill, New York, 1959.
6. Davis, C. V., and K. E. Sorensen: *Handbook of Applied Hydraulics,* 3d ed., McGraw-Hill, New York, 1969.
7. Henderson, F. M.: *Open-Channel Flow,* Macmillan, New York, 1966.
8. Hinds, J., W. P. Creager, and J. D. Justin: *Engineering for Dams,* vols. 1–3, Wiley, New York, 1945.
9. Jaeger, C.: *Engineering Fluid Mechanics,* Blackie, London, 1956.
10. Joint Committee of the American Society of Civil Engineers and the Water Pollution Control Federation: *Design and Construction of Sanitary and Storm Sewers*, ASCE Manual and Report 37, New York, 1969.
11. Metcalf, L., and H. P. Eddy: *American Sewage Practice,* vol. 1, 2d ed., McGraw-Hill, New York, 1928.
12. Metcalf & Eddy, Inc.: *Wastewater Engineering: Collection, Treatment, Disposal,* McGraw-Hill, New York, 1972.
13. Moody, L. F.: Friction Factors for Pipe Flow, *Trans. ASME,* vol. 66, p. 671, 1944.
14. Vennard, J. K., and R. L. Street: *Elementary Fluid Mechanics,* 5th ed., Wiley, New York, 1975.
15. Webber, N. B.: *Fluid Mechanics for Civil Engineers,* SI ed., Chapman and Hall, London, 1971.
16. Williams, G. S., and A. Hazen: *Hydraulic Tables,* 3d ed., Wiley, New York, 1920.

CHAPTER

THREE

WASTEWATER FLOWS AND MEASUREMENT

Determining the rates of wastewater flow is a fundamental step in the design of wastewater collection and pumping facilities. Reliable data on projected flows must be available if these facilities are to be designed properly, and if the associated costs are to be minimized. The purpose of this chapter is to develop a basis for properly estimating wastewater flowrates from a community. The subjects discussed include: (1) the various components of wastewater from a community, (2) wastewater sources and flowrates, (3) flowrate data analysis, and (4) flow measurement by direct-discharge methods and velocity-area methods. Much of the material in this chapter has been taken from Chap. 2 of the companion text of this book [11], and, where appropriate, design data and information have been updated to reflect current practice.

3-1 COMPONENTS OF WASTEWATER

The components that make up the wastewater from a community depend on the type of collection system used and may include:

Domestic (also called sanitary) wastewater. Wastewater discharged from residences and from commercial, institutional, and similar facilities.
Industrial wastewater. Wastewater in which industrial wastes predominate.
Infiltration/inflow (I/I). Extraneous water that enters the sewer system from the ground through various means, and storm water that is discharged from sources such as roof leaders, foundation drains, and storm sewers.
Storm water. Runoff resulting from rainfall and snowmelt.

The types of collection systems for which each of these components becomes important are discussed in Chap. 4.

3-2 WASTEWATER SOURCES AND FLOWRATES

Data for estimating average wastewater flows from various domestic and industrial sources and the infiltration/inflow contribution are presented in this section. Variations in the flows that must be established before sewers are designed are also discussed.

Sources and Rates of Domestic Wastewater Flows

The principal sources of domestic wastewater in a community are the residential and commercial districts. Other important sources include institutional and recreational facilities. Methods for projecting domestic wastewater flows for areas that are being developed are considered in the following discussion.

Residential districts. For small residential districts, wastewater flows are commonly determined on the basis of population density and the average per capita contribution of wastewater. Data on ranges and typical flows are given in Table 3-1. For large residential districts, it is often advisable to develop flowrates on the basis of land use areas and anticipated population densities. Where possible, these rates should be based on actual flow data from selected typical residential areas located near the area being considered. If these data are unavailable, an estimate of 70 percent of the domestic water-withdrawal rate may be used. In many cases, design flows are fixed by federal, state, and local regulatory agencies.

Table 3-1 Average wastewater flows from residential sources

		Flow, L/unit · d	
Source	Unit	Range	Typical
Apartment	Person	200–340	260
Hotel, resident	Resident	150–220	190
Individual dwellings			
Average home	Person	190–350	280
Better home	Person	250–400	310
Luxury home	Person	300–550	380
Semimodern home	Person	100–250	200
Summer cottage	Person	100–240	190
Trailer park	Person	120–200	150

Note: L × 0.2642 = gal

In the past, the preparation of population projections for estimating wastewater flowrates was often the responsibility of the sanitary engineer, but today such data are usually available from local, regional, and state planning agencies. If they are not available and must be prepared, Refs. 9, 10, and 12 may be consulted. Population-projection methods that have been used are described in Table 3-2. Saturation (maximum) population density values that are required for estimating flows from land areas with various-use classifications can be obtained from local planning commissions. Even though such values are given or prescribed, they should be checked and assessed in the light of possible future changes in land use patterns.

Commercial districts. Commercial wastewater flows are generally expressed in cubic metres per hectare per day (gallons per acre per day) and are based on existing or anticipated future development or comparable data from other areas. Unit flows may vary from 14 to more than 1500 $m^3/ha \cdot d$ (1500 to more than 160,000 gal/acre · d). Estimates for certain commercial sources may also be made from the data in Table 3-3.

Institutional facilities. The actual records of institutions are the best sources of flow data for design purposes. When records are unavailable, the flows from institutional facilities can be estimated with the data shown in Table 3-4.

Recreational facilities. Flows from many recreational facilities are highly seasonal. Some typical data are presented in Table 3-5.

Table 3-2 Population projections methods[a]

Method[b]	Description of method
Graphical	Graphical projections of past population-growth curves are used to estimate future population growth.
Decreasing-rate-of-growth	Population is estimated on the basis of the assumption that as the city becomes larger, the rate of growth from year to year becomes smaller.
Mathematical or logistic	Population growth is assumed to follow some logical mathematical relationship in which population growth is a function of time.
Ratio and correlation	The population-growth rate for a given community is assumed to be related to that of a larger region, such as the county or state.
Component	Population is forecast on the basis of a detailed analysis of the components that make up population growth, namely, natural increase and migration. Natural increase represents the increase resulting from the excess of births over deaths.
Employment forecast	Population growth is estimated on the basis of various employment forecasts. In actual practice, the relationship between population and the number of jobs is derived by using the techniques of the ratio and correlation method.

[a] Additional details on these and other population projection methods may be found in Refs. 9, 10, and 12.

[b] Methods are arranged in order of increasing complexity.

Table 3-3 Average wastewater flows from commercial sources[a]

Source	Unit	Flow, L/unit · d	
		Range	Typical
Airport	Passenger	8–15	10
Automobile service station	Vehicle served	30–50	40
	Employee	35–60	50
Bar	Customer	5–20	8
	Employee	40–60	50
Hotel	Guest	150–220	190
	Employee	30–50	40
Industrial building (excluding industry and cafeteria)	Employee	30–65	55
Laundromat	Machine	1800–2600	2200
	Wash	180–200	190
Motel	Person	90–150	120
Motel with kitchen	Person	190–220	200
Office	Employee	30–65	55
Restaurant	Meal	8–15	10
Rooming house	Resident	90–190	150
Store, department	Toilet room	1600–2400	2000
	Employee	30–50	40
Shopping center	Parking space	2–8	4
	Employee	30–50	40
	Employee	30–50	40

[a] Adapted in part from Ref. 5.
Note: L × 0.2642 = gal

Table 3-4 Average wastewater flows from institutional sources[a]

Source	Unit	Flow, L/unit · d	
		Range	Typical
Hospital, medical	Bed	500–950	650
	Employee	20–60	40
Hospital, mental	Bed	300–550	400
	Employee	20–60	40
Prison	Inmate	300–600	450
	Employee	20–60	40
Rest home	Resident	200–450	350
	Employee	20–60	40
School, day			
With cafeteria, gym, and showers	Student	60–115	80
With cafeteria, but no gym and no showers	Student	40–80	60
Without cafeteria, gym, and showers	Student	20–65	40
Schools, boarding	Student	200–400	280

[a] Adapted in part from Ref. 5.
Note: L × 0.2462 = gal

Table 3-5 Average wastewater flows from recreational sources

Source	Unit	Flow, L/unit · d	
		Range	Typical
Apartment, resort	Person	200–280	220
Cabin, resort	Person	130–190	160
Cafeteria	Customer	4–10	6
	Employee	30–50	40
Campground (developed)	Person	80–150	120
Cocktail lounge	Seat	50–100	75
Coffee shop	Customer	15–30	20
	Employee	30–50	40
Country club	Member present	250–500	400
	Employee	40–60	50
Day camp (no meals)	Person	40–60	50
Dining hall	Meal served	15–40	30
Dormitory, bunkhouse	Person	75–175	150
Hotel, resort	Person	150–240	200
Laundromat	Machine	1800–2600	2200
Store, resort	Customer	5–20	10
	Employee	30–50	40
Swimming pool	Customer	20–50	40
	Employee	30–50	40
Theater	Seat	10–15	10
Vistor center	Visitor	15–30	20

Note: L × 0.2462 = gal

Table 3-6 Per capita wastewater flows from conventional domestic devices [13]

Device	Wastewater flow	
	L/capita · d	Percent
Bathtub faucet	30.3	12
Clothes washing machine	34.1	14
Kitchen sink faucet	26.5	11
Lavatory faucet	11.4	5
Shower head	45.4	19
Toilet	94.6	39
Total	242.3	100

Note: L × 0.2642 = gal

Reduction of Domestic Wastewater Flows

Because of the importance of conserving both resources and energy, various means for reducing wastewater flows from domestic sources are gaining increasing attention. The principal method of reducing wastewater flow is by reducing water use.

Per capita wastewater flows from conventional domestic devices are given in Table 3-6. The principal devices and systems for reducing wastewater flow reductions are described in Table 3-7. The actual wastewater flow reductions and the percentage reductions possible with these devices and systems are reported in Table 3-8. The social acceptability and the ease with which these

Table 3-7 Flow-reduction devices and systems[a]

Device/system	Description and/or application
Batch-flush valve	Used extensively in commercial applications. Can be set to deliver between 1.9 L/cycle for urinals and 15 L/cycle for toilets.
Brick in toilet tank	A brick or similar device in a toilet tank achieves only a slight reduction in wastewater flow.
Dual-cycle tank insert	Insert converts conventional toilet to dual-cycle operation. In new installations, a dual-cycle toilet is more cost effective than a conventional toilet with a dual-cycle insert.
Dual-cycle toilet	Uses 4.75 L/cycle for liquid wastes and 9.5 L/cycle for solid wastes.
Faucet aerator	Increases the rinsing power of water by adding air and concentrating flow, thus reducing the amount of wash water used. Comparatively simple and inexpensive to install.
Level controller for clothes washer	Matches the amount of water used to the amount of clothes to be washed.
Limiting-flow shower head	Restricts and concentrates water passage by means of orifices that limit and divert shower flow for optimum use by the bather.
Limiting-flow valve	Restricts water flow to a fixed rate that depends on household water system pressure.
Pressure-reducing valve	Maintains home water pressure at a lower level than that of the water-distribution system. Reduces household flows and decreases the probability of leaks and dripping faucets.
Recirculating mineral oil toilet system	Uses mineral oil as a water-transporting medium and requires no water. Operates in a closed loop in which toilet wastes are collected separately from other household wastes and are stored for later pickup by vacuum truck. In the storage tank, wastes are separated from the transporting fluid by gravity. The mineral oil is drawn off by pump, coalesced, and filtered before being recycled to the toilet tank.
Reduced-flush device	Toilet tank insert that either prevents a portion of the tank contents from being dumped during the flush cycle or occupies a portion of the tank volume so that less water is available per cycle.
Urinal	Wall-type urinal for home use that requires 5.7 L/cycle.
Vacuum-flush toilet system	Uses air as a waste-transporting medium and requires about 1.9 L/cycle.
Wash-water recycle system for toilet flushing	Recycles bath and laundry wastewaters for use in toilet flushing.

[a] Adapted from Ref. 13.
Note: L × 0.2642 = gal

Table 3-8 Reductions achieved by flow-reduction devices and systems[a]

Device/system[b]	Wastewater flow reduction	
	L/capita · d	Percent of total[c]
Level control for clothes washer	4.5	2
Pressure-reducing valve[d]	60.6	25
Recirculating mineral oil toilet system	94.6	39
Shower		
Limiting-flow valve	22.7	9
Limiting-flow shower head	28.4	12
Sink faucet		
Faucet aerator	1.9	1
Limiting-flow valve	1.9	1
Toilet		
Brick in toilet tank	3.8	2
Dual-batch-flush valve	58.7	24
Dual-cycle tank insert	37.9	16
Dual-cycle toilet	66.2	27
Reduced-flush device	37.9	16
Single-batch-flush valve	28.4	12
Toilet and urinal with batch-flush valves	54.9	23
Urinal with batch-flush valve	26.5	11
Water-saver toilet	28.4	12
Vacuum-flush toilet system[d]	85.2	35
Washwater recycle system for toilet flushing	94.6	39

[a] Adapted from Ref. 13.
[b] See Table 3-7 for descriptions and applications of devices and systems.
[c] Percent of total for conventional devices reported in Table 3-6.
[d] Both single and multiple homes.
Note: L × 0.2642 = gal

devices can be installed are discussed in Ref. 13. Another method of achieving flow reductions is to restrict the installation and use of appliances that tend to increase water consumption, such as automatic dishwashers and garbage grinders.

In many communities, the use of one or more of the flow-reduction devices is now specified for all new residential dwellings; in others, the use of garbage grinders has been limited in new housing developments. Furthermore, many individuals concerned about conservation have installed devices such as those listed in Table 3-8 to reduce water consumption. It will probably be some time before the actual impact of these devices and methods is known.

Sources and Rates of Industrial Wastewater Flows

Industrial wastewater flowrates vary with the type and size of the industry, the degree of water reuse, and the on-site wastewater treatment methods used, if any. Peak flows may be reduced by the use of detention tanks and equalization basins. A typical design value for estimating the flows from industrial districts that have no wet-process industries is about 30 m^3/ha · d (~3000 gal/acre · d). If the water requirements for the industries are known, wastewater flow projections can be based on water-flow projections. For industries without internal reuse programs, about 85 to 95 percent of the water used in the various operations and processes will probably become wastewater. For large industries with internal-water-reuse programs, separate estimates must be made. Average domestic (sanitary) wastewater contributed from industrial activities may vary from 30 to 95 L/capita · d (8 to 25 gal/capita · d).

Reduction of Industrial Wastewater Flows

On October 18, 1972, Congress passed the Federal Water Pollution Control Act Amendments of 1972 (Public Law 92-500). Among the many far-reaching implications of this law is the requirement for secondary treatment for all discharges regardless of receiving-water quality. As more and more communities move to comply with the requirement for secondary treatment, the cost to industry in terms of user charges (based on flow and the quantity of organic and suspended material discharged into sewers) has become, in many cases, prohibitive.

To lessen the charges for wastewater treatment in community facilities, most industries have established extensive programs to reduce the quantity and strength of the wastewater discharged. In many instances, discharges to sewers have been eliminated totally. For example, a number of canneries have withdrawn all discharges from community systems in favor of alternative methods of treatment and disposal, usually in some form of land treatment.

The significance of the reduction in industrial discharges is that great care should be taken in estimating these flows when sizing new sewers. In many cases, the staged decrease in industrial discharges may be offset by increased discharges of domestic wastewater. In other cases, it may be necessary to perform a separate analysis to determine if problems will develop in the future when the flow in the sewer decreases.

Infiltration/Inflow

Extraneous flows in sewers have been defined as follows [3]:

Infiltration. The water entering a sewer system, including sewer service connections, from the ground through such means as, but not limited to, defective pipes, pipe joints, connections, or manhole walls. Infiltration does not include, and is distinguished from, inflow.

Inflow. The water discharged into a sewer system, including service connections, from such sources as, but not limited to, roof leaders, cellar, yard, and area drains, foundation drains, cooling-water discharges, drains from springs and swampy areas, manhole covers, cross connections from storm sewers and combined sewers (see Sec. 4-1), catch basins, storm water, surface runoff, street wash water, or drainage. Inflow does not include, and is distinguished from, infiltration.

Infiltration/Inflow. The total quantity of water from both infiltration and inflow without distinguishing the source.

Many extensive programs of sewer system evaluation have been and are being undertaken, because the Federal Water Pollution Control Act Amendments of 1972 require that applicants for treatment-works grants must demonstrate that each sewer system discharging into the proposed treatment works will not be subject to excessive infiltration/inflow. The subject of infiltration and inflow is discussed in detail in Chap. 6.

Infiltration into sewers. A portion of the precipitation in a given area runs quickly into the storm sewers or other drainage channels; another portion evaporates or is absorbed by vegetation; and the remainder percolates into the ground, becoming groundwater. The proportion that percolates into the ground depends on the character of the surface and soil formation and on the rate and distribution of the precipitation according to seasons. Any reduction in permeability, such as that caused by buildings, pavements, or frost, decreases the opportunity for precipitation to become groundwater and increases the surface runoff correspondingly.

The amount of groundwater flowing from a given area may vary from a negligible amount for a highly impervious district or a district with a dense subsoil, to 25 or 30 percent of the rainfall for a semipervious district with a sandy subsoil permitting rapid passage of water into it. The percolation of water through the ground from rivers or other bodies of water sometimes has a considerable effect on the groundwater table, which rises and falls continuously.

The presence of high groundwater results in leakage into the sewers and in an increase in the quantity of wastewater and the expense of disposing of it. This leakage from groundwater, or infiltration, may range from 0.01 to more than 1.00 $m^3/d \cdot mm\text{-}km$ (100 to 10,000 $gal/d \cdot in\text{-}mi$). The number of millimeter-kilometers (inches-miles) in a sewer system is the sum of the products of sewer diameters, in millimeters (inches), times the lengths, in kilometers (miles), of sewers of corresponding diameters. Expressed another way, infiltration may range from 0.2 to 30 $m^3/ha \cdot d$ (20 to 3000 $gal/acre \cdot d$). During heavy rains, when inflow may occur through manhole covers and other connections to sewers, the infiltration/inflow rate may exceed 500 $m^3/ha \cdot d$ (50,000 $gal/acre \cdot d$). Infiltration/inflow is a variable part of the total wastewater flow, depending on the quality of the material and workmanship in the sewers and building connections, the character of the maintenance, and the elevation of the groundwater compared with that of the sewers.

The sewers first built in a district usually follow the watercourses in the bottoms of valleys, close to (and occasionally below) the beds of streams. As a result, these old sewers may receive comparatively large quantities of groundwater, whereas sewers built later at higher elevations will receive relatively smaller quantities of groundwater. With an increase in the percentage of area in a district that is paved or built over, comes (1) an increase in the percentage of storm water that is conducted rapidly to the storm sewers and watercourses, and (2) a decrease in the percentage of the storm water that can percolate into the earth and may infiltrate into the sanitary sewers (see Sec. 4-1). A sharp distinction is to be made between maximum and average rates of infiltration into the sewer systems. The maximum rates are necessary to determine required sewer capacities; the average rates are necessary to estimate such factors as annual costs of pumping and treatment of wastewater.

The rate and quantity of infiltration depend on the length of sewers, the area served, the soil and topographic conditions, and, to a certain extent, the population density (which affects the number and total length of house connections). Although the elevation of the water table varies with the quantity of rain and snowmelt percolating into the ground, the leakage through defective joints, porous concrete, and cracks has been large enough, in many cases, to lower the groundwater table to the level of the top of the sewer or lower.

Most of the sewers built during the first half of this century were laid with cement mortar joints or hot-poured bituminous compound joints. Manholes were almost always constructed of brick masonry. The deterioration of pipe joints and sewer-to-manhole joints and the lack of waterproofing of brickwork used in the construction of these old sewers have resulted in a high potential for infiltration. In modern sewer design, high-quality pipe with dense walls, precast manhole sections, and joints sealed with rubber or synthetic gaskets are standard. The use of these improved materials has greatly reduced infiltration into newly constructed sewers, and the increase of infiltration rates with time will probably be much slower than has been the case with the older sewers.

Inflow into sewers. For the purpose of analyzing sewer-flow measurements (gagings) and because of the measuring techniques available, inflow is usually subdivided into two categories. The first category includes cellar and foundation drainage, cooling-water discharges, and drainage from springs and swampy areas. This type of inflow causes a steady flow that cannot be identified separately and so is included in the measured infiltration. The second category consists of inflow that is related directly to storm-water runoff and, as a result of rainfall, causes an almost immediate increase in flows in sewers. Possible sources are roof leaders, yard and areaway drains, manhole covers, cross connections from storm drains and catch basins, and combined sewers (see Sec. 4-1).

Infiltration design allowances for sewers. When designing for unsewered areas or for the relief of overtaxed existing sewers, allowance must be made for unavoidable infiltration/inflow as well as for the expected wastewater. For existing sewers, infiltration allowances should be determined on the basis of flowrate measurements, with appropriate modifications to account for expected future leakage. For new sewers, or existing sewers for which no data are available, average rates may be determined from data derived from similar existing sewers, with appropriate modifications to account for differences in materials and construction and in expected future conditions.

If relevant flow data are unavailable, average infiltration allowances presented in Fig. 3-1 may be used for new sewers or recently constructed sewer systems having precast manholes and pipe joints made with gaskets of rubber or rubberlike material. In all cases, the infiltration allowances for design should reflect the expected condition of the sewer system at the end of the period for which it is being designed.

The average flowrates for designing wastewater-treatment plants and pumping stations may be estimated by adding average domestic and industrial flows and average infiltration allowances. Inflow rates, because of their episodic nature, do not appreciably affect the average design flows.

The infiltration design allowance discussed here has little or no relationship to the allowances used for the acceptance of newly constructed sewers. The acceptance allowances are designed to measure how well the construction job was done, whereas the design allowance is used to account for what may ultimately happen to the sewer, including the construction of building sewers on private property.

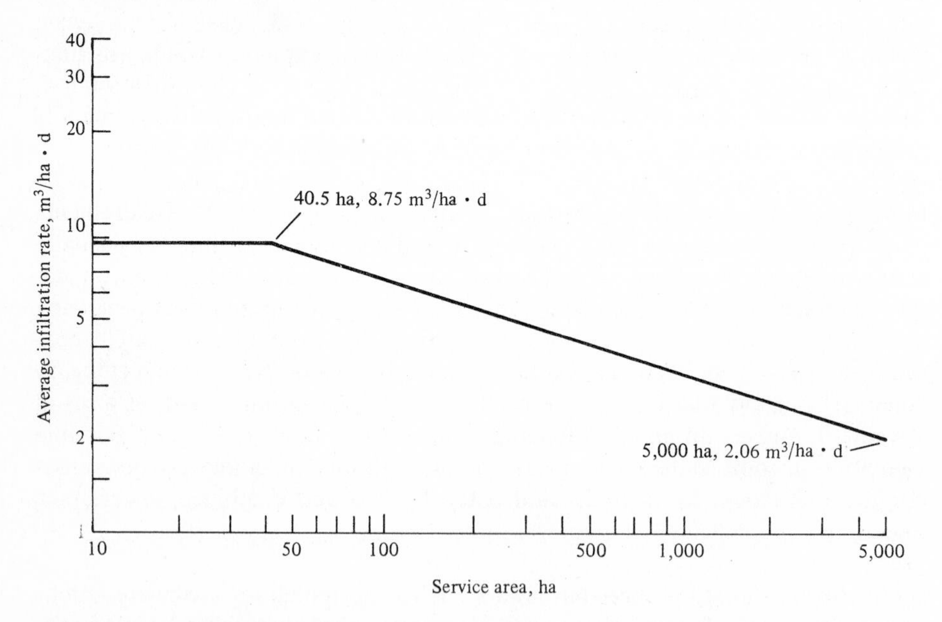

Figure 3-1 Average infiltration rate allowance for new sewers. Note: ha × 2.4711 = acre; m³/ha·d × 106.9 = gal/acre·d.

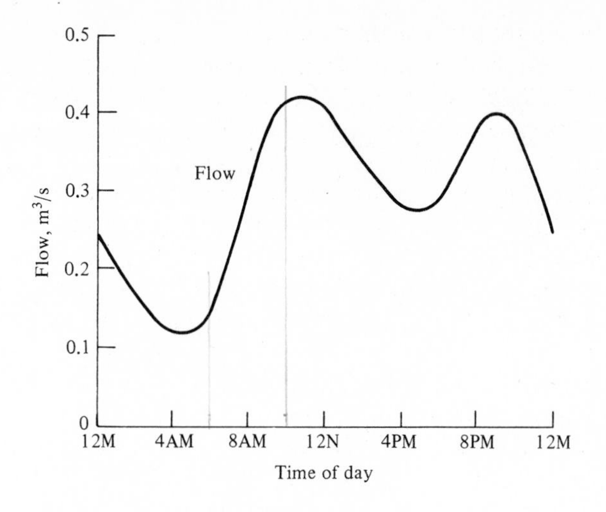

Figure 3-2 Typical hourly variation in domestic wastewater flows. Note: $m^3/s \times 22.8245$ = Mgal/d.

Variations in Wastewater Flows

Short-term, seasonal, and industrial variations in wastewater flows are briefly discussed here. The analysis of flowrate data with respect to peak flows to be expected is discussed in Sec. 3-3.

Short-term variations. The variations in wastewater flows tend to follow a somewhat diurnal pattern, as shown in Fig. 3-2. Minimum flows occur during the early morning hours when water consumption is lowest and when the base flow consists of leakages, infiltration, and small quantities of wastewater. The first peak flow generally occurs just after the peak late-morning water use. A second peak flow generally occurs in the early evening.

When extraneous flows are minimal, the wastewater-discharge curves closely parallel water-consumption curves, but the time lag depends on the distance of the sewer from the initial source of wastewater. If home laundering is not done on a typical day, the variation in weekday flows is negligible. Typical weekly flows for both wet and dry periods are plotted in Fig. 3-3.

Seasonal variations. Seasonal variations in wastewater flows occur in resort areas, in small communities with college campuses, and in communities with seasonal commercial and industrial activities. The expected magnitude of the variations depends on both the size of the community and the seasonal activity. An extreme example is the variation in the city of Modesto, Calif., which occurs because of the substantial amount of industrial wastes from canneries and other activities related to agriculture.

Infiltration/inflow quantities also vary seasonally. Storm water and groundwater can enter the system through cracks, malformed or broken joints, unauthorized drainage connections, and poorly constructed house connections.

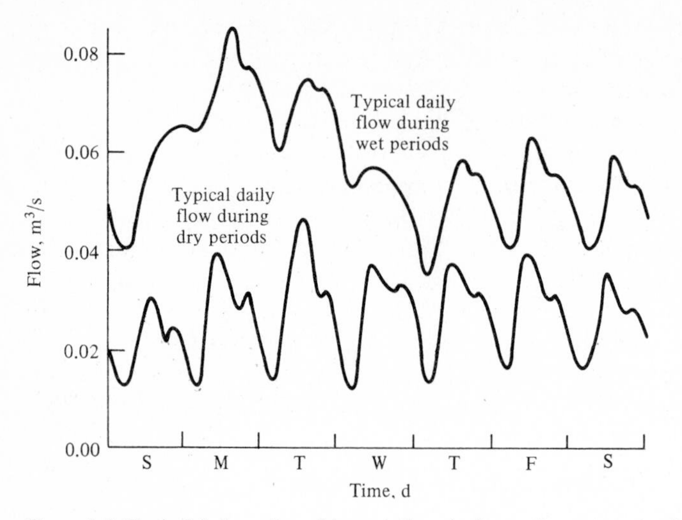

Figure 3-3 Typical daily and weekly variations in domestic wastewater flows. Note: $m^3/s \times 22.8245 = Mgal/d$.

The magnitude of this effect depends on the type of collection system and on location. For example, in the western United States, the rainfall pattern tends to be cyclical; there is little or no rainfall during the summer. In the eastern United States, the rainfall pattern tends to be more uniform.

Industrial variations. There is no foolproof procedure for predicting industrial wastewater discharges. Although internal process changes may reduce discharges, plant expansion may increase discharges. Where joint treatment facilities are to be constructed, special attention should be given to industrial discharge projections, whether they are prepared by the industry or jointly with the city's staff or engineering consultant. Industrial discharges are most troublesome in smaller wastewater-treatment plants where there is limited capacity to absorb shock loadings.

3-3 ANALYSIS OF WASTEWATER FLOWRATE DATA

Because the hydraulic design of sewer facilities is affected by variations in wastewater flows, design values for the expected peak flows must be developed. The best current design practice calls for estimating peaking factors for domestic and industrial wastewater flow and for infiltration and inflow separately.

Peaking Factors for Wastewater Flows

Ideally, peaking factors (the ratio of peak flow to average flow) would be derived or estimated for each major establishment or for each category of flow in the system. The individual average flows are multiplied by these factors to obtain the peak flows. The resulting peak flows from several areas would be combined to obtain the total expected peak flows. Unfortunately, this degree of refinement is seldom possible; therefore, peaking factors usually must be estimated by more generalized methods.

If flow-measurement records are inadequate to establish peaking factors, the curve given in Fig. 3-4 may be used. This curve was developed from analyses of the records of numerous communities throughout the United States. It is based on average domestic wastewater flows, exclusive of infiltration/inflow, and may be used for estimating peak flows from residential areas. It also applies to wastewater that contains small amounts of commercial flows and industrial wastes as well as residential wastewater.

When commercial, institutional, or industrial wastewaters make up a significant portion of the average flows (say 25 percent or more of all flows, exclusive of infiltration), peaking factors for the various flow categories should be estimated separately. If possible, peaking factors for industrial wastewater should be estimated on the basis of average water use, number of shifts worked, and pertinent details of plant operations.

Many state agencies have also set peak design flowrates to be used when no actual measurements are available. Typically, these flowrates are about 1500 L/capita · d (400 gal/capita · d) for laterals and 900 L/capita · d (240 gal/capita · d) for trunk sewers (assuming no extraneous flows other than normal infiltration) [7].

Peak Infiltration Flows

Peak infiltration allowances for sewer design are often related to the sizes of the areas served, as shown in the curves presented in Fig. 3-5. In the absence of contradictory measurements, these curves may be considered conservative for most sewer designs. Curve *A* relates to areas with old sewers; curve *B* relates to areas with either old or new sewers. The choice between curves *A* and *B* for old sewers depends on the present and expected future conditions of the sewers, the elevation of the groundwater table, and the method of joint construction. For example, if sewer joints are known or believed to have been formed with cement mortar, and the presence of a high-groundwater table is known or expected, curve *A* or higher rates should be used. In addition to

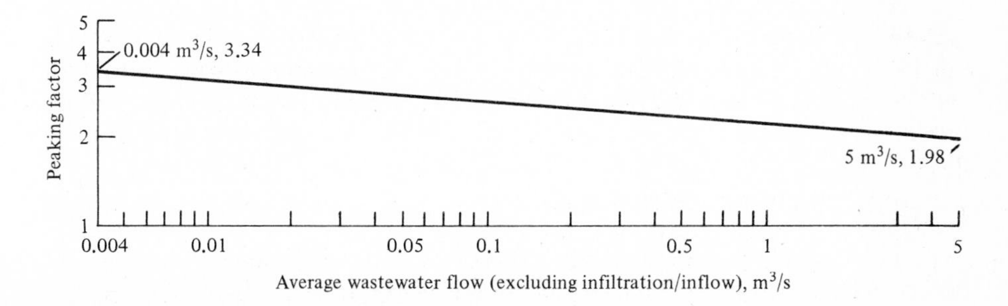

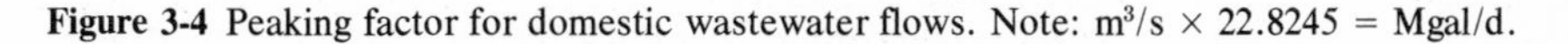

Figure 3-4 Peaking factor for domestic wastewater flows. Note: $m^3/s \times 22.8245 = Mgal/d$.

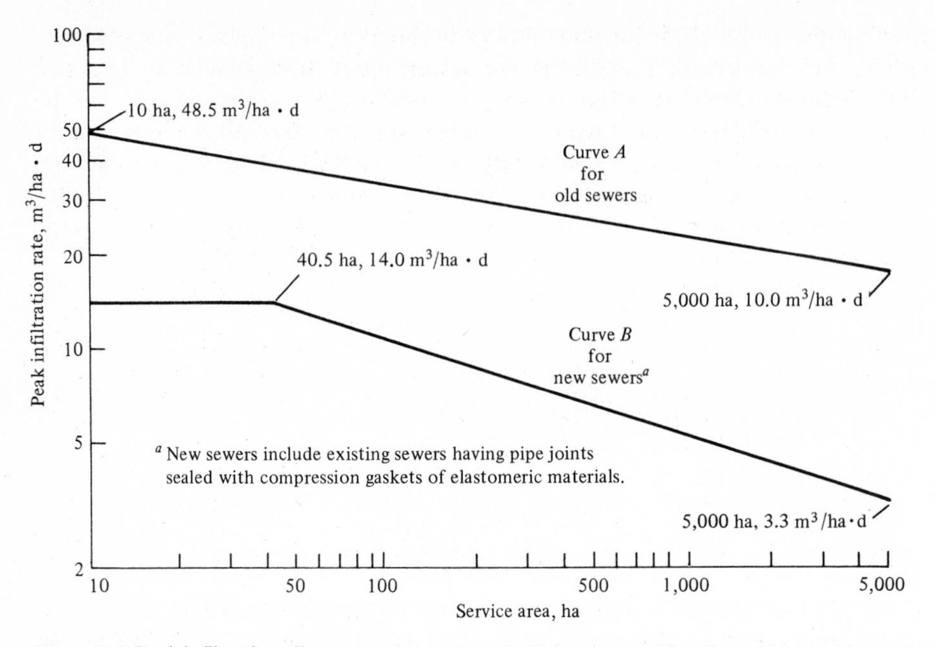

Figure 3-5 Peak infiltration allowances. Note: ha × 2.4711 = acre; m^3/ha · d × 106.9 = gal/acre · d.

newly designed sewers, the category of "new sewers" includes those recently constructed sewer systems in which precast concrete manholes were used and in which the pipe joints were sealed with compression gaskets or rubber or rubberlike materials.

When curves of average infiltration rates, justified by relevant measurements, are available, the peak infiltration rates for sewer design may be obtained by multiplying an appropriate peaking factor by the average values obtained from these curves. Peaking factors for infiltration are properly derived from flow measurements; common values range from 1.5 to 2.0.

Peak Inflow Design Allowance

Separate design allowances for peak inflow rates should be made when designing relief for, or extensions of, existing sewer systems. These allowances should be based on analyses of sewer flow measurements (gagings) where possible, with appropriate reductions attributable to corrective measures proposed for the existing system.

Although properly designed and constructed new sewers should be free from rainfall-related inflow, entry points for such inflow will develop eventually because of loose-fitting manhole covers, inadvertent connection of sump pumps, roof leaders, catch basins, and other drains, or other causes. However, conservative infiltration rates would normally be sufficient to allow for such possible occurrences. If it is customary, in the area to be sewered, to connect areaway drains or other drains to the sewers, and there is little prospect of eliminating this practice in the future, appropriate inflow allowances must be made in the design of new sewers. Also, the use of perforated manhole covers would indicate the need for added inflow allowance. Leakage through a manhole cover submerged under 25 mm (1 in) of water can vary from 75 to 265 L/min (20 to 70 gal/min), depending on the size and number of openings in the cover [16].

Example 3-1 Estimation of flowrates Estimate the expected average and peak domestic and industrial wastewater flows from the Covell Park Development shown in Fig. 3-6.

Data on the expected saturation population densities and wastewater flows for the various types of housing in the Covell Park Development were derived from actual records of similar nearby developments and are given in Table 3-9. The commercial (including the shopping area) and industrial wastewater flowrate allowances were estimated to be 20 and 30 m^3/ha · d (2100 and 3200 gal/acre · d), respectively. These estimates were derived from an analysis of the individual types of facilities to be included within the two zones. On the basis of actual flow records of similar activities, the average peaking factors are 1.8 for the commercial flows and 2.1 for the industrial flows.

The planned school within the Covell Park Development is to serve 2000 students at ultimate capacity. The average flow is 75 L/student · d (20 gal/student · d), and the peaking factor for the school is 4.0.

SOLUTION

1. Set up a computation table for estimating domestic and industrial wastewater flows. The required computations are summarized in Table 3-10. Although Table 3-10 is self-explanatory, the following comments clarify some of the specific entries.
   a. In cols. 3 and 6, a simple unweighted average of the data in Table 3-9 is used to determine the population and unit flow figures for the area of Covell Park with mixed residential dwellings.
   b. The peaking factor for the residential areas was obtained from Fig. 3-4 by using the total residential average flow because the flow from the entire area is being determined. Where sewers are to serve the individual areas, the average flow of each individual area contributing to a section of sewer would be included in estimating the appropriate peaking factor.
2. Summarize the domestic and industrial wastewater flow.

	Flow, m^3/d	
	Average	Peak
Domestic	14,231	34,683
Industrial	3,300	6,930
Total	17,531	41,613

Comment To determine the required capacity of the sewer to serve the area, an allowance for infiltration should be added to the domestic and industrial wastewater flows determined in this example.

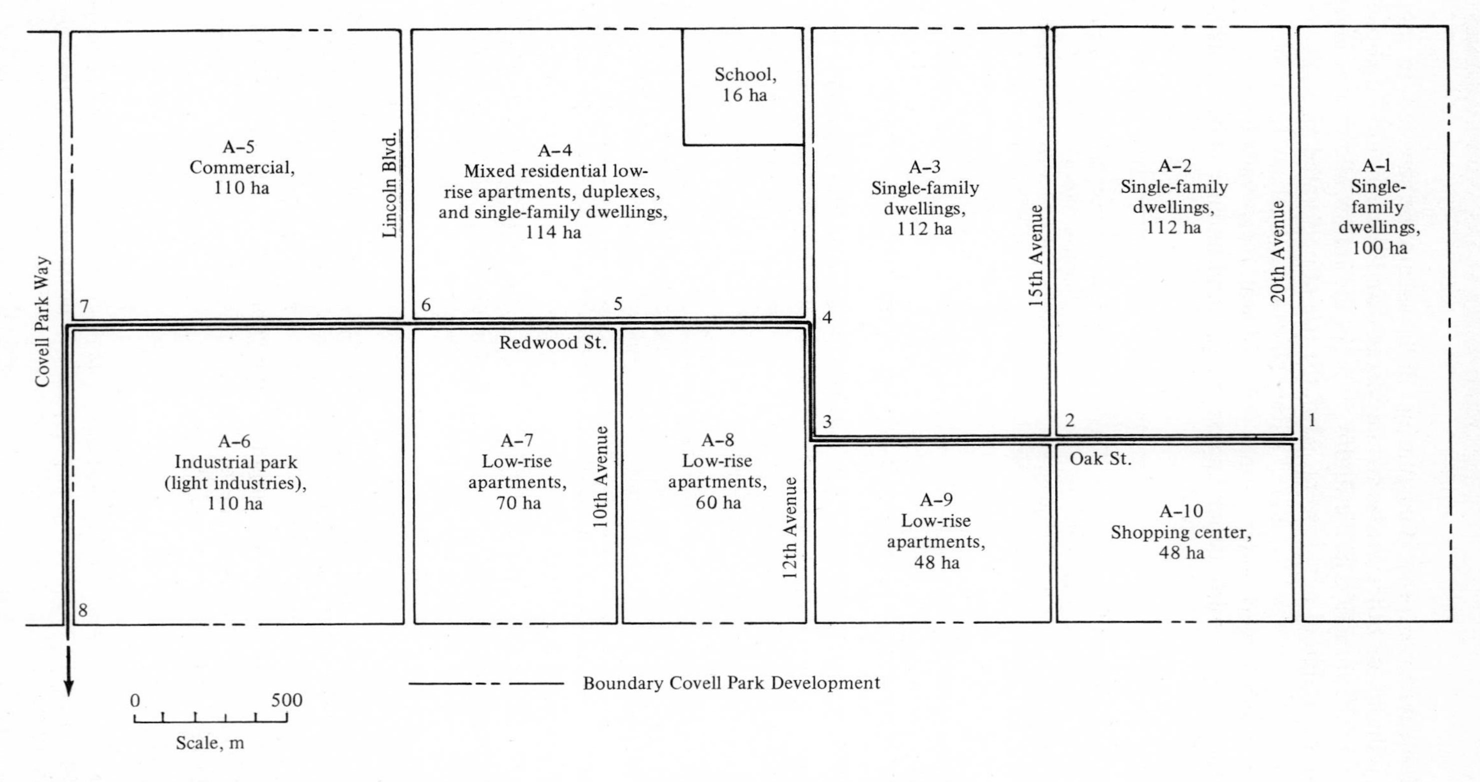

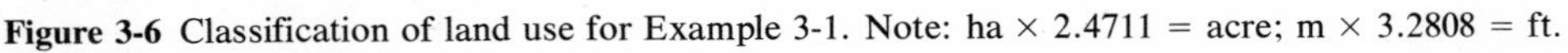

Figure 3-6 Classification of land use for Example 3-1. Note: ha × 2.4711 = acre; m × 3.2808 = ft.

Table 3-9 Saturation population densities and wastewater flows for residential area of the Covell Park Development for Example 3-1

Type of development	Saturation population density		Wastewater flows	
	persons/ha	(persons/acre)	L/capita · d	(gal/capita · d)
Single-family dwellings	38	(15)	300	(80)
Duplexes	60	(24)	280	(75)
Low-rise apartments	124	(50)	225	(60)

Note: ha × 2.4711 = acre
L × 0.2642 = gal

3-4 FLOW MEASUREMENT BY DIRECT-DISCHARGE METHODS

The ability to measure wastewater flows is of fundamental importance in designing all wastewater-management facilities. The two principal methods for measuring flowing fluids are (1) direct-discharge methods and (2) vlocity-area methods. Direct-discharge methods are discussed in this section; velocity-area methods are discussed in Sec. 3-5.

For direct-discharge methods, the rate of discharge relates to one or two easily measured variables. Frequently, if numerous flowrates are to be determined, rating curves are developed to simplify the work involved. The principal methods and applications are described in Table 3-11. Sketches of some of the apparatuses are shown in Fig. 3-7. Weirs, Venturi meters, Parshall flumes, and Palmer-Bowlus flumes are discussed in the following section. Details on the other methods may be found in the references given in Table 3-11.

Weirs

One of the most accurate devices for measuring water is a weir, provided the conditions under which the discharge coefficients of given types of weirs were determined are approximately duplicated in the actual gagings. The three most common types of weirs—rectangular, triangular, and trapezoidal—and submerged weirs are discussed here. To determine the discharge over broad-crested weirs and dams having different types of crests, the reader may consult Refs. 2, 4, . ., and 8.

Rectangular weirs. A rectangular weir is a notched overflow structure placed across the channel perpendicular to the direction of flow (see Fig. 3-8*a*). If the stream issuing from the weir is contracted as shown in Fig. 3-8*a*, the weir is a contracted weir. If the length of the crest is extended so that the notch is

Table 3-10 Determination of average peak domestic and industrial wastewater flow from the Covell Park Development for Example 3-1

Land-use classification	Total area, ha	Population density, persons/ha	Total population	Average unit flow		Average flow, m^3/d	Peaking factor	Peak flow, m^3/d
				Basis	Value			
(1)	(2)	(3)	(4)	(5)	(6)	(7)	(8)	(9)
Single-family dwellings	324	38	12,312	L/person · d	300	3,694	2.6[a]	9,604
Mixed residential dwellings	114	74[b]	8,436	L/person · d	268[b]	2,261	2.6[a]	5,879
Low-rise apartments	178	124	22,072	L/person · d	225	4,966	2.6[a]	12,912
School	16		2,000	L/person · d	75	150	4.0	600
Shopping center	48			M^3/ha · d	20	960	1.8	1,728
Commercial	110			m^3/ha · d	20	2,200	1.8	3,960
Industrial park	110			m^3/ha · d	30	3,300	2.1	6,930
Total	900					17,531		41,613

[a] From Fig. 3-4 based on a total residential average flow of 10,921 m^3/d (3694 + 2261 + 4966).

[b] Based on unweighted average of data reported in Table 3-9.

Note: $ha \times 2.4711 = \text{acre}$

$m^3/d \times 2.6417 \times 10^{-4} = \text{Mgal/d}$

$L \times 0.2642 = \text{gal}$

Table 3-11 Direct-discharge methods for flow measurement

Method/apparatus	Description/application	References
California pipe	In this method the flowrate is related to the depth of flow from an open end of a partially filled horizontal pipe that is discharging freely to the atmosphere. The discharge pipe should be horizontal and should have a length of at least six pipe diameters. If the pipe is expected to flow almost full, an air vent should be installed upstream of the pipe entrance to ensure free circulation of air in the unfilled portion of the discharge pipe.	8, 17
Computation	This method requires field measurements of the depth of flow and slope of the sewer. A value for the coefficient of roughness must also be selected. The method, at best, is an approximation dependent on the steadiness of the flow at the time of observation and the precision with which the coefficient of roughness is assumed for the existing conditions. This method is also based on the assumption that flow is occurring at normal depth. Despite these limitations, this method is used frequently for making wastewater-flow measurements.	
Direct weighing	In this method, which is used to measure small flows, the mass of fluid discharged over a specified time period is weighed and converted to a flowrate using the specific weight of the fluid. This is essentially a laboratory method; it is not readily adaptable to field use.	
Flow nozzles	Nozzle flow meters in pipes make use of the Venturi principle, but use a nozzle inserted in the pipe instead of the Venturi tube to produce the pressure differential. The form of the nozzle, the method of inserting it in a pipe, and the method of measuring the difference in pressure vary with the manufacturer. Open flow nozzles attached to the ends of pipe are usually of the Kennison type shown in Fig. 3-7*b*. Because nozzles placed at the ends of pipes are essentially proportional weirs, only a single pressure connection is needed to measure the head.	12
Magnetic flow meters	When an electric conductor passes through an electromagnetic field, an electromotive force or voltage is induced in the conductor that is proportional to the velocity of the conductor. This statement of Faraday's law serves as the basis of design for electromagnetic flow meters, as shown in Fig. 3-7*d*. In actual operation, the liquid in the pipe (usually water or wastewater) serves as the conductor. The electromagnetic field is generated by placing coils around the pipe. The induced voltage is then measured by electrodes placed on either side of the pipe. The electrodes must penetrate the wall of the	12

Table 3-11 Direct-discharge methods for flow measurement *(Continued)*

Method/apparatus	Description/application	References
	pipe to come into direct contact with the flowing liquid. When necessary, an automatic electrode cleaning system may be incorporated in which ultrasonic waves are continuously introduced into the liquid in the immediate vicinity of the electrodes. The resulting wave motion prevents a buildup of film or foreign particles from forming on the electrodes. Magnetic flow meters are usually available for pipe sizes varying from 50 to 900 mm (2 to 36 in) in diameter; larger sizes require special order.	
Orifice	An orifice is a cylindrical or prismatic opening through which fluid flows. The standard orifice, as generally defined, is one in which the edge of the orifice that determines the jet is such that the jet, upon leaving it, does not again touch the wall of the orifice. Practically, this result is obtained by having the outside of the orifice beveled. The flowrate is determined using Torricelli's theorem.	6, 8, 10, 12, 18
Orifice, pipe	A plate with a cylindrical opening in the center is usually inserted into closed pipelines. The flowrate is determined from differential pressure readings.	6, 8, 10, 12, 18
Sonic meters	Sonic meters are used to measure wastewater flows in pipes. Two basic types of meters are available; the first type uses two transducers mounted angularly to the flow line on opposite sides of the pipe, and the second type uses a single transponder mounted on the pipe wall. Each type uses a different sonic principle for flow measurement, but both types are usually full line size and thus do not add any head loss to the piping system. A number of sonic level meters have been adapted for use with Venturi flumes and weirs for level (and flow) measurement.	1
Tracers, chemical and radioactive	In chemical or radioactive gagings, a known concentration of a chemical or radioactive substance is added continuously, at a constant rate, to the stream in which the discharge is to be determined. At a distance downstream sufficient to ensure complete mixing of the tracer and stream, the stream is sampled and the concentration of the chemical or radioactive substances is determined. The flow in the stream can then be determined by using a mass balance.	12
Venturi flumes	Venturi flumes use the critical depth principle to measure flows in open channels. The two best known types are the Parshall and the Palmer-Bowlus flumes. The Parshall flume (see Fig. 3-7*c*) is usually fixed and is often used to measure flows at treatment plants. The Palmer-Bowlus flume is small and movable and is commonly used to measure wastewater flows in sewers.	12, 14, 15, 18, 20

Table 3-11 Direct-discharge methods for flow measurement *(Continued)*

Method/apparatus	Description/application	References
Venturi meters	The Venturi meter (see Fig. 3-7*a*), which is used to measure flows in closed conduits, consists of three parts: (1) the inlet cone, in which the diameter of the pipe is gradually reduced; (2) the throat or constricted section; and (3) the outlet cone, in which the diameter increases gradually to that of the pipe in which the meter is inserted. The throat in standard meter tubes is from one-fifth to three-fourths the diameter of the pipe. Its length is approximately equal to its diameter. A single pressure tap is provided both in the throat and in the straight portion of the inlet. The determination of the quantity of water flowing is based on the difference between pressures observed or indicated at the inlet and at the throat of the meter.	8, 10, 12, 18
Volumetric measurement	The volume of fluid discharged over a specified time period is measured. Generally, this can be done only with very low flows.	
Weir, sharp	A sharp weir is a barrier (usually a metal or plastic plate) over which the fluid to be measured is made to flow. Rectangular, triangular, and trapezoidal weirs are the three types most commonly used. The flowrate is determined by measuring the observed head on the crest of the weir (rectangular or trapezoidal weirs) or over the vertex of the weir notch (triangular weirs). The head is the difference in elevation between the crest and the surface of the water in the channel, at a point upstream taken just beyond the beginning of the surface curve. The flowrate is determined from a rating curve in which the flowrate is plotted versus the observed head.	4, 6, 8, 10, 12, 18, 20

Note: mm × 0.03937 = in

coincident with the sides of the channel (see Fig. 3-8*b*), the end contractions will be suppressed and the weir is a suppressed weir.

The general equation used to compute the discharge from a rectangular weir is

$$Q = C_d \tfrac{2}{3}\sqrt{2g}Lh^{3/2} \tag{3-1}$$

where Q = discharge, m^3/s (ft^3/s)
C_d = discharge coefficient
g = acceleration due to gravity, 9.81 m/s^2 (32.2 ft/s^2)
L = weir crest length, m (ft)
h = head on weir crest (difference in elevation between the crest of the weir and the water-surface elevation at a point upstream of the local drawdown curve), m (ft)

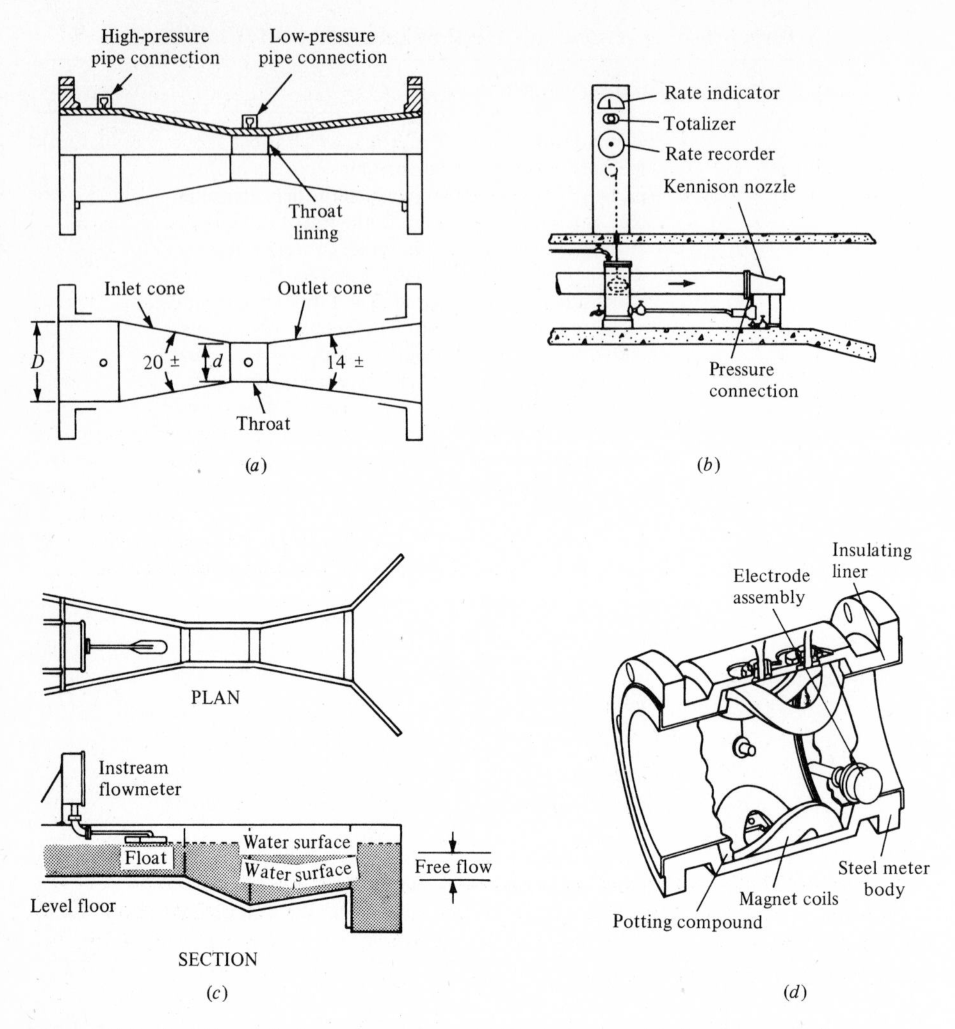

Figure 3-7 Typical direct-discharge flow meters. (*a*) Venturi meter. (*b*) Kennison open-flow nozzle installation (from BIF). (*c*) Parshall flume-metering installation. (*d*) Magnetic flowmeter (from Fischer & Porter).

All of the equations now in use may be reduced to this form. Because several methods of correcting for the velocity of approach have been proposed, it is recommended that the form of the equation as proposed by various researchers be used.

The Francis equation, developed in 1823, is the most common equation for estimating the flow over rectangular weirs. For contracted weirs (see Fig. 3-8*a*), neglecting the approach velocity, the Francis equation is

$$Q = 1.84(L - 0.1nh)h^{3/2} \qquad \text{(SI units)} \qquad (3\text{-}2)$$

$$Q = 3.33(L - 0.01nh)h^{3/2} \qquad \text{(U.S. customary units)} \qquad (3\text{-}2a)$$

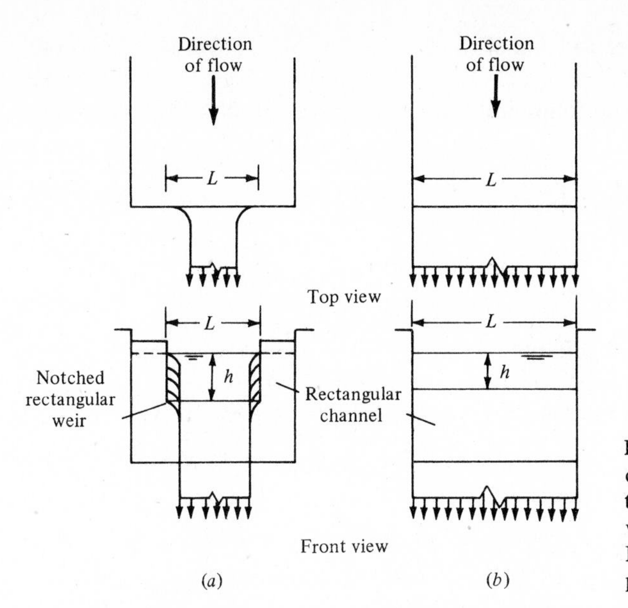

Figure 3-8 Definition sketch for contracted and suppressed rectangular weir. (*a*) Rectangular weir with end contractions. (*b*) Rectangular weir with suppressed end contractions.

where Q = discharge, m^3/s (ft^3/s)
C_d = discharge coefficient = 0.622
g = acceleration due to gravity, 9.81 m/s^2 (32.2 ft/s^2)
L = length of crest of weir, m (ft)
n = number of end contractions
h = head on weir crest, m (ft)

When the flow approaching the weir has appreciable velocity, the discharge will be greater than that calculated by Eq. 3-2. Under these conditions, the head term h should be corrected to account for the approach velocity. For contracted weirs, Eq. 3-2 corrected for the approach velocity is

$$Q = 1.84(L - 0.1nh)[(h + h_v)^{3/2} - h_v^{3/2}] \qquad \text{(SI units)} \qquad (3\text{-}3)$$

$$Q = 3.33(L - 0.1nh)[(h + h_v)^{3/2} - h_v^{3/2}] \qquad \text{(U.S. customary units)} \qquad (3\text{-}3a)$$

In Eq. 3-3, h_v is equal to the head due to the mean approach $V^2/2g$. When a weir has one or more end contractions, the approach velocity should be calculated by using the entire cross-sectional area of the approach channel; discharge should be determined by Eq. 3-3. Usually an estimate of the approach velocity based on an approximate value of discharge from Eq. 3-2 will be sufficiently accurate for use in Eq. 3-3.

For suppressed weirs (see Fig. 3-8*b*), neglecting the approach velocity, the Francis equation is

$$Q = 1.84Lh^{3/2} \qquad \text{(SI units)} \qquad (3\text{-}4)$$

$$Q = 3.33Lh^{3/2} \qquad \text{(U.S. customary units)} \qquad (3\text{-}4a)$$

To avoid the evaluation and use of the term h_v in Eq. 3-3, which makes the calculation more tedious, Eq. 3-1 may be used with the value of the discharge coefficient determined by the following relationship first proposed by Rehbock in 1911 [10].

$$C_d = 0.605 + \frac{1}{1050h - 3} + \frac{0.08h}{z} \qquad \text{(SI units)} \tag{3-5}$$

$$C_d = 0.605 + \frac{1}{320h - 3} + \frac{0.08h}{z} \qquad \text{(U.S. customary units)} \tag{3-5a}$$

where C_d = coefficient of discharge
h = head on weir crest, m (ft)
z = distance from bottom of channel to crest of weir, m (ft)

The weir equations (Eqs. 3-2, 3-3, and 3-4) apply only to vertical sharp-crested rectangular weirs with free overall when:

1. The head h is not greater than one-third the length L.
2. The head is no less than 0.15 m (0.5 ft) and no more than 0.6 m (2 ft).
3. The approach velocity is 0.3 m/s (1 ft/s) or less.
4. The height of the weir is at least three times the head.
5. The space under the water flowing over the weir is fully aerated so that the pressure on the underside of the water is atmospheric.

From a practical standpoint, these equations are probably usable with heads higher than 0.6 m (2 ft) but not much lower than 0.15 m (0.5 ft), and with approach velocities up to 1 m/s (3.3 ft/s).

Triangular weirs. A typical triangular weir is shown in Fig. 3-9. The basic expression used to compute the discharge through a triangular weir is

$$Q = C_d \frac{8}{15} \sqrt{2g} \tan \frac{\theta}{2} h^{5/2} \tag{3-6}$$

where Q = discharge, m^3/s (ft^3/s)
C_d = discharge coefficient (experimentally determined)
g = acceleration due to gravity, 9.81 m/s^2 (32.2 ft/s^2)
θ = angle of notch (see Fig. 3-9)
h = head over vertex of weir notch, m (ft)

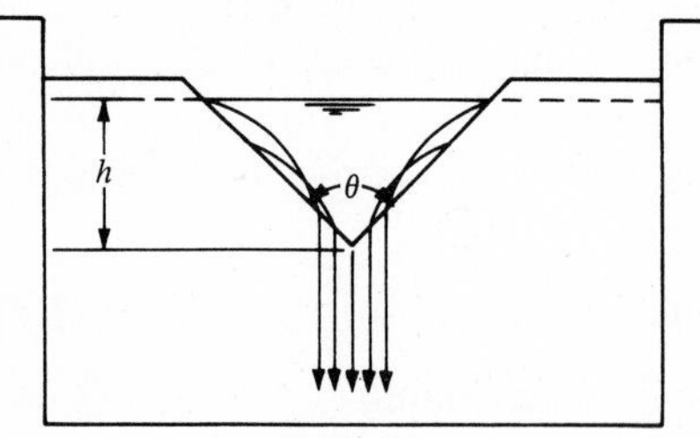

Figure 3-9 Definition sketch for triangular weir.

For a notched weir in which θ is equal to 90° [that is, $\tan(\theta/2) = 1$], it has been found experimentally that the discharge may be computed by the following equation:

$$Q = 0.55h^{5/2} \qquad \text{(SI units)} \tag{3-7}$$

$$Q = 2.5h^{5/2} \qquad \text{(U.S. customary units)} \tag{3-7a}$$

For heads lower than about 0.3 m (1 ft), the value of C_d begins to increase, depending on the notch angle. Such depths are not uncommon. If accuracy greater than that of Eq. 3-7 is desired, Eq. 3-6 may be used. A plot of C_d versus h is shown in Ref. 2.

Trapezoidal weirs. The trapezoidal weir differs from the rectangular weir in that the sides are inclined rather than vertical. Usually the sides are given an inclination of 1 (horizontal) to 4 (vertical), because at this angle the slope is just about sufficient to offset the effect of the end contractions. When this is done the weir is known as the Cipolletti weir (see Fig. 3-10). The equation for the Cipolletti weir is

$$Q = 1.859Lh^{3/2} \qquad \text{(SI units)} \tag{3-8}$$

$$Q = 3.367Lh^{3/2} \qquad \text{(U.S. customary units)} \tag{3-8a}$$

where L = the length of the crest (see Fig. 3-10), m (ft).

Submerged weirs. When the water surface in the channel below the weir is higher than the crest, the weir is said to be submerged or drowned (see Fig. 3-11). Measurements by submerged weirs are much less certain than measurements by weirs with free discharge, but their use is sometimes unavoidable.

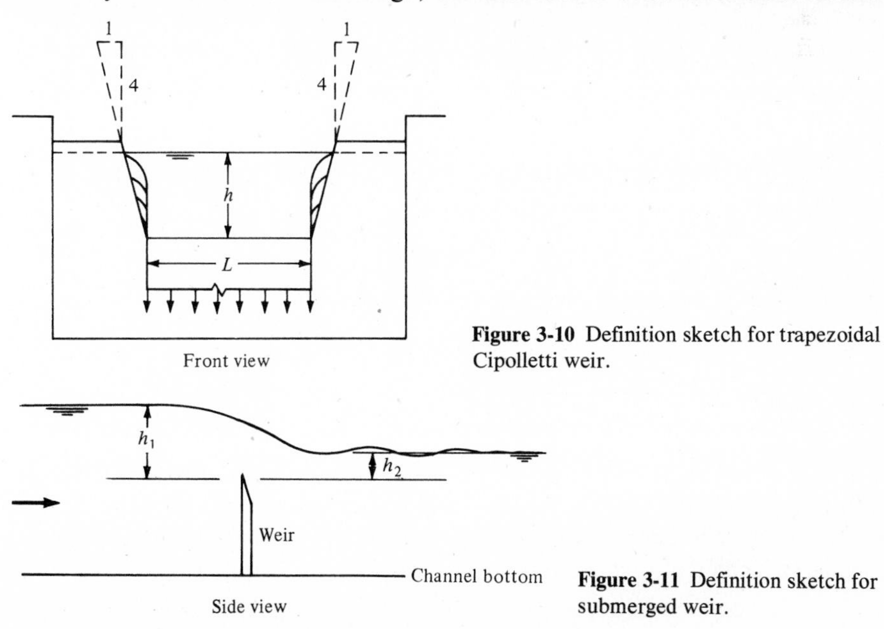

Figure 3-10 Definition sketch for trapezoidal Cipolletti weir.

Figure 3-11 Definition sketch for submerged weir.

On the basis of a series of experiments conducted on rectangular, triangular, parabolic, and proportional weirs, Villemonte [19] found that for all types of submerged weirs the discharge could be computed using the following:

$$\frac{Q}{Q_1} = \left[1 - \left(\frac{h_2}{h_1}\right)^n\right]^{0.385} \tag{3-9}$$

where Q = submerged discharge, m^3/s (ft^3/s)
Q_1 = free discharge, m^3/s (ft^3/s)
h_1 = upstream head, m (ft)
h_2 = downstream head, m (ft)
n = exponent in free-discharge equation $Q_1 = CH_1^n$ for each particular weir

Venturi Meter

A typical Venturi meter for measuring flow is shown in Fig. 3-7*a*. The equation for computing the discharge through a Venturi meter is derived from Bernoulli's equation (see Eq. 2-6, Chap. 2). For a horizontal meter, without allowance for energy losses, the appropriate equation is

$$Q = \frac{A_1A_2\sqrt{2g(h_1 - h_2)}}{\sqrt{A_1^2 - A_2^2}} = \frac{A_1A_2\sqrt{2gH}}{\sqrt{A_1^2 - A_2^2}} \tag{3-10}$$

where A_1 = area at upstream end, m^2 (ft^2)
A_2 = area at throat of meter, m^2 (ft^2)
g = acceleration due to gravity, 9.81 m/s^2 (32.2 ft/s^2)
h_1, h_2 = pressure heads, m (ft)
$H = h_1 - h_2$

Under actual operating conditions and for standard meter tubes, including allowances for losses, Eq. 3-10 reduces to

$$Q = CA_2\sqrt{2gH} \tag{3-11}$$

The coefficient written C is made up of two parts, or

$$C = C_1C_2$$

where $C_1 = A_1/\sqrt{A_1^2 - A_2^2}$
C_2 = coefficient of energy loss

For standard meter tubes in which the diameter of the throat is between one-third and one-half that of the pipe, the values of C_1 range between 1.0062 and 1.0328, and the energy-loss coefficient C_2 varies from 0.97 to 0.99. Thus, the range of values of C is from 0.98 to 1.02.

When Venturi meters are used to measure wastewater, there should be a manual means of periodically cleaning each pressure-sensing tap. Cleanout rods are used which are designed so that, on closing with a twisting action, a rod is forced through the opening to clean out any matter that may have clogged it. In addition, a continuous flushing system is usually necessary if Venturi meters for wastewater are to be maintained in good operating condition. Such a continuous flushing system is shown in Fig. 3-12. A photograph of a Venturi meter is shown in Fig. 3-13.

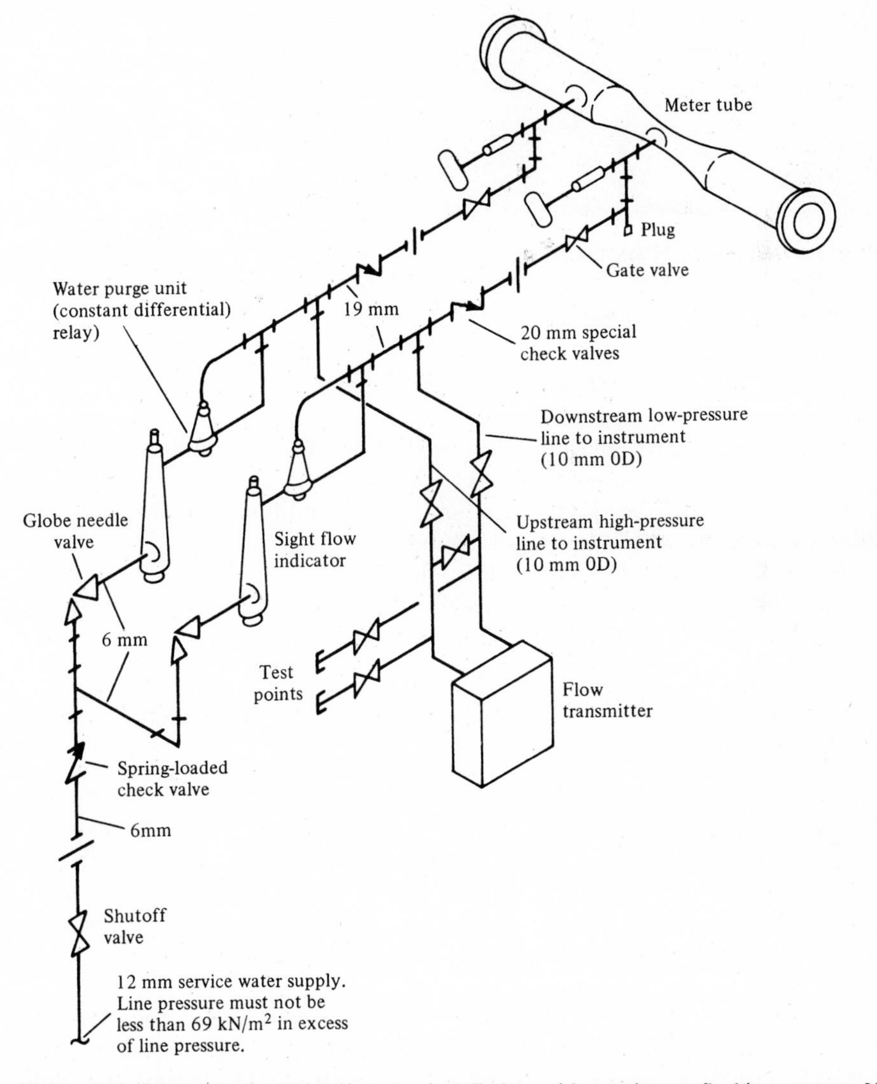

Figure 3-12 Schematic for Venturi meter installation with continuous-flushing system. Note: mm × 0.03937 = in; $kN/m^2 \times 0.0145 = lb_f/in^2$.

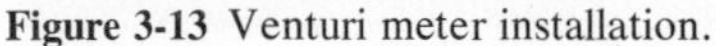

Figure 3-13 Venturi meter installation.

Parshall Flume

The Venturi type of meter is applicable only to closed pipes under pressure. The Venturi principle is used, however, to measure water flow in open channels by means of the Parshall flume [15], as shown in Fig. 3-7*c*. When operating under free-flow conditions, the flow passes through critical depth in the throat, followed by a hydraulic jump. Under some conditions of flow in the downstream channel, the jump may be submerged.

Because the throat width is constant, the discharge under free-flow conditions can be calculated by a single upstream measurement of depth. If the flume is operating under submerged conditions, the downstream head must also be measured to determine the discharge.

The Palmer-Bowlus Flume

The Palmer-Bowlus flume was developed for measuring flow in various open channels [14]. The principle of its operation is similar to that of the Parshall flume. The meter is usually placed in the sewer at a manhole, as shown in Fig. 3-14. To function properly, the flume must act as a hydraulic control in which critical flow is developed. This is usually assured when wastewater is backed up in the sewer above the flume, as a result of its installation, and when discharge from the flume is supercritical.

With critical flow on the flume assured, and with little energy loss, the rate of discharge may be related to the upstream depth. Thus, by measuring the upstream depth, the discharge can be read from a calibration curve that is usually supplied with each unit.

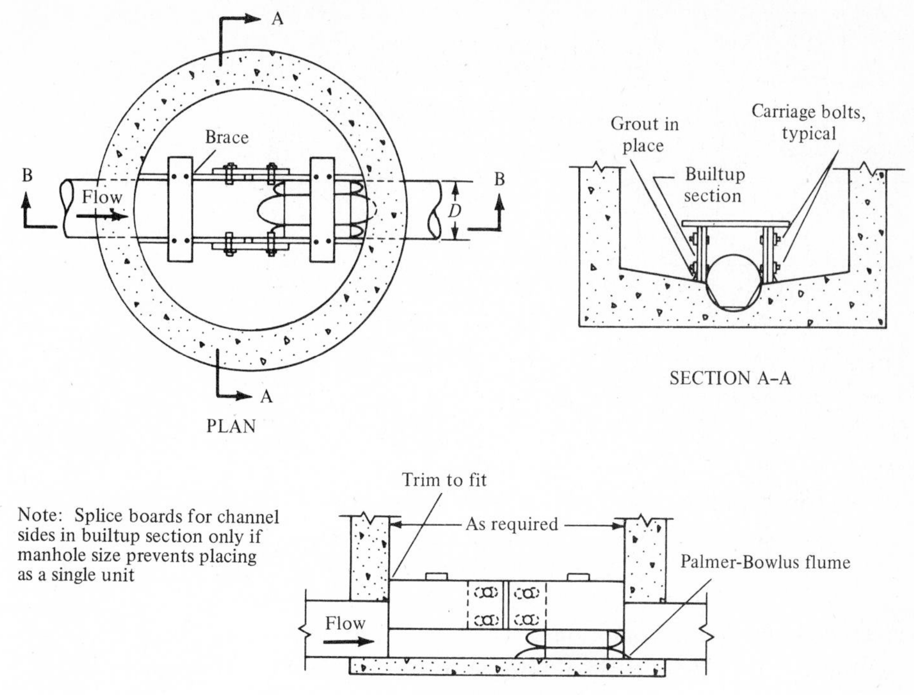

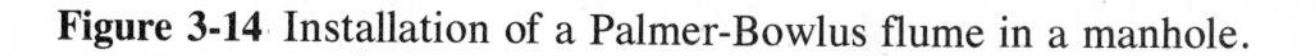

Figure 3-14 Installation of a Palmer-Bowlus flume in a manhole.

The advantages of the Palmer-Bowlus flume are that it can be installed in existing systems, head loss is minor, and it is self-cleansing. Care must be taken to avoid leakage under the flume and conditions in which the flume will be "drowned out." To keep the method accurate, the depth of flow in the upstream should not exceed 0.9 of the pipe diameter, and the point of upstream measurement should be about 0.5 of the pipe diameter from the entrance to the flume. A method of developing a rating curve for Palmer-Bowlus flumes is presented in the following discussion.

Sectional views of a typical flume are shown in Fig. 3-15. Neglecting the friction losses and equating the energies at points 1 and 2 by using Bernoulli's equation, the following equation is obtained:

$$y_1 + \frac{V_1^2}{2g} = y_2 + \frac{V_2^2}{2g} + y_t \tag{3-12}$$

where y_1 = depth of flow in upstream section, m (ft)
V_1 = flow velocity in upstream section, m/s (ft/s)
g = acceleration due to gravity, 9.81 m/s^2 (32.2 ft/s^2)
y_2 = depth of flow in flume, m
V_2 = flow velocity in flume, m/s (ft/s)
y_t = height of flume bottom above channel bottom, m (ft)

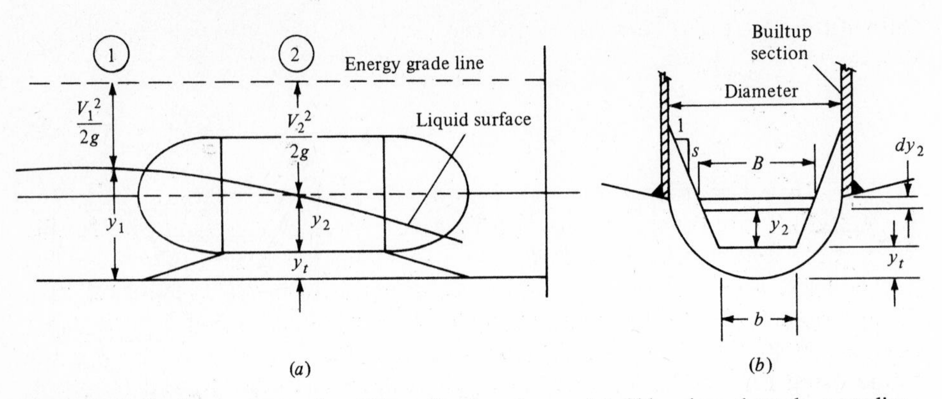

Figure 3-15 Sectional views of a Palmer-Bowlus flume. (*a*) Side view through centerline. (*b*) End view.

Solving for y_1 yields

$$y_1 = y_t + y_2 + \frac{V_2^2}{2g} - \frac{V_1^2}{2g} \tag{3-13}$$

Under free-discharge conditions, the flow will pass through the critical-flow condition in the throat of the flume, in which case y_2 will equal y_c.

The specific energy at any section in the throat is given by the following equation:

$$E = y_2 + \frac{V_2^2}{2g} = y_2 + \frac{Q^2}{A_2^2 2g} \tag{3-14}$$

where Q = discharge, m^3/s (ft^3/s)
A = area of cross section through which flow is occurring, m^2 (ft^2)

Differentiating Eq. 14 with respect to depth y_2 yields

$$\frac{dE}{dy_2} = 1 - \frac{Q^2}{A_2^3 g} \frac{dA_2}{y\, dy_2} \tag{3-15}$$

Because dA is equal to $B \times dy_2$, Eq. 3-15 may be written as follows:

$$\frac{dE}{dy_2} = 1 - \frac{Q^2 B}{A_2^3 g} \tag{3-16}$$

At critical flow, the energy is at a minimum, and dE/dy_2 is equal to zero. This yields

$$\frac{Q_c^2}{g} = \frac{A_c^3}{B_c} \tag{3-17}$$

from which the velocity head is

$$\frac{V_c^2}{2g} = \frac{Q_c^2}{A_c^2 2g} = \left(\frac{A}{2B}\right)_c \tag{3-18}$$

Substituting y_c for y_2 and Eq. 3-18 into Eq. 3-13, the following equation is obtained:

$$y_1 = y_t + y_c + \left(\frac{A}{2B}\right)_c - \frac{V_1^2}{2g} = y_t + y_c + \left(\frac{A}{2B}\right)_c - \frac{Q^2}{A_1^2 2g} \quad (3\text{-}19)$$

In Eq. 3-19, y_1 depends only on the flowrate because A_1 depends on y_1, and y_t, y_c, A_c, and B_c are fixed for a given flume and flowrate. A rating curve for the flume can be developed by solving Eq. 3-19 for y_1 for various flowrates. The procedure may be outlined as follows: (1) select a depth of flow in the flume; (2) determine the corresponding values of A and B in the flume for the depth selected in step 1; (3) determine the corresponding rate of flow through the flume using Eq. 3-17; and (4) by using the rate of flow determined in step 3, solve Eq. 3-19 for y_1 by trial and error. This four-step procedure is repeated for several depths of flow through the flume. The required rating curve is obtained by plotting the computed values of y_1 versus the corresponding flowrates.

To simplify the computations involved in developing the rating curve, either numerical or graphical methods are used. A graphical method for developing a rating curve based on the use of an Arredi diagram [14] is illustrated in Example 3-2. An Arredi diagram is a graphical device for solving Eq. 3-19 without the need for a trial-and-error solution.

Example 3-2 Development of rating curve for Palmer-Bowlus flume Develop a rating curve for a Palmer-Bowlus flume, such as the one shown in Fig. 3-15, with the use of the Arredi diagram method for an estimated range of flow of 0.025 to 0.15 m^3/s (0.57 to 3.4 Mgal/d). The flume is to be placed in a manhole that connects two 500-mm (20-in) pipes. Referring to Fig. 3-15, the critical dimensions of the flume are: b = 160 mm (6.3 in), y_t = 40 mm (1.6 in), and s = 2. Determine the flowrate when the depth in the upstream section is 250 mm (10 in).

SOLUTION

1. Develop the Arredi diagram for the flow installation. The completed Arredi diagram for this problem is shown in Fig. 3-16; its development is described in the following discussion.
   a. Compute values of $Q^2/A^2 2g$ ($=V^2/2g$) for selected values of Q and A, and plot against A in the upper portion of a piece of millimeter graph paper (see Fig. 3-16). The values selected for Q should cover the anticipated range of flowrates to be measured, and the area values should span a range equal to about twice the area of the sewer.
   b. Determine the corresponding cross-sectional areas for selected values of depth in the throat section of the flume and in the upstream section above the flume. The required computations are summarized in Tables 3-12 and 3-13. In the upstream section before the throat, the sides are vertical above a depth of 250 mm, and the zero area in the throat of the flume occurs at depth y_t, with reference to the bottom of the pipe. For depths less than 250 mm, the cross-sectional areas in the upstream pipe section before the throat are determined using A/A_{full} values from Fig. 2-16 for the appropriate d/D values.
   c. Plot separate curves of depth versus area, in the lower portion of the graph for the throat and upstream sections (see Fig. 3-16).
   d. Compute values of $A/2B$ ($= V^2/2g$) for the values of depth in the throat of the flume used in step 2 (see Table 3-12), and plot the computed values against the area in the upper portion of the graph. The intersection of this curve and the velocity head curves plotted in step a represents the solution of Eq. 3-18 for a given flume geometry. In

addition, the vertical distance from the point of intersection of these curves to the curve of throat area versus depth in the lower portion of the graph represents the total energy in the throat section ($y_t + y_c + V_c^2/2g$).

2. Prepare a rating curve for the flume.
   a. Starting with the lowest rate of flow to be measured, scale from the graph the value in millimeters for the total energy head ($y_t + y_c + V_c^2/2g$) for the throat section. Because the total energy in the upstream section (neglecting frictional and other minor losses) must be equal for the same flowrate, the value of $y_1 + V_1^2/2g$ is equal to the measured value of $y_t + y_c + V_c^2/2g$. An example is shown in Fig. 3-16. This procedure is repeated for the range of flowrates to be measured. A summary of the results is presented in Table 3-14.
   b. The rating curve is prepared by plotting the depth y_1 versus the corresponding value of flowrate (see Fig. 3-17).
3. Find the flowrate for a depth of flow of 250 mm. From Fig. 3-17, the flowrate is 0.04 m^3/s (0.9 Mgal/d).

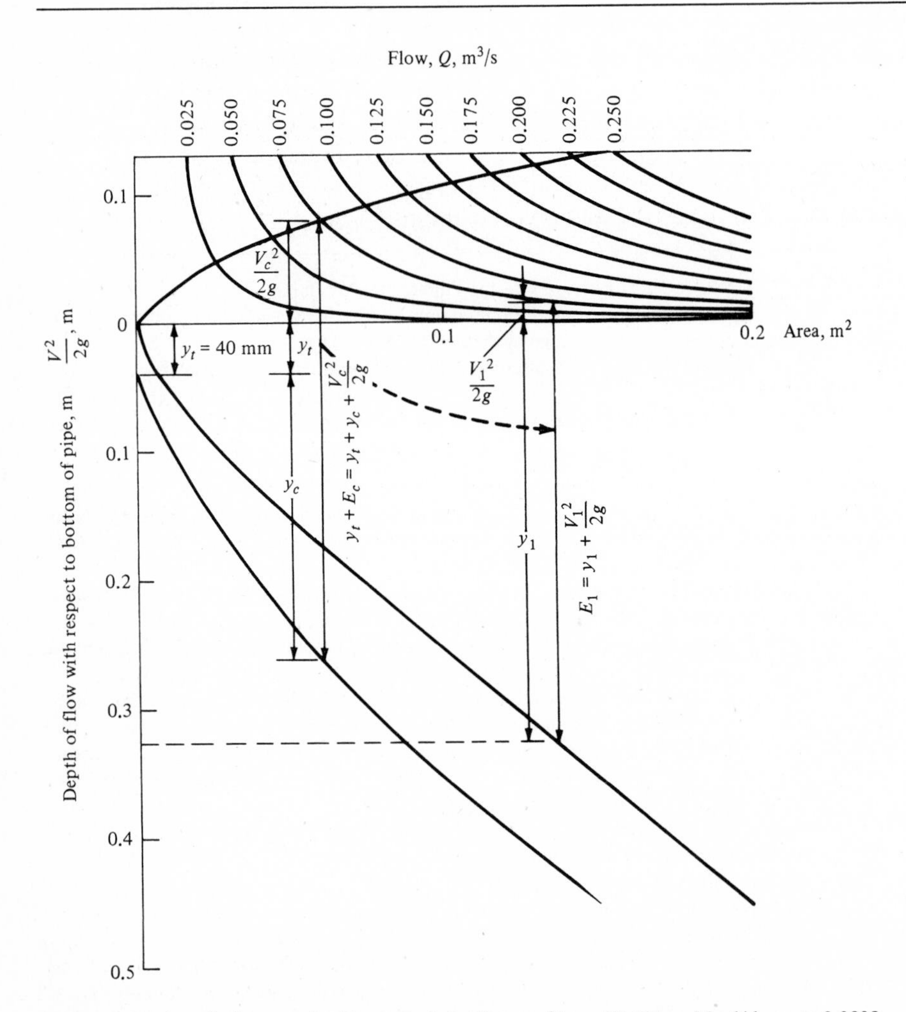

Figure 3-16 Arredi diagram for Example 3-2. Note: $m^3/s \times 22.824$ = Mgal/d; m × 3.2808 = ft; $m^2 \times 10.7639 = ft^2$.

Table 3-12 Computation table for development of the Arredi diagram for the throat section of the Palmer-Bowlus flume for Example 3-2

y_c, mm	$y_c + y_t$, mm	$b + \frac{y_c}{2}$, mm	$A = y_c\left(b + \frac{y_c}{2}\right)$, m^2	$2B = 2b + 4\frac{y_c}{2}$, mm	$\frac{V_c^2}{2g} = \left(\frac{A}{2B}\right)_c$, mm
50	90	185	0.0093	420	22
100	140	210	0.0210	520	40
150	190	235	0.0353	620	57
200	240	260	0.052	720	72
250	290	285	0.0713	820	87
300	340	310	0.093	920	101
340	380	330	0.112	1,000	112
400	440	360	0.144	1,120	129

Note: mm × 0.03937 = in
m^2 × 10.7639 = ft^2

Table 3-13 Computation of area in upstream section for Example 3-2

Depth y_1, mm	Area, m^2
50	0.010
100	0.028
150	0.050
200	0.073
250[a]	0.098
450	0.198

[a]Rectangular section above this depth.
Note: mm × 0.03937 = in
m^2 × 10.7639 = ft^2

Table 3-14 Data for the Palmer-Bowlus flume rating curve derived from Fig. 3-16

Depth, m	Flowrate, m^3/s
0.200	0.025
0.273	0.05
0.327	0.075
0.370	0.10
0.409	0.125
0.440	0.150

Note: m^3/s × 22.8245 = Mgal/d
mm × 0.03937 = in

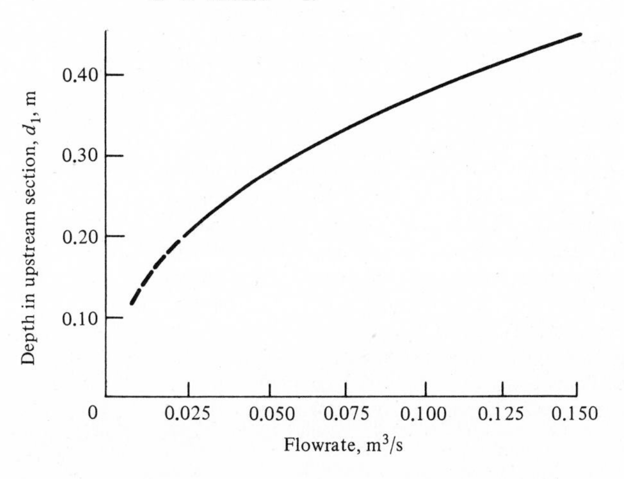

Figure 3-17 Rating curve for Palmer-Bowlus flume for Example 3-2. Note: m^3/s × 22.8245 = Mgal/d; m × 3.2808 = ft.

3-5 FLOW MEASUREMENT BY VELOCITY-AREA METHODS

Using the velocity-area methods, the flowrate is determined by multiplying the velocity of flow, m/s (ft/s), by the cross-sectional area, m^2 (ft^2), through which flow is occurring. The principal methods and apparatuses used to determine velocities are summarized in Table 3-15. Details may be found in the corresponding references given in Table 3-15.

Table 3-15 Methods for velocity measurement

Method/apparatus	Description/application	References
Current (propeller) meters	Current meter measurements may be used to determine accurately the velocity of flow in large sewers or open channels, provided there is not too much paper or other suspended matter present to clog the meter. Gagings of flow may be made by several methods: the one-, two-, and multiple-point methods; the method of integrating in sections; and the method of integrating in one operation. In the one-point method, the meter is held at 0.6 of the depth, measured from the water surface and in the center of the stream. The result is assumed to represent the mean velocity of the stream. This is a rough approximation, suitable only for hasty observations with no pretense of accuracy. In the two-point method, the velocity is observed at 0.2 and 0.8 of the depth, and the average of these two figures is taken to represent the average velocity in the vertical section. The stream can be divided into a number of vertical sections, and the average velocity in each section is approximately determined by this method. In the multiple-point method, sufficient readings are taken so that velocity contours are developed.	8, 10, 12, 18, 20
Electrical methods	Electrical methods to measure the quantity of water flowing in a stream involve the use of equipment such as conductivity cells, hot-wire anemometers, and warm-film anemometers. Although some of these methods have been used in the field, they are not ideally suited for measuring wastewater flows because the floating and suspended material commonly found in wastewater interferes with their position.	10, 12, 18
Float measurements	Float measurements of the flow in sewers are not made routinely, except in rectangular channels or for approximations of the velocity of flow between two manholes. In studies of tidal currents or of wastewater currents in bodies of water into which wastewater may be discharged, floats are used universally. Three types of floats may be used: surface, subsurface, and rod or spar. Only surface velocities can be obtained by the use of surface floats. Owing to the modifying effects of the wind, the results can be considered only as approximations. Subsurface floats consist of relatively large bodies slightly heavier than water, connected by fine wires to surface	10, 12

	floats of sufficient size to furnish the necessary flotation, and carrying markers by which their courses may be traced. The resistance of the upper float and connecting wire is generally so slight that combination may be assumed to move with the velocity of the water at the position of the submerged float.	
	Rod floats have been used to measure flow in open flumes with a high degree of accuracy. They generally consist of metal cylinders so loaded as to float vertically. The velocity of the rod has been found to correspond very closely with the mean velocity of the water in the course followed by the float.	
Pitot tubes	The Pitot tube, which has proved so useful in water pipe gagings, is impractical in extended sewer gagings because the suspended matter in wastewater tends to clog the tube.	8, 10, 12, 18
Tracers, chemical and radioactive	Where velocity measurements are to be made, the chemical or radioactive tracers are usually injected into the stream upstream of two control points. The time of passage of the prism of water containing the tracer is noted at these control points, and the velocity is then computed by dividing the distance between the control points by the travel time.	12
	When salt (NaCl) is used as the tracer, the time of passage between control points is measured by using electrodes connected to an ammeter or recorder. When radioactive tracers are used, the time of passage is noted by radioactive counters attached to the outside of the pipe. The time of passage is the difference between the times when the peak counts were recorded at each counting station.	
Tracers, dye	The use of dyes for measuring the velocity of flow in sewers, particularly in small-pipe sewers, is one of the simplest and most successful methods that have been used. A section of sewer is selected in which the flow is practically steady and uniform. The dye is thrown in at the upper end, and the time of its arrival at the lower end is determined. If a bright-colored dye, such as eosin, is used and a bright plate is suspended horizontally in the sewer at the lower end, the time of appearance and disappearance of the dye at the lower end can be noted with considerable precision, and the mean between these two observed times may be taken as representative of the average time of flow. Other dyes that have been successfully used in tracer studies include flourescein, congo red, potassium permanganate, rhodamine B, and Pontacyle Brilliant Pink B. Pontacyl Brilliant Pink B is especially useful in the conduct of ocean outfall dispersion studies.	8, 10, 12, 18

DISCUSSION TOPICS AND PROBLEMS

3-1 Sewers are to be installed in a recreational camping area that has a developed campground for 300 persons, lodges and cabins for 150 persons, and resort apartments for 50 persons. Assume that persons staying in lodges use the dining hall for 3 meals per day and that a 50-seat cafeteria with 4 employees and an estimated 150 customers per day has been constructed. Daily attendance at visitor centers is expected to be 50 percent of the campground capacity. Other facilities include a 10-machine laundromat, a 20-seat cocktail lounge, and 3 gas stations (7.5 $m^3/d \cdot$ station). Determine the average wastewater flow in cubic meters per day by using the design unit flows reported in Table 3-5.

3-2 Estimate the percentage reduction that can be achieved in the domestic flowrate in the Covell Park Development considered in Example 3-1 if the developer is required to comply with a new water-conservation ordinance that is currently under consideration. The proposed city ordinance will require the use of (1) single-batch flush valve toilets, (2) flow-limiting devices for shower heads, and (3) level controls for washing machines in all single-family dwellings, duplexes, and low-rise apartmens.

3-3 Land use in an area served by a single trunk sewer is given below.

Type of development	Area, ha
Residential	125
Commercial	11
School	4
Industrial	8

The school has 1500 students. The average flowrate is 75 L/student · d, and the peaking factor is 4.0. Average flowrate allowances and peaking factors for the other developments are shown in the table below.

Type of development	Average flowrate, $m^3/ha \cdot d$	Peaking factor
Residential	40	Use Fig. 3–4
Commercial	20	2.0
Industrial	30	2.5

Determine the peak wastewater flowrate in the trunk sewer.

3-4 A residential area having the following housing types and population densities is served by a single trunk sewer. Determine the peak wastewater flowrate including infiltration. Assume that the per capita flowrate is 300 L/d throughout. Use Fig. 3-4 to obtain the peaking factor and Fig. 3-5, curve *B*, to estimate peak infiltration.

Housing type	Area, ha	Population density, persons/ha
A	10	40
B	25	55
C	15	60

3-5 Estimate the wastewater flows from a large industrial development covering an area of 200 ha. From water-meter readings, it has been determined that the annual use of water within the area is $4.24 \times 10^6 m^3$. Twenty percent of the gross area of the development has been landscaped. The average water demand for irrigation of landscape areas is estimated to be 1.3 m/yr.

Assuming that 85 percent of the nonirrigation water consumption ultimately reaches the sewer, estimate the annual wastewater production within the area. Assuming that all industries within the area operate concurrently for 12 h/d, 5 d/wk, throughout the year, and that the wastewater production during the hours of operation is essentially constant, estimate the maximum wastewater flowrate in the trunk sewer that serves the entire area. Also, compute the average annual wastewater production in cubic meters per day, and determine the value of the peaking factor that relates peak flow to average annual flow. Neglect infiltration and inflow.

3-6 Estimate the ratio of the peak hourly flowrate to the average daily flowrate for the curve given in Fig. 3-2.

3-7 Obtain the design curves used by the engineering department in your community to estimate (1) the average and peak infiltration rate allowance for new sewers, and (2) the peaking factors for domestic wastewater flows. How do these curves compare with those given in Figs. 3-1, 3-4, and 3-5?

3-8 A sharp-crested rectangular weir is used to measure flow in a rectangular channel that is 2 m wide. The weir extends completely across the channel and its crest is 1 m above the channel floor. Determine the flowrate in the channel when the depth of water upstream from the weir is 1.3 m. Use the Francis equation with and without correction for approach velocity (Eqs. 3-2 and 3-3, respectively) and Eq. 3-1 with the Rehbock correction (Eq. 3-5) for the discharge coefficient. Compare results.

3-9 A suppressed weir, 8.0 m in length, is to discharge 12 m^3/s of wastewater. To what height will the water rise behind the weir if the height of the weir is 1.25 m?

3-10 To what height may the weir in Prob. 3-9 be built if the height of wastewater behind the weir is not to exceed 2 m?

3-11 Derive the basic equation (Eq. 3-6) governing the discharge over a triangular weir.

3-12 Five triangular weirs with $\theta = 90°$ are used to meter the overflow from a rectangular tank 5 m wide and 10 m long. The bottoms of the weir notches are 7 m above the floor of the tank. Determine the flowrate over the weirs when the tank contains 360 m^3 of water.

3-13 Demonstrate that the side slopes of a Cipolletti weir (see Fig. 3-10) must be equal to 4:1. (*Hint:* Equate the loss of discharge in a rectangular weir caused by the end contractions to the discharge through a triangular weir.)

3-14 Determine the flowrate Q in a Venturi meter when the pressures upstream and in the throat are 196 kN/m^2 and 170 kN/m^2, respectively. The upstream diameter is 150 mm and the throat diameter is 75 mm.

3-15 Derive Eq. 3-10 for a Venturi meter.

3-16 Determine the flowrate Q in the Venturi meter shown in Fig. 3-18. Neglect all losses. Note that Eq. 3-10 (with the terms as defined) cannot be applied directly because the meter shown in Fig. 3-18 is not horizontal.

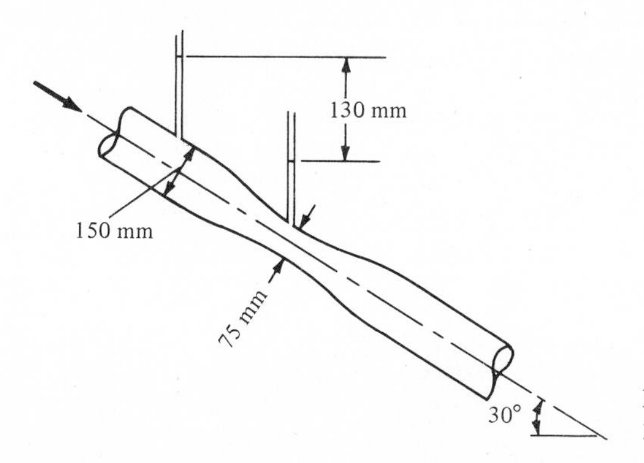

Figure 3-18 Definition sketch for Prob. 3-16. Note: mm × 0.03937 = in.

3-17 Select the throat diameter for a Venturi meter that is to be used in a 250-mm effluent-discharge-force main from a wastewater-treatment plant. The average flow through the plant is 5000 m^3/d and it is desired to have a minimum head differential of 200 mm at low flow. Assume that the flow variation of the treatment plant is as shown in Fig. 3-19, and that the friction coefficient C_2 is 0.982. What is the head differential at maximum flow for the selected Venturi meter?

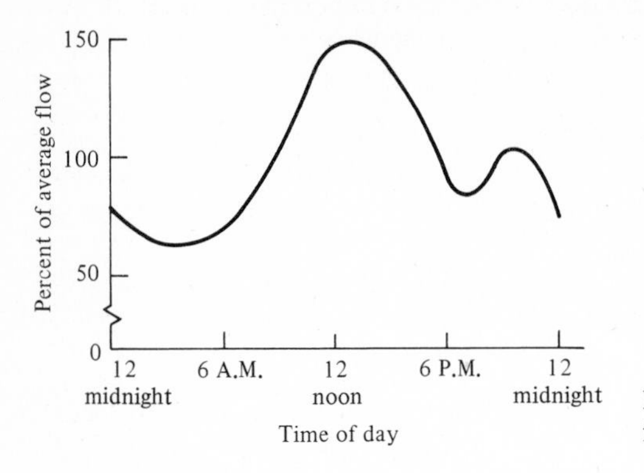

Figure 3-19 Hourly variation of wastewater flow for Prob. 3-17.

3-18 Dye is injected into the upstream end of a reach of sewer with constant slope and a diameter of 0.5 m. The dye is first observed 95 s later at a point 70 m downstream. The depth of flow is determined to be 0.4 m throughout the reach. Determine the flowrate and the slope of the sewer. Assume that $n = 0.015$ for flow at the existing depth.

3-19 A flow-measuring device for a rectangular channel, which operates on the same principle as a Palmer-Bowlus flume for sewers, consists of a short section of channel in which the floor has been raised sufficiently to produce critical flow. Develop a rating curve in which the upstream depth is plotted versus the flowrate for a 3-m-wide rectangular channel in which the floor has been raised 0.75 m for a short distance. The estimated range of flows is 1 m^3/s to 8 m^3/s. Use the Arredi diagram method and neglect all losses.

3-20 Rating curves for Palmer-Bowlus flumes and similar flow-measuring devices can be developed without an Arredi diagram; however, such derivations can be difficult. To understand more fully the benefits of the Arredi diagram, derive, but do not attempt to solve, a relationship which could be used directly for developing a rating curve for the measuring device described in Prob. 3-19. The only variables in the relationship are to be the upstream depth y and the flowrate Q.

REFERENCES

1. Considine, D. M. (ed.): *Process Instruments and Controls Handbook,* 2d ed., McGraw-Hill, 1974.
2. Daugherty, R. L., and J. B. Franzini: *Fluid Mechanics: With Engineering Applications,* 7th ed., McGraw-Hill, New York, 1977.
3. *Federal Register,* vol. 39, n. 29, sec. 35.905, Feb. 11, 1974.
4. Horton, R. E.: *Weir Experiments, Coefficients and Formulas,* U.S. Geological Survey Water Supply and Irrigation Paper 200, 1907.
5. Hubbell, J. W.: Commercial and Institutional Wastewater Loadings, *J. Water Pollut. Control Fed.,* vol. 34, no. 9, 1962.
6. Jaeger, C.: *Engineering Fluid Mechanics,* Blackie, London, 1956.
7. Joint Committee of the American Society of Civil Engineers and the Water Pollution Control Federation: *Design and Construction of Sanitary and Storm Sewers,* ASCE Manual and Report 37, New York, 1969.

8. King, H. W., and E. F. Brater: *Handbook of Hydraulics,* McGraw-Hill, New York, 1963.
9. McJunkin, F. E.: Population Forecasting by Sanitary Engineers, *J. Sanit. Eng. Div., ASCE,* vol. 90, no. SA4, 1964.
10. Metcalf, L., and H. P. Eddy: *American Sewerage Practice,* vol. 1, 2d ed., McGraw-Hill, New York, 1928.
11. Metcalf & Eddy, Inc.: *Wastewater Engineering: Treatment, Disposal, Reuse,* 2d ed., McGraw-Hill, 1979.
12. Metcalf & Eddy, Inc.: *Wastewater Engineering: Collection, Treatment, Disposal,* McGraw-Hill, New York, 1972.
13. Metcalf & Eddy, Inc.: *Report to National Commission on Water Quality on Assessment of Technologies and Costs for Publicly Owned Treatment Works,* vol. 2, prepared under Public Law 92-500, Boston, 1975.
14. Palmer, H. K., and F. D. Bowlus: Adaption of Venturi Flumes to Flow Measurement in Conduits, *Trans. ASCE,* vol. 101, p. 1195, 1936.
15. Parshall, R. L.: The Improved Venturi Flume, *Trans. ASCE,* vol. 89, p. 841, 1926.
16. Rawn, A. W.: What Cost Leaking Manhole? *Waterworks and Sewage Work,* vol. 84, no. 12, 1937.
17. Van Leer, B. R.: The California-Pipe Method of Water Measurement, *Eng. News-Rec.,* Aug. 3, 1922, Aug. 21, 1924.
18. Vennard, J. K., and R. L. Street: *Elementary Fluid Mechanics,* 5th ed., Wiley, New York, 1975.
19. Villemonte, J. R.: Submerged-Weir Discharge Studies, *Eng. News-Rec.,* Dec. 26, 1947.
20. *Water Measurement Manual,* U.S. Department of the Interior, Bureau of Reclamation, United States Printing Office, Washington, D.C., 1953.

CHAPTER

FOUR

DESIGN OF SEWERS

The collection and conveyance of wastewater from the various sources where it is generated is the first step in the effective management of a community's wastewater. The pipes that collect and transport the wastewater away from its sources of generation are called *sewers,* and the network of sewer pipes in a community is known as a *collection system.*

Because such systems must function properly and without creating a nuisance, it is imperative that the fundamental principles involved in their design and implementation be understood clearly. The purpose of this chapter is (1) to describe the types of collection systems and sewers now used; (2) to illustrate the design of both gravity-flow sanitary sewers and gravity-flow storm-water sewers; and (3) to discuss pressure and vacuum sanitary sewers.

The principal appurtenances for gravity-flow sewers and their functions and some of the special structures and appurtenances often used in the design of large sewers are described in Chap. 5. The analysis of infiltration and inflow into sewers and the means to limit their occurrence are discussed in Chap. 6, and the analysis of the biological transformations that can occur in sewers are discussed in Chap. 7. The structural design of sewer pipelines and appurtenances is not presented in this book. For this aspect of design refer to Ref. 3 and to standard structural-design texts.

4-1 TYPES OF COLLECTION SYSTEMS AND SEWERS

Over the years, three types of collection systems have been developed: sanitary, storm-water, and combined. These three terms, which refer to the contents of the sewers, also refer to the sewers themselves. The hydraulic characteristics and applications of each of these systems are identified in Table 4-1.

Table 4-1 Classification of wastewater collection systems

Type of system	Hydraulic characteristics	Purpose
Sanitary[a]	Gravity	Sanitary (gravity) sewers are used to collect wastewater from residential, commercial, industrial, and institutional sources. Allowances must be made for groundwater infiltration and unavoidable inflow.
	Pressure	Sanitary (pressure) sewers are used principally to collect wastewater from residential sources in locations unsuitable for the construction and/or use of gravity sewers; they are also used to collect wastewater from commercial sources, but only some wastewater from industrial sources because of the large volumes that may be involved. These systems are usually small and are designed to exclude groundwater infiltration and storm-water inflow.
	Vacuum	Same as above for pressure systems.
Storm water[b]	Gravity	Storm-water sewers are used to collect storm water from streets, roofs, and other sources. Sanitary wastewater is excluded totally.
Combined	Gravity	Combined sewers are used to collect wastewater from residential, commercial, institutional, and industrial sources and storm water. Additional flows come from groundwater infiltration and storm water inflow. Combined sewers are rarely designed and built in the United States today.

[a]Sanitary sewers are also known as separate sewers.

[b]Pressure and vacuum sewers are seldom used for storm-water flows because of the large quantities involved.

Sanitary sewers, which are often called *separate sewers,* originally were intended solely for the collection of wastewater from residential districts as a means of improving the general sanitation of the community. As shown in Table 4-1, sanitary sewers may be conventional gravity systems, in which the wastewater is transported by gravity, or they may be pressure or vacuum systems in which wastewater is transported under pressure or in a vacuum. Although pressure and vacuum sewers are not common in the United States at the present time (1980), more of these systems will probably be constructed in the future.

Storm-water sewers, as the name implies, are intended solely to collect storm water. With a few exceptions, the design of storm-water sewers is similar to that of sanitary sewers, although storm-water sewers are generally larger. Originally, separate systems of sanitary sewers and storm-water sewers were constructed to avoid the pollution problems associated with discharging untreated wastewater from combined sewers into watercourses. As the effects of storm-water pollution have become better understood, the treatment of storm-water discharges has received more attention [5, 6, 14].

Combined sewers, as noted in Table 4-1, are used for the collection of both wastewater and storm water. To reduce or eliminate pollution problems associated with the overflows from combined sewers, the trend is to construct separate sewers if new or replacement sewers are needed.

The types, sizes, and lengths of sewers in a wastewater collection system vary, depending on the layout of the community and the location of the treatment facilities. The various types of sewers found in most wastewater collection systems are described by function in Table 4-2 and shown graphically in Fig. 4-1.

The sizes of sewers are determined by the quantity of flow and the local building regulations governing the minimum sizes allowed. A typical distribution of sewer sizes in a community, derived from an analysis of 97 communities in 21 states, is given in Table 4-3. Pipe sizes in Table 4-3 are in metric units (see Sec. 2-3).

The total length of sewer, regardless of size, depends on the layout of the community and the location of the treatment facilities. For example, for communities with populations between 25,000 and 50,000, it has been estimated that the total average length of sewer per capita varies from 4.3 to 4.9 m (14 to 16 ft) [9]. Yet, for the city of Flagstaff, Ariz., which has a population of about 35,000, the length of sewer per capita is about 11.5 m (38 ft) [13]. Hence, average data can be misleading.

The principal appurtenance associated with most wastewater collection systems is the manhole, which interconnects sewer pipes and allows entry for

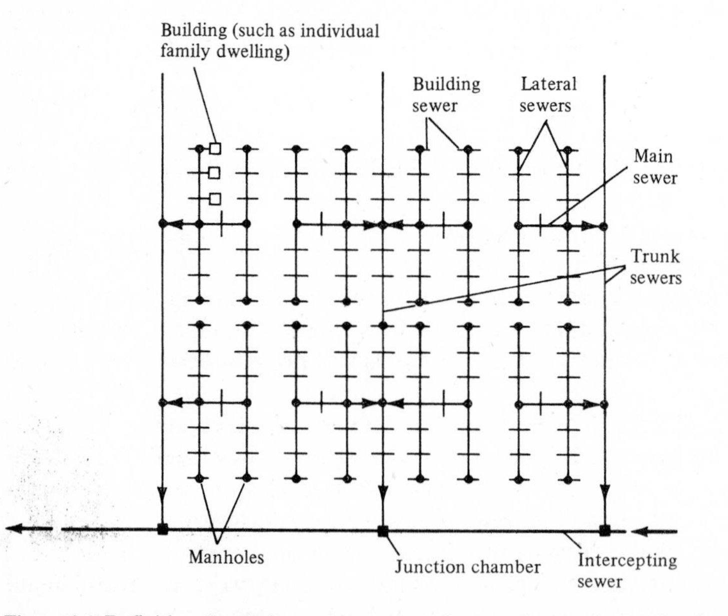

Figure 4-1 Definition sketch for various types of sewers in a typical collection system (See Table 4-2 for descriptions of sewers).

Table 4-2 Types of sewers in a typical collection system as shown in Fig. 4-1[a]

Type of sewer	Purpose
Building	Building sewers, sometimes called *building connections*, connect to the building plumbing and are used to convey wastewater from the buildings to lateral or branch sewers, or any other sewer except another building sewer. Building sewers normally begin outside the building foundation. The distance from the foundation wall to where the sewer begins depends on the local building regulations.
Lateral or branch	Lateral sewers form the first element of a wastewater collection system and are usually in streets or special easements. They are used to collect wastewater from one or more building sewers and convey it to a main sewer.
Main	Main sewers are used to convey wastewater from one or more lateral sewers to trunk sewers or to intercepting sewers.
Trunk	Trunk sewers are large sewers that are used to convey wastewater from main sewers to treatment or other disposal facilities or to large intercepting sewers.
Intercepting	Intercepting sewers are large sewers that are used to intercept a number of main or trunk sewers and convey the wastewater to treatment or other disposal facilities. (The term *intercepting sewer* originally was applied to a sewer that ". . . receives dry-weather flow from a number of transverse sewers or outlets and frequently additional predetermined quantities of stormwater." [3])

[a] Adapted in part from Refs. 3 and 4.

sewer cleaning. A photograph of a manhole being installed is shown in Fig. 4-2. Manholes are used at all major pipe junctions, but not at building connections (see Fig. 4-1). When the flows in very large sewers are to be joined, special junction chambers are used (see Fig. 4-1). Further details on these and other appurtenances are presented in Chap. 5.

4-2 DESIGN OF GRAVITY-FLOW SANITARY SEWERS

In designing a gravity-flow sanitary system, the designer must (1) conduct preliminary investigations; (2) review design considerations and select basic design data and criteria; (3) design the sewers, which includes preparation of a preliminary sewer system design and design of the individual sewers; and (4) prepare contract drawings and specifications. A considerable amount of knowledge and experience is required to perform these tasks. Sometimes a project is done in two steps: the preparation of an engineering report or facilities plan followed by the final design. The report or plan will include much of the activities covered by items (1) and (2) above plus the preparation of a preliminary sewer system design. Final design mainly entails the completion of the field work and preparation of the final contract documents.

The purpose of this section is to provide some of the necessary background knowledge and to illustrate its application in the design of sanitary wastewater collection systems.

Table 4-3 Typical distribution of pipe sizes in cities in the United States[a]

Pipe diameter, mm (in)	Distribution, %	Pipe diameter, mm (in)	Distribution, %
100, 125, 150 (4, 5, 6)	3.6	600 (24)	1.7
200 (8)	73.1	675 (27)	0.2
250 (10)	5.5	750 (30)	1.9
300 (12)	4.5	900 (36)	0.9
350, 375, 400 (14, 15, 16)	3.4	1,050 (42)	1.4
450 (18)	2.2	Total	100.0
500, 530 (20, 21)	1.6		

[a]Adapted from Ref. 9. Based on data from 97 cities in 21 states.
Note: mm × 0.03937 = in

Figure 4-2 Precast manhole base being lowered into position in the sewer trench. (Note that the worker's right hand is on the pipe-to-manhole connector and his left hand is grasping the bottom of the manhole base.)

Preliminary Investigations

Comprehensive preliminary investigations of the area to be sewered are required not only to obtain the data needed for design and construction but also to record pertinent information about the local conditions before construction begins.

At the outset of the engineering work, all pertinent maps and other drawings of the area should be obtained. Municipal and county engineers and surveyors, regional planning agencies, local planning boards, assessment boards, land-title and insurance companies, and public utility officials often have such maps and permit their duplication. For large sewer projects, useful maps may be obtained from the U.S. Geological Survey, various state agencies, or the Soil Conservation Service of the U.S. Department of Agriculture.

Field work. If satisfactory maps are not available, surveys must be made. The degree of precision required depends on the project. The surveys should show the locations of streets, alleys, railways, public parks and buildings, ponds, streams, drainage ditches, and other features and structures which may influence or may be influenced by the sewer system. In some cases, it is necessary to show property lines.

An accurate, permanent, and complete system of bench levels is needed throughout the area to be covered by the proposed sewer system. A bench mark should be established on each block of every street in which a sewer is to be laid and where topographic details are to be obtained subsequently. Profiles should then be made of all existing streets and alleys, and if the existing and "established" grades differ, information about the latter must be obtained.

Where appropriate, the adjacent area(s) where sewers may be needed in the future ought to be considered. In some cases, topographic notes can be used for the plotting of a map with contours at suitable intervals. For preparing preliminary designs, elevations of streets and alleys at intersections, high and low points, and changes in surface slope usually are sufficient; thus surface contours may not be required. However, contour maps often can be prepared economically by aerial photogrammetric methods. Important information to be determined include the elevations of the beds of streams, ditches, canals, and culverts, and the maximum expected and ordinary water-surface elevations.

Information on existing structures and utilities should include:

1. Elevations of the sills of buildings and depths of their basements
2. Character, age, and condition of the pavements of streets in which sewers will be laid
3. Location of water and gas mains, electric conduits, drain lines, and other underground structures

Where good information is lacking on underground facilities, it may be advisable to have pits excavated in the streets to obtain the required data.

Local rainfall and runoff data should be located or, where these are inadequate, measurements in the field should be taken if possible. Information that builders and contractors can supply regarding groundwater should be recorded and, in the case of low-lying areas, it may be desirable to excavate pits or make borings to indicate the groundwater conditions.

It is important to ascertain the character of the soil in which the sewers are to be constructed so that the cost can be estimated with fair accuracy. For this purpose, subsurface explorations with soil borings should be made, with samples and blow counts on the sampler being taken at 1.5-m (5-ft) intervals and at every change in soil type. Soil borings should extend down at least to 1.5 m (5 ft) below the estimated bottoms of the excavations, or to refusal when conventional drilling equipment can no longer penetrate the soil strata.

Refusal does not necessarily mean that bed rock has been reached. Therefore, at these locations it is often required that up to 3 m (10 ft) of core drilling be done to ascertain whether the refusal was caused by bed rock, a boulder, or extremely hard-packed earth. Where structures such as pumping stations, large junction chambers, or the like are to be constructed, core drilling should always be done when refusal has been reached with the soil boring. Special attention should be paid to the locations and the number of the borings to be made in areas where construction difficulties are likely to develop. Example of such areas are stream, railroad, and highway crossings; the sites of deep excavations or high-groundwater table; and where trenches or other excavations will be close to existing structures.

Other important information includes the local wages of unskilled and skilled labor, the cost of construction materials and supplies, the cost of construction of similar work previously done, and freight rates and rental charges for trucks and equipment. This information is useful in preparing reliable cost estimates.

Preparation of maps and profiles. Work on preliminary maps and profiles should begin as soon as possible during the field work, so that other studies preliminary to design may be started before the field work is finished. As a rule, maps on a scale of 25 m to 10 mm (200 ft to 1 in) are large enough to permit the data to be shown in adequate detail for preliminary system design; but where there are many subsurface structures, a scale of 5 m or less to 10 mm (40 ft or less to 1 in) may be necessary for clarity.

For use in the preparation of design profiles, street centerline elevations should be shown at least every 15 m (50 ft) and at abrupt changes in surface slope. Contours, if available, should be plotted at 0.5-m (2-ft) intervals. Summits in streets should be marked, and the elevations should be given to hundredths of a meter, as should also points of depression or "pockets." An example of such a map for a residential area, where elevations are shown but not pipes or manholes, is given in Fig. 4-3. Profile sheets of the ground surface along the proposed sewer routes should also be prepared in advance of the computations.

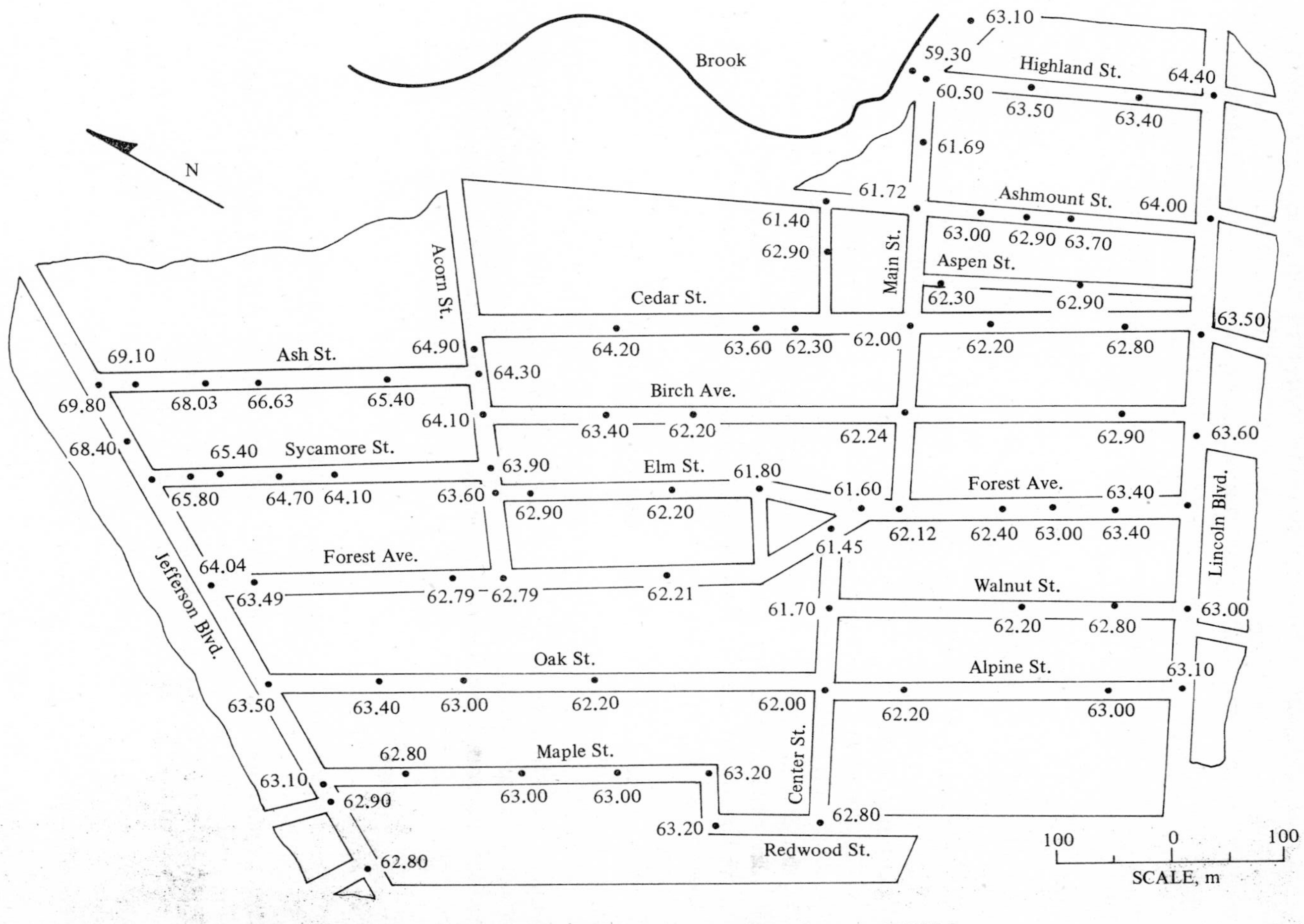

Figure 4-3 Typical map used for the design of sanitary sewers (elevations in meters). Note: m × 3.2808 ft.

System maps for planning or feasibility reports are developed from preliminary designs; these maps are usually presented in the reports at scales of 100 to 200 m to 10 mm (800 to 1600 ft to 1 in). Report profiles, when used, are presented at scales as small as feasible that will permit essential information to be shown. Detailed plans and profiles for construction and bidding purposes are discussed later in this section.

Basic Design Considerations

Designing a sanitary sewer involves:

1. The estimation of wastewater flowrates for the design date and the evaluation of any local conditions that may affect the hydraulic operation of the system
2. The selection of the hydraulic-design equation, alternative sewer pipe materials and minimum sizes, minimum and maximum velocities, and slopes
3. The evaluation of alternative alignments or designs
4. The evaluation of the use of curved sewers
5. The selection of appropriate sewer appurtenances
6. A review of the need for sewer ventilation

These design details are usually determined by the information obtained from the preliminary investigations.

Design flows. In most situations, as noted previously, the total wastewater flow consists of three components: wastewater from residential, commercial, and institutional sources; industrial wastewater; and infiltration. Thus new sanitary sewers are designed for the following flows expected at the design date.

1. Peak hourly flows from residential, commercial, institutional, and industrial sources for the entire service area
2. Peak infiltration allowance for the entire service area

Hydraulic-design equation. As discussed in Chap. 2, the Manning equation is commonly used in sewer design. It is recommended that a Manning n value of 0.013 be used to analyze well-constructed existing sewers and to design new sewers, and that a Manning n value of 0.015 be used to analyze most older existing sewers. Higher values of n should be used for existing sewers if deterioration, departures from line and grade, variations of inside dimensions, deposits, or inferior workmanship is evident.

The n value of 0.013 for new and well-constructed existing sewers is based on the use of individual pipe sections—not less than 1.5 m (5 ft) long with true and smooth inside surfaces—and on the assumption that only first-class construction procedures have been or will be allowed.

Some pipes made from the various plastics reportedly are smoother initially and retain their initial capacities longer than pipes made of traditional materials. Their standard lengths are longer than those of some of the older kinds of pipe. Therefore, some manufacturers have advocated smaller n values for plastic pipes ($n = 0.011$ or $n = 0.010$). But the number of building connections, manholes, and other flow-disturbing appurtenances in a given sewer remains the same, regardless of the pipe material. For this reason, and considering the uncertainties inherent in sewer design and construction, the value of n for sewer design should not be less than 0.013.

Sewer pipe materials and sizes. The principal sewer pipe materials are asbestos cement, ductile iron, reinforced concrete, prestressed concrete, polyvinyl chloride, and vitrified clay. The size ranges and information on the sewers made of these materials are presented in Table 4-4. Other sewer pipe materials include cast iron, corrugated metal, steel, nonreinforced concrete, and various plastics, either plain or reinforced with glass fibers.

In sewer design, a minimum size of sewer pipe must be established because large objects sometimes enter sewers, and clogging is less likely if sewers are not smaller than 200 mm (8 in). Obviously, the smallest sewers should be larger than the building sewer connections in general use, so that articles that pass through the building connections may as readily pass through the sewer. A minimum size of 200 mm (8 in) is recommended for gravity-flow sanitary sewers. The most common size of building connection is 150 mm (6 in), but connections of 125 and 100 mm (5 and 4 in) have been used successfully in some areas.

Minimum and maximum velocities. If wastewater flows for an extended time at low velocities, solids may be deposited in the sewer. Sufficient velocity should be developed regularly to flush out any solids that may have been deposited during low flow periods. The usual practice is to design the slopes for sanitary sewers to ensure a minimum velocity of 0.6 m/s (2.0 ft/s) with flow at one-half full or full depth. The velocity at less than one-half full depth will be less than 0.6 m/s; the velocity for depth between one-half full and full will be slightly greater than 0.6 m/s. Often, minimum and maximum velocities are specified in state and local standards.

Although the velocity near the bottom of the sewer significantly affects how quickly the wastewater flows, a mean velocity of 0.3 m/s (1.0 ft/s) is usually sufficient to prevent the deposition of the organic solids in wastewater. To prevent deposition of mineral matter, such as sand and gravel, a mean velocity of 0.75 m/s (2.5 ft/s) is generally adequate in sanitary sewers. These are minimum figures. In depressed sewers (sometimes called *inverted siphons*), where access for cleaning is difficult, the minimum velocity should be about 1.0 m/s (3.0 ft/s) (see "Depressed Sewers," Chap. 5, Sec. 5-3). Slopes corresponding with mean velocities as low as 0.5 m/s (1.5 ft/s) have been used successfully in some special cases, but sewers at such slopes must be constructed with great care and will probably require frequent cleaning.

Repeated removal of deposited material from sewers is expensive and, if such deposits are not cleaned out, they may cause increasingly troublesome

Table 4-4 Available size ranges and descriptions of commonly used pipe for gravity-flow sewers

Type of pipe	Available size range, mm (in)[a]	Description
Asbestos cement (AC)	100–900 (4–36)	Weighs less than other commonly rigid pipes. May be susceptible to acid corrosion and hydrogen sulfide attack, but if properly cured with steam at high pressure (autoclave process), may be used even in environments with moderately aggressive waters of soils with high-sulfate content.
Ductile iron (DI)	100–1350 (4–54)	Often used for river crossings and where the pipe must support unusually high loads, where an unusually leakproof sewer is required, or where unusual root problems are likely to develop. Ductile-iron pipes are susceptible to acid corrosion and hydrogen sulfide attack, and therefore should not be used where the groundwater is brackish, unless suitable protective measures are taken.
Reinforced concrete (RC)	300–3600 (12–144)	Readily available in most localities. Susceptible to corrosion of interior if the atmosphere over wastewater contains hydrogen sulfide, or from outside if buried in an acid or high-sulfate environment.
Prestressed concrete (PC)	400–3600 (16–144)	Especially suited to long transmission mains without building connections and where precautions against leakage are required. Susceptibility to corrosion (the same as reinforced concrete).
Polyvinyl chloride (PVC)	100–375 (4–15)	A plastic pipe used for sewers as an alternative to asbestos-cement and vitrified-clay pipe. Lightweight but strong. Highly resistant to corrosion.
Vitrified clay (VC)	100–900 (4–36)	For many years the most widely used pipe for gravity sewers; still widely used in small and medium sizes. Resistant to corrosion by both acids and alkalies. Not susceptible to damage from hydrogen sulfide, but is brittle and susceptible to breakage.

[a]Sizes listed are readily available without the need for special design or special manufacturing machinery or equipment. Some sizes within the listed ranges may not be available in all localities. Larger sizes than those listed have been made on special order. For example, one instance of the manufacture of 5200-mm (17-ft) -diameter reinforced concrete has been reported, more than one manufacturer has supplied 2750-mm (108-in) filament-wound reinforced resin pipe, and polyvinyl chloride pipe has been made in sizes up to 675 mm (27 in) in the United States and 1000 mm (39 in) in Europe.

Note: mm × 0.03937 = in

conditions. It is therefore desirable that slopes have self-cleaning velocities even though the cost of constructing steeper slopes may be greater than the added cost of maintaining sewers laid on flatter slopes. This is recommended because, if such maintenance work is neglected, a substantial deposit can result. Then the sewer cannot perform properly, and it may fail to transport the wastewater at the design rate, resulting in damage to property.

The erosive action of the material suspended in the wastewater depends not only on the velocity at which it is carried along the invert of a sewer but also on its nature. Because this erosive action is the most important factor in determining the safe maximum velocities of wastewater, the character of the suspended material must be considered. In general, maximum mean velocities of 2.5 to 3.0 m/s (8 to 10 ft/s) at the design depth of flow will not damage the sewer.

High velocities in small-pipe sewers and the corresponding low depths of flow may allow large objects, which at times enter all sanitary sewer systems, to remain on the inverts, where they may become lodged so firmly that the next rush of wastewater will not detach them.

Minimum slopes. Sewers with flat slopes are often required to avoid excessive excavation where surface slopes are flat or the changes in elevation are small. In such cases, the sewer sizes and slopes should be designed so that the velocity of flow will increase progressively, or at least will be steady throughout the length of the sewer. Then solids washed into the sewer and transported by the flowing stream may be carried through the sewer and not deposited at some point because of a decrease in velocity. In general, the minimum slopes given in Table 4-5 for small-pipe sewers in a sanitary sewer system have been satisfactory.

In long, flat sewers a buildup of hydrogen sulfide may take place. When released to the atmosphere above the wastewater, it may cause serious odor

Table 4-5 Minimum slopes for gravity-flow sanitary sewers[a]

Size, mm	(in)	Slope, m/m[a] $n = 0.013$	$n = 0.015$
200	(8)	0.0033	0.0044
250	(10)	0.0025	0.0033
300	(12)	0.0019	0.0026
375	(15)	0.0014	0.0019
450	(18)	0.0011	0.0015
525	(21)	0.0009	0.0012
600	(24)	0.0008	0.0010
675	(27)	0.0007[b]	0.0009
750	(30)	0.0006[b]	0.0008[b]
900	(36)	0.0004[b]	0.0006[b]

[a]Based on Manning's equation with a minimum velocity of 0.6 m/s. Where practicable steeper slopes should be used.

[b]The minimum practicable slope for construction is about 0.0008 m/m.

Note: mm × 0.03927 = in
m × 3.2808 = ft

problems and the deterioration of materials containing cement, such as reinforced-concrete pipe, asbestos-cement pipe, concrete walls of manholes and other structures, and mortar of brickwork. These subjects are discussed more fully in Chap. 7.

When designing a large sewer (particularly a trunk or intercepting sewer), it is important to consider the conditions that may develop at minimum flow during the first few years after its construction. It should be made certain that the velocities will not be so low, for significant periods of time, as to result in objectionable deposits in the sewer, because removal of these deposits would involve excessive cost. An alternative to the construction of a single large sewer to serve for a long period is to build a smaller sewer initially and to add another sewer when the service area has developed further. Although the construction cost of two smaller sewers may be greater than that of a single large sewer, the high cleaning cost of a large sewer during the early years of operation might more than offset the increased cost of construction of the two smaller sewers.

The concept just described probably is not valid for smaller sewers because the relative costs of construction and cleaning are not the same as for large sewers. For instance, the difference in the construction cost of a 200-mm (8-in) sewer and a 300-mm (12-in) sewer is only the difference in the cost of the pipe itself; pipe costs for these sizes of sewers typically range from 5 to 8 percent of the cost of the completed sewer. Other factors that should be considered are the costs, both tangible and intangible, and the public irritation at the inconveniences and disruptions caused by constructing another sewer in the same area after only a few years have passed.

Alternative alignments and designs. Often two or more alternative pipeline routes may be possible for a given project. Usually it will be necessary to prepare design details for each alternative and to make comparative cost estimates before a decision can be reached. Some otherwise desirable alternative routes may involve the placement of sewers across private property. However, unless there is a significant advantage in cost or other condition resulting from a placement across private property, it is generally not advisable to construct ordinary sewers outside public rights of way. Interceptors are often constructed in private easements because the most favorable locations for interception are usually in valleys near natural drainage channels.

Use of curved sewers. Alternative alignments or designs might include the use of curved sewers. Traditionally, sewers have been laid out in straight lines between manholes so that maintenance personnel can inspect and clean the sewers. Straight alignments were also considered necessary to maintain desirable flow characteristics. However, curved, instead of straight, sewer alignments in curved streets sometimes allow for convenient construction with public rights of way, interfere less with other utilities, require fewer manholes, and possibly provide more convenient sewer service to all the building sites. Many communities have had curved sewers for a sufficiently long time to demonstrate the feasibility of their use.

For smaller sewers, the curvature is usually obtained by laying a succession of straight pipes with the joints deflected slightly. Only the most watertight jointing systems should be specified for these sewers. However, the likelihood of leaks developing or roots entering the sewer will be increased by the use of the deflected joints. For larger sewers, curvature usually is obtained with beveled-end pipe, preferably without joint deflection. Other details of design, including minimum velocity and slope requirements, should be the same as for straight sewers.

If curved sewer alignments are contemplated, the types of cleaning equipment available for sewer maintenance should be studied. Water jets and cleaning balls usually operate in curved sewers with no more difficulty than in straight sewers. However, if heavy grit deposits are expected and bucket cleaners are needed, curvatures should be flat enough so that the bucket can be pulled through the sewer without damaging the sewer walls by either the bucket or its cables.

Some disadvantages associated with curved alignments are: laser equipment for maintaining line and grade during construction cannot be used; sharply curved sewers are difficult to inspect by closed-circuit television or still-photography; and leaks cannot be visually located except near the manholes.

Sewer appurtenances. The principal appurtenances of sanitary sewers are manholes, drop inlets to manholes, building connections, and junction chambers. In addition, many special appurtenant structures may be required, depending on local topography. Detailed descriptions of appurtenances and special structures are presented in Chap. 5. For this reason, the following discussion will be limited to a few specific design considerations concerning manholes.

For sewers that are 1200 mm (48 in) and smaller, manholes should be located at changes in size, slope, or direction. For larger sewers, these changes may be made without installing a manhole. If possible, vertical drops in the flowing wastewater should be avoided to minimize splashing. Where such drops are necessary, drop inlets or other means of conveying the wastewater to a lower elevation should be installed. At such points, vitrified-clay brick or other wear-resistant lining may be provided in concrete structures to prevent erosion of the concrete.

The number of manholes must be adequate so that the sewers can be easily inspected and maintained. Recommended maximum spacing between manholes is discussed in Chap. 5, Sec. 5-1.

Sewer ventilation. When wastewater collection systems were first built, sewer ventilation received much attention, and a variety of theories offered for the best way to provide it. It was at one time thought that sewers should be ventilated at manholes to prevent sewer gases from accumulating and entering dwellings, thus affecting the health of the residents. However, the real needs for ventilation arise from the danger of asphyxiation of workers in sewers and manholes, from the explosion hazard, and from the development of gaseous odors. The subject of ventilation and air relief is discussed in more detail in Chap. 7, Sec. 7-3.

Design of Sanitary Sewers

After a careful review of population and land-use projections, state standards, local building, construction, and sewer codes, the results of the preliminary field investigations, and the various design considerations discussed previously, the next step is to design the sewer. The selection of design criteria and the preparation of sewer design computations for a gravity-flow trunk sewer and for a gravity-flow sanitary sewer for a small residential area are illustrated in Examples 4-1 and 4-2. In Example 4-1, the discussion focuses on how to obtain the average and peak design flowrates for different types of areas. The details of analysis are illustrated in Example 4-2.

Example 4-1 Design of a gravity-flow trunk sewer Develop a preliminary design, including flowrates, pipe sizes, and pipe slopes, for the trunk sewer to be constructed in the Covell Park Development shown in Fig. 4-4. Assume that the ground slope is flat. The general location of the trunk sewer was established when the development was laid out originally. The design of the trunk sewer is to be consistent with the design considerations discussed previously and the following specific data that are applicable to this development. (This same development was discussed in Example 3-1 to illustrate the procedures for estimating the domestic and industrial wastewater flows for a new development.)

Data on the expected saturation population densities and wastewater flows for the various types of housing in the Covell Park Development were derived from actual records of similar nearby developments and are given in Table 4-6. The average commercial (including the shopping area) and industrial wastewater flowrate allowances were estimated to be 20 to 30 $m^3/ha \cdot d$ (2100 and 3200 $gal/acre \cdot d$), respectively. These estimates were determined by analyzing the individual types of facilities to be included within the two zones. On the basis of actual flow measurements of similar activities, the average peaking factors are 1.8 for the commercial flows and 2.1 for the individual flows. Peaking factors for the residential flows are obtained from Fig. 3-4.

The planned school within the Covell Park Development is to serve 2000 students at ultimate capacity. The average flow is 75 $L/student \cdot d$ (20 $gal/student \cdot d$), and the peaking factor for the school is 4.0.

SOLUTION

1. Lay out the trunk sewer. Use the layout given in Fig. 4-4.
2. Locate the manholes. Because this is a preliminary design, only the flows, pipe sizes, and pipe slopes will be established between these points. The details of locating manholes, cleanouts, and other appurtenances are not considered (see Example 4-2). However, to aid in the preliminary computation, number the major junctions or points where the trunk sewer might change size, starting with the upper end of the trunk sewer that terminates in the area of single-family dwellings (see Fig. 4-4).
3. Establish the limits of drainage areas and the points of wastewater contribution. Assume that the boundary of the Covell Park Development also defines the limits of the drainage area and that the points of contribution of both wastewater and infiltration for each subarea are as shown in Fig. 4-4.

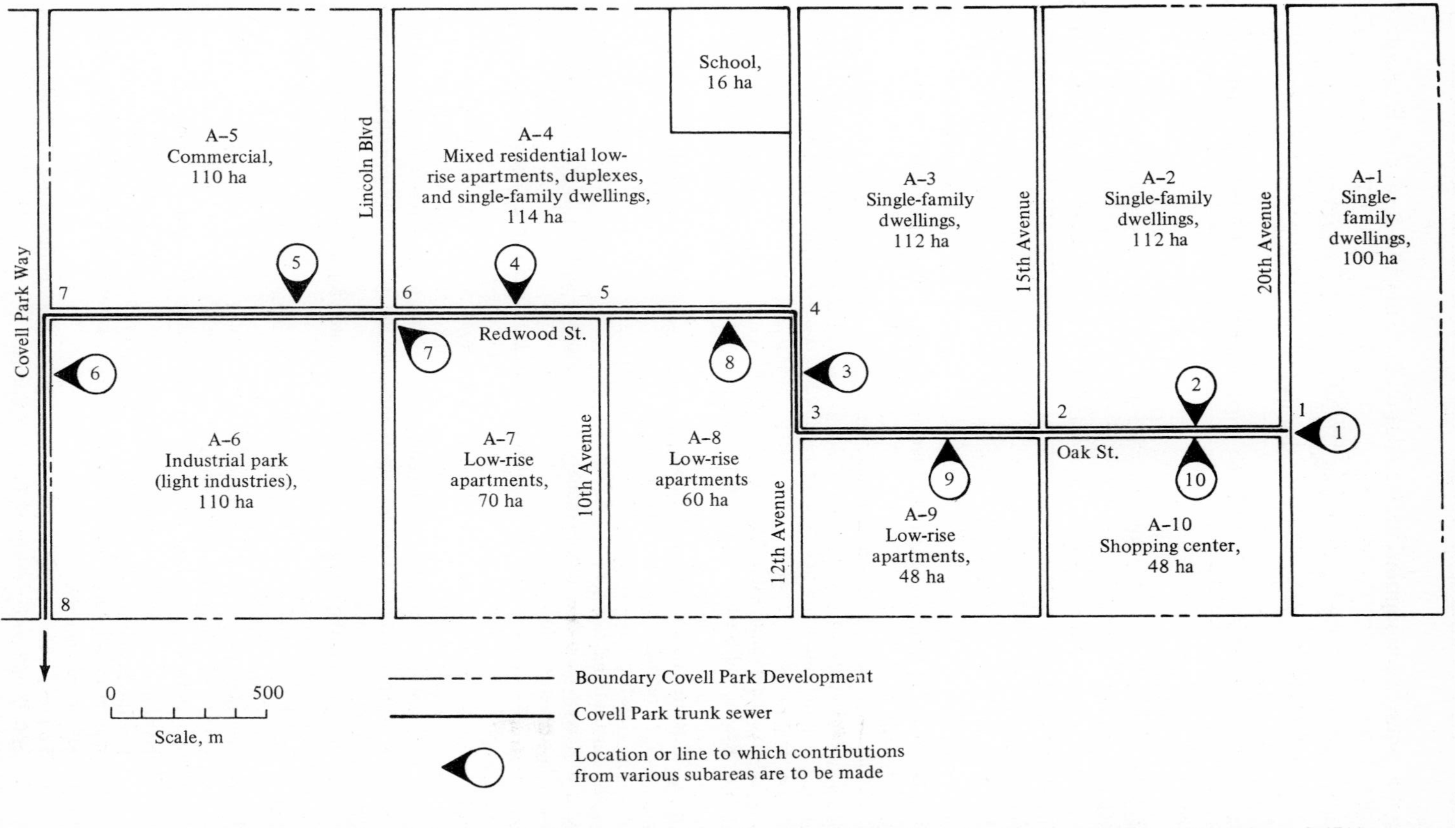

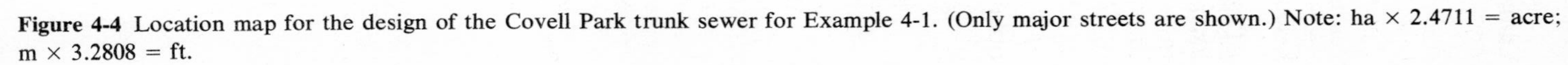

Figure 4-4 Location map for the design of the Covell Park trunk sewer for Example 4-1. (Only major streets are shown.) Note: ha × 2.4711 = acre; m × 3.2808 = ft.

Table 4-6 Saturation population densities and wastewater flows for Example 4-1

Zoning	Type of development	Saturation population density, persons/ha	(persons/acre)	Wastewater flows, L/capita·d	(gal/capita·d)
Residential	Single-family dwellings	38	(15)	300	(80)
Residential	Duplexes	60	(24)	280	(75)
Residential	Low-rise apartments	124	(50)	225	(60)

Note: ha × 2.4711 = acre
L × 0.2642 = gal

4. Determine the area of each of the drainage subareas. The required data are given in Fig. 4.4.
5. Summarize the basic design criteria.
   a. Design period:
   Use the saturation period (time required to reach saturation population).
   b. Population densities:
   Use the data in Table 4-6.
   c. Residential wastewater flows:
   Use the data in Table 4-6.
   d. Commercial and industrial flows:
      i. Commercial—20 m^3/ha · d.
      ii. Industrial—30 m^3/ha · d.
   e. Institutional flows:
   School—150 m^3/d (2000 students × 75 L/student · d)/(1000 L/m^3).
   f. Infiltration allowance:
      i. For residential areas, obtain the peak values from curve *B* for new sewers in Fig. 3-5.
      ii. For commercial, industrial, and institutional areas, also obtain the peak values from curve *B* in Fig. 3-5. However, to take into account that the total length of sewers in these areas will generally be less than that in residential areas, use only 50 percent of the actual area to compute the infiltration allowance.
   g. Inflow allowance:
   Assume that the steady-flow inflow is accounted for in the infiltration allowance.
   h. Peaking factors:
      i. Residential—use the curve given in Fig. 3-4.
      ii. Commercial—1.8.
      iii. Industrial—2.1.
      iv. Institutional (school)—4.0.
      v. Infiltration—1.6.
   i. Hydraulic design equation:
   Use the Manning equation with an n value of 0.013. To simplify the computations, use Figs. 2-12 and 2-13.
   j. Minimum velocity:
   To prevent the deposition of solids at low wastewater flows, use a minimum velocity of 0.75 m/s during the peak flow conditions.
   k. Minimum cover (minimum depth of cover over the top of the sewer):

As established by the local community building code (the actual value of the minimum cover is not required for this example).

6. Prepare a tabulation form for completing the preliminary design. The computations for the trunk sewer are shown in Table 4-7. As shown, set up a separate column for each step in the computation. Although Table 4-7 is self-explanatory for the most part, the following comments clarify its development and use.
   a. The line under consideration is identified in col. 1.
   b. Columns 2 and 3 usually are used for the manhole numbers at each end of the line in col. 1. In this example, the numbers identified in Fig. 4-4 are entered in these columns. In a detailed design, additional manholes would be necessary.
   c. The subarea is identified in col. 4. When a number of subareas contribute to a given line, it is best to use separate entries for each subarea.
   d. The area, in hectares, for each of the residential subareas is entered in col. 5. The population density data from Table 4-6 are entered in col. 6. The population density value for subarea A-4 was obtained by assuming that the areas devoted to each type of residential development would be equal.

      Incremental and cumulative population data are summarized in cols. 7 and 8, respectively. The average unit flow data from Table 4-6 are entered in col. 9. The flow area A-4 was estimated by averaging the data given in Table 4-6. Although a weighted average could be used, a simple average was considered conservative.

      The cumulative average residential flow is given in col. 10. For example, for line 2, the cumulative average flow is equal to 3756 m^3/d (2417 m^3/d + 5952 persons × 225 L/capita · d ÷ 1000 L/m^3). The peaking factor derived from Fig. 3-4, based on the cumulative average flow data given in col. 10, is given in col. 11. The corresponding cumulative peak flow, obtained by multiplying the cumulative average flow (col. 10) by the peaking factor, is given in col. 12.
   e. The commercial area, the corresponding unit flow, and the cumulative average flows are entered in cols. 13, 14, and 15, respectively. The given peaking factor for the commercial area is entered in col. 16, and the computed cumulative peak commercial flows are entered in col. 17.
   f. The entries in cols. 18 through 22 for the industrial flows are the same as described for the commercial flows (cols. 13 through 17).
   g. In a similar manner to the commercial and industrial flows, the institutional flows are entered in cols. 23 through 27, respectively.
   h. The cumulative average and cumulative peak domestic and industrial wastewater flows are summarized in cols. 28 and 29, respectively.
   i. The first step in determining the infiltration allowance (cols. 30 through 33) is to enter the area for each subarea in col. 30. The cumulative area is summarized in col. 31. The unit infiltration allowance, obtained from curve *B* in Fig. 3-5 by using the cumulative area data (col. 31), is entered in col. 32. The peak cumulative infiltration allowance in col. 33 is obtained by multiplying the corresponding values in cols. 31 and 32.
   j. The total cumulative average and cumulative peak design flows for the various lines are given in cols. 34 and 35, respectively. These values include the domestic, industrial, and infiltration/inflow contributions. The peak design flows, expressed in cubic meters per second, are given in col. 36.
   k. The data in col. 36 are used to determine the sewer design data in cols. 37 through 40. The trial-and-error procedure required to determine the sewer pipe diameter, the slope, and the capacity and velocity when flowing full, given in cols. 37 through 40, is discussed below. The description only illustrates the trial-and-error procedure involved in selecting sewer sizes: the cardinal rule in sewer design is to work from a ground-surface profile in sizing and laying out sewers (see Example 4-2).
      i. To find the required sewer sizes, locate the value of the peak flow on the abscissa in Fig. 2-13 and move vertically up the diagram to the minimum acceptable velocity.

Table 4-7 Computation table for design of the trunk sewer in Example 4-1

	Location					Residential flow					
						Design population					
Line	From	To	Subarea[a]	Area, ha[a]	Pop. density, persons/ha	Incre-ment	Cum. total	Avg. unit flow, L/capita · d	Cum. avg. flow, m³/d (8 × 9)	Peaking factor[b]	Cum. peak flow, m³/d (10 × 11)
(1)	(2)	(3)	(4)	(5)	(6)	(7)	(8)	(9)	(10)	(11)	(12)
1	1	2	A-1	100	38	3,800	3,800	300	1,140	. . .	. . .
			A-2	112	38	4,256	8,056	300	2,417	. . .	. . .
			A-10	. . .	. . .	. . .	8,056	. . .	2,417	2.9	7,009
2	2	3	A-9	48	124	5,952	14,008	225	3,756	2.8	10,517
3	3	4	A-3	112	38	4,256	18,264	300	5,033	2.7	13,589
4	4	5	A-8	60	124	7,440	25,704	225	6,707	2.7	18,109
5	5	6	A-4	114	74	8,436	34,140	268	8,968	2.6	23,317
6	6	7	A-7	70	124	8,680	42,820	225	10,921	2.6	28,395
			A-5	. . .	. . .	. . .	42,820	. . .	10,921	2.6	28,395
7	7	8	A-6	. . .	. . .	. . .	42,820	. . .	10,921	2.6	28,395

Commercial flow					Industrial flow					Institutional flow				
Area, ha[a]	Avg. unit flow, m³/ha · d	Cum. avg. flow, m³/d	Peak-ing factor	Cum. peak flow, m³/d (15 × 16)	Area, ha[a]	Avg. unit flow, m³/ha · d	Cum. avg. flow, m³/d	Peak-ing factor	Cum. peak flow, m³/d (20 × 21)	Area, ha[a]	Avg. flow, m³/d	Cum. avg. flow, m³/d	Peak-ing factor	Cum. peak flow, m³/d (25 × 26)
(13)	(14)	(15)	(16)	(17)	(18)	(19)	(20)	(21)	(22)	(23)	(24)	(25)	(26)	(27)
. . .	. . .	. . .	. . .	. . .	. . .	. . .	. . .	. . .	. . .	. . .	. . .	. . .	. . .	. . .
. . .	. . .	. . .	. . .	. . .	. . .	. . .	. . .	. . .	. . .	. . .	. . .	. . .	. . .	. . .
48	20	960	1.8	1,728	. . .	. . .	. . .	. . .	. . .	. . .	. . .	. . .	. . .	. . .
. . .	. . .	960	1.8	1,728	. . .	. . .	. . .	. . .	. . .	. . .	. . .	. . .	. . .	. . .
. . .	. . .	960	1.8	1,728	. . .	. . .	. . .	. . .	. . .	. . .	. . .	. . .	. . .	. . .
. . .	. . .	960	1.8	1,728	. . .	. . .	. . .	. . .	. . .	. . .	. . .	. . .	. . .	. . .
. . .	. . .	960	1.8	1,728	. . .	. . .	. . .	. . .	. . .	16	150	150	4.0	600
. . .	. . .	. . .	. . .	. . .	. . .	. . .	. . .	. . .	. . .	. . .	. . .	. . .	. . .	. . .
110	20	3,160	1.8	5,688	. . .	. . .	. . .	. . .	. . .	. . .	. . .	150	4.0	600
. . .	. . .	3,160	1.8	5,688	110	30	3,300	2.1	6,936	. . .	. . .	150	4.0	600

Cumulative subtotal		Infiltration				Total design flow			Sewer design[c]			
Cum. avg. flow, m^3/d (10 + 15 + 20 + 25)	Cum. peak flow, m^3/d (12 + 17 + 22 + 27)	Area, ha[a,d]	Cum. area, ha[e]	Peak unit infil. allowance, $m^3/ha \cdot d$[f]	Cum. peak infil allowance, m^3/d	Cum. avg. flow, m^3/d[g]	Cum. peak flow, m^3/d (29 + 33)	Cum. peak flow, m^3/s	Sewer diameter, mm	Slope, m/m	Capacity full, m^3/m	Velocity, full, m/s
(28)	(29)	(30)	(31)	(32)	(33)	(34)	(35)	(36)	(37)	(38)	(39)	(40)
...	...	100	...	...	...	...	...	...	...	...	...	...
...	...	112	...	...	...	...	...	...	...	...	...	...
3,377	8,737	0.5(48)	236	8.3	1,959	4,601	10,969	0.124	450	0.0019	0.124	0.78
4,716	12,245	48	284	7.8	2,215	6,100	14,460	0.167	525	0.0015	0.167	0.77
5,993	15,317	112	396	7.1	2,812	7,751	18,129	0.210	600	0.0012	0.213	0.75
7,667	19,837	60	456	6.8	3,101	9,605	22,938	0.265	675	0.0011	0.279	0.78
10,078	25,645	114 + 0.5(16)	578	6.3	3,641	12,345	29,286	0.339	750	0.0010	0.352	0.80
...	...	70	...	...	...	...	...	...	...	...	...	...
14,231	34,683	0.5(110)	703	6.0	4,218	16,867	38,901	0.450	900	0.0007	0.479	0.75
17,531	41,619	0.5(110)	758	5.8	4,396	20,279	46,015	0.533	900	0.0009	0.543	0.85

[a] See Fig. 4-4.
[b] Peaking factor taken from Fig. 3-4.
[c] Sewer design was completed using Fig. 2-14.
[d] Use of the 0.5 factor applied to commercial, industrial, and institutional areas is documented in Step *f* of the solution.
[e] Cumulative area contributing to a given line.
[f] Data from Fig. 3-5, curve *B* for new sewers.
[g] To obtain average infiltration allowance, peak values were divided by 1.6.

Note:
ha × 2.4711 = acre
L × 0.2642 = gal
$m^3/d \times 2.6417 \times 10^{-4}$ = Mgal/d
$m^3/ha \cdot d \times 106.9064$ = gal/acre · d
$m^3/s \times 35.3147 = ft^3/s$
mm × 0.03937 = in
m/s × 3.2808 = ft/s

ii. Next, select the pipe diameter line that is nearest to the point found in part (i). Move along the diameter line to a point that just satisfies the minimum velocity and the peak discharge requirement. If it is desirable and practical to use a larger or smaller pipe and/or a flatter or steeper slope, the designer may choose adjacent pipe diameters as long as the velocity and discharge requirements can be satisfied. In actual sewer design problems, depth of cover requirements and other constraints may necessitate the use of different slopes and possibly different sizes than would otherwise be chosen.

iii. Once a suitable combination is found, enter the required information in cols. 37 through 40 of Table 4-7.

iv. The final step in the design of a sanitary sewer is to establish the invert elevations of the individual sewers. The procedures involved in doing this are detailed in Example 4-2.

Comment A computation table, such as Table 4-7, not only saves time but also displays both the data and the computed results in an orderly sequence for subsequent use. The specific columns in a given computation table depend on the factors that must be considered in arriving at the peak design flows. For example, an additional set of columns would be required for sewer inflow data if inflow were calculated separately. Most sanitary and civil engineering consulting firms have developed tabulation forms of their own for sewer design computations. Although the forms may differ in specific details and in the order of presentation from that of Table 4-7 and the computation form used in Example 4-2, the same information is usually presented. Some engineering firms have developed computer programs for sewer design.

Example 4-2 Design of a gravity-flow sanitary sewer for a small residential area Design a gravity-flow sanitary sewer system for the residential district shown in Fig. 4-3. The future average saturation population density is estimated to be 40 persons/ha (16 persons/acre). The average flowrate of residential wastewater is estimated to be 300 L/person · d (80 gal/person · d). Use Fig. 3-4 to estimate the peak wastewater flow. Use curve *B* of Fig. 3-5 to estimate the maximum rate of groundwater infiltration.

The minimum size of sewer pipe is 200 mm (8 in). The wall thickness of the 200-mm pipe is 20 mm (0.8 in). The minimum velocity of flow in the sewer when full is 0.6 m/s (2.0 ft/s). A value of 0.75 m/s (2.5 ft/s) is more desirable if it can be achieved without unnecessary cost. Use Manning's equation with an n value of 0.013 to determine the capacity of the sewers. Because the homes in this area have basements, the minimum depth below the street surface to the top of the sewers is 2.75 m (9 ft). In areas where basements are not usually constructed, the depth of cover to the top of the sewer may be less.

SOLUTION

1. Lay out the sewers. Draw a line to represent the proposed sewer in each street or alley to be served (see Fig. 4-5). Near or on the line, indicate by an arrow the direction in which the wastewater is to flow. Except in special cases, the sewer should slope with the surface of the street. It is usually more economical to plan the system so that the wastewater from any street will flow to the point of disposal by the most direct (and, consequently, the shortest) route. In general, the laterals connect with the mains and these, in turn, connect with the trunk sewer, which leads to the points of discharge or to an intercepting sewer.
2. Locate the manholes. Locate a manhole at (1) changes in direction; (2) changes in slope; (3) at pipe junctions, with the exception of building connections (see Fig. 4-1); (4) at the upper ends of all laterals for cleaning and flushing the lines; and (5) at intervals from 90 to 120 m, or less, as required. Give each manhole an identification number (see Fig. 4-5).
3. Establish the limits of the service area. Sketch the limits of the service areas for each lateral, as shown in Fig. 4-5. If a single lateral will be required to accommodate an area larger than can be served by the minimum size of sewer with the minimum slope, the area should be subdivided further. Where the streets are laid out, assume that the limits are

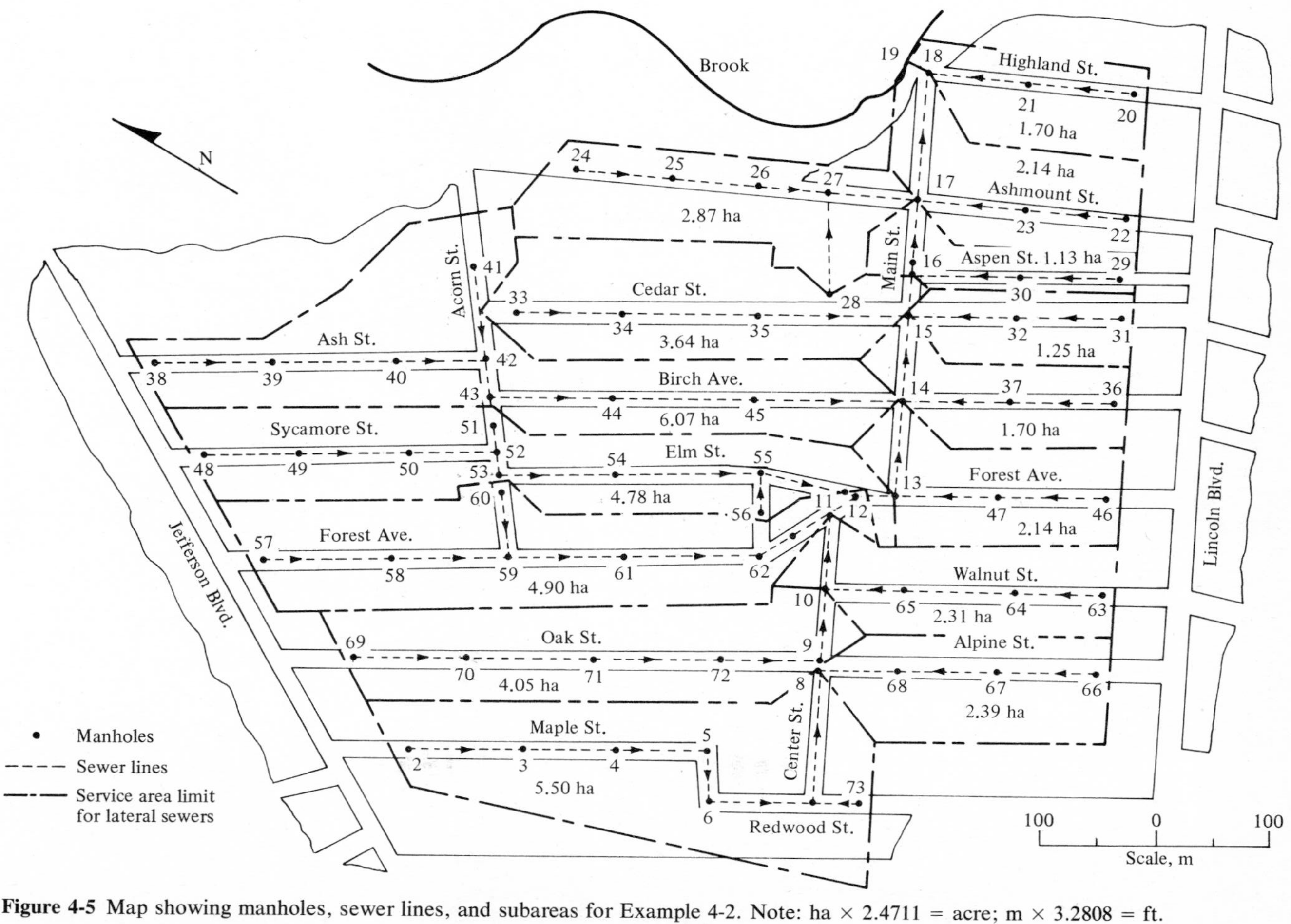

Figure 4-5 Map showing manholes, sewer lines, and subareas for Example 4-2. Note: ha × 2.4711 = acre; m × 3.2808 = ft.

Table 4-8 Computation table for design of the gravity-flow sewer in Example 4-2

Location				Area		Domestic flow					
Line	From manhole	To manhole	Length of sewer, m	Incre-ment, ha	Cum. total, ha	Pop. density, persons/ha	Cum. design pop., persons (6 × 7)	Avg. unit flow, L/capita · d	Cum. avg. flow, m^3/d (8 × 9)	Peaking factor[a]	Cum. peak flow, m^3/d (10 × 11)
(1)	(2)	(3)	(4)	(5)	(6)	(7)	(8)	(9)	(10)	(11)	(12)
1	57	58	116	...	...	...	...	...	...	...	...
2	58	59	105	...	...	...	...	...	...	...	...
3	59	61	105	...	...	...	...	...	...	...	...
4	61	62	120	...	...	...	...	...	...	...	...
5	62	11	73	...	4.90	40	196	300	58.8	3.8	223
6	11	12	28	14.25	19.15	40	766	300	229.8	3.45	793
7	12	13	37	4.78	23.93	40	958	300	287.4	3.4	977
8	13	14	85	2.14	26.07	40	1043	300	312.9	3.35	1048
9	14	15	80	7.77	33.84	40	1354	300	406.2	3.3	1340
10	15	16	37	4.89	38.73	40	1550	300	465.0	3.3	1535
11	16	17	72	1.13	39.86	40	1595	300	478.5	3.25	1555
12	17	18	118	5.01	44.87	40	1795	300	538.5	3.2	1723
13	18	19	23	1.70	46.57	40	1863	300	558.9	3.2	1788

Infiltration/inflow		Design flows		Sewer design				Sewer layout			
								Ground surface elevation		Invert elevation	
Peak I/I allowance, $m^3/ha \cdot d$[b]	Cum. peak I/I allowance, m^3/d (6 × 13)	Cum. peak flow, m^3/d (12 + 14)	Cum. peak flow, m^3/s	Pipe diameter mm	Slope, m/m	Capacity full, m^3/s	Velocity full, m/s	At upper manhole	At lower manhole	Upper end	Lower end
(13)	(14)	(15)	(16)	(17)	(18)	(19)	(20)	(21)	(22)	(23)	(24)
...	...	...	...	200	0.004	0.021	0.66	63.49	63.12	60.52	60.06
...	...	...	...	200	0.004	0.021	0.66	63.12	62.79	60.06	59.64
...	...	...	...	200	0.004	0.021	0.66	62.79	62.37	59.64	59.22
...	...	...	...	200	0.004	0.021	0.66	62.37	61.82	59.22	58.74
14.0	69	292	0.0034	200	0.004	0.021	0.66	61.82	61.45	58.74	58.44
14.0	268	1061	0.0123	200	0.004	0.021	0.66	61.45	61.60	57.10	56.99
14.0	335	1312	0.0152	200	0.004	0.021	0.66	61.60	62.12	56.99	56.84
14.0	365	1413	0.0164	200	0.004	0.021	0.66	62.12	62.24	56.84	56.50
14.0	474	1814	0.0210	200	0.004	0.021	0.66	62.24	62.00	56.50	56.18
14.0	542	2077	0.0240	250	0.0025	0.030	0.61	62.00	61.90	56.13	56.04
14.0	558	2113	0.0245	250	0.0025	0.030	0.61	61.90	61.72	56.04	55.86
13.7	615	2338	0.0271	250	0.0025	0.030	0.61	61.72	60.50	55.86	55.56
13.5	629	2417	0.0280	250	0.0025	0.030	0.61	60.50	59.30	55.56	55.51

[a]Peaking factors were taken from Fig. 3-4.
[b]Peak infiltration values were obtained from curve *B* of Fig. 3-5.

Note:

m × 3.2808 = ft
ha × 2.4711 = acre
L × 0.2642 = gal
$m^3/d \times 2.6417 \times 10^{-4}$ = Mgal/d
$m^3/ha \cdot d \times 106.9064$ = gal/acre · d
$m^3/s \times 35.3147$ = ft^3/s
mm × 0.03937 = in
m/s × 3.2808 = ft/s

midway between them. If the street layout is not shown on the plan, the limits of the different service areas cannot be determined as closely, and the topography may serve as a guide.

4. Determine the area of each service area. Measure the area of each service area by using a planimeter, and enter the values on the map (see Fig. 4-5).
5. Summarize the basic design criteria.
   a. Design period:
   Use the saturation period.
   b. Population density:
   The average population at saturation development is 40 persons/ha.
   c. Residential wastewater flows:
   i. Use a value of 300 L/person · d.
   ii. Obtain peaking factors from Fig. 3-4.
   d. Infiltration allowance:
   Obtain the peak values from curve *B* for new sewers given in Fig. 3-5.
   e. Inflow allowance:
   Assume that steady-flow inflow is accounted for in the infiltration allowance.
   f. Hydraulic-design equation:
   Use the Manning equation with an n value of 0.013. To simplify the computation, use Figs. 2-12 and 2-13.
   g. Minimum pipe size:
   As fixed by the local code, the minimum pipe size is 200 mm.
   h. Minimum velocity:
   To prevent deposition of wastewater solids in the sewer, use a minimum velocity of 0.6 m/s.
   i. Minimum cover:
   As established by the local community building code, the minimum cover equals 2.75 m since the homes in this area have basements.
6. Prepare a tabulation form to record the data and steps in the computations for each section of sewer between manholes.

   A computation table similar to the one in Example 4-1, with separate columns for each step in the computation and a separate line for each section of sewer between manholes, is given in Table 4-8. Because there are no commercial, industrial, and institutional flows involved, the number of columns in the table can be reduced considerably. For illustration, use the sewer starting with manhole 57 and terminating with manhole 19, just beyond the end of Main Street. With the exception of the sewer layout data in cols. 21 through 24, the data and computations in Table 4-8 are similar to those described for Table 4-7.

   The necessary layout data for the sewer design given in Table 4-8 are obtained as follows:
   a. The ground surface elevations at the manhole locations in cols. 21 and 22 are obtained by interpolation with the elevation data given in Fig. 4-3.
   b. The sewer invert elevations shown in cols. 23 and 24 are obtained by trial and error with a sewer profile work sheet (see Fig. 4-6). The first step in preparing a work sheet is to plot the ground-surface elevations given in cols. 21 and 22, working backwards from a convenient point. For the profile work sheet shown in Fig. 4-6, station 0 + 00 was taken at manhole 19. The stationing continues until station 9 + 99 is reached, which corresponds to the total length of sewer in this example (the sum of col. 4 in Table 4-8).

After the ground-surface profile is drawn, the next step is to begin sketching the invert and crown (inside bottom and inside top of the pipe, respectively) of each sewer section as the necessary elevation data are developed. The method for establishing the invert elevations will be illustrated by analyzing selected sewer lines starting with line 1, which connects manholes

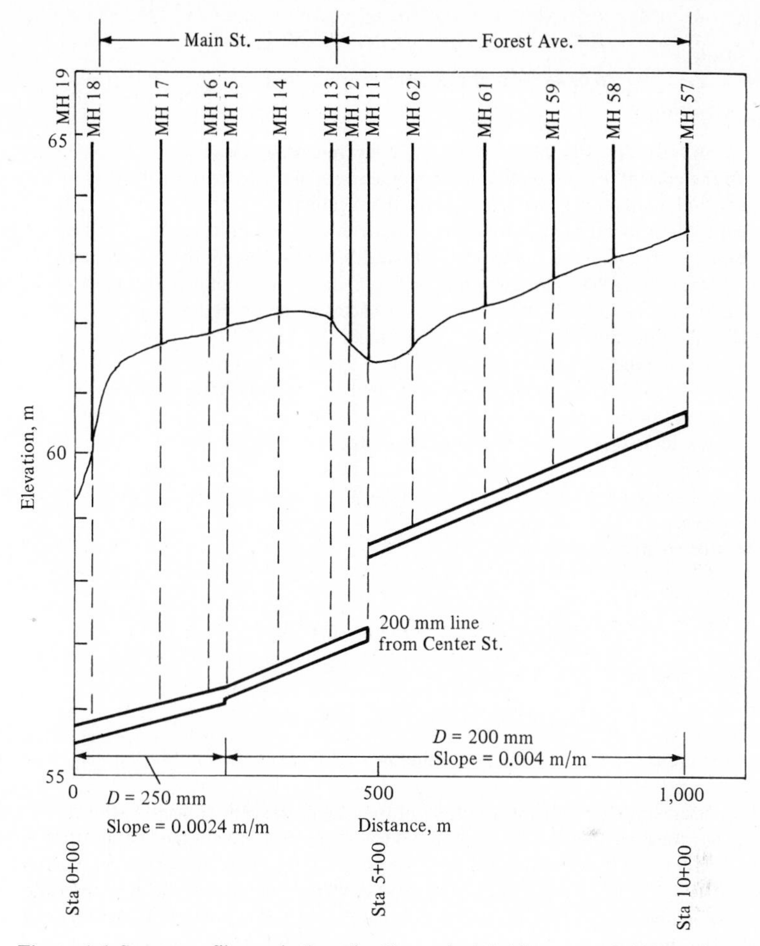

Figure 4-6 Sewer profile work sheet for Example 4-2. Note: m × 3.2808 = ft; mm × 0.03937 = in

57 and 58. The first step is to locate the invert of the upper end of the pipe at such an elevation that the minimum cover requirement is satisfied, taking into account both the inside diameter of the pipe and its wall thickness. In Fig. 4-6, the upper invert elevation of the 200-mm (8-in) pipe is set initially at elevation 60.52 m (ground surface − cover − pipe wall thickness − pipe diameter = 63.49 m − 2.75 m − 0.02 m − 0.20 m). The lower elevation is computed by subtracting the fall as follows:

$$\begin{matrix}\text{Lower}\\ \text{invert}\\ \text{elevation}\end{matrix} = \begin{matrix}\text{upper}\\ \text{invert}\\ \text{elevation}\end{matrix} - \begin{pmatrix}\text{slope}\\ \text{of}\\ \text{sewer}\end{pmatrix}\begin{pmatrix}\text{length}\\ \text{of}\\ \text{sewer}\end{pmatrix}$$

For line 1:

$$\begin{matrix}\text{Lower}\\ \text{invert}\\ \text{elevation}\end{matrix} = 60.52\text{ m} - (0.004\text{ m/m})(116\text{ m}) = 60.06\text{ m}$$

For line 2:

$$\text{Lower invert elevation} = 60.06 \text{ m} - (0.004 \text{ m/m})(105 \text{ m}) = 59.64 \text{ m}$$

If it is apparent from Fig. 4-6 that the depth of cover (remember to allow for the pipe wall thickness above the crown) for any section has become too shallow, repeat the process with a lower initial invert elevation or a steeper slope for that section.

When a manhole is located at a sewer junction, the outlet sewer elevation is fixed by the lowest inlet sewer. The junction at manhole 11 illustrates this. Although the calculations are not shown in this example, a 200-mm sewer section beginning with minimum cover at manhole 2 and continuing at a slope of 0.004 m/m to manhole 11 has an invert elevation of 57.10 m at manhole 11. Thus the upper invert elevation of line 6 must also be 57.10 m. (If line 6 had been larger than the 200-mm pipe with the invert elevation of 57.10 m, the crowns of the two pipes would have been matched at manhole 11, as discussed in the next paragraph.) Often, the location of the sewers in the cross streets will control the design.

If the pipe size increases, the crowns of the two pipes must be matched at the manhole. This is done to avoid the backing up of wastewater into the smaller pipe. An example of this is the increase in size from 200 to 250 mm at manhole 8. For this case, the calculations are as follows: Lower invert elevation of the 200-mm sewer = 56.18 m. (This value is obtained by continuing the process just illustrated.)

Upper invert elevation for the 250 mm-sewer (line 8) is:

$$56.13 \text{ m } (56.18 \text{ m} + 0.20 \text{ m} - 0.25 \text{ m})$$

Lower invert elevation for the 250-mm sewer is:

$$56.13 - (0.0025 \text{ m/m})(37 \text{ m}) = 56.04 \text{ m}$$

These procedures are repeated until the entire line is established.

Comment It is interesting to compare the effect of the size of the service area on the overall peaking factor. For the area shown in Fig. 4-4, the ratio of the peak flow to the average flow is 2.27. To obtain the average flow for the area shown in Fig. 4-5, the cumulative peak infiltration value is divided by 1.6, and the resulting value is aded to the cumulative average domestic flow. The corresponding ratio of the peak to the average flow is 2.54 [(2,417 m^3/d)/(558.9 m^3/d + (629 m^3/d)/1.6)].The corresponding ratio of areas is 19.3 (900 ha/46.57 ha).

Preparation of Contract Drawings and Specifications

Detailed contract drawings, including plans, profiles, and details of the works to be constructed, and specifications should be completed before bids are requested, so that all of the data will be available to prospective bidders.

Construction contractors are justified in making lower bids when they are supplied with complete information about the conditions they will encounter; otherwise, they must estimate many of them. All facts ascertained and interpretive studies made, whether advantageous or not, should be made public, because the judgment of the engineer must be entirely unbiased, and the contractor should, in all fairness, have all the information available to assist in formulating a bid. Also, if any facts or studies are suppressed, contractors' claims for extra costs based on such suppression may be allowed by the courts.

Contract drawings. The construction plans must indicate the locations of the works to be constructed. Base plans on which these locations are drawn should show surface contours, when available; street centerline elevations; all streets, railroads, buildings, pipes, conduits, manholes, gate boxes, and catch basins; and the names of streets, parks, public buildings, and watercourses. The magnetic or true north, or both, should be indicated.

A small-scale plan indicating the extent of the works, a key plan showing how the detailed plans fit together, and a location map should be prepared. In many projects, the first two plans are combined. The preliminary design maps often will be suitable for the base map for these purposes.

Profiles should show the ground surface, the location of existing buildings and the elevations of their sills, the proposed sewer, its slope and size, and the invert elevation at each manhole as well as the size and elevation of the sewer into which the flow of the sewer under construction is to discharge. As far as practicable, the profiles should show all available information bearing on subsurface features and the location, size, and character of structures likely to be found during excavation, together with details of the works to be constructed.

Engineering practice in displaying data from soil borings varies. In some projects, all the data are shown on the drawings; in others, only the locations are shown on the drawings, and the details of the test borings are in the bound-contract documents. Such data consist of copies of the detailed logs or reports prepared by the contractor who made the borings.

Prospective bidders should be encouraged to make their own appraisals of subsurface conditions from the driller's logs. However, because of many court decisions, any interpretations of subsurface data that may have been made, whether in the form of notes, soil or geologic profiles, or otherwise, should appear on the contract drawings or in the other contract documents.

The scales of the plans and profiles are determined by the number of obstacles to construction and, therefore, the amount of detail required. The profile should be drawn either directly above or below the location plan. The plan should be of the same scale as the horizontal scale of the profile and should show all structures, both above and below ground surface, which may influence the sewer's location or which may affect the construction operations.

Scales of plans and horizontal scales of the corresponding profiles are commonly 5 m, or less, to 10 mm (40 ft, or less, to 1 in). Vertical-profile scales are usually exaggerated, often as much as 10 times the horizontal. An example would be a horizontal scale of 5 m to 10 mm (40 ft to 1 in) with a vertical scale of 0.5 m to 10 mm (4 ft to 1 in).

An example of a typical contract drawing is shown in Fig. 4-7. The data usually required for preparing contract drawings are illustrated in this figure, and the method of showing the data with scales may be seen. The greater amount of detail required on contract drawings, as compared with sewer-profile work sheets, and the difference in scale and arrangement that this necessitates are apparent when Fig. 4-7 is compared with the profile shown in Fig. 4-6. Note that both a plan and a profile are required for indicating the work to be done.

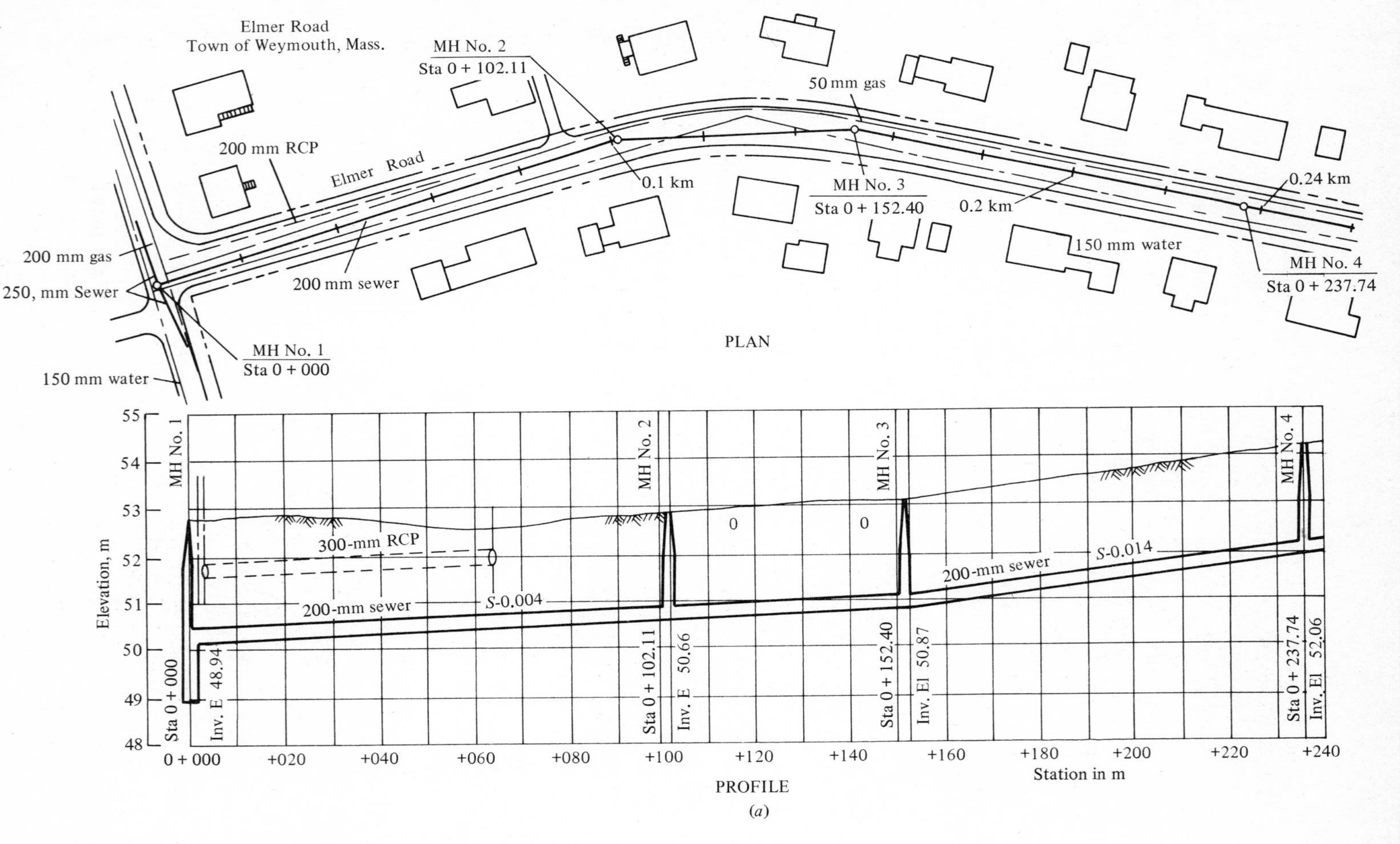

Figure 4-7(*a*) Typical sewer contract drawing in metric units.

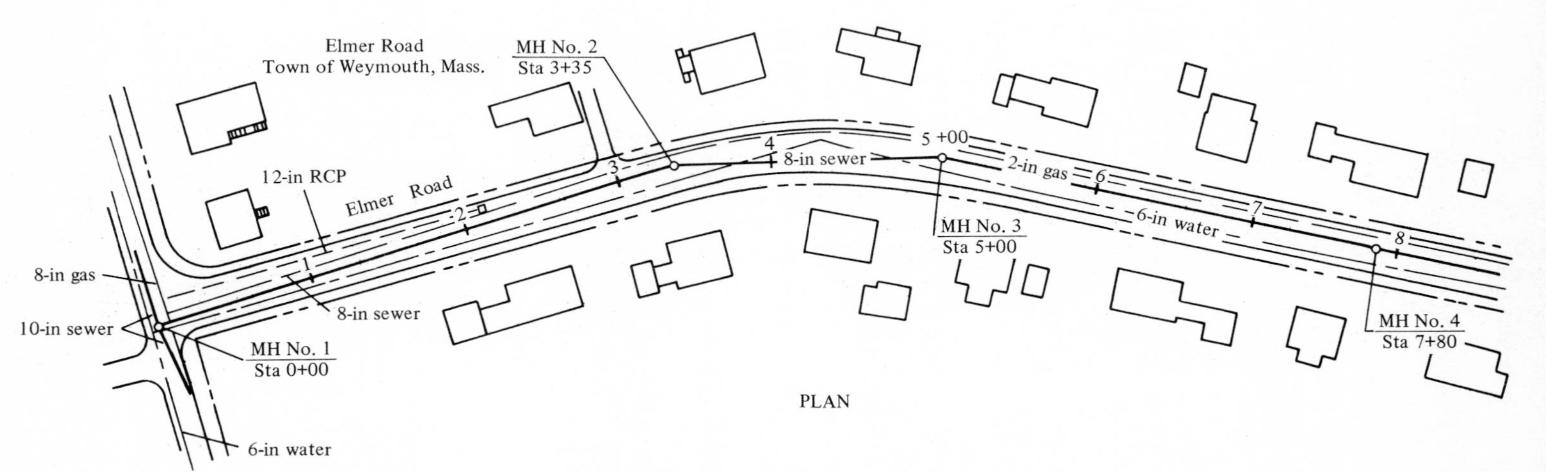

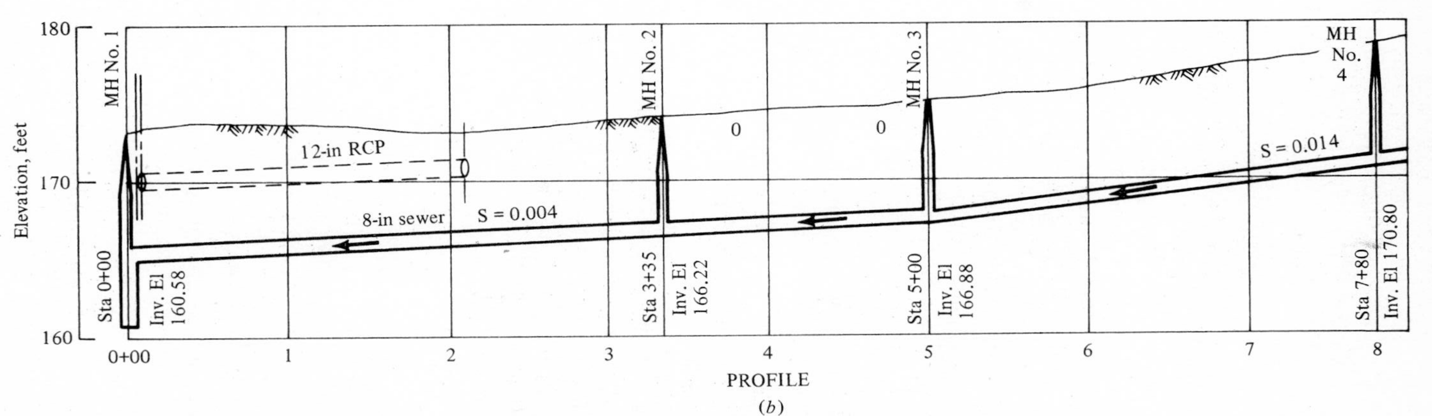

(b)

Figure 4-7(b) Typical sewer contract drawing in U.S. customary units.

Special details of the sewers and appurtenances may be included as a part of the contract drawings or as an appendix to the contract specifications.

Contract specifications. The contract and specifications should describe clearly, and as completely as possible, all work, requirements, and conditions included in the contract or that affect the contract. Although the preparation of these details increases the cost of the engineering work, the total cost of the project will usually be less than when the drawings merely show in a general way what is to be done or when the specifications are incomplete or obscure.

The contract specifications include (1) the special instructions to the bidder, (2) the bid forms for inserting the prices for doing the work, (3) the contract agreement that stipulates the legal provisions, (4) special and general conditions related to the work to be done, and (5) technical specifications. The technical specifications describe the types and quality of material to be furnished, construction and installation procedures, and performance requirements. Performance requirements in sewer construction are evidenced by acceptance tests that demonstrate the ability of the newly constructed sewer to retain wastewater within acceptable limits. The types of test specified and acceptable performance criteria are discussed in the following section.

Acceptance Tests for New Sewers

For newly built gravity-flow sewers, provision for tests as a basis for acceptance of the sewers should be included in all contract documents. Testing should be done, in sections of not more than about 300 m (1000 ft), immediately after completion of the trench backfill, so that any necessary reconstruction may be accomplished expeditiously. Water testing and low-pressure air testing are the principal methods now used to assess the water tightness of new sewers.

Water testing. Direct measurements of leakage may be made by conducting infiltration or exfiltration tests on the newly completed sections of sewers and manholes.

The infiltration test usually is preferred when the groundwater level is at least 0.3 to 0.6 m (1 to 2 ft) above the crown of the sewer. The upstream end of the section to be tested is plugged, and a V-notched weir is installed in the manhole at the lower end. Sufficient readings of the head on the weir are made to permit calculation of the average rate of leakage into the sewer.

When the groundwater level is too low for infiltration testing, the leakage may be determined by measuring the exfiltration (outward flow) from the sewer. In this test, the ends of the section of sewer to be tested, including a manhole at each end, are plugged, and all stoppers and plugs are braced or otherwise secured to resist the internal pressure resulting from the test. The section is then filled with water to a predetermined level above the crown of the sewer. The rate of leakage is computed on the basis of the observed drop in water level over a reasonably long period of time or by metering the volume of water required to be supplied to the system to maintain the original water level.

The criterion for acceptance of newly constructed sewers is a rate of leakage (whether in or out) usually expressed as liters per millimeter or diameter per kilometer per day (gallons per inch of diameter per mile per day). With modern materials and methods of construction, a readily attainable rate is 20 L/mm of diam · km · d (200 gal/in of diam · mi · d). Individual pipeline sections sometimes test out at 10 L/mm of diam · km · d (100 gal/in of diam · mi · d) or less, but it is doubtful that such a low value is reasonable or desirable when manholes and stubs for building connections are included in the test section.

Low-pressure air testing. Methods of testing sewers for leaks by using air under low pressure have been developed in recent years and are now favored by many engineers. Although there is no direct correlation between air leakage and water leakage, it is generally believed that a sewer that passes a specified air test would also pass a water test if one were made. For this air test procedure, a section of sewer between manholes is plugged, as shown in Fig. 4-8 [2]. All plugs and stoppers are secured to resist the expected internal pressure. Air is then introduced into the test section at a pressure above the maximum pressure exerted by any groundwater that may be present outside the pipe. After the pressure in the pipe has stabilized, the air supply is disconnected quickly, and the time required for the pressure in the sewer to drop a specified amount is observed. Additional details beyond those presented in the following discussion may be found in Refs. 4, 10, and 11.

The air testing of sewers is accomplished with air pressures in the range of 20 to 35 kN/m^2 (3 to 5 lb$_f$/in^2) above any outside pressure on the pipe. The most commonly used pressure is 27.5 kN/m^2 (4 lb$_f$/in^2). After a line or section to be tested has been pressurized and the pressure has stabilized (minimum of 2 min), the air is shut off. Then, starting at a pressure of 24 kN/m^2 (3.5 lb$_f$/in^2), the time required (in minutes) for the pressure to drop by 6.9 kN/m^2 (1 lb$_f$/in^2) is measured.

It has been established that a section of sewer is performing satisfactorily if the time required (in seconds) for the pressure to drop from 24 to 17.1 kN/m^2 (3.5 to 2.5 lb$_f$/in^2) is greater than or equal to the least of the two times computed with the use of the following equations. The equation that yields the shortest time is the controlling one. The application of these equations is illustrated in Example 4-3.

$$t_Q = \frac{0.032}{Q}(d_1^2L_1 + d_2^2L_2 + \cdots + d_n^2L_n) \quad \text{(SI units)} \tag{4-1}$$

$$t_Q = \frac{0.022}{Q}(d_1^2L_1 + d_2^2L_2 + \cdots + d_n^2L_n) \quad \text{(U.S. customary units)} \tag{4-1a}$$

$$t_q = \frac{1.0184}{q}\,\frac{d_1^2L_1 + d_2^2L_2 + \cdots + d_n^2L_n}{d_1L_1 + d_2L_2 + \cdots + d_nL_n} \quad \text{(SI units)} \tag{4-2}$$

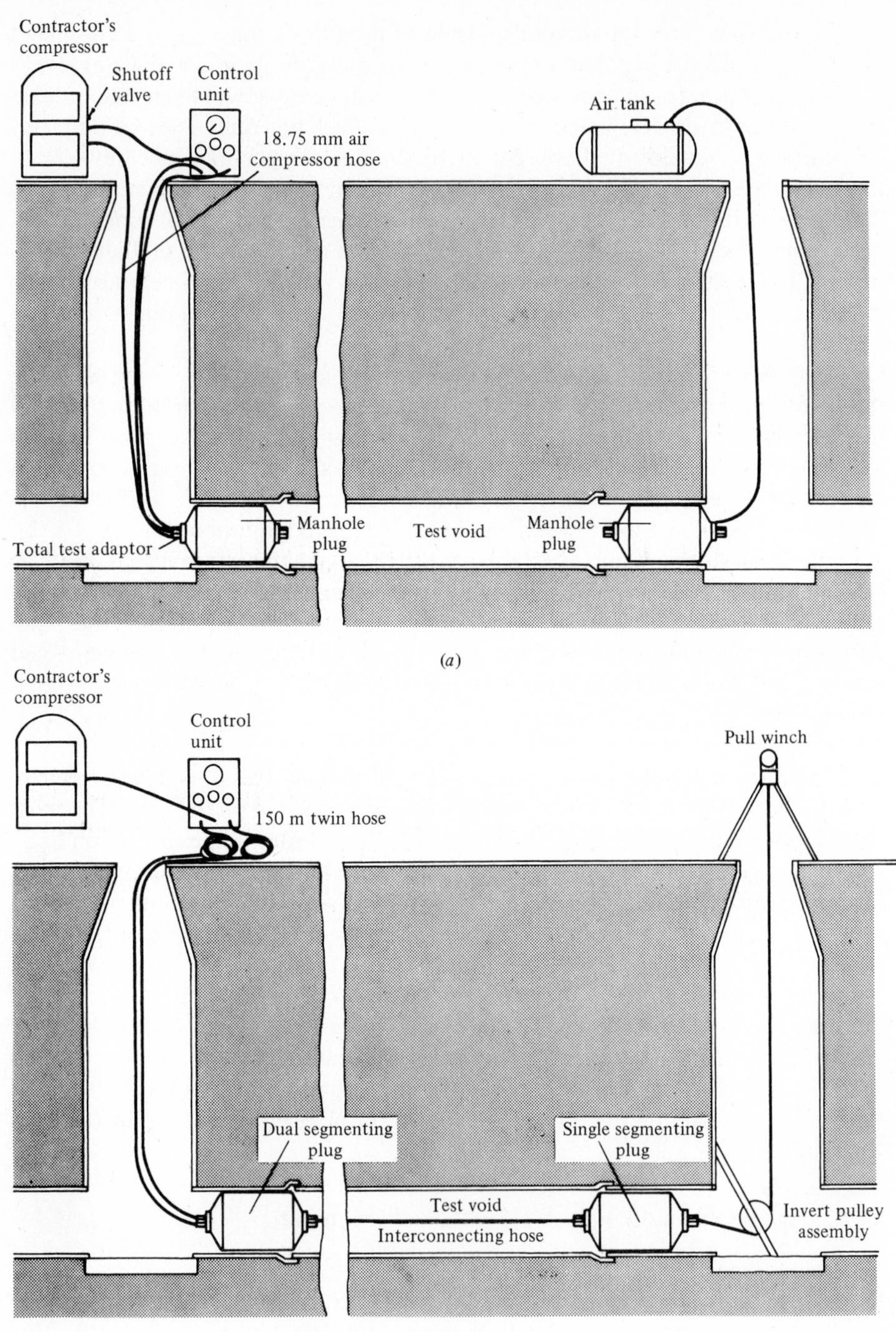

Figure 4-8 Typical setups for low-pressure air testing of sewers. (*a*) Manhole-to-manhole air test. (*b*) Segment of pipe air tests. Note: mm × 0.3937 = in; m × 3.2808 = ft.

$$t_q = \frac{0.085}{q} \frac{d_1^2L_1 + d_2^2L_2 + \cdots + d_n^2L_n}{d_1L_1 + d_2L_2 + \cdots + d_nL_n} \quad \text{(U.S. customary units)} \qquad (4\text{-}2a)$$

where $t_{Q'}$, t_q = allowable time for a decrease in pressure from 24 to 17.1 kN/m² (3.5 to 2.5 lb_f/in^2), s

Q = 56.7 L/min (2.0 ft³/min) or air loss

q = 0.913 L/min · m² of internal pipe surface (0.0030 ft³/min · ft² of internal pipe surface) or air loss

d = diameter of sewer pipe in section of system being tested, mm (in)

L = length of sewer pipe being tested, m (ft)

Typically, a tight sewer will hold the required pressure much longer than the minimum time required, but a defective joint or other significant leak will cause an almost immediate loss of pressure. In the interest of safety, the testing equipment should be provided with an air-regulator valve or air-safety valve set so that the internal air pressure in the pipeline cannot exceed 55 kN/m² (8 lb_f/in^2).

When low-pressure air procedures are used for testing the sewer, manholes should be tested separately for leakage because they will not have been included in the air test. For example, the state of New Hampshire requires that manhole leakage shall not exceed 12.4 L/vertical m · d (1.0 gal/vertical ft · d) when tested with no external groundwater and the manhole is full to the top of the cone section.

Example 4-3 Evaluation of air testing results An air test has been performed on a recently completed section of 300-mm (12 in) sewer 100 m (328 ft) in length. Also included were 150 m (492 ft) of 150-mm (6-in) building connection pipes. The air pressure was equalized at 27.5 kN/m² (4 lb_f/in^2) for 6 min. The time required for the pressure to drop from 24 to 17.1 kN/m² (3.5 to 2.5 lb_f/in^2) was observed to be 375 s. Does the completed sewer meet acceptance requirements?

SOLUTION

1. Determine the acceptance time t_Q by using Eq. 4-1.

$$t_q = \frac{0.0032}{Q} (d_1^2L_1 + d_2^2L_2)$$

$$= \frac{0.0032}{56.7} [(300)^2(100) + (150)^2(150)]$$

$$= 698 \text{ s}$$

2. Determine the acceptance time t_q by using Eq. 4-2.

$$t_q = \frac{1.0184}{q} \frac{d_1^2L_1 + d_2^2L_2}{d_1L_1 + d_2L_2}$$

$$= \frac{1.0184}{0.913} \frac{(300)^2(100) + (150)^2(150)}{(300)(100) + (150)(150)}$$

$$= 263 \text{ s}$$

3. Because the observed time (375 s) is greater than the controlling time t_q (263 s), computed in step 2, the section of sewer being tested meets the acceptance requirement.

4-3 DESIGN OF GRAVITY-FLOW STORM-WATER SEWERS

The procedure used for the design of a storm-water sewer system is similar to that needed for a sanitary sewer system, as discussed in the preceding section on sanitary sewer design. Only those preliminary activities and design features which are different for storm water sewers are discussed in this section. The design of storm water sewers is illustrated at the end of this section in Example 4-4.

Storm-Water Design Flows

The method for determining the rates of storm-water runoff transported by storm-water sewers is still imprecise, although much progress has been made during recent years [5, 6, 14]. Such progress has come about with increased technical knowledge of hydrologic events and characteristics in relation to rainfall and runoff, particularly the large amount of data available for study and correlation of rainfall, runoff, topography, and development. The extensive use of flow hydrograph data, operation of flood-control reservoirs, pumping of storm water runoff, and the requirements for substantially flood-free highways and airfields have resulted in various methods of estimating rainfall runoff rates and amounts. However, much is yet to be learned about these relationships, and research in this area must be a continuing process.

Methods of computation. The quantity of storm-water runoff may be determined for storm-water sewer design by any of the following methods: (1) empirical formulas, (2) the rational method, (3) rainfall-runoff correlation studies, (4) hydrograph methods, (5) the inlet method, and (6) digital computer models. The method selected depends on the local geographic and hydrologic conditions, the availability of past rainfall and runoff data, the size of the drainage area, and the degree of protection needed. Detailed data on these methods are presented in Refs. 1, 3, 7, 8, and 12. Of the methods listed, the rational method is the most common.

The rational method In the rational method, the following relationship is used [1, 7]:

$$Q = 240CiA \qquad \text{(SI units)} \qquad (4\text{-}3)$$

$$Q = CiA \qquad \text{(U.S. customary units)} \qquad (4\text{-}3a)$$

where Q = maximum rate of runoff, m^3/d (acre-in/h $\simeq$ ft^3/s)

C = average runoff coefficient
i = average rainfall intensity, mm/h (in/h)
A = drainage area tributary to the point under study, ha (acre)

Briefly, the computation of storm-water runoff rates, using the rational method, requires that the following basic data be determined:

1. The time-intensity rainfall relation to be used as the basis of design.
2. The probable future condition of the drainage area, i.e., the percentage of impervious surface expected when the district is developed to the extent assumed.
3. The runoff coefficient, relating the peak rate of runoff at any location to the average rate of rainfall during the concentration time (see item 6) for that location.
4. The probable time required for water to flow over the surface of the ground to the first inlet, called the *inlet time* or *time of entrance*.
5. The area tributary to the sewer at the location at which the size is to be determined.
6. The time required for water to flow in the sewer from the first inlet to the location in question which, added to the inlet time, gives the time of concentration at that location.

When estimating storm-water runoff quantities by the rational method, the average rainfall intensity in the computations is assumed to be the value that corresponds to a duration equal to the time of concentration.

The inlet time will vary from 5 to 10 min for high-density areas with closely spaced street inlets, from 10 to 20 min for well-developed areas with relatively flat slopes, and from 20 to 30 min for residential areas with widely spaced street inlets. If reliable information for a given location is unavailable, a value of 20 min is often assumed for inlet time.

A note of caution about the use of the rational method: Because it is assumed in the application of this method that the entire area is contributing, the computed flowrates will tend to be increasingly conservative as the size of the drainage area increases. (Remember that rainfall does not fall in a uniform pattern, and some places in a large drainage area will receive no rainfall.) For this and other reasons, the rational method should be applied only to small drainage areas.

Storm-Water Sewer Pipe Materials and Sizes

Reinforced concrete is the most common pipe material in storm-water sewers. Other pipe materials are used, as for sanitary sewers when the special needs and economics of the project require them. The effects of corrosive waters and aggressive soil environments should be considered when selecting pipe materials for storm-water sewers as well as sanitary sewers.

In general, the minimum size of a storm-water sewer should be 300 to 375 mm (12 to 15 in). Building connections are preferably at least 150 mm (6 in) in diameter.

Minimum Velocities

Because storm-water sewers usually receive larger quantities of grit, they should be designed with higher minimum velocities (and corresponding minimum slopes than sanitary sewers). A minimum velocity of 1.0 m/s (3.0 ft/s) is common. Special circumstances may require higher minimum velocities.

Example 4-4 Design of a storm-water sewer Design a storm-water sewer system for the area shown in Fig. 4-9. Elevations are shown in Fig. 4-3. The location of the proposed trunk storm-water sewer that is to receive the storm water from the district is also shown in Fig. 4-9. (In most designs, the invert elevation where a main or trunk sewer would be connected to a trunk or intercepting sewer would be known.)

Compute the rainfall intensity for the area by the following equation:

$$i = 518.2/t^{0.61}$$

where i = rainfall intensity, mm/h
t = rainfall duration, min

This equation represents the average rate of rainfall for a duration of t min that may be expected to be equaled or exceeded, on the average, once in a 5-yr period. The rate was derived from precipitation records for the area in which the drainage district is located.

From a careful study of local conditions, including the present and probable future development of the district, it was concluded that, at ultimate development, 70 percent of the surfaces in the district would be impervious. On the basis of analyses in similar areas, the coefficients of runoff for the pervious and impervious portions of the district can be assumed to vary with rainfall duration, as shown in Fig. 4-10. This variation occurs partly because much of the early part of the rainfall soaks into the soil or is stored in local depressions before appreciable runoff develops.

Design the storm-water sewers with the crown at a depth of at least 1.8 m (6 ft) below the surface of the street. In many communities, especially in cold climates, the minimum depth would be greater than 1.8 m (6 ft) to allow for clearance under existing and future water mains. The minimum size of the sewer is to be 300 mm. The assumed minimum velocity is to be 0.9 m/s when flow is at full depth. Determine the capacities of the sewers by using Manning's equation with an n value of 0.013.

SOLUTION

1. Lay out the storm-water sewers. Draw a line to represent the storm-water sewer in each street or alley to be served. Place an arrow on or near each sewer to show the direction of flow. The sewers should, in general, slope with the street surface. However, it usually is more economical to lay out the system so that the water reaches the trunk or intercepting storm-water sewer by the most direct route.

   In some communities, roof water is allowed to discharge on the ground and flow over the surface to the gutter inlets. Under such circumstances, sewers may be provided only to the last gutter inlet rather than to a point opposite the last house lot, thereby effecting some saving in cost. This practice is open to the criticism that it does not give equal service to all property and is, therefore, unfair. In the example discussed here, the intention is to provide drainage facilities for all property within the district.

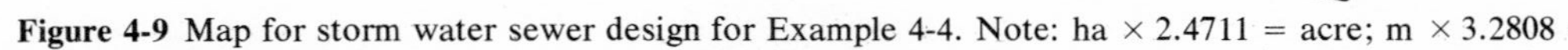
Figure 4-9 Map for storm water sewer design for Example 4-4. Note: ha × 2.4711 = acre; m × 3.2808 = ft.

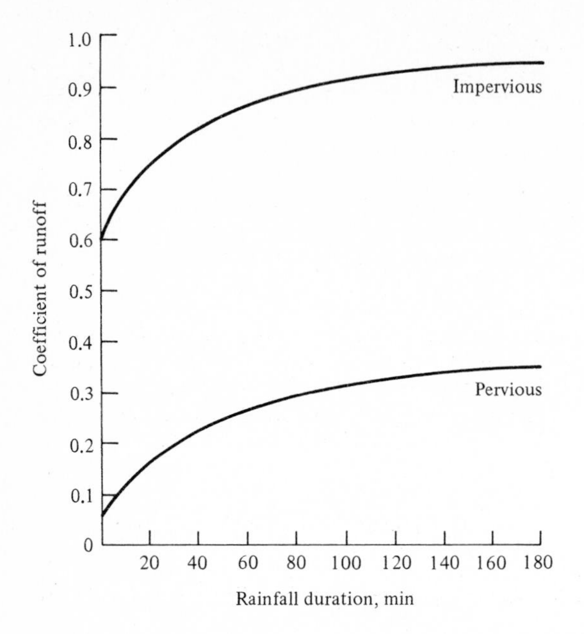

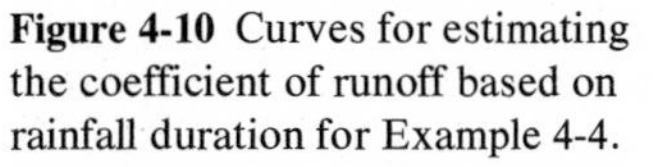
Figure 4-10 Curves for estimating the coefficient of runoff based on rainfall duration for Example 4-4.

2. Locate the manholes. Locate the manholes tentatively, giving to each an identification number. (Manholes have been numbered only on the selected storm-water sewer lines for which design calculations are performed in this example.) In this example, a manhole is placed at each bend or angle, at all junctions of storm-water sewers, at all points of change in size or slope, and at intermediate points where the distance exceeds 100 m for 300- to 600-mm sewers and 120 m for larger sewers.

   Where adequate velocity can be maintained during practically all flow conditions, and where the sewer is large enough for maintenance personnel to walk without stooping, intervals between manholes up to 200 m may be used. Sufficient manholes should be built to allow access for inspection and cleaning. Later, when the profiles are drawn and the final slopes are being fixed, the locations for some manholes may be changed so that the sewers would be at the most advantageous depth, particularly where the slope of the street surface is not substantially uniform. Other considerations, such as underground obstacles, may require the installation of additional manholes because of a change in alignment or special forms of construction involved in junctions or connections with other sewers.
3. Establish the limits of the drainage area. Sketch the limits of the drainage-area tributary at each manhole. The assumed character of future development and the topography determines the proper limits.
4. Determine the area of drainage subareas. Measure each individual area by planimeter or other suitable method.
5. Summarize the basic design data and criteria.
   a. Rainfall intensity relationship:

$$i = \frac{518.2}{t^{0.61}}$$

   b. Runoff coefficient:
   Use Fig. 4-10 to obtain the runoff coefficients for the pervious and impervious areas. A weighted average of these values can be used to obtain the runoff coefficient for the overall area.

c. Inlet time:

$$t_i = 20 \text{ min}$$

d. Hydraulic-design equation:
Use the Manning equation with an n value of 0.013.

e. Minimum pipe size:
As specified, the minimum pipe size is 300 mm.

f. Minimum velocity:
To prevent the deposition of solids, the minimum velocity is 0.9 m/s.

g. Minimum cover:
As specified, the minimum cover is 1.8 m.

6. Prepare a runoff curve. To make the determination of runoff volumes more convenient, the runoff curve shown in Fig. 4-11 is developed. To illustrate the method used to develop this curve, sample calculations for a storm duration of 60 min are as follows:

a. Determine the runoff coefficient for pervious and impervious surfaces for a storm duration of 60 min. The following values were obtained from Fig. 4-10:

$$C\text{, impervious} = 0.86$$

$$C\text{, pervious} = 0.26$$

b. Determine the overall runoff coefficient:
A weighted average is used based on 70 percent of the overall area being impervious and 30 percent pervious.

$$C\text{, overall} = (0.70)(0.86) + (0.30)(0.26) = 0.68$$

c. Determine the rainfall intensity i:

$$i = \frac{518.2}{60^{0.61}}$$

$$i = 42.64 \text{ mm/h}$$

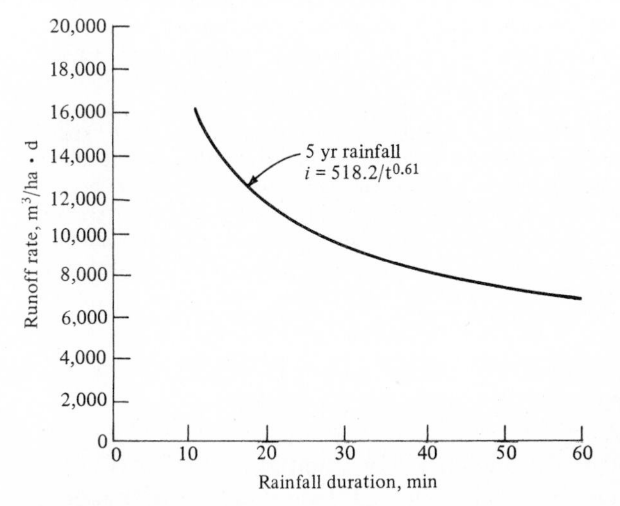

Figure 4-11 Curve for estimating storm water runoff rates for storms of various duration for Example 4-4. Note: $\text{m}^3/\text{ha}\cdot\text{d} \times 106.9064 = \text{gal/acre}\cdot\text{d}$.

d. Determine the runoff per unit area:

$$\frac{Q}{A} = 240Ci$$

$$\frac{Q}{A} = 240(0.68)(42.64)$$

$$= 6{,}959 \text{ m}^3/\text{ha} \cdot \text{d}$$

Similar calculations are performed for other storm durations until sufficient data have been generated to plot the complete runoff curve (see Fig. 4-11).

7. Prepare a tabulation form to record the data and steps in the computations for each section of sewer between manholes. The computations for a selected line are shown in Table 4-9. The first six columns of Table 4-9 require no explanation. Columns 7 and 8 record the travel time from the upper end of the drainage area (col. 7) and the time spent in each pipe section. The first value entered in col. 7 is the assumed inlet time, 20 min. Using this value, the rate of runoff from a storm of this duration is obtained from Fig. 4-11, and the value is entered in col. 9. The cumulative storm-water flowrate is obtained by multiplying the cumulative total area (col. 6) by the unit runoff rate (col. 9), and entering the resulting value in col. 10. The cumulative runoff rate, expressed in cubic meters per day in col. 10, is expressed in cubic meters per second in col. 11.

   The next four columns record the sewer design parameters. The numerical values for these parameters are obtained in exactly the same way as the corresponding values were obtained in Table 4-8 in Example 4-2 (see step 6 of the solution for Example 4-2).

   After the necessary sewer design values for one line have been entered in cols. 12 to 15, the time of travel in the sewer (col. 8) is obtained by dividing the length of sewer (col. 4) by the flow velocity (col. 15). The resulting time of travel is added to the inlet time to obtain the total flow time for the next sewer section. The unit runoff rate is now determined by using the total travel time, and the process is repeated until the entire table is completed.

   Each lateral is designed in a similar way. If necessary, the first design of the main or submain is subsequently modified to serve the laterals properly. It is possible in some cases to omit some manholes on lateral storm-water sewers by using the inlet substructures, at which junctions any changes in size, direction, or slope may be made.

Comment Because the flow velocities at design conditions are much higher for storm-water sewers than for sanitary sewers, the actual design must provide for additional head to compensate for minor losses, such as those caused by bends, manholes, transitions, and velocity changes.

The storm water sewers designed in this example might be overtaxed on the average of once in about 5 yr, but the cost of providing for storms of greater intensity would be higher. During the earlier years after construction of the storm water sewers, they will be able to carry the runoff from higher rates of rainfall, but they will not be able to carry it later because impervious surface and runoff will progressively increase as the service area becomes more developed, e.g., more paved areas. In the future, when the district is more built up and funds are available, relief sewers can be constructed to handle the higher runoff rates if flooding has become serious enough to warrant the expenditure.

4-4 PRESSURE AND VACUUM SANITARY SEWERS

Where the topography is suitable, gravity-flow sanitary sewers have been selected and will probably continue to be selected. However, where the topog-

Table 4-9 Computation table for design of the gravity-flow storm water sewer in Example 4-4

	Location			Area		Flow time			Design flows		Sewer design			
Line	From manhole	To manhole	Length of sewer, m	Incre-ment, ha	Cum. total, ha	To upper end, min	In section, min. (4 ÷ 15)	Unit rate of runoff, $m^3/ha \cdot d$[a]	Cum. flow, m^3/d (6 × 9)	Cum. flow, m^3/s	Pipe diam., mm	Slope, m/m	Capac-ity full, m^3/s	Veloc-ity full, m/s
(1)	(2)	(3)	(4)	(5)	(6)	(7)	(8)	(9)	(10)	(11)	(12)	(13)	(14)	(15)
1	1	2	85	0.93	0.93	20[b]	1.6	11,500	10,695	0.124	450	0.0026	0.145	0.91
2	2	3	91	0.97	1.90	21.6	1.6	11,100	21,090	0.244	600	0.0018	0.261	0.92
3	3	4	83	0.89	2.79	23.2	1.4	10,700	29,853	0.346	675	0.0017	0.347	0.97
4	4	5	45	0.61	3.40	24.6	0.8	10,400	35,360	0.409	750	0.0014	0.417	0.94
5	5	6	99	0.89	4.29	25.4	1.8	10,200	43,758	0.506	900	0.0010	0.572	0.90
6	6	7	122	1.25	5.54	27.2	2.0	9,900	54,846	0.635	900	0.0013	0.653	1.03
7	7	8	11	2.43	7.97	29.2	0.2	9,600	76,512	0.886	1,050	0.0011	0.906	1.05
8	8	9	67	4.12	12.09	29.4	1.1	9,500	114,855	1.329	1,350	0.0008	1.510	1.05
9	9	10	70	2.30	14.39	30.5	1.0	9,400	135,266	1.566	1,350	0.0009	1.601	1.12
10	10	11	34	4.80	19.19	31.5	0.5	9,200	176,548	2.043	1,500	0.0009	2.121	1.20
11	11	12	29	4.50	23.69	32.0	0.4	9,200	217,948	2.532	1,650	0.0008	2.578	1.21
12	12	13	86	2.27	25.96	32.4	1.1	9,100	236,236	2.734	1,650	0.0009	2.734	1.28
13	13	14	75	6.97	32.93	33.5	0.9	9,000	296,370	3.430	1,800	0.0009	3.448	1.36
14	14	15	44	5.55	38.48	34.4	0.5	8,900	342,472	3.964	1,950	0.0008	4.025	1.35
15	15	16	62	1.18	39.66	34.9	0.7	8,800	349,008	4.039	1,950	0.0009	4.269	1.43
16	16	17	110	5.35	45.01	35.6	1.3	8,800	396,088	4.584	2,100	0.0008	4.904	1.42
17	17	18	50	1.66	46.67	36.9	. . .	8,600	401,362	4.645	2,100	0.0008	4.904	1.42

[a] Unit runoff rates were obtained from Fig. 4-10.
[b] Assumed inlet time.

Note:

$m \times 3.2808 = ft$ $\qquad$ $mm \times 0.03937 = in$

$ha \times 2.4711 = acre$ $\qquad$ $m/s \times 3.2808 = ft/s$

$m^3/ha \cdot d \times 106.9064 = gal/acre \cdot d$

$m^3/d \times 2.6417 \times 10^{-4} = Mgal/d$

$m^3/s \times 35.3147 = ft^3/s$

raphy is unfavorable, and where the water tables are high or structurally unstable soil or rocky conditions occur, pumping stations and force main sewers (a force main is a pipe under pressure) may be needed. To overcome these difficulties, both pressure and vacuum sewers have been developed as alternatives. Because these systems are continually being improved, it is advisable that current manufacturers' literature and operating data be obtained when considering the use of these sewers. If possible, visits should be made to existing installations.

Pressure Sewers

The main components of a pressure sewer system are shown in Fig. 4-12*a*. In most pressure sewer systems, wastewater from individual buildings is collected in a gravity sewer and discharged into a holding tank. From the holding tank the wastewater is discharged periodically into a pressure main by means of a grinder pump that can grind the solids in the wastewater. A holding tank and pump are required at each inlet point to the pressure main. To reduce capital and maintenance costs, a single holding tank and pump can be used for several homes. Wastewater from the pressure main is discharged either into a gravity line or to the influent facilities of the treatment plant. Typical design data on pressure sewer systems are reported in Table 4-10.

A pressure system eliminates the need for small pumping stations and makes it possible to substitute a small-diameter plastic pipe for a much larger-diameter conventional pipe [9]. However, all of this is accomplished at the expense of having to install a grinder pump at each inlet to the pressure main. In addition to the initial cost of the pumps, the associated power and maintenance expenses must be considered.

Vacuum Sewers

The principal features of a vacuum sewer system are shown in Fig. 4-12*b*, and operational data are reported in Table 4-10. In these systems, wastewater from an individual building flows by gravity to the location of a vacuum ejector (vacuum valve of special design). The valve seals the line leading to the main so that the required vacuum levels can be maintained in the main. When a given amount of wastewater accumulates behind the valve, the valve is programmed to open and close after the wastewater enters as a liquid plug. Vacuum pumps in a central pumping station maintain the vacuum in the system. The pumping station is usually near the treatment facilities or any other convenient discharge point.

Application of Pressure and Vacuum Sewers

In a report prepared for the National Commission on Water Quality [9], it was concluded that pressure and vacuum systems offer some advantages over con-

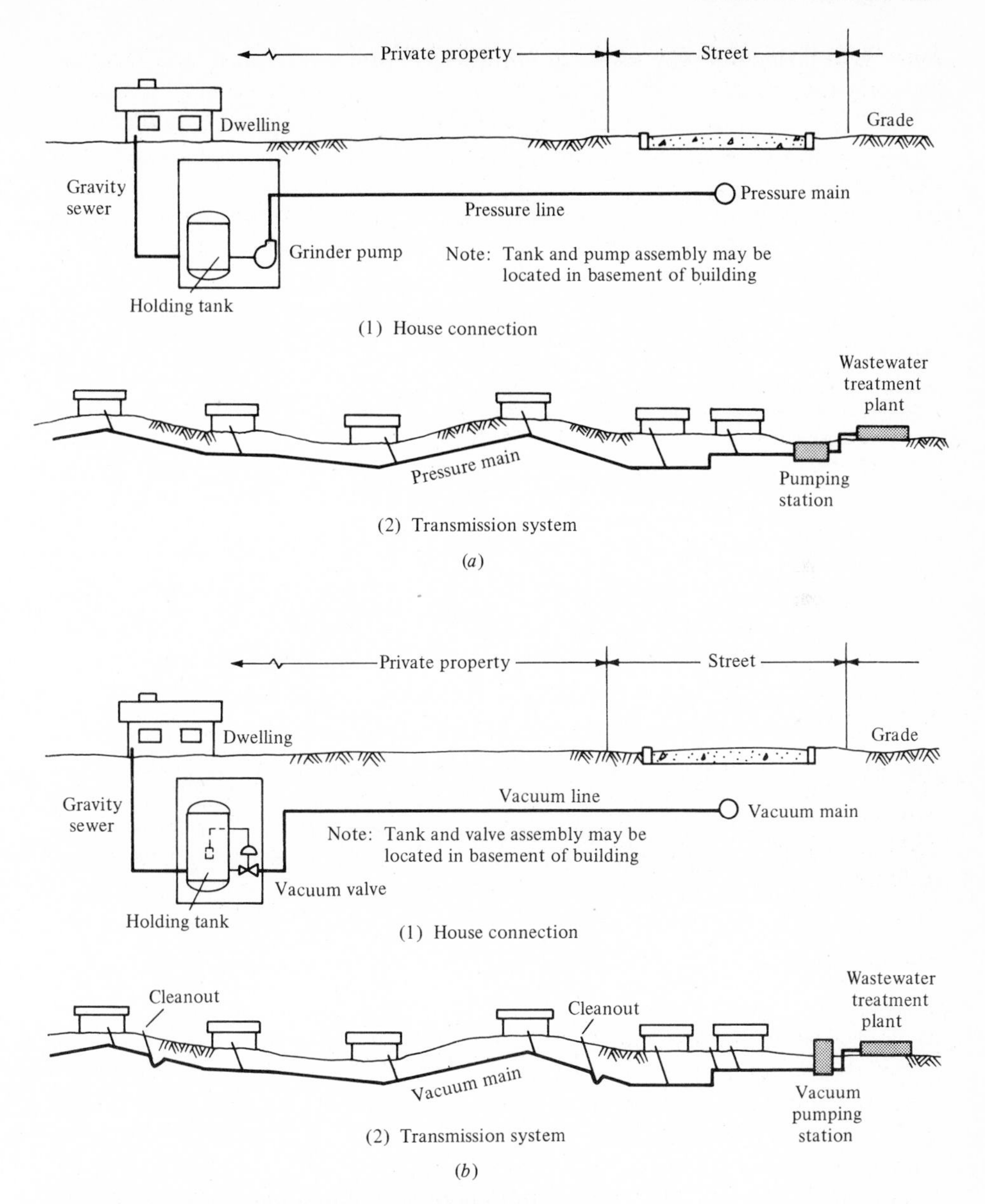

Figure 4-12 Identification sketch for the principal components of pressure and vacuum sewer systems (*a*) Pressure sewer. (*b*) Vacuum sewer.

ventional gravity sewers when certain soil types and terrain conditions are present. A pressure sewer connected to a residence too low to be served by a gravity sewer is illustrated in Fig. 4-13. As more and more areas are served by sewers, the need to find more cost-effective alternatives will continue to increase, and these systems cannot be overlooked.

Table 4-10 Typical design data for pressure and vacuum sewer systems

	Range	Typical
Pressure sewers		
Grinder pump, kW	0.75–1.5	1.12
Grinder pump discharge pressure (gage), kN/m^2	200–275	240
Size of line from pump to pressure main, mm	25–50	30
Size of pressure main, mm	50–300	[a]
Vacuum sewers		
Height of water level on vacuum discharge valve, mm	75–1,000	750
Size of line from discharge line to vacuum main, mm	75–125	100
Vacuum maintained in collection tank at pumping station, mm Hg	300–500	400

[a]Varies with location in system.
Note: $kN/m^2 \times 0.1450 = lb_f/in^2$
$mm \times 0.03937 = in$

When a pressure or vacuum sewer system is proposed, the question of ownership, maintenance, and repair of the equipment should be discussed. A stock of spare parts and the services of qualified personnel should be readily available, whether from the municipal sewer department or from an independent agency. Responsibilities and liabilities for the failure of the equipment and the adequacy of maintenance and repair should be defined clearly.

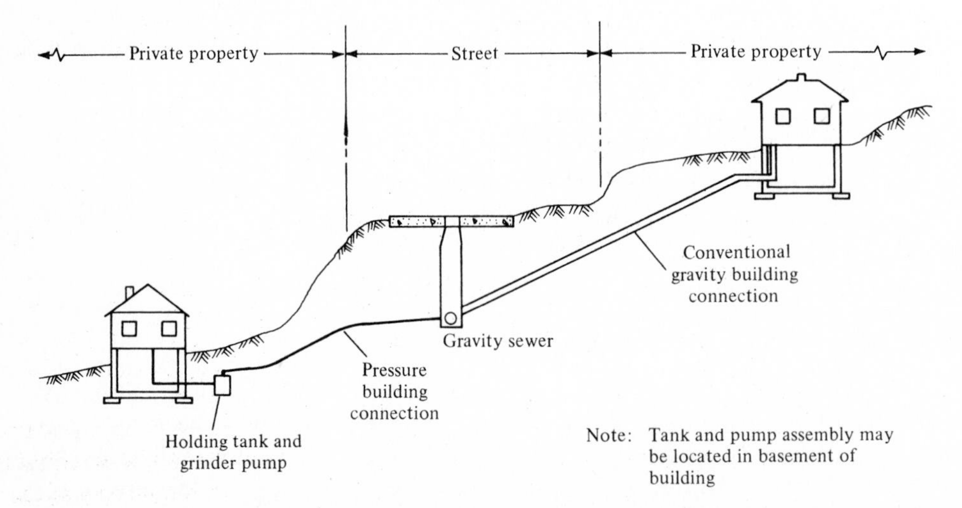

Figure 4-13 Typical pressure connection to gravity sewer from isolated low building site.

One problem with these systems is the lack of knowledge about them. According to the National Commission report [9]:

> What is required is (1) the awareness that alternatives do exist so that appropriate cost-effectiveness comparisons can be made, and (2) the willingness (or incentive) to make these comparisons.

DISCUSSION TOPICS AND PROBLEMS

4-1 Trace the path of the sewer leading from your home or school building to the local wastewater-treatment plant. Prepare a sketch (not to scale) of the layout and classify the types of sewers in the collection system you have identified.

4-2 Summarize the steps involved in designing a separate sanitary sewer and discuss their relative importance.

4-3 Develop a preliminary design, including flowrates, pipe sizes, and pipe slopes for the trunk sewers to be constructed in the development shown in Fig. 4-14. Flowrate determinations and design conditions are to be based on the information from Examples 3-1 and 4-1. Assume that the average park attendance is 350 persons/d and the flow is 60 L/person · d. The peaking factor for the park is 4.0. Also assume that the hospital is a 400-bed facility employing about 150 people and has a peaking factor of 4.0.

4-4 Develop a preliminary design, including flowrates, pipe sizes, and pipe slopes, for the trunk sewer to be used in the development shown in Fig. 4-14. Flowrate determinations and design conditions are to be based on the following information. Expected saturation population densities and wastewater flows for the various types of housing are given in Table 4-11. Commercial and industrial wastewater flow is estimated to be 28 and 56 $m^3/ha \cdot d$, respectively. Average peaking factors are 1.5 for the commercial flows and 3.0 for the industrial flows. The school is to serve 2500 students at ultimate capacity, with an average flow of 80 L/students · d and a peaking factor of 4.0. Hospital and park flows are the same as those listed in Prob. 4-3.

4-5 Develop a preliminary design, including flowrates, pipe sizes, and pipe slopes, for the trunk sewer to be used in the Covell Park Development shown in Fig. 4-4. However, underground facilities prevent line 7 (point 7 to point 8) from being considered. An alternative plan with a sewer down 20th Avenue and out of the development is proposed. Also estimate the wastewater flows from the information presented in Prob. 4-4.

4-6 Develop a preliminary design, including flowrates, pipe sizes, and pipe slopes, for the trunk sewer to be used in the development shown in Fig. 4-15. Assume that the ground slopes down toward the southwest to point *a*. Estimate the wastewater flowrates from the information presented in Example 4-1.

4-7 Develop a preliminary design, including flowrates, pipe sizes, and pipe slopes, for the trunk sewer to be used in the development shown in Fig. 4-15. Assume that the ground slopes down

Table 4-11 Saturation population densities and wastewater flows for Prob. 4-4

Zoning	Type of development	Saturation population density, person/ha	Wastewater flow, L/capita · d
Residential	Single-family dwellings	30	380
Residential	Duplexes	50	310
Residential	Low-rise apartments	100	260

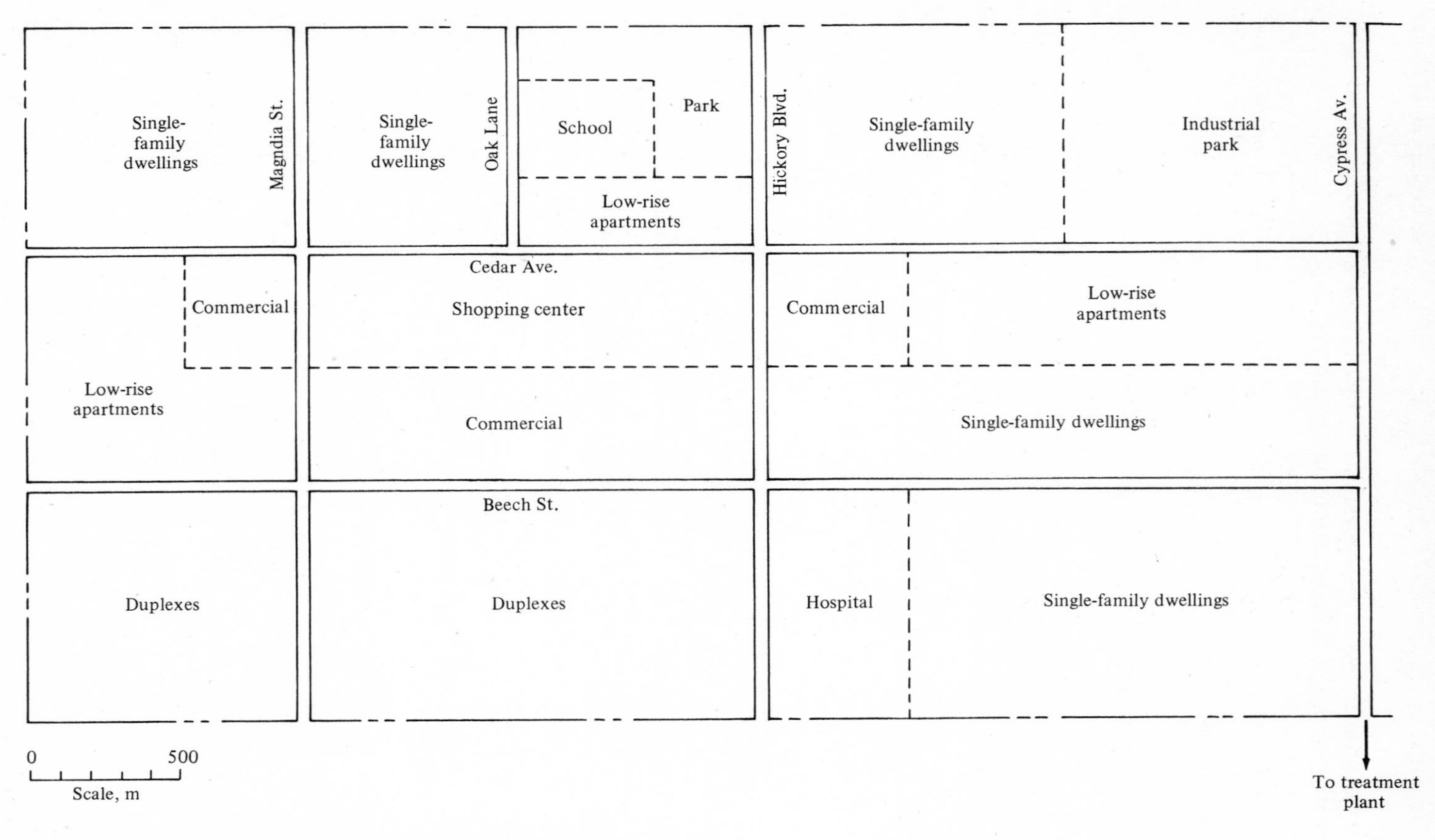

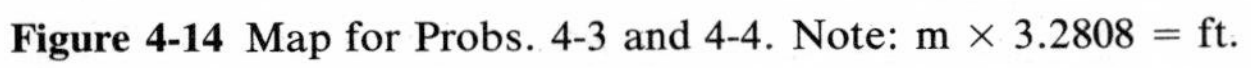

Figure 4-14 Map for Probs. 4-3 and 4-4. Note: m × 3.2808 = ft.

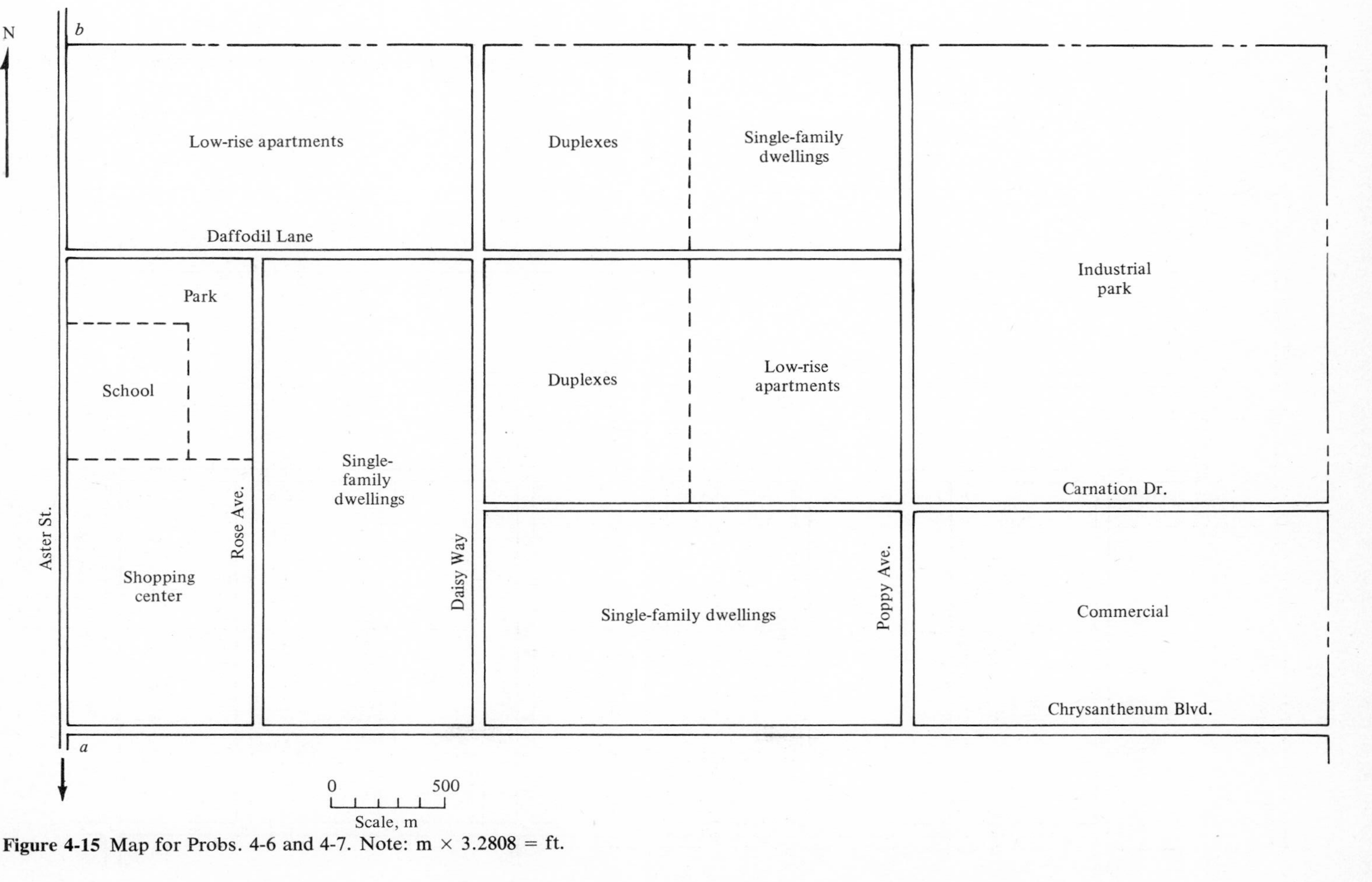

Figure 4-15 Map for Probs. 4-6 and 4-7. Note: m × 3.2808 = ft.

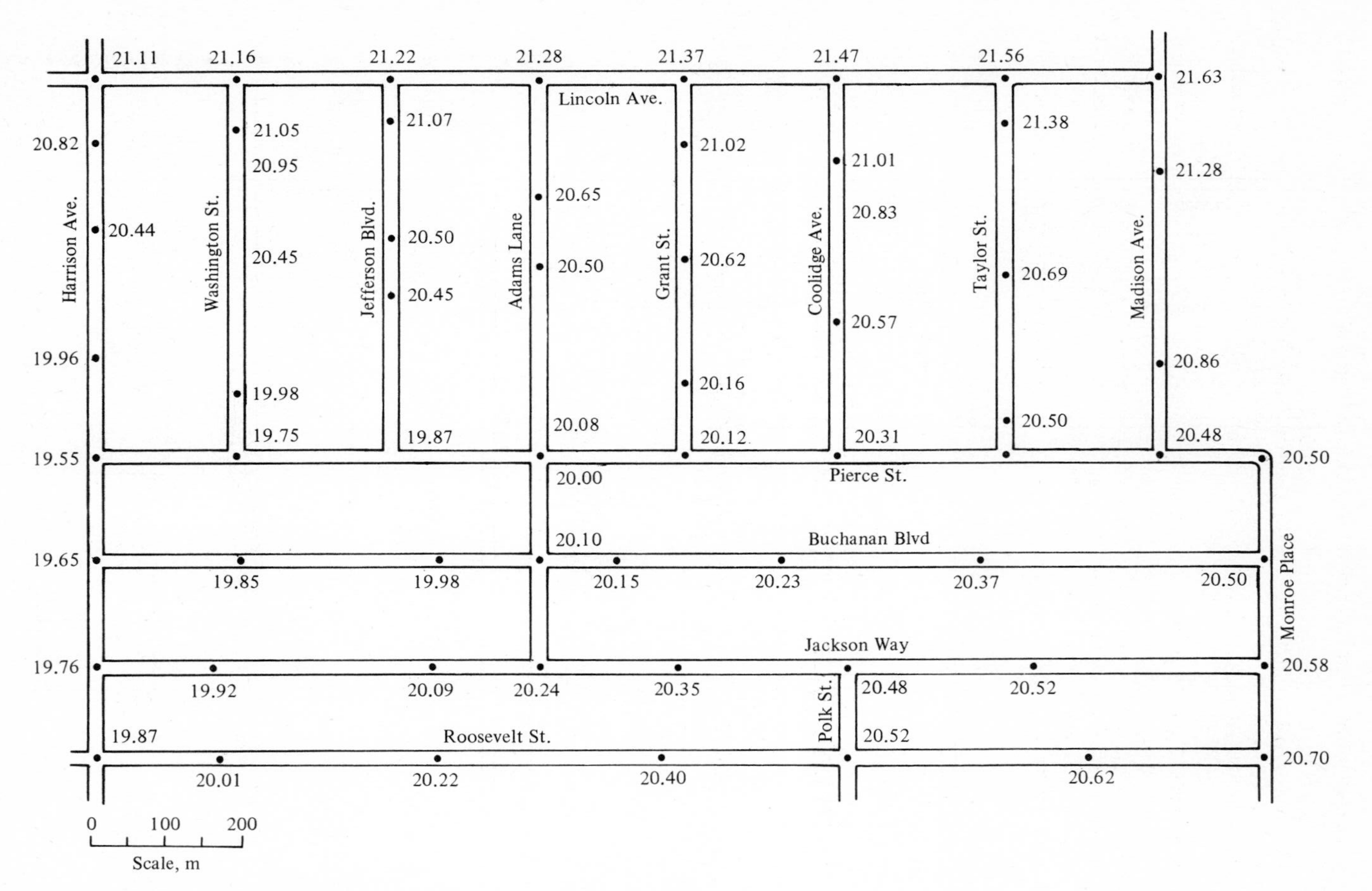

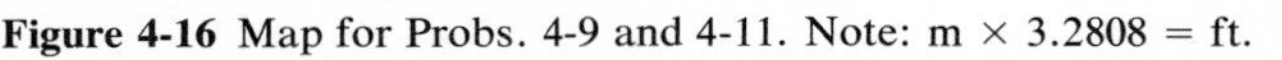

Figure 4-16 Map for Probs. 4-9 and 4-11. Note: m $\times$ 3.2808 = ft.

toward the northwest to point b. Estimate the wastewater flowrates from the information presented in Prob 4-4.

4-8 Referring to Fig. 4-5, design a sewer starting at manhole 2 in Maple Street and ending at manhole 11 in Center Street. Prepare a design table similar to the one used in Example 4-2 and draw a profile. Flowrate determinations and design constraints are the same as those given in Example 4-2.

4-9 Referring to Fig. 4-16, design a gravity-flow sanitary sewer to serve Pierce Street starting at Madison Avenue and ending at Harrison Avenue. Prepare a sewer design table and draw a profile. The future average saturation population density is estimated to be 32 persons/ha and the average flowrate is estimated to be 380 L/person · d. Minimum size of sewer pipe is 150 mm. Pipe thickness is 20 mm and the minimum depth of cover is 1 m.

4-10 Referring to Fig. 4-17, design a gravity-flow sanitary sewer starting at the intersection of 4th

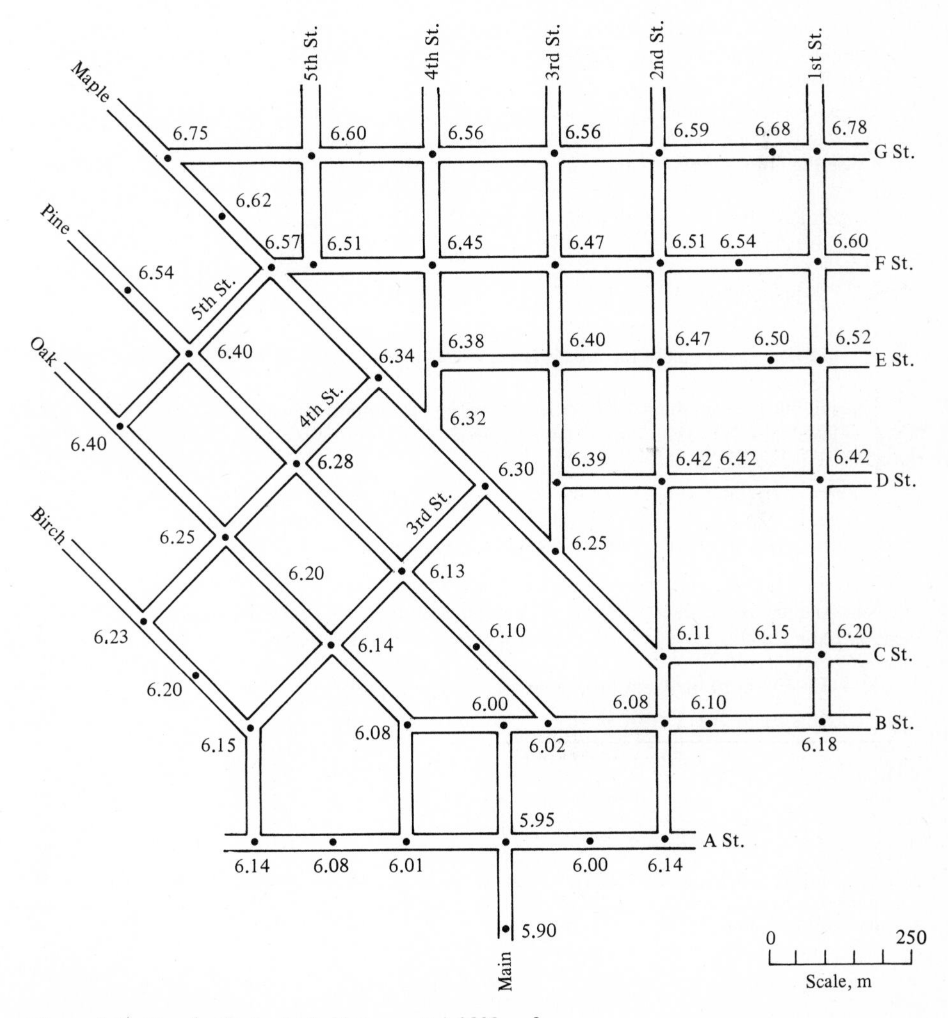

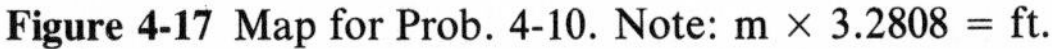
Figure 4-17 Map for Prob. 4-10. Note: m × 3.2808 = ft.

and G streets and ending at the intersection of 2d and B streets. Prepare a sewer design table and draw a profile. Estimate wastewater flows from the data given in Prob. 4-9.

4-11 Compute the volume of excavation and length of pipe of various diameters required for the sewer shown in Fig. 4-6. Assume that the width of the trench is 1.4 times the inside diameter plus 0.3 m. The minimum width of the trench is 1 m, and, to allow for pipe-bedding material, the depth of excavation is to be 0.2 m below the invert of the sewer.

4-12 Compute the volume of excavation and length of pipe of various diameters required for the sewer designed in Prob. 4-8. Use the design constraints given in Prob. 4-11.

4-13 Compute the volume of excavation and length of pipe of various diameters required for the sewer designed in Prob. 4-9. Use the design constraints given in Prob. 4-11.

4-14 Compute the volume of excavation and length of pipe of various diameters required for the sewer designed in Prob. 4-10. Use the design constraints given in Prob. 4-11.

4-15 List six methods of determining storm-water quantities and briefly discuss their distinguishing characteristics.

4-16 A parking lot at a large shopping center has an area of 4.3 ha. During a storm with an average rainfall intensity of 12 mm/h, the runoff from the parking lot is 0.12 m^3/s. Calculate the average runoff coefficient.

4-17 Prepare a design table like the one shown in Table 4-9 for the storm-water sewer in Forest Avenue from the manhole near Jefferson Boulevard to the manhole at the intersection with Center Street (see Fig. 4-9). Manholes are to be positioned as shown in Fig. 4-9 and may be numbered as desired.

Compute the rainfall intensity for the area with the following equation:

$$i = 300/t^{0.5}$$

where i = rainfall intensity, mm/h
t = rainfall duration, min

Assume that 55 percent of the area is impervious and use Fig. 4-10 to obtain the coefficients of runoff.

The minimum sewer diameter is to be 300 mm and the minimum velocity is to be 0.9 m/s when flow is at full depth. Use Manning's equation with $n = 0.013$ for design calculations. The inlet time can be assumed to be 20 min.

4-18 Prepare a preliminary gravity-flow storm-water sewer design for the Covell Park Development shown in Fig. 4-4. Use the information given in Table 4-12 to determine the runoff flows. Compute the rainfall intensity with the following equation: $i = 200/t^{0.65}$

4-19 Referring to Fig. 4-16, design the gravity-flow storm-water sewer for Pierce Street starting at Madison Avenue and ending at Harrison Avenue. Determine the runoff quantities with the data given in Example 4-4. Prepare a sewer design table and draw a profile.

Table 4-12 Percent imperviousness in each zoning area for Prob. 4-18

Zoning	Percent of impervious area
Residential	
Single family	55
Duplexes	60
Low-rise apartments	75
Commercial	95
Shopping center	100
Industrial	98

4-20 If the storm-water sewer designed in Prob. 4-19 operated as a combined sewer until the separate sanitary sewer were constructed, determine the velocity in the storm-water (combined) sewer under peak dry-weather-flow conditions by using the design data given in Prob. 4-9 or the design flows if Prob. 4-9 has been solved.

4-21 Obtain some sewer contract drawings from the engineering department of your community or school and compare them with Fig. 4-7. How are they similar? What are the major differences, if any?

4-22 A low-pressure air test is to be conducted on a 450-mm sewer. If the test section is 150 m long and has no building connections, what is the minimum allowable time for a decrease in pressure from 24 to 17.1 kN/m^2?

4-23 If the sewer section mentioned in Prob. 4-22 also has 100 m of 150-mm building connections and 200 m of 100-mm building connections, what would be the minimum allowable time for the specified pressure drop?

4-24 A sewer main with an inside diameter of 600 mm has just been constructed in a high-groundwater area and is to be tested by the infiltration test. What is the maximum allowable flow at the downstream end of a 150-m test section for compliance with a specification of 20 L/(mm of diam · km · d)?

4-25 A new sewer main with an inside diameter of 200 mm was tested using the exfiltration test. A section of sewer 100 m long and one manhole on each end was isolated and filled with water for the test. The manholes have an inside diameter of 1.3 m. After 24 hr, the water level measured in the upstream manhole dropped 0.52 m. The contract specifications require a maximum exfiltration rate of 20 L/(mm of diam. · km · d). Was the specification met?

4-26 Referring to Fig. 4-13, estimate the power required to pump the wastewater from the holding tank to the gravity sewer located in the street. The static head between the low level in the holding tank and gravity sewer is 4 m. The inside diameter and the length of the plastic pressure line from the holding tank to the gravity sewer are 30 mm and 60 m, respectively. Assume that the minor losses and pump losses together will be about 1.5 m. The minimum scouring velocity required to avoid the deposition of solids in the plastic pipe can be estimated by the following relationship:

$$V_s = 0.061\sqrt{\frac{d}{2}}$$

where V_s = minimum scouring velocity, m/s
d = inside diameter of pipe, mm

4-27 Estimate the power required in Prob. 4-26 if the inside diameter of the plastic pipe is 25 mm and the length of the pipe is 80 m.

REFERENCES

1. American Society of Civil Engineers: *Hydrology Handbook,* Manual 28, 1949.
2. Hobbs, S. H., and L. G. Cherne: Air Testing Sanitary Sewers, *J. Water Pollut. Control Fed.,* vol. 40, no. 4, 1968.
3. Joint Committee of the American Society of Civil Engineers and the Water Pollution Control Federation: *Design and Construction of Sanitary and Storm Sewers,* ASCE Manual and Report 37, New York, 1969.
4. Kerri, K. D., and J. Brady (eds.): *Operation and Maintenance of Wastewater Collection Systems,* prepared for the U.S. Environmental Protection Agency, Office of Water Program Operation, by Department of Civil Engineering, California State University, Sacramento, 1976.
5. Lager, J. A. and W. G. Smith: *Urban Stormwater Management and Technology: An Assessment,* U.S. Environmental Protection Agency, Report 670/2-74-040, NTIS no. PB 240 687, 1974.

6. Lager, J. A., et al.: *Urban Stormwater Management and Technology: Update and Users Guide,* U.S. Environmental Protection Agency, Office of Research and Development, Report 600/8-77-014, Cincinnati, Ohio, 1977.
7. Linsley, R. K., Jr., M. A. Kohler, and J. L. H. Paulhus: *Hydrology for Engineers,* 2d ed., McGraw-Hill, New York, 1975.
8. Metcalf & Eddy, Inc.: *Storm Water Management Model,* U.S. Environmental Protection Agency, Water Quality Reports nos. 11024 DOC 07/71 through 11024 DOC 10/71, 4 vols., 1971.
9. Metcalf & Eddy, Inc.: *Report to National Commission on Water Quality on Assessment of Technologies and Costs for Publicly Owned Treatment Works,* vol. 1, prepared under Public Law 92-500, Boston, 1975.
10. Ramseier, R. E., and G. C. Riek: Experience in Using the Low-Pressure Air Test for Sanitary Sewers, *J. Water Pollut. Control Fed.,* vol. 38, no. 10, 1966.
11. Ramseier, R. E.: Testing New Sewer Pipe Installations, *J. Water Pollut. Control Fed.,* vol. 44, no. 4, 1972.
12. Storm Drainage Research Committee: *The Design of Storm Water Inlets,* The Johns Hopkins University Storm Drainage Research Project, 1956.
13. Tchobanoglous, G.: *Review of Proposed Modification of Wildcat Hill Wastewater Treatment Facility,* report prepared for the city of Flagstaff, Ariz., Davis, Calif., 1975.
14. Wanielista, M. P.: *Stormwater Management: Quantity and Quality,* Ann Arbor Science Publishers, Inc., Ann Arbor, Mich., 1978.

CHAPTER

FIVE

SEWER APPURTENANCES

All wastewater and storm-water collection systems are composed of two major components: (1) sewer pipe and (2) various structures, castings, and hardware items. The latter group of components is identified collectively as sewer appurtenances. These are used to ensure that the collection system will function as designed and so that it can be inspected and maintained in good working order.

The purpose of this chapter is to present and discuss the design and application of the principal appurtenances used with collection systems including (1) manholes, building connections, and flushing devices; (2) street inlets and catch basins; (3) junctions and transitions, depressed sewers (commonly called) *inverted siphons*), and vertical drops and energy dissipators; (4) overflow and diversion structures; (5) regulating devices; and (6) outlets.

5-1 MANHOLES, BUILDING CONNECTIONS, AND FLUSHING DEVICES

The principal appurtenances of gravity-flow sanitary sewers are manholes, building connections, and flushing devices. These structures are discussed in this section. Street inlets and catch basins as well as common appurtenances for storm-water sewers are discussed in Sec. 5-2.

Manholes

Although manholes are the most familiar appurtenance of a wastewater collection system, they were not used extensively until some time after many large sewers had been constructed (about 1880). Manholes were installed to facilitate the removal of grit and silt that collected on the inverts of sewers with low velocities. Before that time, when a sewer became so badly clogged that it needed cleaning, it was customary to excavate down to the sewer, break through its walls, remove the obstruction, and then close in the sewer again.

After the value of manholes on small sewers was recognized, it was determined that there should be no change of grade or alignment in a sewer between points of access to it, unless the sewer was large enough to enable a worker to pass through it readily. However, in some cases, the manholes were placed too close together. This practice is objectionable because the cost is unnecessary and traffic can damage pavement where the roadway has manhole frames.

Manhole size. Manholes should be large enough to provide easy access to the sewer. The clearance opposite the steps should be sufficient for a worker to pass up or down without difficulty. In small sewers—less than 600 mm (24 in) in diameter—there should be enough room for a worker to handle a shovel, and the bottom should afford footing for a person working in the manhole, but still drain to the sewer. The manholes in small sewers are usually about 1.2 m (4 ft) in diameter when the sewers are circular cross sections.

For sewers larger than 600 mm, larger manhole bases are needed, although the same size of manhole barrel is usually used for all sewers. Special conditions sometimes require a larger size, such as when a large gaging or cleaning device must be lowered through it for use at the bottom of the manhole. Examples of a manhole with cast-in-place base and a precast reinforced-concrete manhole base for vitrified-clay sewer pipe are shown in Figs. 5-1 and 5-2, respectively. Manholes used for other types of pipe are similar.

Access structures on large sewers are sometimes constructed so that a boat or scraper can be lowered into the sewer. The boat may be used for inspection, and the scraper may be used to loosen material that forms a coating on the inside surface of a sewer. Where the sewer is much larger than the diameter of the manhole, the outside of the manhole is usually set tangent to one side of the sewer; otherwise, a worker would have difficulty entering the sewer and would need a ladder to reach the top of the sewer. On very large sewers, manholes are sometimes separate from the sewer, and shafts are built to lead into it.

Manhole spacing. Manholes for smaller sewers—600 mm (24 in) in diameter and less—should be placed at intervals not greater than 100 m (350 ft). For sewers 700 to 1200 mm (27 to 48 in) in diameter, the maximum interval should be 120 m (400 ft). For sewers larger than 1200 mm (48 in), manholes may be placed at somewhat greater intervals, depending on local circumstances, such as breaks in grade and the location of street intersections. In any case, the length between any two manholes should not exceed the length that can be cleaned with the equipment expected to be used by the local sewer department or agency responsible for maintaining the collection system.

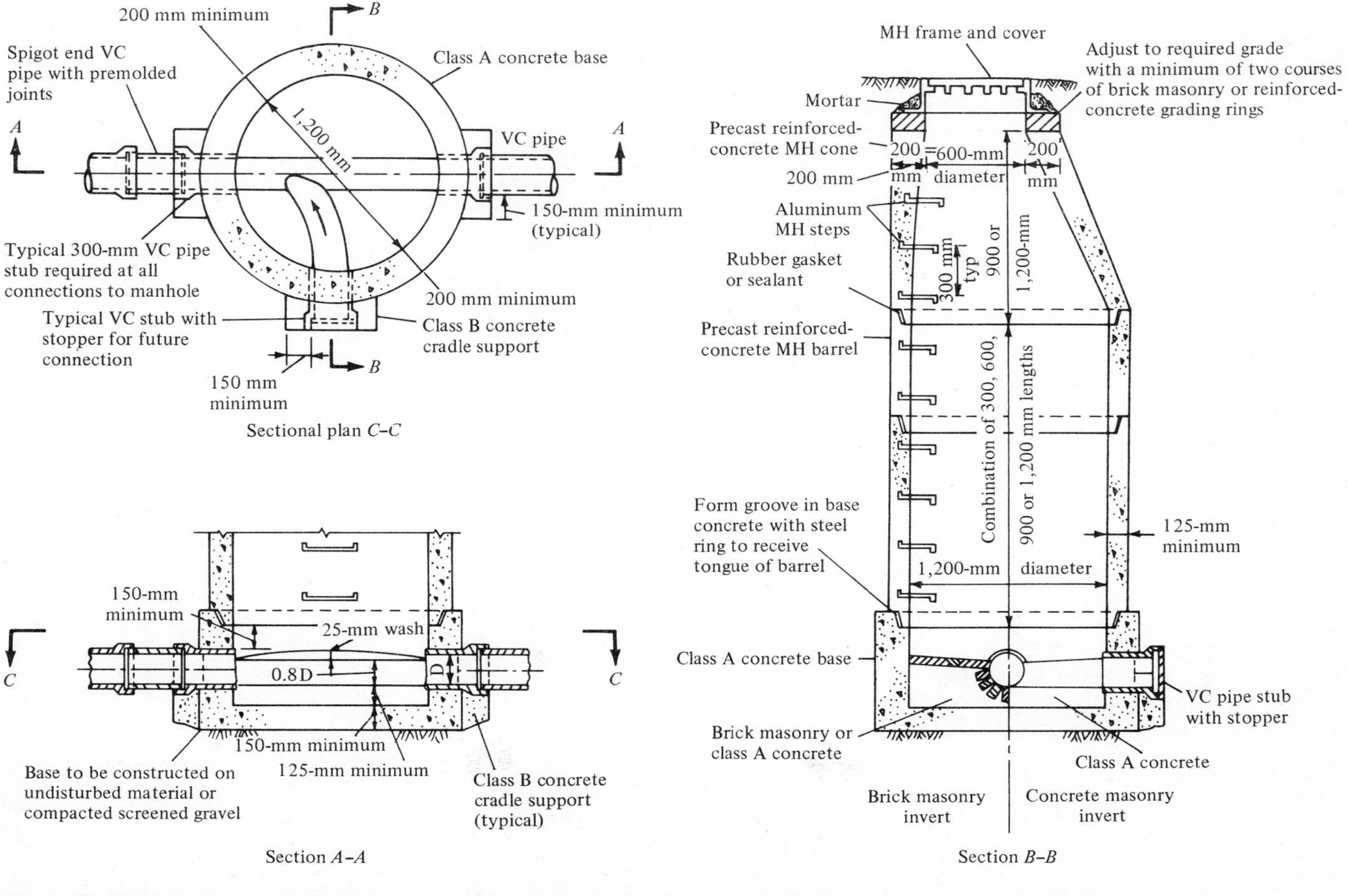

Figure 5-1 Manhole with cast-in-place base for vitrified-clay sewers 600 mm in diameter and smaller. Note: mm × 0.03937 = in.

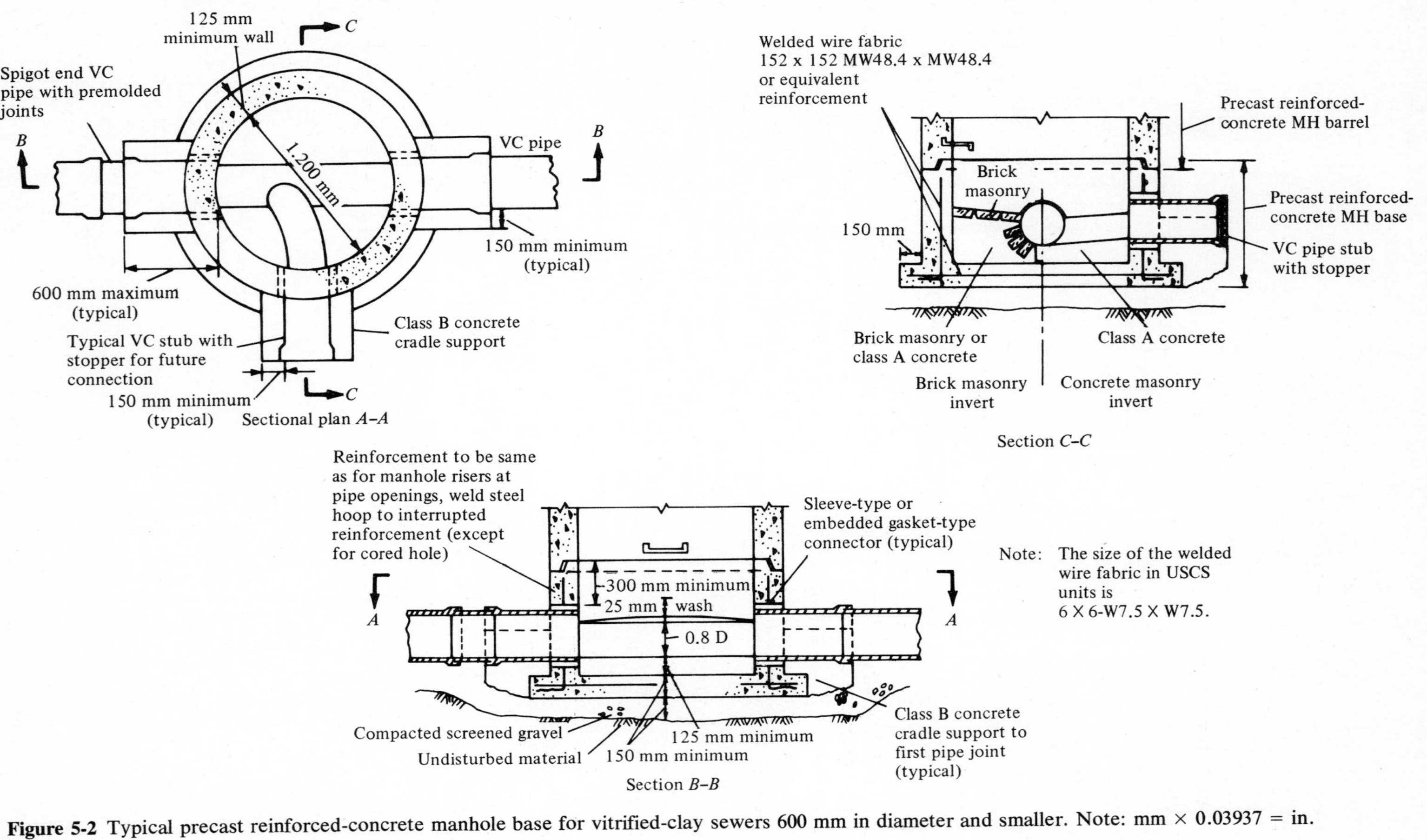

Figure 5-2 Typical precast reinforced-concrete manhole base for vitrified-clay sewers 600 mm in diameter and smaller. Note: mm × 0.03937 = in.

Transitions and turns in sewer manholes. Changes in size or shape of the cross section of the sewer at a manhole produce disturbances in flow with accompanying loss of head. Gradual transitions can reduce these head losses. If the sidewalls of the channels in manholes are extended to the crown of the sewer, lower head losses will result at high flows. There is no uniform practice in this detail, however; the berm at the top of the sidewall has been placed from a point mid-depth of the sewer to a point level with the crown.

In sewers 600 mm (24 in) and smaller, a turn of 90° or less may be made in one standard manhole. In sewers ranging from 700 to 1200 mm (27 to 48 in), a 90° turn may be apportioned between two manholes, each located at least two manhole diameters from the point of intersection, with a straight alignment from manhole to manhole. The more common practice is to place one large manhole with a specially designed base at the point of intersection. Curves may be used for sewers larger than 1200 mm, by employing pipe sections cast with a bevel on one end or monolithic (cast-in-place) structures. The centerline radius for monolithic concrete sewers should be from four to eight times the diameter of the sewer to reduce energy losses to a minimum. The minimum practical centerline radius for a sewer made of bevel-end pipe is about 15 m (50 ft).

Where a turn is made in a manhole, compensation for the curve loss is desirable. Where curves in a sewer alignment are used, compensation for the required additional head loss should be made. However, because of the approximate nature of all hydraulic computations for sewer design, these refinements are seldom included. One method of compensating for the curve loss is to compute the hydraulic grade line along each tangent (to the point of intersection of the forward and back lines) and to apply a steeper slope to the sewer along the length of the curve, thus using the additional available head. In other words, the total change of elevation around the curve would be the same amount computed along the tangents.

Manhole construction. Practically all sewer manholes constructed in the past were made of brick masonry, often with cast-in-place concrete bases. Many of these older manholes are still in place, and they frequently become sources of significant volumes of groundwater infiltration into the sewer.

The need for reducing this infiltration, as well as the high cost of labor for brickwork and the large amount of time required for in-place construction, has led to the development of precast concrete manhole sections (see Fig. 5-3). The sections are now used almost exclusively for sewer manholes. When the sections are manufactured by the best modern methods and assembled properly in the trench, the manholes should be nearly leak-free, especially if modern elastomeric pipe-to-manhole connectors join the sewer pipes to precast manhole bases. Factory-built manholes of glass-reinforced plastics are also available and may be suitable for some applications.

Manhole steps. In the past, brick manholes had forged or cast-iron steps that were built into the brickwork or concrete of the manhole during construction. Current practice is to make the steps of aluminum alloy, plastic-coated

Figure 5-3 Unassembled precast-concrete manhole components.

steel, or other suitable material. The steps are either embedded in the walls of precast sections during manufacture or inserted into the walls immediately after casting, preferably with plastic inserts. The details of securing steps to the walls of precast manhole sections depend on the method used for casting the concrete sections.

Whenever manholes have aluminum steps, the step portions in the wall must be insulated from the concrete to prevent deterioration of the metal by interaction with the concrete. This insulation may be accomplished by coating the portions of the steps to be embedded with a heavy bituminous material. Plastic inserts will also serve the same purpose. Zinc chromate, which has sometimes been used for this purpose, has proved unsatisfactory because it chips easily, thus leaving areas of metal unprotected, and because it tends to saponify when in contact with cement in the presence of water.

Regardless of the techniques used to affix the steps to the manhole walls, they must be securely embedded and capable of supporting the person using them. Steps must be set in a continuous vertical alignment to form a ladder, with rungs uniformly placed from 300 to 400 mm (12 to 16 in) apart vertically. In some communities, manhole steps are not allowed, and portable ladders are required for access to the sewers through the manhole.

Manhole frames and covers. Factors that should be considered in selecting manhole frames and covers include:

1. Safety, so that covers will not slip off
2. Convenience of repair and replacement, necessitated by the wear from traffic

3. Strength sufficient to withstand the impact from wheel loads
4. Freedom from rattle and noise
5. Cost
6. Possibility of adjustment when a new wearing course is added to existing pavements to correct for unevenness
7. Appearance
8. Protection against entrance of storm-water runoff and lighted cigars and cigarettes
9. Protection by locking devices against removal to prevent the dumping of liquid or solid wastes into the opening

The cover should be flat and even with the pavement so that it will neither interfere with traffic nor cause excessive wear of the pavement. Its design should be standardized so that it can be readily replaced if it is stolen or broken. The cover should be corrugated or provided with bosses (raised parts) to have a nonskid surface. Circular tops are in almost universal use for sewer manholes. They are inherently stronger than rectangular tops and have the advantage that they cannot be dropping into the manhole.

Frames are usually 150 to 300 mm (6 to 12 in) high, depending partly on the type of pavement in which they are installed. Design practice as to clear opening varies widely; a 600-mm (24-in) cover, allowing a 550-mm (21-in) clear opening, generally is satisfactory. In some cases, the opening may be 750 mm (30 in) in diameter. This size allows workers to enter the sewer with a portable ladder. The weight of frames generally ranges from 115 to 230 kg (250 to 500 lb); the weight of covers generally ranges from 45 to 70 kg (100 to 150 lb).

Manhole frames and covers are usually of cast iron, but semisteel or cast steel may be used. At one time it was common to install perforated manhole covers for ventilation. To avoid plugging with sticks and dirt, the holes were usually larger on the inside than on the outside. Manholes for sanitary sewers should not have perforated covers because of the possible inflow of storm water as well as the possible escape of odors and visible vapors.

Drop manholes. Where the difference in elevation between inflow and outflow sewers exceeds about 0.5 m (1.5 ft), flow from the inflow sewer may be dropped to the elevation of the outflow sewer by an outside or inside connection known as a *drop manhole* (also known as a *drop inlet*), as shown in Figs. 5-4*a* and 5-4*b*. The purpose of drop manholes is to protect workers when they enter the manhole and to avoid the splashing of wastewater with the consequent liberation of objectionable gases, many of which may be odorous.

The dimensions of the pipe fittings establish the minimum vertical drop possible at a manhole. This minimum distance is about 530 mm (21 in) for a 200-mm (8-in) sewer. Because of unequal earth pressures, resulting from backfilling operations near the manhole, the entire outside drop connection is usually encased in concrete.

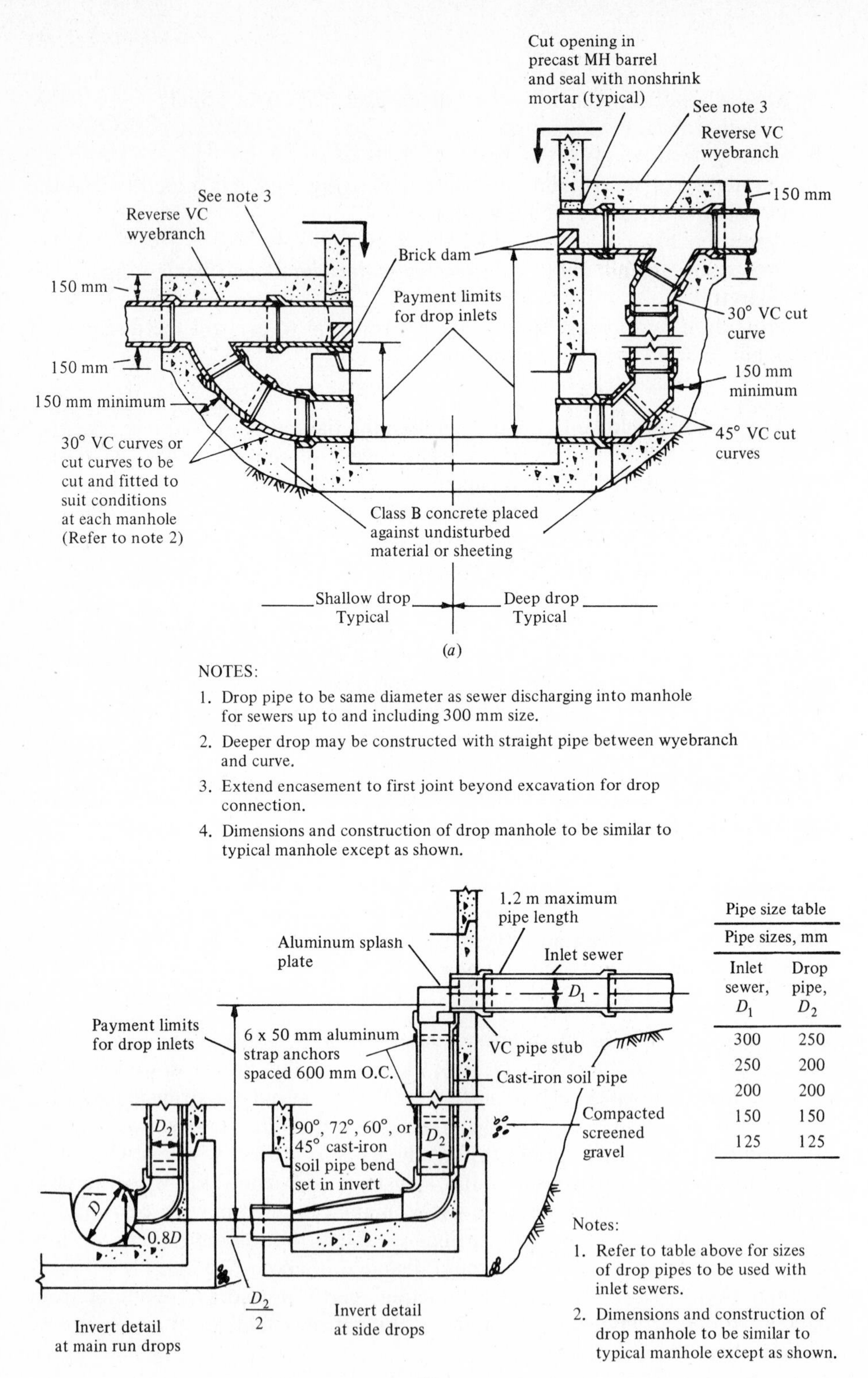

Pipe size table

Pipe sizes, mm	
Inlet sewer, D_1	Drop pipe, D_2
300	250
250	200
200	200
150	150
125	125

Building Connections

Building connections, also called *house* or *building sewers* or *house connections,* are small-pipe sewers leading from buildings to the public sewer in the street (see Fig. 5-5). In some places, the builder or owner is responsible for the construction of these sewers and the connection to the public sewers. In others, the municipality installs the portion of the work in the public street or even the entire connection from the street sewer to the building.

Most cities require the use of cast-iron pipe for the house sewer for a distance of 1 or 2 m outside the wall of the building. Regardless of the type of pipe used, care must be taken to support the pipe properly so that it will not be damaged by subsequent settling. If the pipe passes through the foundation wall, a joint should be provided at the outside face of the wall.

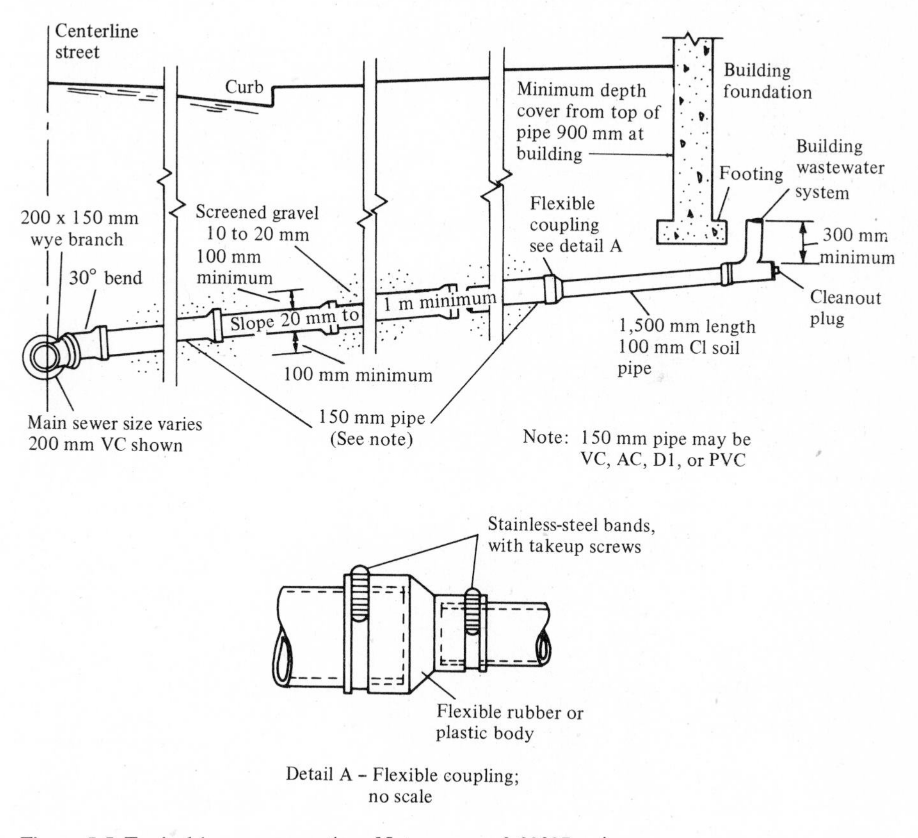

Figure 5-5 Typical house connection. Note: mm × 0.03937 = in.

Figure 5-4 Typical drop inlets for vitrified-clay pipe. (*a*) Outside drop. (*b*) Inside drop for sewers 600 mm in diameter and smaller. Note: mm × 0.03937 = in.

The minimum size of house connection pipes should be 100 mm (4 in); 125 or 150 mm (5 or 6 in) sizes are preferable. The minimum slope for a connection is usually fixed by local regulation, and less than 20 mm/m (0.25 in/ft) is rarely permitted or desirable.

The building connection enters the sewer at a wye-branch fitting for the smaller size sewers or at a T-branch fitting for the larger size sewers (see Fig. 5-6). Where the sewer is in a deep trench, a vertical pipe encased in concrete, called a *chimney,* is sometimes used to allow the house sewer to be placed at a depth that can be constructed economically. The house sewer is then connected to the branch at the top of the chimney. Typical details of a sewer chimney are shown in Fig. 5-7.

Flushing Devices

In the past, flushing devices were used so that sewers could be laid on relatively flat slopes which were inadequate to produce scouring velocities. Such devices are almost never used now. Current design practice is to construct sewers on

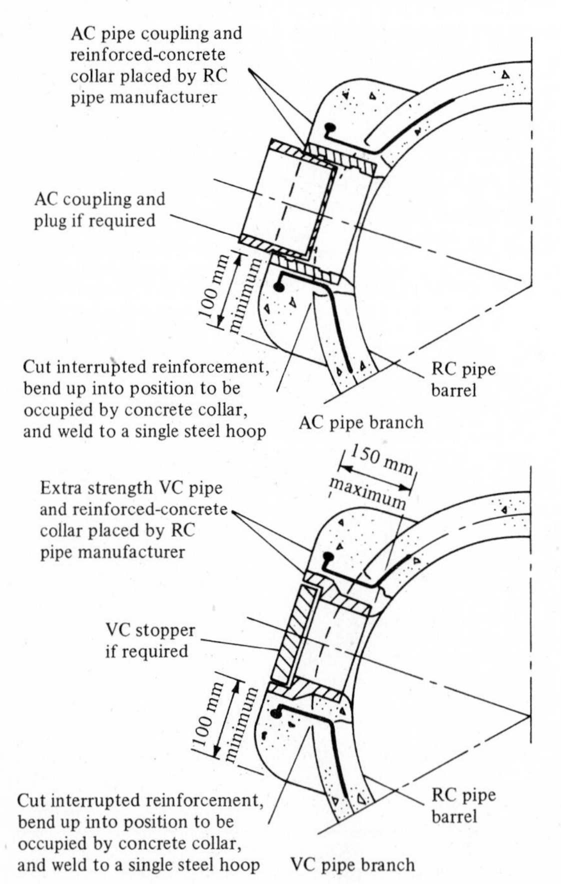

Figure 5-6 T-branch for building or chimney connection in reinforced-concrete pipe. Note: mm × 0.03937 = in.

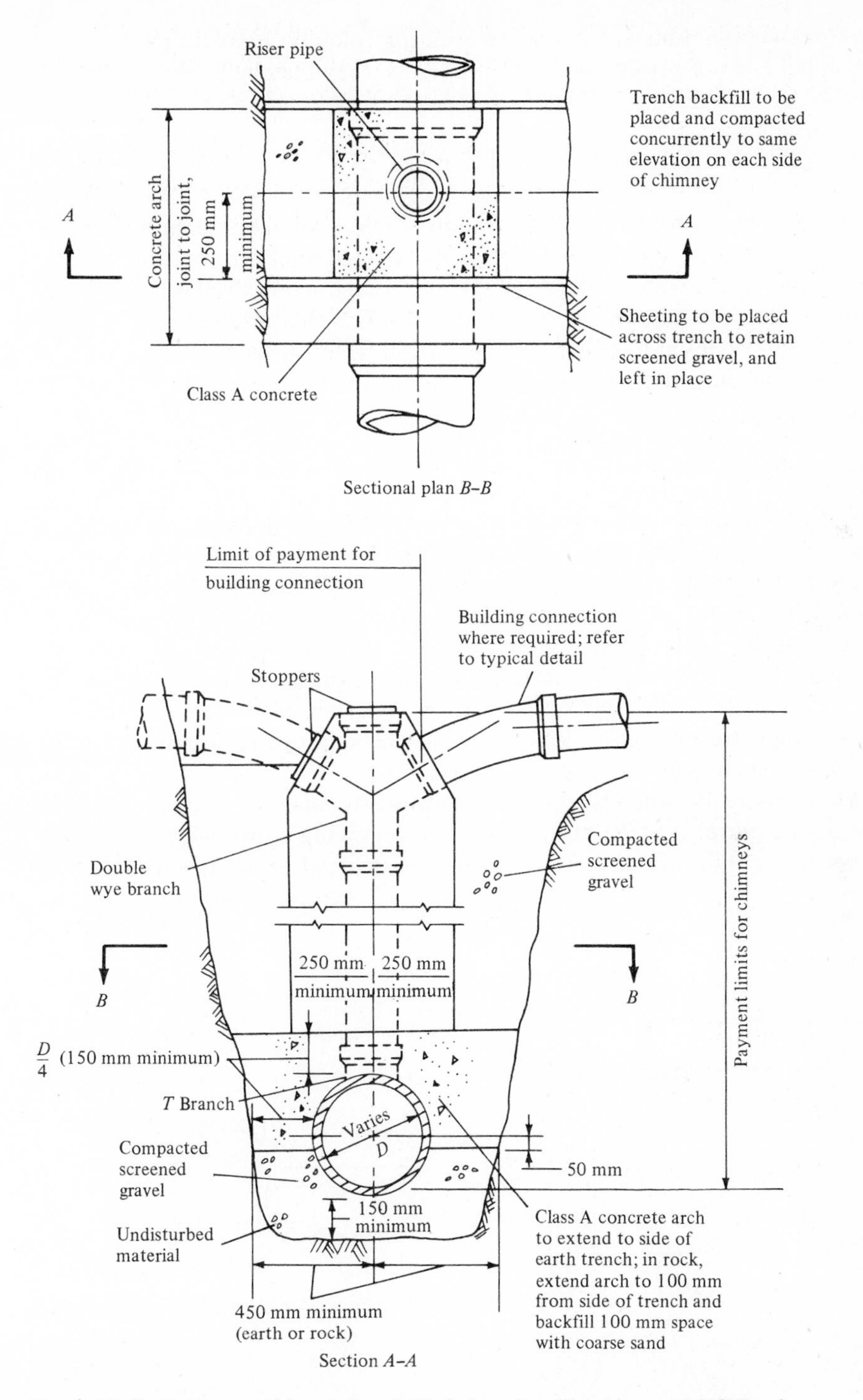

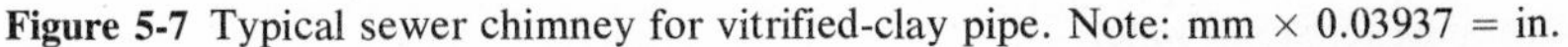

Figure 5-7 Typical sewer chimney for vitrified-clay pipe. Note: mm × 0.03937 = in.

slopes that will function with adequate scouring velocity. If solids are deposited in the upper ends of sewers or in sewer laterals, they may be flushed out with a fire hose.

5-2 STREET INLETS AND CATCH BASINS

Storm water that is not lost by evaporation or percolation drains into the street gutters and is removed at suitable intervals through inlets connecting with the underground storm-water collection system. These inlets discharge either directly into storm-water sewers or into catch basins that are intended to intercept the solid wastes and sediment flushed from street surfaces. They should be located so that they interfere as little as possible with pedestrian and vehicular traffic.

Street Inlets

Street inlets are used to transfer storm water to the storm sewer. Because the inlets can be clogged by the debris carried by the storm water, care should be taken to minimize the potential for clogging. Clogging can occur at the opening through which water enters the street inlet or in the trap or pipe connecting the inlet to the sewer. Objects that cause the most trouble are sticks, wastepaper, and leaves. If sticks become lodged against the opening, the leaves and wastepaper drawn to it by the next flush of storm water are likely to cause a blockage. Openings presenting the least possible obstacle to the entrance of these three classes of wastes have been widely used.

Where there is much accumulation of wastepaper or leaves, a grating placed in the gutter may be of little value in removing storm water because it may become either partially or completely obstructed at the first runoff from

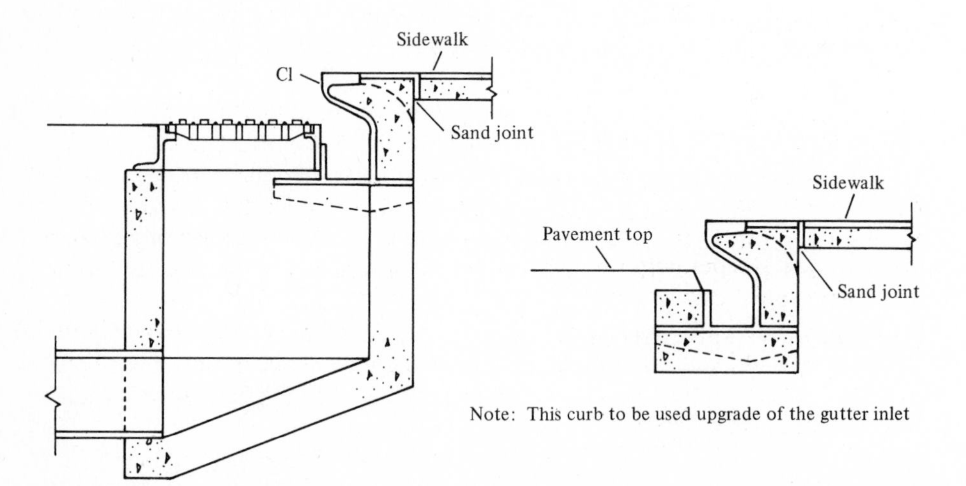

Figure 5-8 Street inlet for storm water.

the storm. In most cases, it is advisable to provide an opening in the curb adequate to remove the total flow accumulating at the inlet. Cleanliness of streets and adjacent sidewalks is essential to the successful functioning of inlets.

Types of inlets. There are three general types of storm-water inlets: (1) curb inlets, (2) gutter inlets, and (3) combination inlets. A curb inlet is a vertical opening in the curb through which the flow passes. A gutter inlet is an opening in the gutter beneath one or more grates through which the gutter flow falls. A combination inlet includes both a curb opening and a gutter opening, with the gutter opening directly in front of the curb opening. The two openings may be overlapping or offset.

At low points, the clear opening of an inlet is critical. Since minimizing the clogging is the most important design consideration, curb inlets are more desirable in such locations. In addition, two hydraulic conditions can occur at low points: inflow caused by weir action or orifice action. The rate of inflow varies with the depth of the gutter flow to the two-thirds power in the weir action and to the one-half power in the orifice action. Consideration should be given to this fact when selecting the type of inlet at low points. As a general rule, gutter inlets are used on steep grades, and curb inlets are used on flat grades with steep cross slopes.

Depressed gutter inlets may interfere with vehicular traffic and present a hazard to pedestrians. Moreover, their capacity is rarely much greater than that of undepressed gutter inlets. Gutter inlets, especially those with depressions, are more likely to clog than curb inlets. Curb inlets with depressions close to the curb present only slight interference to vehicular traffic and slight hazard to pedestrian traffic, yet they have greater capacity than undepressed curb inlets. When designing storm-water systems, the allowable capacities of curb and gutter inlets are sometimes taken as 10 to 30 percent smaller than the nominal capacities to allow for reduced capacity caused by clogging [11].

Inlet castings and assembled gratings. As with manhole frames and covers, the configuration of inlet castings varies considerably in different communities, but the tendency is to adopt a few standard types which meet the conditions usually encountered in practice. Typical examples of inlet castings are shown in Fig. 5-9*a*. Where a larger inlet opening may be required, either because of a higher runoff rate or a steeper gutter grade, assembled gratings (see Fig. 5-9*b*) are used rather than typical cast grates.

The inlet gratings shown in Fig. 5-9*a* are suitable for highways where bicycles are not allowed. For city streets and other areas where bicycle riding is anticipated, grates with crossbars or diagonal bars spaced to limit dropdown of a bicycle wheel should be used (see Fig. 5-9*b*).

Hydraulics of inlets. Hydraulic conditions in street gutters, the various types of inlets, and the methods of construction make hydraulic design of such facilities

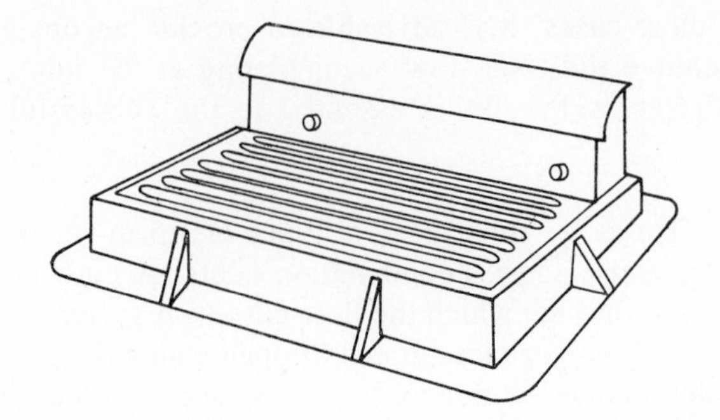

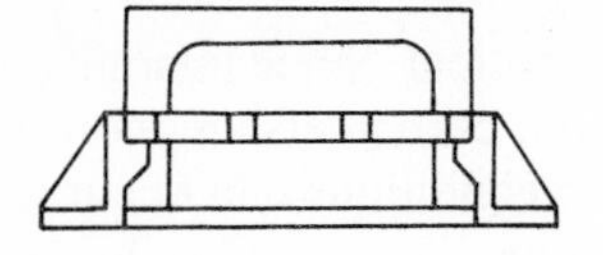

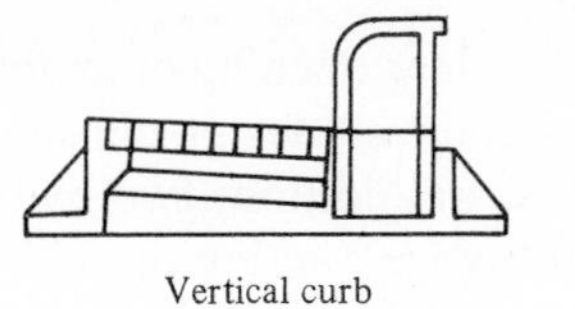

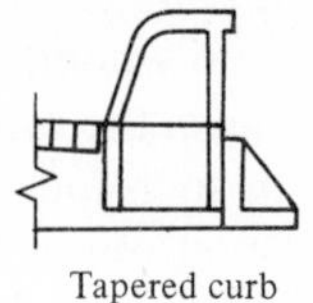

(*a*)

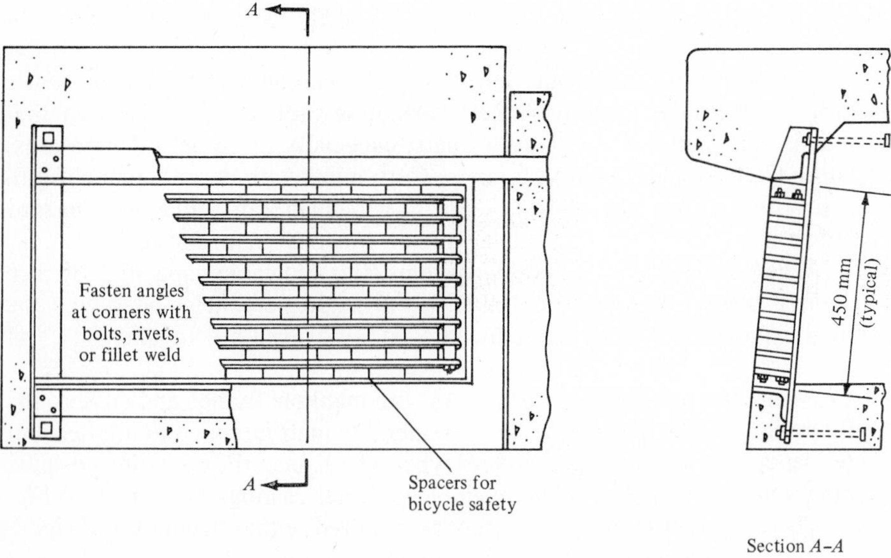

(*b*)

Figure 5-9 Typical manufactured gutter inlets. (*a*) Various inlet castings. (*b*) Assembled grating. Note: mm × 0.03937 = in.

difficult. Field tests are necessary for accurate determinations. Several studies have been conducted to provide design factors for special conditions that may be extrapolated, with due caution, for general design. Refs. 3, 4, 5, and 16 provide data on inlet design.

Catch Basins

The catch basin (see Fig. 5-10) was once considered an essential part of a combined wastewater or storm-water collection system. The velocity of the wastewater in many sewers was often insufficient to prevent the formation of sludge deposits, and it was much more expensive to remove this sludge from the sewers than from catch basins. This situation was especially critical when streets were crude and uneven, and little attention was given to keeping them clean. The sewers themselves were not laid with the present regard for self-cleansing velocities. Under such conditions it was natural that catch-basins were more common than now. Durable, smooth pavements, more efficient street cleaning, and sewers laid on steeper slopes have reduced the need for catch basins to a few special situations.

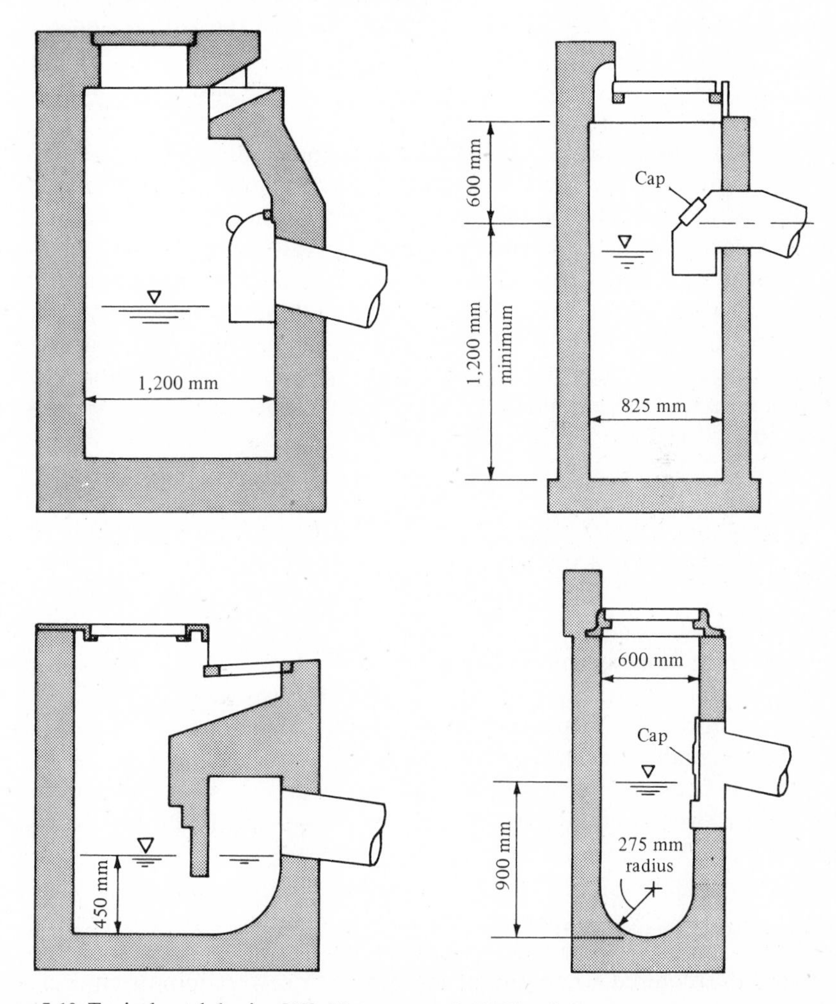

Figure 5-10 Typical catch basins [10]. Note: mm $\times$ 0.03937 = in.

Where the slopes of storm-water sewers are sufficient to ensure self-cleansing velocities, street flushing is a common method of cleaning. Where streets are cleaned by flushing, catch basins become filled with solid wastes more quickly. The low flowrates associated with sewer flushing would have little effect on the flushing of solids from the catch basin. However, when the catch-basin sump is approximately one-half full of solids, the solids removal efficiency is reduced considerably, even for small storm flows, and solids are carried into the sewer.

The required frequency of catch-basin cleaning depends on several local conditions, such as sump capacity, quantity of accumulated street solids, antecedent dry period, meteorological conditions, street-cleaning methods and practices, surrounding land use, topography, and the erodibility of the soils subject to washoff. It could be more expensive to remove the wastes accumulating in the catch basins than to leave them in the gutters and remove them by mechanical street-sweeping or vacuuming methods. Since a large portion of the solid wastes will reach the sewer if a catch basin is allowed to fill, self-cleaning slopes are still essential.

Although catch basins are not needed at frequent intervals as was formerly believed, their use may be advantageous where a large quantity of grit will probably be washed into the inlet. Cleaning should not be neglected until the inlet is clogged and the attendant flooding occurs; it should be done regularly. Catch basins need not be cleaned when there is little accumulation in them unless the nature of the deposit creates offensive odors and annoys persons passing by or living nearby. Under certain conditions, catch basins may become breeding places for mosquitoes. Additional information derived from a detailed evaluation of catch basins and their performance may be found in Ref. 10.

5-3 JUNCTIONS AND TRANSITIONS, DEPRESSED SEWERS, AND VERTICAL DROPS AND ENERGY DISSIPATORS

In addition to the common appurtenances used in collection systems, there is often a need for more specialized appurtenances. Junction and transition structures are required on large sewers so that the wastewater will flow with minimum disturbance. Depressed sewers (inverted siphons) are used in stems to overcome conditions imposed by local topography, such as crossing depressions or river beds [8]. Vertical drops and energy dissipators are used to reduce the excess energy contained in wastewater, such as at the bottom of a steep sewer.

Junctions and Transitions

Junctions are required when one or more branch sewers join or enter a main sewer. If the sewers are small, the changes in size, direction, and slope may be made at manholes (see Fig. 5-11). Junction chambers provided with manholes should be used when the sewers to be joined are too large to be accommodated in a manhole (see Fig. 5-12).

Regardless of the type of junction, great care must be taken to avoid conditions which may contribute to the development of excessive turbulence (eddy formation) and accumulations of sludge, rags, and other debris found in waste-

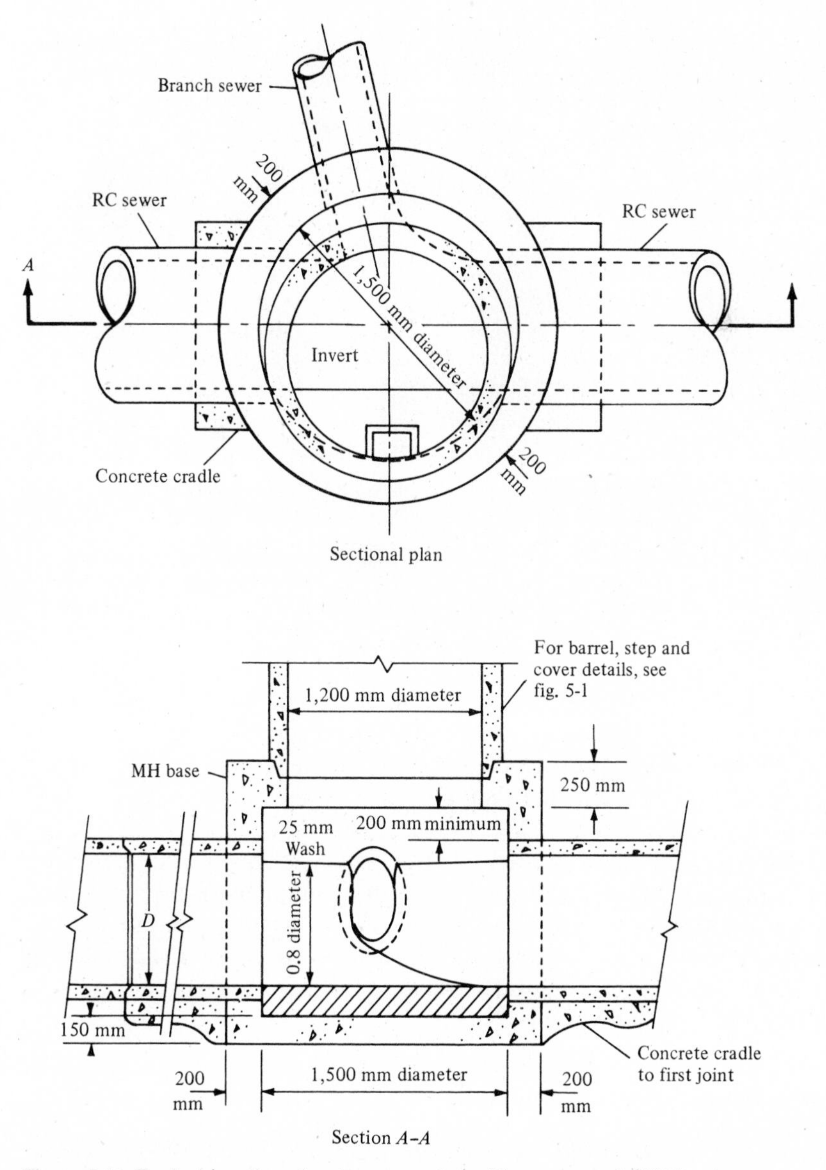

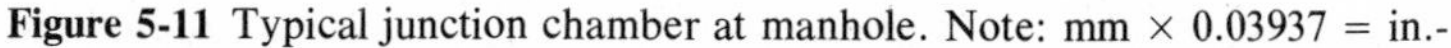

Figure 5-11 Typical junction chamber at manhole. Note: mm × 0.03937 = in.-

Figure 5-12 Construction of large sewer junction chamber (*Courtesy Sacramento County, CA*). Note: mm × 0.03937 = in.

water. The reduction of turbulence is important if the release of dissolved gases is to be minimized. To limit these problems, all junction transitions should be smooth, channels should be constructed in the manhole bottoms, and the invert of smaller sewers should be higher than the invert of the larger sewer to which they join.

Transition structures are needed where changes in the channel geometry (hence the state of the flow) must be made with a minimum of turbulence and energy loss. For example, a transition structure would be required to go from supercritical to subcritical flow and the reverse, as encountered in sudden expansions and contractions. If a transition structure is to be designed, Refs. 4, 5, 7, and 8 in Chap. 2 should be consulted.

Depressed Sewers (Inverted Siphons)

Any dip or sag introduced into a sewer to pass under structures, such as conduits or subways, or under a stream or across a valley, is often called an *inverted siphon*. It is misnamed, for it is not a siphon. The term *depressed sewer* is suggested as more appropriate. Since the pipe constituting the depressed sewer is below the hydraulic grade line, it is always full of water under pressure, although there may be little flow in the sewer. A depressed sewer and its associated inlet and outlet chambers are shown in Fig. 5-13. The design of the depressed sewer shown in Fig. 5-13 is given in Example 5-1.

Because of practical considerations, such as the increased danger of blockage in small pipes, minimum diameters for depressed sewers are usually about the same as for ordinary sewers: 150 or 200 mm (6 or 8 in) in sanitary sewers, and about 300 mm (12 in) in storm-water sewers. Since obstructions are much more difficult to remove from a depressed sewer than from a sewer, special care should be taken to prevent their formation. The velocity in the depressed sewer should be as high as is practicable, that is, about 0.9 m/s (3 ft/s) or more for domestic wastewater, and 1.25 to 1.5 m/s (4 to 5 ft/s) for storm water. The use of several pipes, instead of one pipe, for the depressed sewer is also advantageous. To maintain reasonable velocities at all times, the pipes are arranged so that additional pipes are brought into service progressively as wastewater flows increase.

Design features. Catch basins or grit chambers have sometimes been built upstream of depressed sewers, but they are difficult to clean, and the material removed from them usually has an offensive odor. Depressed sewers should be flushed frequently and inspected regularly to ensure prompt removal of obstructions.

Smaller pipes are sometimes clogged by sticks caught in the bends. When clogging occurs or is likely to occur, racks should be arranged in front of the inlets to the smaller pipes so that the material collected on them is washed off and carried through the larger pipes during high flows.

A depressed sewer may be flushed or cleaned in various ways, depending on the available facilities and surrounding conditions. The wastewater may be backed up and then released; clean water may be admitted at the head of the sewer; or the sewer may be drained and cleaned by hand, using joint rods with suitable scrapers or other tools.

Manholes or cleanout chambers should be installed at each end of a depressed sewer to provide access for rodding and pumping, and, in the case of pipes of larger size, for entrance. Placing intermediate manholes on a depressed sewer is objectionable if the wastewater will be free to rise in them, since grease and other scum tend to accumulate in the manhole. However, manholes are advantageous if the wastewater is confined within the pipes as it passes through them. They allow access or are a means of removing the deposit in the depressed sewer through a gated connection or similar device.

Since a depressed sewer is subject at all points of its cross section to an internal pressure, the pipe walls will be in tension, although the magnitude of the tension will be affected by external water and soil pressures. Because of these tensile stresses, depressed sewers are usually constructed of ductile-iron,

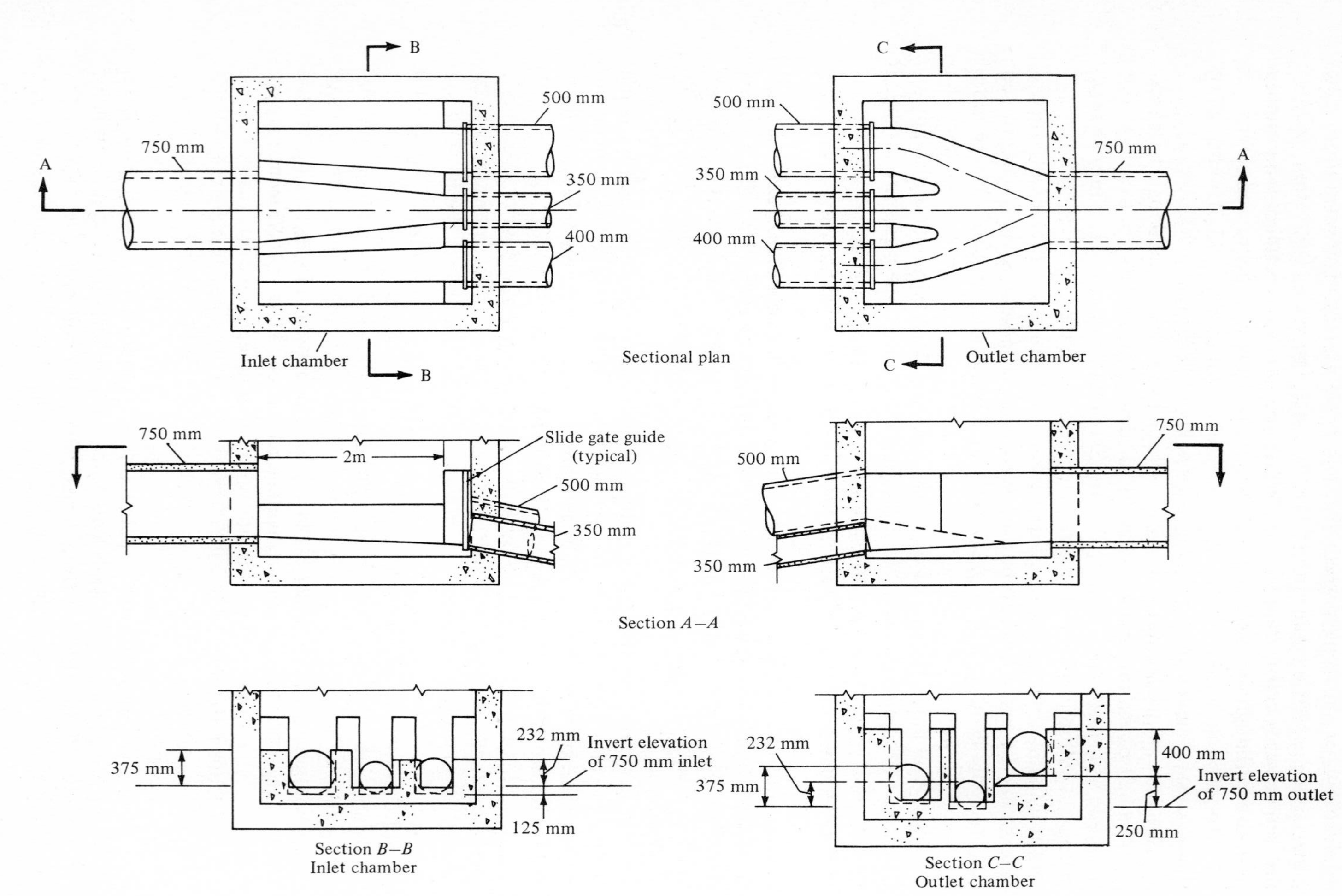

Figure 5-13 Definition sketch for depressed sewer. Note: mm × 0.03937 = in.

reinforced-concrete, or other pressure-rated pipe. Depressed sewers built on or under river beds should have sufficient weight or anchorage to prevent their flotation when empty, a condition that might occur during the construction period or when a pipeline has been dewatered for repairs.

Design computations. The computation of the pipe sizes for depressed sewers is the same as that for sewers and water mains. The diameter depends on the hydraulic grade line and the maximum flow of wastewater to be carried. The head loss or drop in the hydraulic grade line actually required at any time for the flow will be the difference in level in the free water surfaces at the two ends. It will equal the sum of the friction head and other minor losses. The losses are relatively small for low velocities, but increase roughly with the square of the velocity. For a clean 300-mm (12-in) depressed sewer that is 15 m (50 ft) long and has a velocity of 0.9 m/s (3 ft/s), a total loss of 150 mm (6 in) would probably be a maximum figure. For a velocity of 0.6 m/s (2 ft/s), the total loss would be 75 mm (3 in); but for velocity of 1.8 m/s (6 ft/s), the total loss would be 600 mm (2 ft). The head loss caused by the friction would be about one-half of these values and the remainder would be attributable to minor losses.

Example 5-1 Design of a depressed sewer Design a depressed sewer, consisting of several pipes and an inlet and outlet chamber (see Fig. 5-13), to replace an existing single-pipe depressed sewer that has not performed properly because low velocities in the depressed sewer have increased sedimentation. Use ductile-iron pipes for the depressed sewers. The basic data and assumptions are as follows:

1. Length of depressed sewer = 135 m (440 ft)
2. Maximum sewer depression = 2.75 m (9.0 ft)
3. Diameter of reinforced-concrete gravity sewer to be connected by depressed sewer = 750 mm (30 in)
4. Slope of existing 750-mm (30-in) sewer = 0.0022 m/m (ft/ft)
5. Minimum velocity in depressed sewer = 0.9 m/s (3.0 ft/s)
6. Design flows:
   (*a*) Minimum flow = 0.110 m^3/s (4 ft^3/s)
   (*b*) Maximum dry-weather flow = 0.255 m^3/s (9 ft^3/s)
   (*c*) Ultimate maximum flow equals the capacity of the gravity sewer
7. Available fall (invert to invert) = 1100 mm (3.6 ft)
8. Assumed head loss at inlet = 120 mm (0.4 ft)
9. Available head loss for friction in depressed sewer = 980 mm (3.2 ft)
10. Available hydraulic grade line (approximate) = 0.98 m/135 m = 0.0073 m/m

Because depressed sewers can become coated with grease or other materials that tend to reduce the sewer capacity, it is generally advisable to use a value of 0.015 for Manning's n. The pipe sizes selected should be particularly adapted to the minimum flow, the maximum dry-weather flow, and the ultimate maximum flow.

Three depressed-sewer pipes are often selected to meet these requirements, with regulation at the inlet so that the minimum flow will be confined to one pipe, the maximum dry-weather flow will be confined to two pipes, and the combined capacity of the three pipes will be equivalent to that of the influent sewer (see Fig. 5-13).

SOLUTION

1. Design the depressed sewer.
   a. Determine the capacity of the existing 750-mm sewer. The capacity of the 750-mm concrete gravity sewer laid on a slope of 0.0022 m/m is 0.522 m^3/s, as computed by using Manning's equation with an n value of 0.013. The velocity at this flowrate is about 1.18 m/s.
   b. Determine the size of the depressed-sewer pipe for low-flow conditions. The required capacity of the first pipe is 0.110 m^3/s. From Fig. 2-14, the capacity of a 350-mm ductile-iron pipe ($n = 0.015$) with a hydraulic grade line of 0.0073 m/m is 0.108 m^3/s. The velocity would be 1.12 m/s.
   c. Determine the size of the second depressed-sewer pipe required for the maximum dry-weather flow. Because the capacity of the 350-mm pipe is 0.108 m^3/s, the required capacity of the second pipe is 0.147 m^3/s (0.255 m^3/s − 0.108 m^3/s). From Fig. 2-14, the capacity of a 400-mm ductile-iron pipe ($n = 0.015$) with a hydraulic grade line of 0.0073 m/m is 0.154 m^3/s. The velocity would be 1.23 m/s. The capacity of the 350-mm pipe and the 400-mm pipe is 0.262 m^3/s, beyond which a third pipe begins to operate.
   d. Determine the size of the third depressed-sewer pipe required to handle the ultimate maximum flow. The ultimate maximum flow is 0.522 m^3/s. Since the combined capacity of the 350- and 400-mm pipes is 0.262 m^3/s, the required capacity of the third pipe is 0.260 m^3/s. From Fig. 2-14, it will be found that the required pipe diameter is between 450 and 525 mm.

      A diameter of 500 mm is chosen because this is the standard ductile-iron pipe size in this range. On the basis of Manning's equation, the capacity of a 500-mm ductile-iron pipe ($n = 0.015$) with a hydraulic slope of 0.0073 m/s is 0.280 m^3/s. The velocity would be 1.42 m/s. The combined capacity of 350- and 400-mm pipes is 0.262 m^3/s, and the total available capacity is 0.542 m^3/s (0.522 m^3/s is required).
2. Design the inlet chamber.
   a. Referring to Fig. 5-13, the invert of the 350-mm sewer is continuous with that of the 750-mm gravity sewer, with a wall on each side of a central channel extending from the 750-mm sewer to the 350-mm pipe. One wall crest corresponds to the flow depth in the 750-mm sewer with the minimum flow. The other wall crest corresponds to the depth in the 750-mm sewer with the maximum dry-weather flow. Thus the minimum flow will be in the 350-mm pipe, and the flow up to the maximum dry-weather flow will be in the 350- and 400-mm pipes. For higher flows, a portion of the maximum dry-weather flow will occur in the 500-mm pipe.
   b. To determine the elevations of the tops of the walls, which will permit overflow to the 400- and 500-mm pipes, it is necessary to compute the depth of flow in the 750-mm gravity sewer when the flowrate is the same as the capacity (0.108 m^3/s) of the 350-mm pipe, and when the combined capacity (0.262 m^3/s) of the 350- and 400-mm pipes is reached. From the results of these calculations, determined with the use of Table 2-5, the heights are estimated to be about 244 and 375 mm, respectively. The accuracy of these calculations is adequate, provided that the velocity in the gravity sewer is not greater than about 1.5 m/s.
   c. Under maximum flow conditions, the overflow walls will be submerged and, therefore, cannot be considered weirs. Instead, they must be considered obstructions causing head losses that are difficult to predict. If it is assumed that the flow over these walls is approximately at right angles to that in the gravity sewer (conservative assumption), then this loss may be taken as the head required to produce the necessary velocity across the top of the wall. In this computation, it is assumed that the velocity head in the gravity sewer is not available to produce the flow at the right angle, but is lost in the inlet pool.

A maximum of 0.154 m^3/s (0.262 m^3/s − 0.108 m^3/s) must pass across the wall to the 400-mm pipe. The depth on the wall will then be at least 131 mm, or the difference in elevation between the two walls (375 mm − 244 mm), the higher wall having been designed to overflow just as the capacity of the first two pipes is exceeded.

At least 0.260 m^3/s must pass over the higher wall. The depth available is the distance from the top of the wall to the crown of the 750-mm sewer, or 375 mm. Assuming that the length of the wall is 2 m, the inlet losses may be approximated as described below.

In the 350-mm pipe, the loss is negligible because the invert is continuous with that of the 750-mm sewer, and smooth transitions can be made to take advantage of the approach velocity. Even at low flows in the gravity sewer, this velocity is equal to or greater than that in the 350-mm pipe. (For the partly filled sewer, the approach velocity may be estimated using Fig. 2-16.)

With the 400-mm pipe at maximum dry-weather flow, the velocity over the wall between the 350- and 400-mm pipes equals 0.154 $m^3/s \div (0.131\ m \times 2\ m)$ or 0.59 m/s; the corresponding velocity head is 18 mm. The difference in water-surface elevations may be taken as $1.5\ V^2/2g$ = 27 mm. After passing over the wall, a velocity of 1.23 m/s in the 400-mm pipe must be obtained in a new direction (conservative assumption), which requires a velocity head of 77 mm. Thus the total head loss is 104 mm (less than the assumed inlet allowance of 120 mm).

With the 500-mm pipe, the velocity over the wall equals 0.260 $m^3/s \div (0.375\ m \times 2\ m)$ or 0.35 m/s; the corresponding velocity head is 6 mm. The velocity in the 500-mm pipe is 1.32 m/s; the required velocity head is 90 mm. The total head loss is 99 mm [90 + (1.5)(6.)]. The assumed allowance of 120 mm for inlet losses appears sufficient, and no revision of the computation of available slope is required.

The relative elevations of the pipe inverts can now be calculated. When the capacity of the 350-mm pipe is reached, the depth of flow in the 750-mm gravity sewer is 244 mm. Since no allowance is needed for velocity head loss, the crown of the 350-mm sewer is 244 mm above the invert of the 750-mm pipe. The crown of the 400-mm pipe will be at an elevation of 375 mm minus the 104-mm head loss, or 271 mm above the invert of the 750-mm pipe. The crown of the 500-mm pipe is 99 mm below the crown of the 750-mm pipe. From these figures, the relative invert elevations may be determined.

The invert deviations just determined from the required hydraulic grade elevations are the maximum elevations that may be used and still have the depressed sewer operate as intended. Any or all of the pipes may be set lower than the required grades. Accordingly, for convenience of construction, all three pipes leading from the inlet chamber were set at the lowest allowable invert elevation, which is the one for the 400-mm pipe.

3. Design the outlet chamber.
   a. At the depressed-sewer outlet, the junction of the three pipes with the 750-mm gravity sewer should be designed to reduce the opportunity for eddies to carry sediment back into the pipes that frequently have no flow through them, but are full of standing water. This is especially important with the 500-mm pipe, which will not be in operation except at unusual rates of flow.
   b. The design is completed by locating the two larger pipes, or the inverts of corresponding channels within the junction chamber, as high as possible to avoid pooling and reducing the velocity in the chamber. As a further precaution, the outlet of the 500-mm pipe (least frequently required) may be raised so that the invert of its channel has a sharp forward pitch toward the invert of the 750-mm sewer. The crown of the pipe must not be raised above that of the 750-mm sewer, or a portion of it will be above the hydraulic grade line.

Comment In designing the depressed sewers, it was assumed that flexible joint pipe was used. This type of pipe, which allows joint deflections, permitted the three pipes to be depressed the required amount.

If bends amounting to 10 or 20° are required to make the depression, an additional head loss allowance must be made for each bend.

If the n value used had been 0.013, instead of 0.015, the required pipe sizes would be 350 mm (14 in), 350 mm (14 in), and 500 mm (20 in), instead of 350 mm (14 in), 400 mm (16 in), and 500 mm (20 in), on the basis of nominal diameters. Since the actual inside diameters of ductile-iron pipe suitable for depressed sewers, such as the one in this example, are about 4 percent larger than nominal, a built-in extra capacity of about 10 percent is available even without adjusting the n value. However, this extra capacity could be lost because of a buildup of coatings on the pipe walls. A change of n from 0.013 to 0.015 would provide an additional 15 percent capacity. For ductile iron, the use of a larger n value is not recommended since the extra capacity provided is not needed. For pipe where the actual inside diameter is the same as nominal, an increased n may be justified, although sufficient extra capacity may result merely from using standard pipe sizes.

Vertical Drops and Energy Dissipators

It is often necessary to discharge wastewater flowing in small surface sewers to larger trunk and intercepting tunnel sewers that may be constructed at a lower elevation. The flow is usually transferred through some type of vertical drop shaft. Where vertical drops must be used, they should be designed to avoid the entrapment of gases in the shaft. This is usually accomplished by having the wastewater flow in a spiral pattern downward through the shaft and by maintaining an open-air core throughout the shaft length.

Because the falling fluid contains a significant amount of kinetic energy, some form of energy dissipation must be included in the design if excessively turbulent conditions are to be avoided at the point of entry. The Hydro-Brake regulator discussed in Sec. 5-5 can be used for this purpose. The literature should be consulted for other methods that have been used, but it should be recognized that most applications are site-specific.

In the design of conventional sewers, some form of energy dissipation is often required to reduce the energy content (both potential and kinetic) of the flowing wastewater. The most common device for small sewers is the drop manhole (see Fig. 5-4). Where steep slopes must be used and the provision of energy-dissipation facilities is not practicable, it usually will be necessary to construct the sewer invert with special corrosion-resistant materials. Because of the problems caused by vertical drops, especially with the release of sewer gases, they should be avoided where possible.

5-4 OVERFLOW AND DIVERSION STRUCTURES

A combined sewer carries both wastewater and storm-water runoff. Wastewater-treatment facilities are generally sized to handle dry-weather wastewater flows, but not the high flowrates and volumes associated with storm-water runoff. Where existing combined sewers must be used to deliver flow to a treatment facility, it may be necessary to provide a diversion or overflow structure. Storm water may be diverted by side weirs, baffled side weirs, transverse weirs, leaping weirs, and relief siphons. During dry weather such structures allow all dry-weather wastewater to be carried to the treatment facility, but during wet weather, they divert that portion of the total flow which exceeds the capacity of the treatment facility. The flow may be diverted to a point of disposal without treatment or to temporary storage facilities.

Side Weirs

A weir parallel to the wastewater flow in the side of the sewer pipe is known as a *side weir*. The weir should be high enough to prevent any discharge of dry-weather flows, but low and long enough to discharge the required excess flow during storm-water flows. Flow over a side weir depends primarily on the depth of flow above the weir in the adjacent channel. The rate of discharge over the weir varies along the crest because of the change in depth resulting from the diversion of water without appreciable loss of energy.

Some of the earliest experiments to determine the capacity of side weirs were those conducted by Engels in 1917 [11] and by Smith and Coleman in 1923 [15]. Empirical relationships were developed from the experimental results. In addition, other empirical relationships were developed or studied by Parmley in 1905 [14], Babbitt in 1917 [2], and Tyler, Carollo, and Steyskal in 1929 [18].

A theoretical analysis of the discharge over side weirs was presented in 1928 by Nimmo [13]. Nimmo approached the problem of determining the discharge over a side weir by considering changes in momentum in upstream and downstream channel cross sections along the weir. He noted that the change in momentum in the length dL minus the momentum lost in the overflowing water must equal the sum of the external forces, and he applied this theory to a rectangular channel. By assuming a constant channel width and a horizontal channel bottom, and neglecting friction, he arrived at the following equation for the water-surface profile over a weir:

$$\frac{dd}{dL} = \frac{AQ}{BQ^2 - QA^3} \quad \frac{dQ}{dL} \tag{5-1}$$

In Eq. 5-1, dQ/dL is the flow per unit length over the weir and is given by the following normal weir equation:

$$\frac{dQ_w}{dL} = C\sqrt{2g}(d - c)^{3/2} \tag{5-2}$$

where dQ/dL = discharge per unit length of weir, m^2/s (ft^2/s)
C = constant
g = acceleration due to gravity, 9.81 m/s^2 (32.2 ft/s^2)
d = depth of flow in channel, m (ft)
e = height of weir, m (ft)

Working from the assumption of constant total energy along the weir, de Marchi in 1934 [6] published the results of a theoretical investigation in which he arrived at an equation identical to the one developed by Nimmo (Eq. 5-1). De Marchi demonstrated that, if dd/dL in Eq. 5-1 is positive (tranquil flow), there will be a rising water-surface profile over the weir, whereas if dd/dL is negative (shooting flow), a falling water-surface profile results. This analysis by de Marchi explains the previously mentioned results observed by Engels and by Smith and Coleman. In fact, de Marchi showed that three possible profiles existed [6]. Subsequently, it has been found that there are additional flow conditions in which a hydraulic jump develops within the weir section [7]. The three profiles identified by de Marchi are:

1. If the combined sewer has a steep slope, producing uniform supercritical flow upstream of the weir, the profile is as shown in Fig. 5-14*a*. The weir has no effect in the upstream direction because the flow is supercritical. Along the length of the weir, the depth gradually decreases; beyond the weir, the depth increases as the flow is retarded and approaches the normal depth asymptotically.
2. If the combined sewer has a mild slope, with uniform subcritical flow some distance upstream of the weir, the weir only affects the flow in the upstream direction, as shown in Fig. 5-14*b*. Downstream of section 2, the

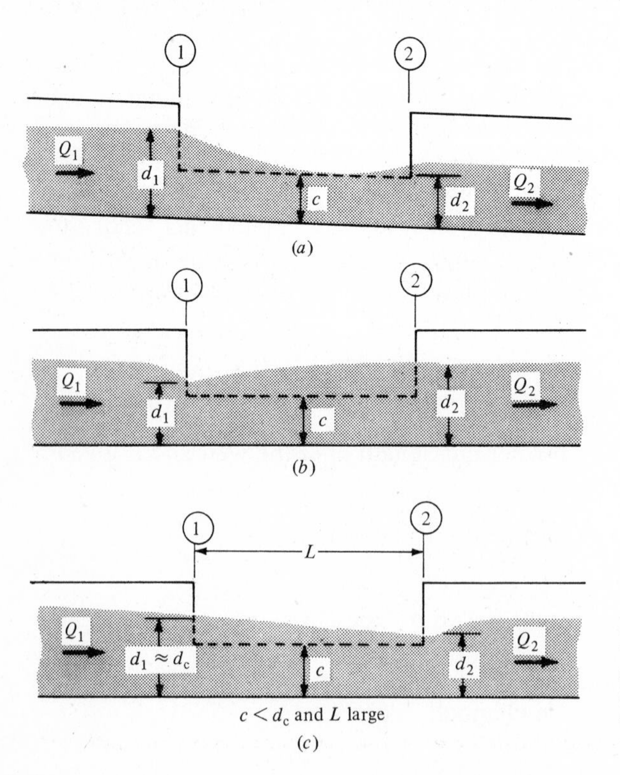

Figure 5-14 Possible types of water-surface profiles in channel adjacent to side weirs [6]. (*a*) Steeply sloping bed; (*b*) bed with mild slope; (*c*) another bed with mild slope.

depth tends toward the nomral depth corresponding to the flow Q_2 remaining in the channel. Along the length of the weir, the depth gradually increases in the direction of flow. Upstream of section 1, the depth approaches the normal depth asymptotically for the initial flow Q_1.

3. If the sewer slope is mild, and if the weir height is less than the critical depth corresponding to the initial flow Q_1 and the weir is sufficiently long, the incoming flow is drawn down upstream of the weir and critical depth develops near the weir entrance. As it passes parallel to the weir, the contained flow becomes increasingly supercritical and its depth decreases, as shown in Fig. 5-14*c*. If the downstream sewer slope remains mild, a transition to subcritical flow (hydraulic jump) may be expected a short way downstream of the weir.

For a theoretical analysis of the flow over side weirs, two basic equations are required: one accounts for either the momentum or the energy along the channel and the other relates the discharge over the weir length. The development of these equations and their application to side weirs are discussed in the text that follows.

Side weirs—falling water surface. The equations and procedures for computing weir lengths when the water-surface profile is falling are taken from an analysis of this problem presented by Ackers in 1957 [1]. A definition sketch for this problem is shown in Fig. 5-15.

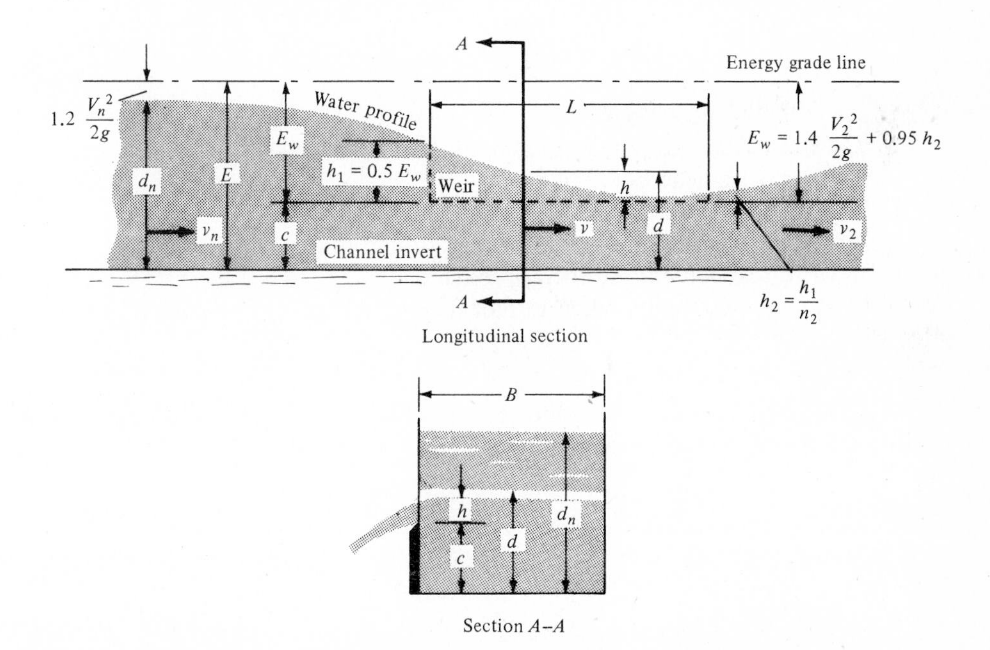

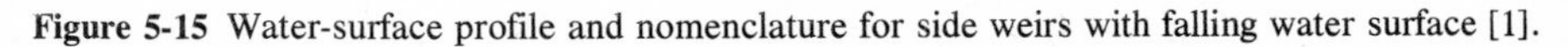

Figure 5-15 Water-surface profile and nomenclature for side weirs with falling water surface [1].

Ackers developed a design equation for computing the required weir length by combining Bernoulli's equation, including an allowance for channel velocity distribution, with a weir-discharge formula. The relationship developed by Ackers is

$$L_2 = 2.03B\left\{2\sqrt{2}\left[\sqrt{n_2 - 0.4}\left(1 - 0.4\frac{c}{E_w}\right) + 0.310\frac{c}{E_w} - 0.948\cos^{-1}\sqrt{\frac{0.4}{n_2} + 0.065}\right]\right\} \tag{5-3}$$

where L = length of weir, m (ft)
B = channel width (or diameter), m (ft)
$n_2 = h_1/h_2$
h_1 = head on weir at upstream end of weir, m (ft)
h_2 = head on weir at downstream end of weir, m (ft)
c = height of weir above channel invert, m (ft)
E_w = specific energy of flow relative to crest of weir, m (ft)

The resulting equations for selected values of n_2 are given in Table 5-1. The relationship between the term c/E_w, which is the height of the weir divided by the specific energy in the channel, and the ratio of the length of weir to the channel width (or diameter) L/B is shown graphically in Fig. 5-16.

As suggested by Ackers, the specific energy relative to the weir crest E_w, including a velocity head correction, may be computed with the following expression:

$$E_w = \alpha\frac{V_n^2}{2g} + \alpha'(d_n - c) \tag{5-4}$$

where α = velocity-head coefficient
V_n = normal velocity in approach channel, m/s (ft/s)
α' = pressure-head coefficient
d_n = normal depth in approach channel, m (ft)

On the basis of his experiments, Ackers found that α and α' may be taken as 1.2 and 1.0, respectively, in the approach channel, and the head of the weir at the upstream end may be taken as half the specific energy referenced to the weir crest.

Table 5-1 Design equations for side weirs with falling water surfaces based on Eq. 5-3

n_2	Equation for weir length
5	$L = 2.03B\,(2.81 - 1.55c/E_w)$
7	$L = 2.03B\,(3.90 - 2.03c/E_w)$
10	$L = 2.03B\,(5.28 - 2.63c/E_w)$
15	$L = 2.03B\,(7.23 - 3.45c/E_w)$
20	$L = 2.03B\,(8.87 - 4.13c/E_w)$

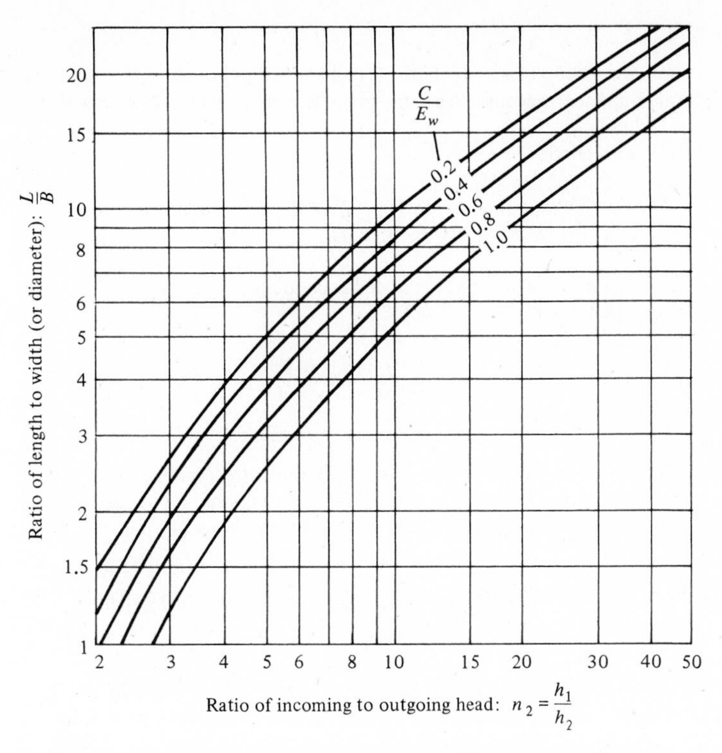

Figure 5-16 Design chart for side weirs with falling water surface [1].

Therefore

$$h_1 = 0.5E_w$$

and

$$h_2 = \frac{E_w}{2n_2}$$

At the downstream end of the weir, Ackers found the correction coefficients for velocity head and pressure head to be 1.4 and 0.95, respectively.

Equations 5-3 and 5-4 apply only if there is a falling water-surface profile along the weir. However, it can be shown that a falling profile results if c/E_w is less than about 0.6. The application of these equations is illustrated in Example 5-2.

Example 5-2 Design of side weir in existing pipe Determine the length of weir that must be placed in an existing 1200-mm (47-in) pipe, which will be used as a combined sewer, if the maximum wet-weather flow is 1.9 m^3/s (43.4 Mgal/d) and the maximum allowable wet-weather flow to the treatment plant is not to exceed 0.7 m^3/s (16.0 Mgal/d). The pipe slope is 0.003 m/m (ft/ft) and Manning's n is 0.013. The maximum dry-weather flow is 0.14 m^3/s (3.2 Mgal/d). All flow is to be treated.

SOLUTION

1. Compute the maximum capacity of the pipelines. Using Manning's equation, Eq. 2-24, the capacity of 1200-mm-diameter pipe laid on a slope of 0.003 m/m with an n value of 0.013 is 2.1 m^3/s. The corresponding velocity is 1.89 m/s.
2. Compute the flow characteristics at 1.9 m^3/s.
   a. Depth of flow:
   From Fig. 2-16, for a Q/Q_{full} value of 0.905 (1.9/2.1), the corresponding d/D value is 0.73. Thus the depth of flow d_n equals (0.73)(1200), or 876 mm.
   b. Velocity:
   From Fig. 2-16, for a d/D value of 0.73, the corresponding V/V_{full} value is 1.14. The velocity V_n equals (1.89)(1.14), or 2.16 m/s.
   c. Type of flow:
   Using Eq. 2-48, the critical depth in the 1200-mm pipe for a flow of 1.9 m^3/s is about 760 mm. The flow is, therefore, subcritical since $d_n > d_c$.
3. Compute the flow characteristics at 0.14 m^3/s.
   a. Depth of flow:
   From Fig. 2-16 for a Q/Q_{full} value of 0.07 (0.14/2.1), the corresponding d/D value is 0.2. Thus the depth of flow d_n equals (0.2)(1200), or 240 mm. Since all the dry-weather flow must be retained, the weir height c will be at least 240 mm above the bottom of the pipe.
   b. Velocity:
   From Fig. 2-16 for a d/D value of 0.2, the corresponding V/V_{full} value is 0.56. The velocity V_n equals 1.89 × (0.56), or 1.06 m/s.
4. Weir calculations.
   a. Compute the specific energy E_w, using Eq. 5-4, at the upstream end of the weir:

$$E_w = \frac{1.2V_n^2}{2g} + (d_n - c)$$

$$= 1.2\frac{(2.16)^2}{2g} + (0.876 - 0.24) = 0.92 \text{ m}$$

   b. Compute c/E_w:

$$\frac{c}{E_w} = \frac{0.24}{0.92} = 0.26$$

   Since $c/E_w < 0.6$, a falling head is possible.
   c. Compute the required weir length. For an assumed n_2 value of 10 in Eq. 5-3 (see Table 5-1), the required length is given by

$$L = (2.03B)(5.28 - 2.63\frac{c}{E_w})$$

$$= (2.03)(1.2)[5.28 - 2.63(0.26)] = 11.2 \text{ m}$$

5. Estimate the peak wet-weather discharge to the treatment plant.
   a. Compute the velocity at the lower end of the weir using Eq. 5-4,

$$1.4\frac{V_2^2}{2g} = 0.92 - 0.95\frac{0.92}{20} = 0.876$$

$$V_2^2 = 12.28 \quad \text{and} \quad V = 3.50 \text{ m/s}$$

   b. Outflow depth:
   Inflow head,

$$h_1 = 0.5, E_w = 0.5(0.92) = 0.46$$

Outflow head,

$$h_2 = \frac{h_1}{n_2} = \frac{0.46}{10} = 0.046$$

Outflow depth,

$$d_2 = 0.24 + 0.046 = 0.286$$

c. Outflow area:
From Fig. 2-16 for a d/D value of 0.2' (286/1200), the corresponding A/A_{full} value is 0.17.

$$\text{Outflow area} = 0.17\ \pi(0.6)^2 = 0.192\ \text{m}^2$$

d. Discharge to the treatment plant:

$$\text{Discharge} = 0.192\ (3.50) = 0.673\ \text{m}^3/\text{s}$$

Since $0.673\ \text{m}^3/\text{s} < 0.7\ \text{m}^3/\text{s}$, the design is satisfactory.

Side weirs—rising water surface. The analysis of weirs under the conditions of a rising-water-surface profile is based on the equation developed by de Marchi for rectangular channels [6]:

$$l - l_0 = \frac{B}{\mu}\left[\frac{2E - 3c}{E - c}\sqrt{\frac{E - d}{d - c}} - (3\ \sin^{-1})\sqrt{\frac{E - d}{E - c}}\right] \tag{5-5}$$

where l = distance from an origin to a point on weir, m (ft)
l_0 = constant of integration
B = channel width, m (ft)
μ = weir constant
$E = d + V^2/2g$, specific energy referred to channel bottom, m (ft)
d = depth of flow in the channel adjacent to weir, m (ft)
V = velocity of flow in channel, m/s (ft/s)
c = height of weir crest above channel bottom, m (ft)

De Marchi replaced the terms in the brackets with $\phi\ (d/E)$, a flow function, and evaluated it for various values of d/E and c/E. The results are shown in Fig. 5-17.

Since l_0 is a constant of integration for any given weir, the length between two points on a weir can be expressed as $l_2 - l_1$, and when $\phi(d/E)$ is evaluated at either end of the weir, $l_2 - l_1 = L_t$ equals the theoretical length of the weir crest. Equation 5-3 can now be written as

$$L_t = \frac{B}{\mu}\left[\phi\left(\frac{d_2}{E}\right) - \phi\left(\frac{d_1}{E}\right)\right] \tag{5-6}$$

The weir constant for use in Eq. 5-6 is given by

$$\mu = \frac{C}{\sqrt{2g}} \tag{5-7}$$

where C = weir coefficient in $Q = CLH^{3/2}$.

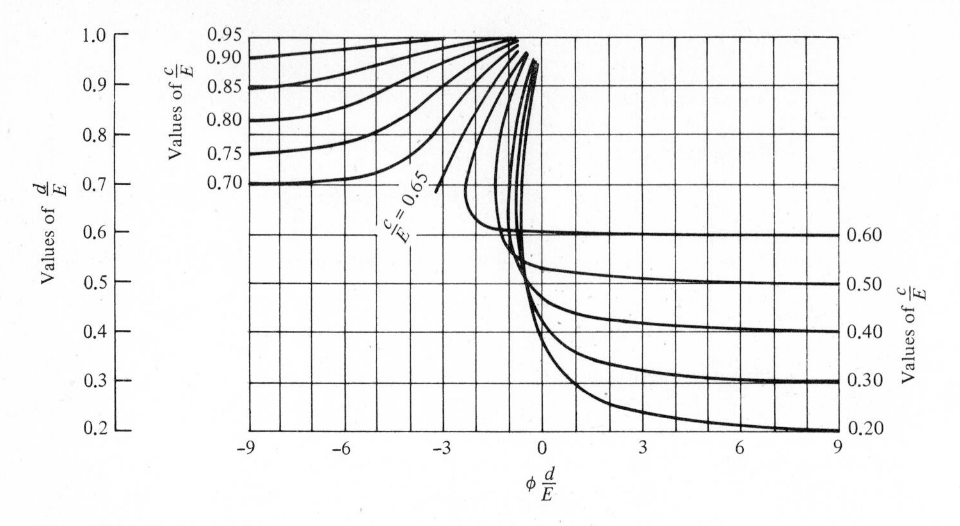

Figure 5-17 Values of $f\,(d/E)$ for various values of c/E for side weirs [6].

For a sharp-crested weir with flow normal to the crest, discharging freely with fully aerated nappe, the weir coefficients are

$$C = 1.84 \qquad \text{(SI units)}$$

$$C = 3.33 \qquad \text{(U.S. customary units)}$$

The corresponding value of the weir constant μ is 0.415.

When using Eq. 5-6, the specific energy E is assumed to be constant along the length of the weir. In a rectangular channel, flow, depth, and specific energy are related by

$$Q = Bd\sqrt{2g(E - d)}$$

where Q = volume rate of flow, m^3/s (ft/s).

When the length of the weir and the depth and rate of flow just downstream of the weir are known, Eq. 5-6 can be used to determine the allowable rate of flow from upstream of the weir. At the downstream end of the weir, d_2 is known, E can be calculated, and the value of $\phi(d_2/E)$ can be read from Fig. 5-18. Using the given length of weir and Eq. 5-6, $\phi(d_1/E)$ can be calculated, and d_1/E can be read from Fig. 5-17. Since E is known (assumed constant along the weir), d_1 can be calculated, and the allowable inflow can be computed by Eq. 5-8. This procedure is illustrated in Example 5-3.

The length of weir necessary to reduce the flow by a specified amount can be determined by a similar computation, except that a trial length of weir is assumed and the corresponding inflow rate is computed. If the computed inflow Q_1 does not equal the design inflow, the computation is repeated with revised trial lengths until the required value of Q_1 is attained, or it is demonstrated that the sought-for reduction in flow is not possible regardless of the length of the weir.

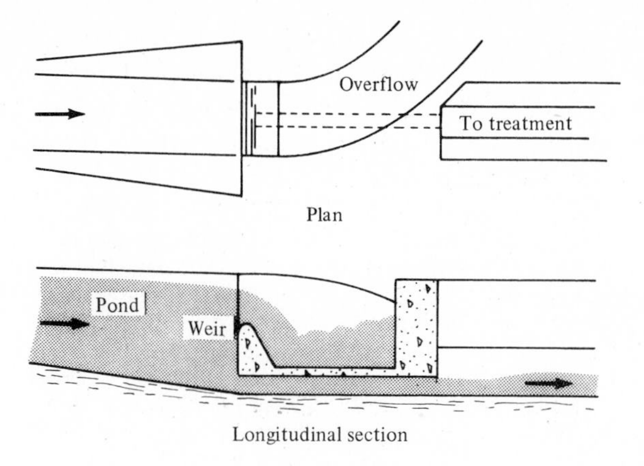

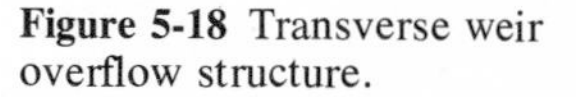

Figure 5-18 Transverse weir overflow structure.

Collinge found that the actual discharge over the weir is less than that calculated on the basis of a standard weir coefficient [6]. This was confirmed by Subramanya and Awasthy [17] who, in 1972, demonstrated that the weir constant is less than 0.415 and is principally a function of the Froude number of the incoming flow. A factor has been derived that can be applied to the theoretical length calculated in Eq. 5-6 to give a close approximation to the required length. This length-correction factor is

$$K = \frac{3.1}{2.8 - F_1} \tag{5-9}$$

where K = length-correction factor

$F_1 = \dfrac{V_1}{\sqrt{gd_1}}$ = Froude number of channel flow at the upstream end of the weir

and the required length is

$$L = KL_t \tag{5-10}$$

where L_t is the theoretical length calculated by Eqs. 5-6 and 5-8, with μ = 0.415. The value of K in Eq. 5-9 was derived from an envelope curve used to determine the maximum values of K [17].

After determining the weir length necessary to effect a desired reduction of flow by using de Marchi's relationships, it is suggested that K be evaluated and the theoretical length be increased in Eq. 5-10.

Similarly, when an allowable rate of inflow has been determined for a side weir of known length, the values of F_1 and K should be evaluated, and a corrected value of L_t for use with Eq. 5-6 should be determined in Eq. 5-10. The allowable inflow can then be recomputed with the new L_t in Eq. 5-6. The process should be repeated until sufficiently close agreement between the values of L_t used in Eq. 5-6 and the values of L_t calculated in Eq. 5-10 is reached.

However, the value of K calculated with Eq. 5-9 is intended to be a maximum value so as to ensure the design of a sufficiently long weir. For an existing side weir, it may be significantly smaller, and the weir may be more effective than the analysis indicates.

Example 5-3 Capacity of a sewer with an existing side weir diversion A 0.5-m (1.6-ft) square combined sewer (n = 0.013) has a mild slope of 0.001 m/m (ft/ft). A flow of 0.12 m³/s is the maximum rate to be allowed to enter a treatment plant. Determine the maximum inflow rate commensurate with a rising head along a 3-m (9.8-ft) length of weir. [*Note:* In this example, the weir length L is known. Therefore a trial value of the corresponding theoretical length L_t must be assumed for use in Eq. 5-6. On that basis, the conditions at the upstream end of the weir will be determined, F_1 and K will be evaluated, and L_t will be calculated from the given length L by use of Eq. 5-10. The resulting computed value of L_t will be compared with the assumed trial value. If agreement is not sufficiently close (5 to 10 percent), the process will be repeated with new trial values of L_t until satisfactory agreement is reached.]

SOLUTION

1. Compute the downstream depth d_2 using Manning's equation (Eq. 2-24).

$$Q = \frac{1.00}{n} AR^{2/3}S^{1/2}$$

where Q = 0.12 m³/s
n = 0.013
$A = d_2(0.5\ \text{m})$
$R = d_2(0.5\ \text{m})/(2d_2 + 0.5\ \text{m})$
S = 0.001 m/m

$$0.12\ \text{m}^3/\text{sb} = \frac{1.00}{0.013} d_2(0.5\ \text{m})\left[\frac{(0.5\ \text{m})d_2}{2d_2 + 0.5\ \text{m}}\right]^{2/3}(0.001)^{1/2}$$

$$0.099 = d_2\left[\frac{d_2(0.5\ \text{m})}{2d_2 + 0.5\ \text{m}}\right]^{2/3}$$

$$d_2 = 0.355\ \text{m}$$

2. Determine the specific energy E.
   a. Calculate V_2:

$$V = \frac{1.00}{n} R^{2/3}S^{1/2}$$

$$= \frac{1.00}{0.013}\left[\frac{(0.5\ \text{m})(0.355\ \text{m})}{2(0.355\ \text{m}) + 0.5\ \text{m}}\right]^{2/3}(0.001)^{1/2}$$

$$= 0.677\ \text{m/s}$$

   b. Calculate E:

$$E = d_2 + \frac{V_2^2}{2g}$$

$$= 0.355\ \text{m} + \frac{(0.677\ \text{m/s})^2}{2(9.81\ \text{m/s}^2)}$$

$$= 0.378\ \text{m}$$

3. Determine the weir height c. Based on a constant value of E, if c is made equal to or less than two-thirds of E, critical flow will occur at section 1 and the water surface will drop while passing along the weir. Since a rising water surface is specified, assume $c = 0.7E$ as a minimum practical value. Therefore let $c = 0.7(0.378 \text{ m}) = 0.265$ m.
4. Determine d, the upstream depth of flow.
   a. Determine $\phi(d_2/E)$ from Fig. 5-18 and with $c/E = 0.7$:

$$d_2/E = \frac{0.355}{0.378} = 0.939$$

$$\phi\left(\frac{d_2}{E}\right) = -1.6$$

   b. Determine $\phi(d_1/E)$ using Eq. 5-6. Assume a trial value of $L_t = 2.7$ m:

$$L_t = \frac{B}{\mu}\left[\phi\left(\frac{d_2}{E}\right) - \phi\left(\frac{d_1}{E}\right)\right]$$

$$2.7 \text{ m} = \frac{0.5 \text{ m}}{0.415}\left[(-1.6) - \phi\left(\frac{d_1}{E}\right)\right]$$

$$\phi\left(\frac{d_1}{E}\right) = -3.84$$

   c. Compute d_1 using Fig. 5-18. With $c/E = 0.7$ and $\phi(d_1/E) = -3.84$, $d_1/E = 0.77$. Thus,

$$d_1 = 0.77(0.378 \text{ m}) = 0.291 \text{ m}$$

5. Determine Q.

$$V_1 = [2g(E - d_1)]^{0.5}$$

$$= [2(9.81 \text{ m/s}^2)(0.378\text{m} - 0.291 \text{ m})]^{0.5}$$

$$= 1.31 \text{ m/s}$$

$$Q_1 = A_1V_1$$

$$= (0.291 \text{ m})(0.5 \text{ m})(1.31 \text{ m/s})$$

$$= 0.190 \text{ m}^3/\text{s}$$

6. Compute and apply the length correction factor.

$$F_1 = \frac{V_1}{\sqrt{gd_1}} = \frac{1.31 \text{ m/s}}{\sqrt{(9.81 \text{ m/s}^2)(0.291 \text{ m})}} = 0.78$$

$$K = \frac{3.1}{2.8 - F_1} = \frac{3.1}{2.8 - 0.78} = 1.53$$

$$L_t = \frac{3.0}{1.53} = 1.96 < 2.7 \quad \text{(trial value)}$$

7. Recompute d_1 with $L_t = 2.0$.

$$2.0 \text{ m} = \frac{0.5 \text{ m}}{0.415}\left[-1.6 - \phi\left(\frac{d_1}{E}\right)\right]$$

$$\phi\left(\frac{d_1}{E}\right) = -3.26$$

Using Fig. 5-18 and $c/E = 0.7$, $d_1/E = 0.82$; thus $d_1 = 0.310$ m.

8. Determine Q_1.

$$V_1 = 2(9.81 \text{ m/s}^2)(0.378 \text{ m} - 0.310 \text{ m})$$
$$= 1.16 \text{ m/s}$$
$$Q_1 = A_1 V_1$$
$$= (0.5 \text{ m})(0.310 \text{ m})(1.16 \text{ m/s})$$
$$= 0.179 \text{ m}^3\text{/s}$$

9. Compute and apply the length correction.

$$F_1 = \frac{1.16}{\sqrt{9.81(0.310)}} = 0.66$$

$$K = \frac{3.1}{\sqrt{2.8 - 0.66}} = 1.45$$

$$L_t = \frac{3.0}{1.45} = 2.07 \cong 2.0 \qquad \text{(trial value)}$$

10. Conclusion. The maximum inflow rate for the prescribed conditions, based on Q_1 from step 8, is approximately 0.18 m³/s.

Comment In this example, the total weir discharge equals 0.06 m³/s, about 33 percent of the inflow. The weir discharge and allowable inflow for a weir twice as long (6 m) would be increased only about 5 percent. Thus a side weir with a rising water surface appears to be a relatively inefficient diversion device.

Design of side weirs. The side weirs most likely to be useful in the design of overflow diversion structures are those illustrated in Fig. 5-14*a* and 5-14*c*; the weir height is less than the critical depth of the incoming flow, and the water-surface profile is falling in the direction of flow. For equal lengths of weir, overflow capacities will be greater with a falling water surface than with a rising water surface.

For the range of values in Fig. 5-17, Eq. 5-6 can be used for problems involving both rising and falling water surfaces. However, Collinge found that the best agreement between theoretical and actual discharges occurs when (1) the Froude number, F_1, of the flow at the upstream end of the weir is between 0.3 and 0.92 (that is, when the inflow is subcritical), and (2) the water surface is rising along the weir [6]. Agreement between theoretical and actual discharges was poor when F_1 was greater than about 1.2 and was extremely poor when F_1 was close to 1.0. It is, therefore, recommended that Eq. 5-6 be used only for conditions of subcritical approach flow and rising water surface. For these conditions, $c/E > 1/3$; for other conditions, Eq. 5-3 (see Table 5-1) is recommended.

Equations 5-3 through 5-8 for side weirs with either rising or falling water surfaces are based on weir discharge with fully aerated nappe. If the nappe adheres to the weir or if there is less than full aeration under the nappe, the weir discharge will be greater than the calculated discharge by an indeterminate amount.

Baffled Side Weirs

A baffled side weir is one in which a baffle is placed across the main channel to increase the side-weir capacity. The baffle is placed across the channel so that its bottom edge and the channel walls and bottom form a rectangular orifice. Depending on flow conditions and controls in the downstream channel, flow under the baffle and into the downstream channel will be a supercritical jet, followed by a hydraulic jump, or it will take the form of discharge from a submerged orifice. In either case, the baffle will increase the head on the side weir, thereby increasing its capacity.

The depth of flow d_2 in the channel at the downstream end of the weir may be determined by calculating the head corresponding to the required flow under the baffle on the basis of flow through a rectangular orifice with contractions suppressed on the bottom and sides. With d_2 known, the length of the weir may be calculated as for a weir without a baffle, except that the specific energy is determined from the depth at the downstream end.

Tyler, Carollo, and Steyskal in their experiments found that a horizontal skimming plate had no advantage. They also found that a bulkhead placed squarely across the sewer, with its bottom edge at the elevation of the weir crest, more efficiently deflected excess flow over a side weir than one set at an angle [18]. They suggested, as a result of their experiments, that, by making the elevation of the lower edge of this bulkhead adjustable, it could be set to pass the desired amount of wastewater down the main sewer.

Transverse Weirs

One method of avoiding the uncertainty involved in the use of a side weir is to place the weir directly across the line of flow and to deflect the sewer to one side, thus reversing the arrangement used with side weirs. It should be possible to determine the discharge of a weir in this position by the usual weir equation (Eq. 3-2) with greater certainty than when the weir is in the side wall of the sewer, because the variation in head along the crest will presumably be smaller.

Another scheme for diverting storm water flow using a transverse weir is shown in Fig. 5-18. Under peak wet-weather flow conditions, the sewer leading to the treatment plant will be surcharged and should be designed accordingly. In addition, a deflector should be installed to avoid the formation of a vortex whenever the sewer leading to the treatment plant becomes surcharged.

Leaping Weirs

A leaping weir is formed by a gap in the invert of a sewer so that the ordinary flow of wastewater falls through the opening and passes to the interceptor.

During storms, the increased velocity of flow causes most of the flow to leap the opening and enter the storm outlet. An example of a leaping weir is shown in Fig. 5-19. The steel weir plate is inclined to give a spouting effect and is designed so that it can be adjusted for various flow conditions.

The design of leaping weirs is usually a trial-and-error process involving computation of the trajectory of flow and determination of the percentage of flow captured at various weir settings.

Relief Siphons

The relief siphon affords a means of regulating the maximum water-surface elevation in a sewer with smaller variations in high-water level than can be obtained with other devices. A siphon works automatically and does not require any auxiliary mechanisms. Since it uses all available head, it discharges at higher velocities than do overflow weirs. Although this device has obvious advantages, its infrequent use for diversion may be due to inadequate information about its design and operation, such as the minimum head required for priming, or to possible noise and vibration from sudden starting and stopping of the siphon, especially with high heads. A typical relief siphon is shown in Fig. 5-20.

The approximate cross section for the siphon throat can be determined by Eq. 5-11.

$$Q = ca\sqrt{2gh} \tag{5-11}$$

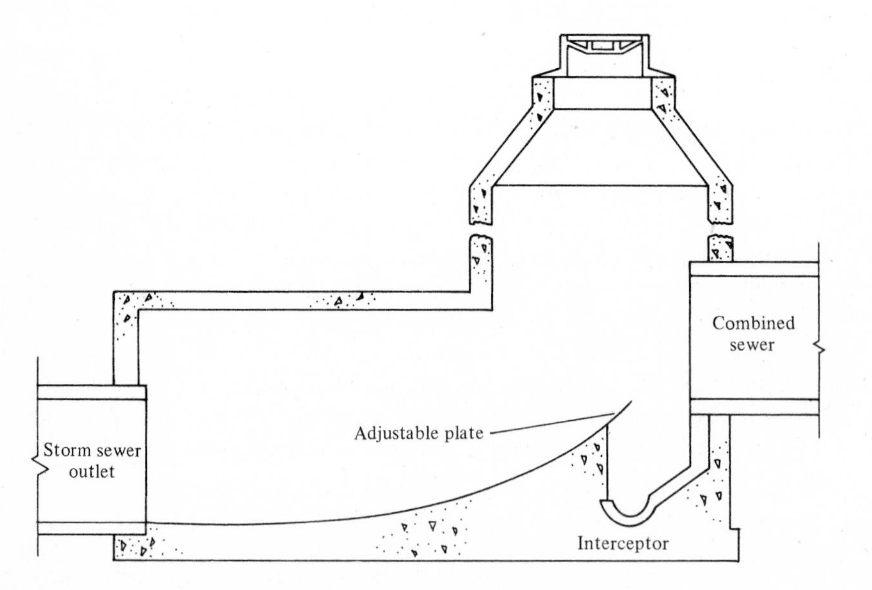

Figure 5-19 Adjustable leaping weir.

where Q = discharge, m^3/s (ft^3/s)
c = coefficient of discharge (0.6 to 0.8)
a = area of cross section of throat, m^2 (ft^2)
h = head, m (ft) = difference in water level above and below siphon

A trial section is then selected and the losses are determined. The sum of all the energy losses, including entrance and exit losses, should equal the total available head. If not, the trial section should be revised and calculations should be repeated until agreement is reached. The design then involves working out such details as the method of venting, the shape of the section to meet the particular requirements, and the fixing of elevations of the inlet, spillway, outlet, and air vent. An air vent with an area equal to the area of the throat divided by 24 ($a/24$) has been found to be ample. The siphon inlet should be

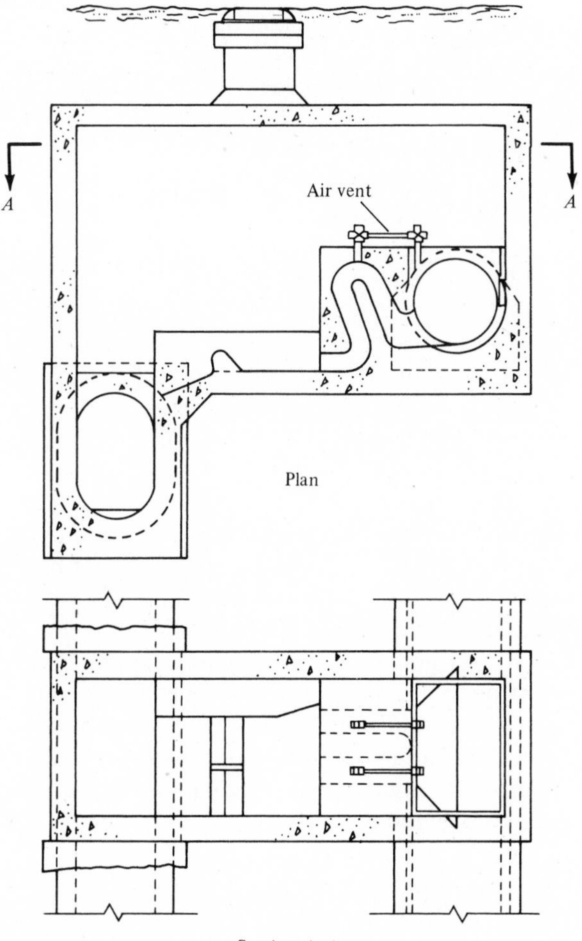

Figure 5-20 Relief siphon.

large so that the entrance velocity will be small, thus preventing large head losses. The section tapers gradually to the throat, and the lower leg, which may be either vertical or inclined, is of uniform section or slightly flared.

When the wastewater rises to the elevation of the crest in the siphon, it will flow over the crest in a thin sheet, falling against the opposite wall and carrying with it some air. This condition will continue until the wastewater has risen sufficiently to seal the air vent, after which the air remaining in the siphon will be exhausted quickly and siphonic action will begin. The siphon will continue to discharge flow from the sewer until the wastewater surface has been lowered sufficiently to uncover the vent, at which point siphonic action and flow diversion will cease.

The wastewater levels at which diversion begins and ends are attained by properly setting the elevation of the vent. Generally, the vent should be lower than the siphon crest, so that only some overflow occurs before the siphonic action begins or after it ceases. Theoretically, the maximum operating head for a siphon is about 10 m (33 ft) at sea level. It decreases at the rate of about 1 m for every 850 m of increase in elevation above sea level. In practice, because of the release of dissolved gases, velocity-head reduction, and flow losses, the maximum feasible head will be about 6 m (20 ft), which is ample for siphon reliefs used in wastewater systems. When deciding whether a siphon should be used to divert a relatively small wastewater flow, the possibility that solids in wastewater will clog the air vents should be considered.

5-5 REGULATING DEVICES

As discussed in Sec. 5-4, a storm overflow is designed to divert the excess wastewater above a definite rate into a relief sewer or other facility. In a sense, these overflow structures are regulators because they limit the amount of wastewater flow that will remain in the combined sewer. However, when dry-weather flow needs to be diverted from a combined sewer to an interceptor and then to a treatment facility, the rate of flow to the interceptor must be limited to an acceptable maximum while the remainder continues to flow in the combined sewer to its usual outlet.

In a sanitary sewer, a limitation of flow from a branch sewer might be needed to prevent unacceptable surcharging of an existing interceptor. In this case, the excess wastewater flow from the branch sewer would be diverted to a treatment facility through a relief sewer built for that purpose. In both cases, the flow would be limited by using regulating devices. Three such devices are the reverse taintor gate, the tipping-plate regulator, and the Hydro-Brake. The application of these devices is also considered.

Reverse Taintor Gate

The reverse taintor gate (also known as an *automatic sewer regulator*) responds to the water level in the combined sewer or branch sewer, as shown in Fig.

5-21, or to the water level or hydraulic grade level in the interceptor. In either case, the float travel and the corresponding gate travel may be adjusted to regulate closely the flow to the interceptor. These regulators are made in a range of gate sizes capable of limiting maximum diverted flows from 0.014 m^3/s (0.5 ft^3/s) to 4.0 m^3/s (140 ft^3/s).

Tipping-Plate Regulator

Tipping-plate regulators (also known as *Milwaukee or Chicago regulators*) have been in use for many years. In these devices, the plate is pivoted off-center, and its motion is controlled by the difference of water levels above and below the gate. One example of a tipping-plate regulator is shown in Fig. 5-22. Maximum diverted flows ranging from 0.01 m^3/s (0.3 ft^3/s) to 0.9 m^3/s (30 ft^3/s) can be accommodated by proper selection of available gate sizes. Multiple gates can be used to increase the capacity of an installation.

Hydro-Brake Regulator

One of the simplest and most effective regulating devices is the recently developed Hydro-Brake shown in Fig. 5-23. This device is manufactured by the Hydro-Storm Sewage Corporation. The patented configuration of the Hydro-

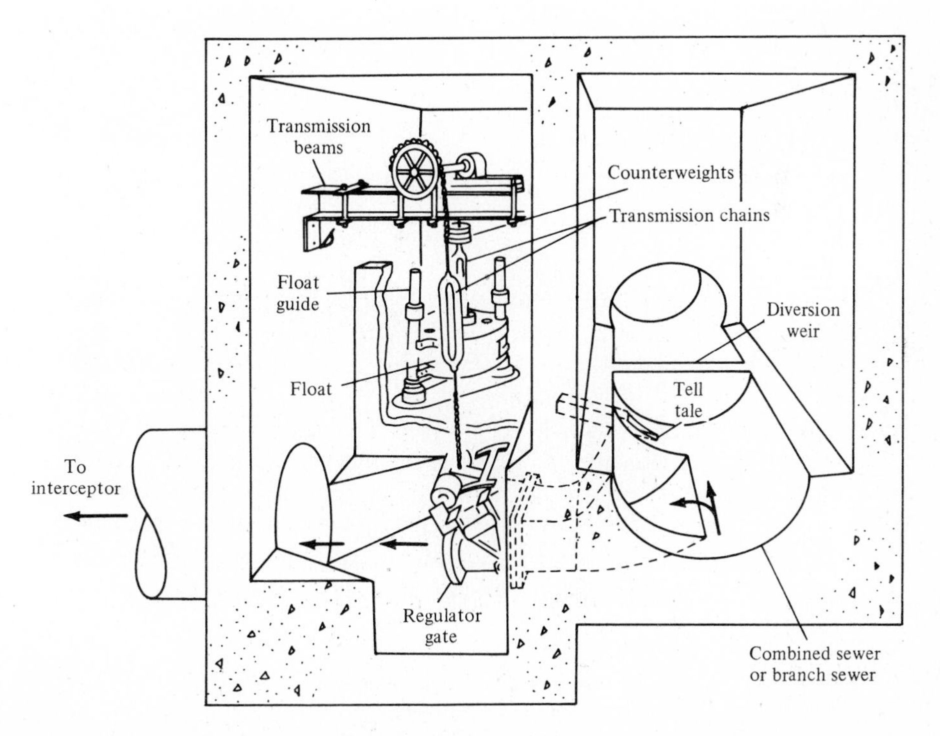

Figure 5-21 Reverse taintor gate (*Courtesy Brown and Brown*).

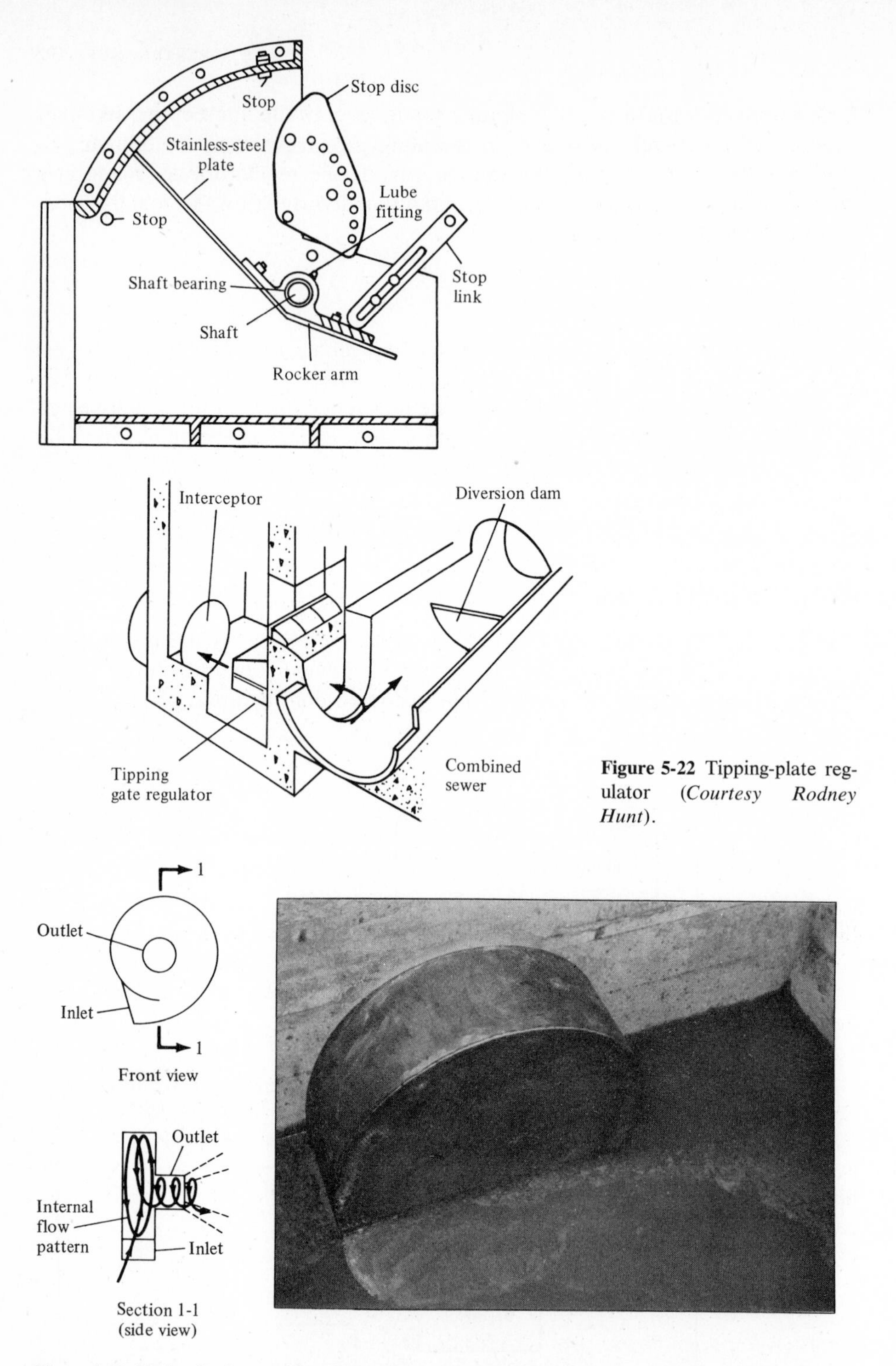

Figure 5-22 Tipping-plate regulator (*Courtesy Rodney Hunt*).

Figure 5-23 Hydro-Brake used for the dissipation of energy in wastewater and storm water collection systems (*Courtesy Hydro-Storm Sewage Corporation*).

Brake imparts a more-or-less centrifugal motion to the entering fluid. This action, which commences when a predetermined liquid head has been reached, effectively reduces the rate of discharge. This device has been used extensively on combined storm-water sewers to limit the flow. Fabricated of stainless steel, this device should have a long, useful life.

Application

Automatic mechanical regulators, either the float-operated type or the flap-gate type, are generally impracticable for small flows because of the small waterways and the consequent danger of clogging. All such devices require periodic inspection, skillful adjustment, and careful maintenance of all parts by workers with some mechanical skill. Lack of necessary periodic inspection and maintenance accounts for an extensive history of failure and abandonment of automatic mechanical regulators with moving parts. By comparison, the Hydro-Brake regulator does not require any adjustment after installation; it needs little or no monitoring or maintenance. Because of its simplicity and effectiveness, this device will probably be used more in the future.

Automatic mechanical regulators have been used successfully in urban areas where the flow in combined sewers is to be controlled. The key to their successful operation is continued monitoring and maintenance on a regular basis. Typical regulator stations using automatic mechanical regulators are shown in Fig. 5-24.

5-6 OUTLETS

Strictly speaking, the outlet of a sewerage system is the end of an outfall sewer where the wastewater is discharged. There may be a number of these outlets where the system has several storm-water outfalls or overflows. In every case, the objective should be to discharge the wastewater at a point where its presence will cause no offense.

If an outlet ends at a body of water subject to considerable fluctuations in level, such as the near-shore discharge of storm water into tidal waters, and it is necessary to prevent this water from entering the sewer, a backwater or tide gate is necessary. This gate consists of a flap hung against an inclined seat. The hinges may be at the top when the gate consists of a single leaf, as is usually the case, or they may be at the side when the gate consists of two leaves. An example of a single-leaf tide gate is shown in Fig. 5-25.

Where treated wastewater is discharged into the sea, into tidal waters, or into a lake or a deep river, the outlet is usually submerged to a considerable depth to disperse the wastewater effluent thoroughly through a large volume of water. Details on the design of submerged (submarine) outfalls are covered in Ref. 12, chap. 14.

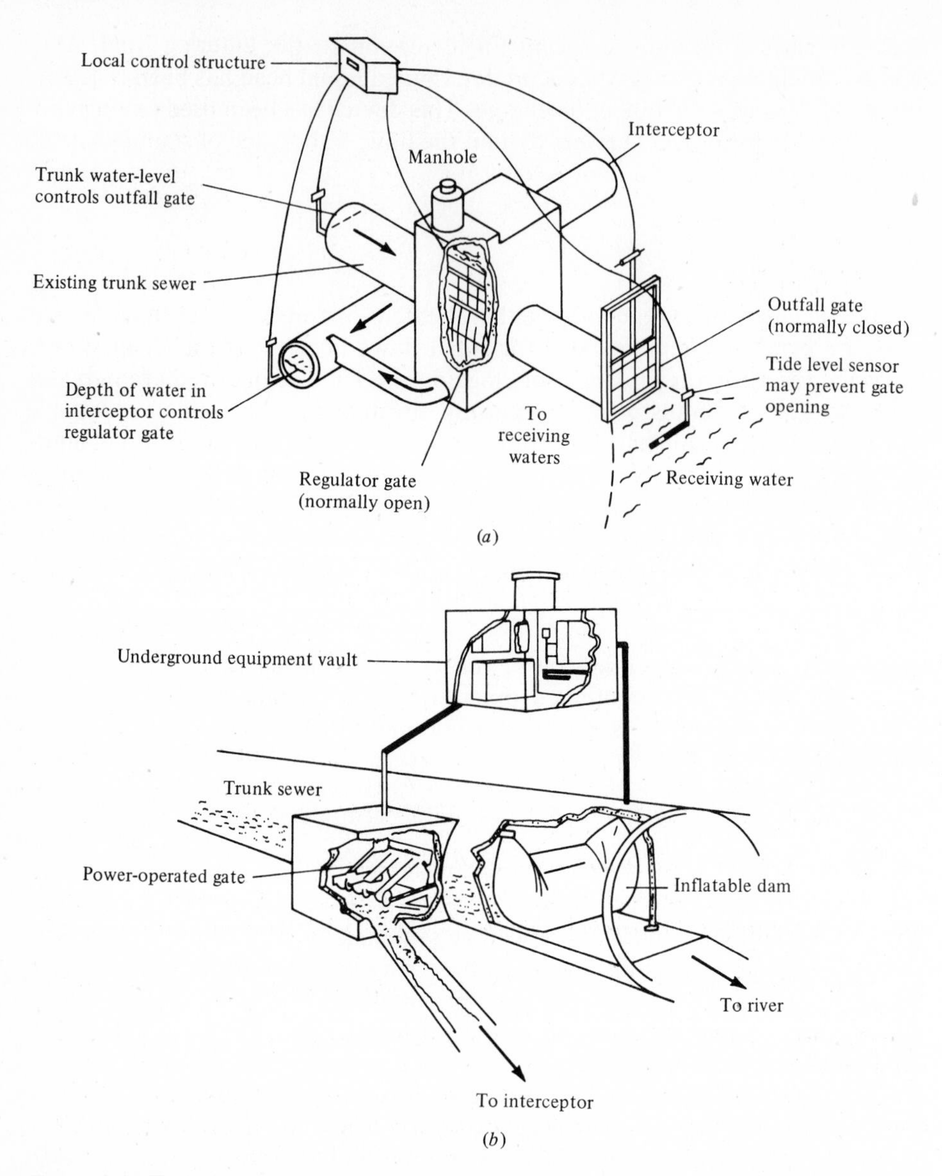

Figure 5-24 Typical regulator stations of inline storage systems. (*a*) Seattle, Wash.; (*b*) Minneapolis-St. Paul, Minn. [8, 9].

DISCUSSION TOPICS AND PROBLEMS

5-1 Obtain a copy of the standard manhole details used by your local community or school to prepare sewer-contract drawings. How do the manhole details compare with the details shown in Fig. 5-1? Identify the similar and dissimilar features.

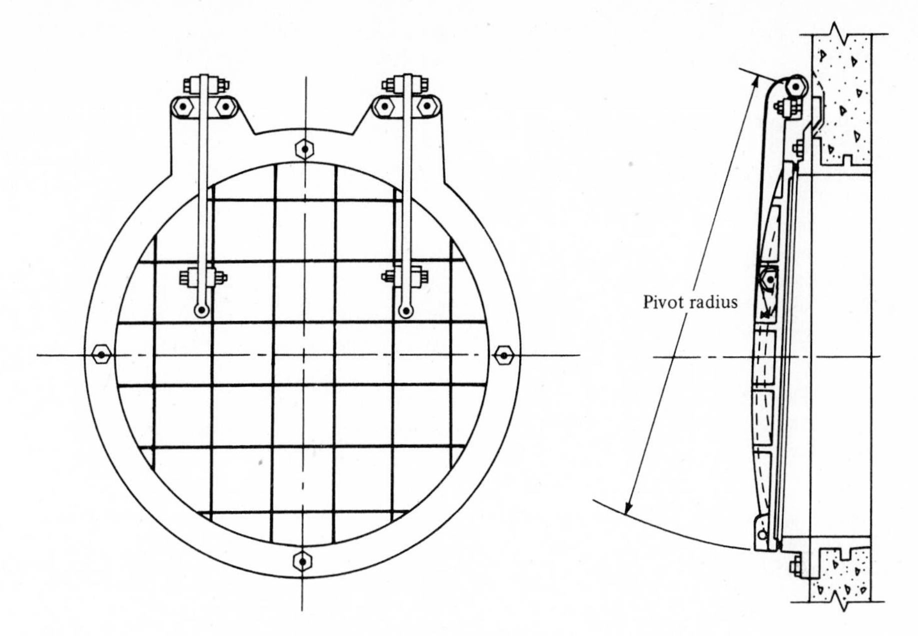

Figure 5-25 A single-leaf tide gate (*Courtesy Armco*).

5-2 Obtain a copy of the standard drop inlet (outside and inside) details used by your local community or school to prepare sewer-contract drawings. How do the drop inlet details compare with those shown in Figs. 5-4*a* and 5-4*b*? Identify the similar and dissimilar features.

5-3 Identify and sketch at least three types of storm-water inlets used in your community. Of the inlets identified, how many have spacers for bicycle safety?

5-4 Identify the types of catch basins used in the storm-water collection system in your community or school. Select three or more catch basins at random and determine if they have been cleaned recently.

5-5 An existing 1500-mm-diameter storm-water sewer has a slope of 0.0009 m/m (n = 0.013). The maximum dry-weather flow is 0.35 m³/s and an average storm will produce 1.0 m³/s. Design a three-pipe depressed sewer for maximum dry-weather flow, average storm-water flow, and maximum storm-water flow (the capacity of the 1500-mm sewer). The available hydraulic slope is 0.007 m/m and n is 0.015. The minimum scouring velocity that must be maintained is 0.9 m/s.

5-6 Design a depressed sewer consisting of three pipes and an inlet chamber for a 900-mm gravity sewer with a slope of 0.001 m/m (n is 0.013). The basic data are as follows:

1. Length of depressed sewer = 160 m
2. Maximum depression of sewer = 2.75 m
3. Minimum velocity in depressed sewer = 0.9 m/s
4. Design flows:
   (*a*) Minimum flow = 0.090 m³/s
   (*b*) Maximum dry-weather flow = 0.30 m³/s
   (*c*) Ultimate maximum flow = capacity of sewer
5. Available fall = 1200 mm
6. Manning's n = 0.015

5-7 Design a depressed sewer consisting of three pipes and an inlet chamber for a 600-mm (n = 0.013) sanitary sewer with a slope of 0.0031 m/m. Basic data are as follows:

1. Length of depressed sewer = 50 m
2. Minimum velocity = 0.9 m/s
3. Maximum depression of sewer = 3 m
4. Design flows:
   (*a*) Minimum flow = 0.034 m^3/s
   (*b*) Average flow = 0.114 m^3/s
   (*c*) Peak flow = 0.342 m^3/s
5. Available fall = 650 mm
6. Manning's n = 0.015

5-8 A large 4200-mm-diameter sanitary sewer must pass under a depressed highway before reaching the wastewater-treatment facilities. Design a depressed sewer with inlet structures. The basic data are as follows:

1. Length of depressed sewer = 100 m
2. Minimum velocity in depressed sewer = 0.9 m/s
3. Maximum depression of sewer = 3 m
4. Design flows:
   (*a*) Minimum flow = 11.42 m^3/s
   (*b*) Average flow = 14.28 m^3/s
   (*c*) Maximum flow = 19.07 m^3/s
5. Available fall = 550 mm
6. Manning's n = 0.015

5-9 A storm-water sewer of 1200-mm diameter has a slope of 0.0015 m/m. The maximum dry-weather flow is 0.3 m^3/s and an average storm will produce 0.7 m^3/s. Design a three-pipe depressed sewer for maximum dry-weather flow, average storm-water flow, and maximum storm-water flow (the capacity of the 1200-mm sewer). The available hydraulic slope is 0.005 m/m and n is 0.013. Minimum scouring velocity is 0.9 m^3/s.

5-10 Design the necessary sewer appurtenances for the proposed treatment-plant influent sewer shown in Fig. 5-26. Excess wastewater is diverted into a holding basin. The holding basin is used for flow equalization. The basic data are:

1. Maximum treatment-plant capacity = 20,000 m^3/d
2. Sewer flow:
   (*a*) Average dry-weather flow = 8000 m^3/d
   (*b*) Maximum dry-weather flow = 20,000 m^3/d
   (*c*) Peak storm-water flow = 30,000 m^3/d
3. Available fall = 500 mm in a siphon length of 100 m
4. Slope of sewer = 0.0018 m/m

5-11 Determine the length of weir that must be placed in a proposed 900-mm combined sewer. The maximum allowable wet-weather flow to the treatment facilities is 42,000 m^3/d. The maximum wet-weather flow is 80,000 m^3/d and the maximum dry-weather flow is 10,000 m^3/d. The slope of the sewer is 0.0028 m/m. Assume that n_2 is 10.

5-12 An off-line holding basin serves a wastewater-treatment plant. Design a side weir to divert any flow greater than 0.50 m^3/s to the basin. Basic data are as follows:

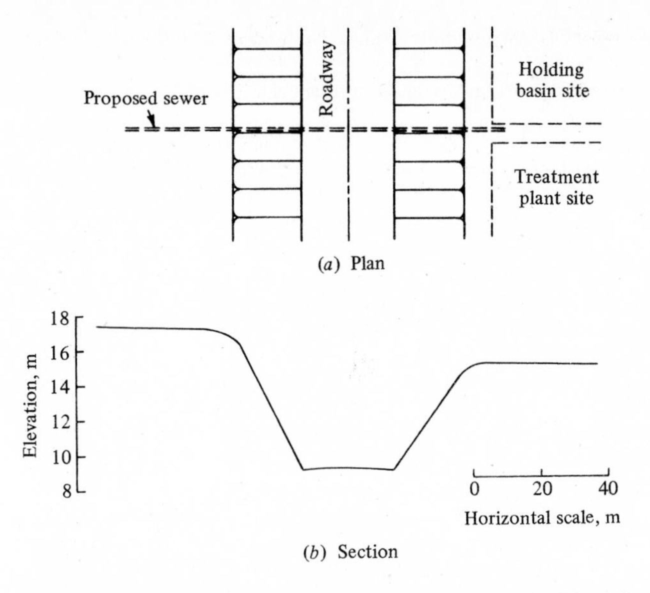

Figure 5-26 Definition sketch for Problem 5-10. m × 3.2808 = ft.

1. Average flowrate = 0.23 m^3/s
2. Sewer diameter = 1050 mm
3. Slope = 0.001 m/m
4. $n_2 = 15$
5. Manning's $n = 0.013$
6. Peak flow = 0.80 m^3/s

5-13 Wastewater flow greater than 1.50 m^3/s must be bypassed around a treatment facility for storm water. Design a side weir bypass system for a 1500-mm sewer with a slope of 0.0012 m/m. The average wastewater flow is 80,000 m^3/d with a peaking factor of 2.5. Assume that $n = 0.013$ and $nm_2 = 20$.

5-14 Determine the size of an incoming sewer and the length of a side weir bypass to supply an off-line equalization basin. Flow in excess of 7000 m^3/d is to be diverted. The average flowrate is 2000 m^3/d and the peaking factor is 6.0. Assume that $n = 0.013$ and slope = 0.003 m/m for the sewer, and use an n_2 value of 15 for the design of the weir.

REFERENCES

1. Ackers, P.: A Theoretical Consideration of Side Weirs as Stormwater Overflows, *Proceedings, Institution of Civil Engineers, London,* vol. 6, no. 2, 1957.
2. Babbitt, H. E., and E. R. Baumann: *Sewerage and Sewage Treatment,* 8th ed., Wiley, New York, 1958.
3. Bauer, W. J., and D. C. Woo: Hydraulic Design of Depressed Curb-Opening Inlets, *Highway Research Record,* Highway Research Board, no. 58, p. 61, 1964.
4. Bureau of Public Roads: *Urban Storm Drainage,* Hydraulic Information Circular 2, Washington, D.C., 1951.
5. Cassidy, J. J.: Generalized Hydraulic Characteristics of Grate Inlets, *Highway Research Record,* Highway Research Board, no. 123, p. 36, 1966.

6. Collinge, V. D.: The Discharge Capacity of Side Weirs, *Proceedings, Institution of Civil Engineers, London,* vol. 6, no. 2, 1957.
7. Frazer, W.: The Behavior of Side Weirs in Prismatic Rectangular Channels, *Proceedings, Institution of Civil Engineers, London,* vol. 6, no. 2, 1957.
8. Lager, J. A., and W. G. Smith: *Urban Stormwater Management and Technology: An Assessment,* U.S Environmental Protection Agency, Report 670/2-74-014, NTIS No. PB 240 687, Cincinnati, Ohio, 1977.
9. Lager, J. A., et al.: *Urban Stormwater Management and Technology: Update and Users Guide,* U.S. Environmental Protection Agency, Office of Research and Development, Report 600/8-77-014, Cincinnati, Ohio, 1977.
10. Lager, J. A., W. G. Smith, and G. Tchobanoglous: *Catchbasin Technology Overview and Assessment,* U.S. Environmental Protection Agency, Office of Research and Development, Report 600/2-77-051, Cincinnati, Ohio, 1977.
11. Metcalf, L., and H. P. Eddy: *American Sewerage Practice,* vol. I, 2d ed., McGraw-Hill, New York, 1928.
12. Metcalf & Eddy, Inc.: *Wastewater Engineering: Treatment, Disposal, Reuse,* 2d ed., McGraw-Hill, New York, 1979.
13. Nimmo, W. H. R.: Side Spillways for Regulating Diversion Canals, *Trans. ASCE,* vol. 92, 1928.
14. Parmley, W. C.: The Walworth Sewer, Cleveland, Ohio, *Trans. ASCE,* vol. 55, 1905.
15. Smith, D., and G. S. Coleman: The Discharging Capacities of Side Weirs, *Proceedings, Institution of Civil Engineers, London,* 1923.
16. Storm Drainage Research Committee: *The Design of Storm Water Inlets,* The Johns Hopkins University Storm Drainage Research Project, 1956.
17. Subramanya, K., and S. C. Awasthy: Spatially Varied Flow Over Side Weirs, *J. Hydraulics Div., ASCE,* vol. 98, no. HY1, 1972.
18. Tyler, R. G., J. A. Carollo, and N. A. Steyskal: Discharge Over Side Weirs With and Without Baffle, *J. Boston Soc. Civ. Eng.,* vol. 16, 1929.

CHAPTER

SIX

INFILTRATION/INFLOW

Those involved in the design and management of sanitary wastewater collection systems have long recognized the benefits of a leak-free or "tight" system in which extraneous flows (groundwater leakage or infiltration, and stormwater sources or inflow) are either excluded or held to a minimum. Obvious benefits of a tight system include (1) no overloaded or surcharged sewers and the associated problems of wastewater backups and overflows; (2) more efficient operation of wastewater-treatment facilities; and (3) the use of the sewer hydraulic capacity for wastewater instead of infiltration/inflow. The less apparent but usually more significant benefit of a tight system can be lower overall capital and operating costs. Savings in costs result from (1) the extended life of existing system components; (2) smaller future expansions of hydraulically sized components, such as sewers, pumping stations, and sedimentation basins; and (3) lower operating costs realized by a decrease in energy requirements (pumping) and certain chemicals.

Perhaps the single most important development that has highlighted the need for controlling infiltration/inflow has been the requirement that an infiltration/inflow (I/I) analysis be performed before funds can be obtained from the U.S. Environmental Protection Agency (EPA) for construction of treatment facilities. In view of this recent emphasis on infiltration/inflow, the purpose of this chapter is to discuss the EPA evaluation and rehabilitation program for sewer systems, to describe how to conduct an I/I analysis, and to discuss the key factors involved in a detailed sewer system evaluation survey (SSES). Other important topics covered are the various corrective measures available for I/I reduction (called *rehabilitation*) and the design standards for the prevention and control of infiltration/inflow. To understand the terms used in this chapter, definitions of infiltration and inflow are presented first.

6-1 DEFINITIONS

General definitions for infiltration and inflow were provided in Chap. 3. However, to identify and correlate these extraneous flows with observed flow measurements (sometimes called *flow hydrographs*), they are defined in greater detail in the following and are illustrated graphically in Fig. 6-1.

Infiltration. This consists of the water entering a sewer system, including sewer service connections, from the ground through such means as defective pipes, pipe joints, connections, or manhole walls. Infiltration does not include, and is distinguished from, inflow.

Steady inflow. This includes the water discharged from cellar and foundation drains, cooling-water discharges, and drains from springs and swampy areas. This type of inflow causes a steady flow and is identified and measured along with infiltration.

Direct inflow. This consists of those types of inflow that have a direct stormwater runoff connection to the sanitary sewer and cause an almost immediate increase in wastewater flows. Possible sources are roof leaders, yard and areaway drains, manhole covers, cross connections from storm drains and catch basins, and combined sewers.

Total inflow. This is the sum of the direct inflow at any point in the system plus any flow discharged from the system upstream through overflows, pumping station bypasses, and the like.

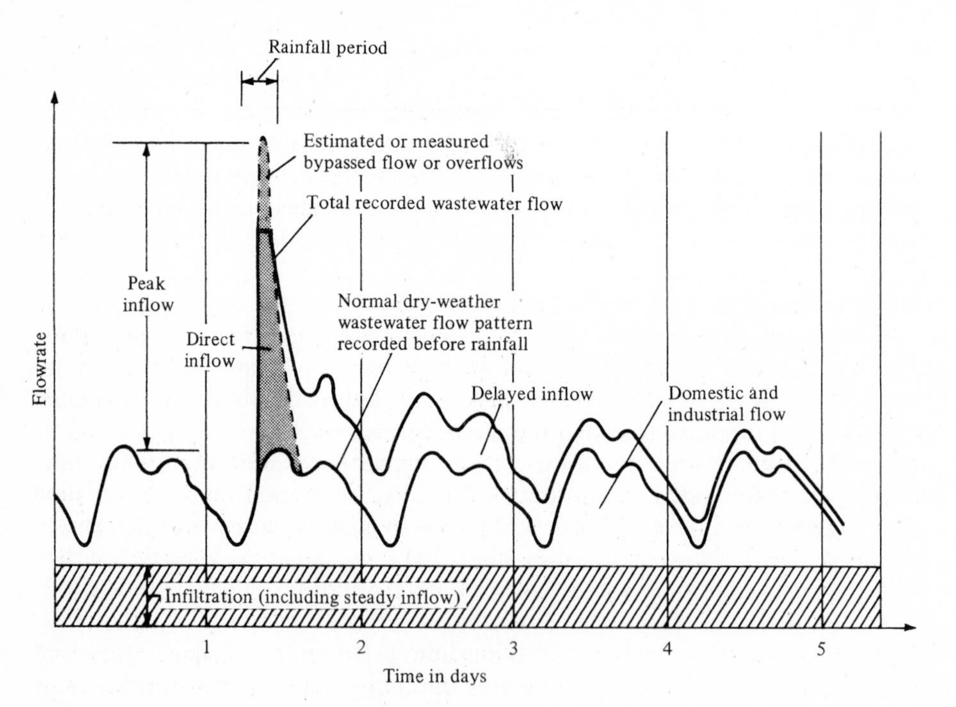

Figure 6-1 Graphic identification of infiltration/inflow.

Delayed inflow. This consists of storm water runoff that may require several days or more to drain through the sewer system. This category of inflow can include the discharge of sump pumps from cellar drainage as well as the slowed entry of surface water through manholes in ponded areas.

6-2 AN OVERVIEW OF SEWER SYSTEM EVALUATION AND REHABILITATION

Reduction of extraneous flows involves (1) definition of extraneous flow sources, (2) preparation of cost-effectiveness analyses to assess the feasibility of making any needed repairs, and (3) subsequent rehabilitation of the sewers if justified. More recently, with the passage of the Federal Water Pollution Control Act Ammendment of 1972 (Public Law 92-500), a structured program of sewer system evaluation and rehabilitation became a prerequisite for obtaining funds under the EPA construction grants program.

Elements of EPA Program

The current EPA sewer evaluation and rehabilitation program has three phases:

Phase I—infiltration/inflow analysis
Phase II—sewer system evaluation survey
Phase III—sewer system rehabilitation

The primary objective of the program is to ensure that, in sizing planned pollution control projects, capacity is not provided to convey or treat excessive infiltration, inflow, or both. Because excessive infiltration/inflow may not be specifically identifiable until the completion of phase II (sewer system evaluation survey), the program has often delayed the planning and/or design and construction process. The three phases of this program are illustrated in Fig. 6-2.

The purpose of phase I (analysis), conducted during the planning or preliminary engineering phase, is to determine whether the infiltration/inflow could be excessive. A cost-effectiveness analysis is conducted as the final step in phase I to determine if each identified I/I source is excessive. An I/I source is excessive if it is less costly to reduce it by repairing the sewer system than transporting and treating the excess flow. Phase II (evaluation survey) only begins if excessive infiltration/inflow is identified in phase I. In the evaluation survey, a detailed investigation is made of the collection system to determine the actual flowrate, type, and location of the I/I flows. In phase III (rehabilitation), only those sources determined excessive are recommended for elimination or reduction. Depending on the type of I/I source, rehabilitation can include plugging the storm-water connections to sewers (and making other provisions for the storm water), sealing leaking pipe joints and manholes, and completely replacing defective sewers.

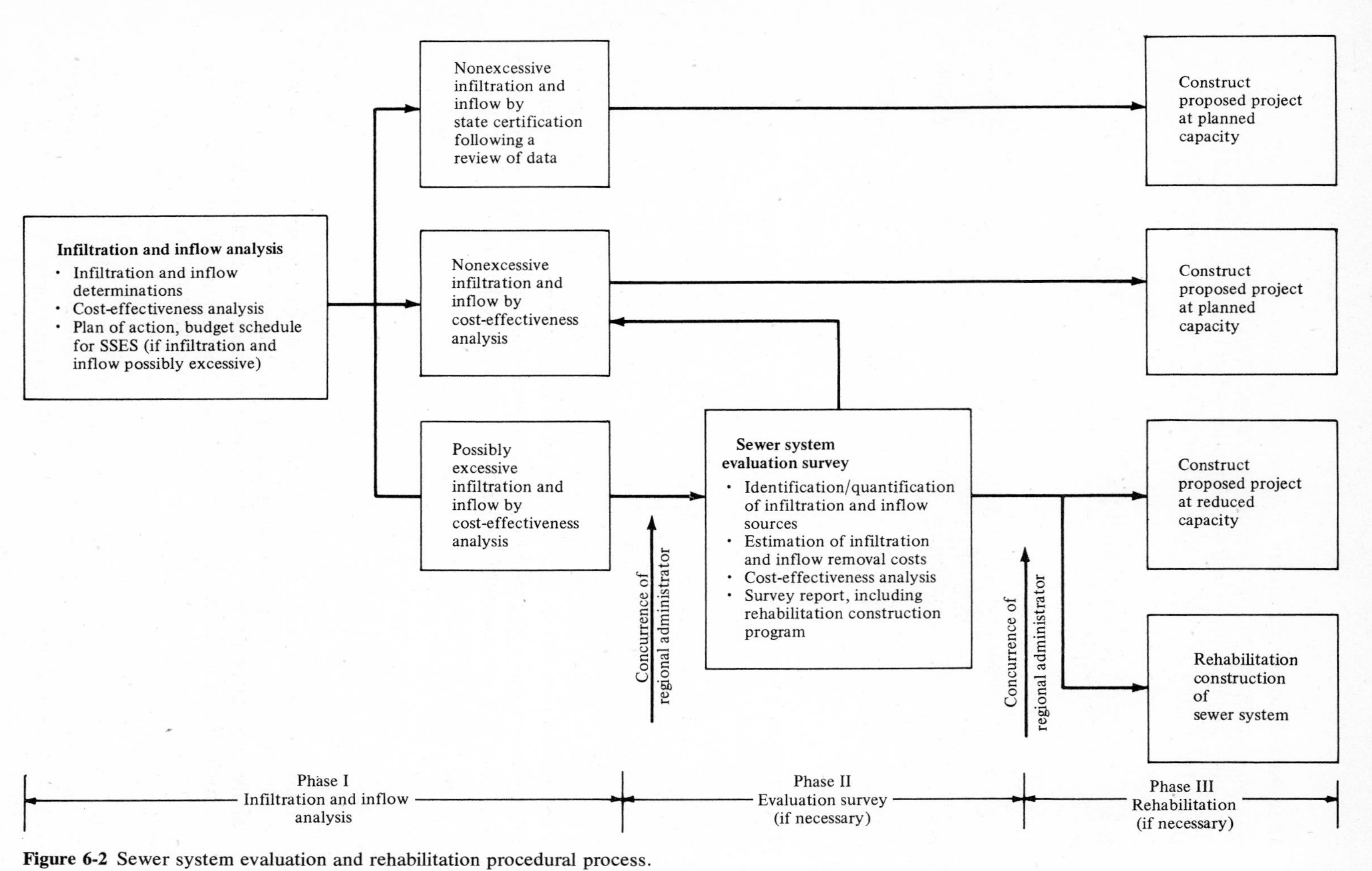

Figure 6-2 Sewer system evaluation and rehabilitation procedural process.

I/I Analysis in the Future

Although the current EPA program has some limitations, especially with respect to the emphasis placed on potential cost savings realized by I/I reduction, it has stressed the importance and proper use of the collection system. More recently several modifications have been made to the program. The most important are (1) simplification in the phase-I analysis, and (2) allowance of a "minor rehabilitation" (sealing of sewers) work within the phase-II survey. It is anticipated that additional changes will be made to the program as more field data are gathered and practical experience is gained.

The control and reduction of extraneous flows should be an ongoing effort. It is anticipated that progress will be made in the following areas and improve future efforts: better diagnostic methods for identifying I/I sources; improved flow-measurement techniques; improved methods of repair and materials for reducing leakage into sewers; and more reliable information on the success and duration of flow-reduction techniques, particularly sealing.

6-3 I/I ANALYSIS (PHASE I)

The I/I analysis should be considered a study of limited scope to determine if infiltration/inflow could be excessive. Thus only the data necessary to reach a conclusion should be collected. For example, if reliable metered flow data are available for a particular tributary area, a brief review of this information should be sufficient to disclose whether infiltration/inflow is excessive.

If an I/I analysis is necessary, it must be conducted systematically. The basic steps involved in such an analysis are discussed in this section. In addition, because of their importance, the determination of infiltration/inflow and the preparation of a cost-effectiveness analysis are discussed.

Basic Steps in I/I Analysis

The following steps are usually part of an I/I analysis:

1. Preparation of a sewer map indicating major system elements and subdivided into appropriate subsystems to facilitate the study.
2. Determination of the wastewater flows within the system, including flows bypassed at a treatment facility or through system overflow points. Flow-metering may be necessary to augment existing flow records.
3. Estimation of the portion of the flows that come from residential, commercial, industrial, and institutional sources.
4. Determination of the flows at key gaging points (if early studies indicate the possible existence of excessive extraneous flows) to isolate key areas for further study. Flow measurement at these key gaging points is preferably

done simultaneously so that the data collected represent I/I information obtained during the same storm event(s).

5. Identification and quantification of the extraneous portion of measured wastewater flows in the collection system and key portions thereof.
6. Evaluation of the wastewater collection system, including such factors as age, length, and materials of construction, and response to storms (from flow hydrographs).
7. Study of the geographical and geological conditions that may affect present and future infiltration rates.
8. Determination of whether infiltration/inflow is excessive by an economic analysis in which the costs of transporting and treating these flows are compared with the costs of various levels of I/I reduction (through rehabilitation). A cost-effectiveness analysis is used to make this assessment.
9. Preparation of a study plan for conducting a detailed evaluation survey (SSES) if excessive infiltration/inflow is identified.

Determination of Infiltration/Inflow

Infiltration is a steady 24-hr flow that usually varies during the year in relation to the groundwater levels above the sewers. It can be estimated by subtracting the 24-hr domestic and industrial flow component from the total 24-hr wastewater flow during dry-weather periods.

Infiltration rates are normally estimated from wastewater flows measured in the sewers during the early morning hours when water use is at a minimum and the flow is essentially infiltration. For small residential subareas (or subsystems) where the time required for wastewater to travel from the most remote part of the system upstream to the gaging point is about 3 to 4 hr or less, usually a high percentage of the early morning flowrate is infiltration. For larger systems where the travel time exceeds 6 hr, a lower percentage of the early morning flow is infiltration, particularly if a large part of the system is remote from the metering point. Only "order-of-magnitude" infiltration rates can be determined for larger systems with similar travel times.

Inflow rates are usually determined by using a network of continuous-flow meters that operate before and during a significant storm. The inflow rate can be determined from the flow hydrographs recorded with the flow meters by subtracting the normal dry-weather domestic and industrial flow and the infiltration (including steady flow) from the measured flowrate (see Fig. 6-1).

When the infiltration and inflow rates are determined for each subarea, it will usually be found that only a small part of a collection system contributes most of the infiltration/inflow. Generally, about 75 percent of the inflow comes from 20 to 30 percent of the system, whereas 75 percent of the infiltration comes from 40 percent of the area. This relationship is illustrated by the results of an evaluation survey in Houston, Texas, shown in Fig. 6-3. An example of an I/I determination summary tabulation taken from an I/I analysis report for Greenfield, Mass., is presented in Table 6-1.

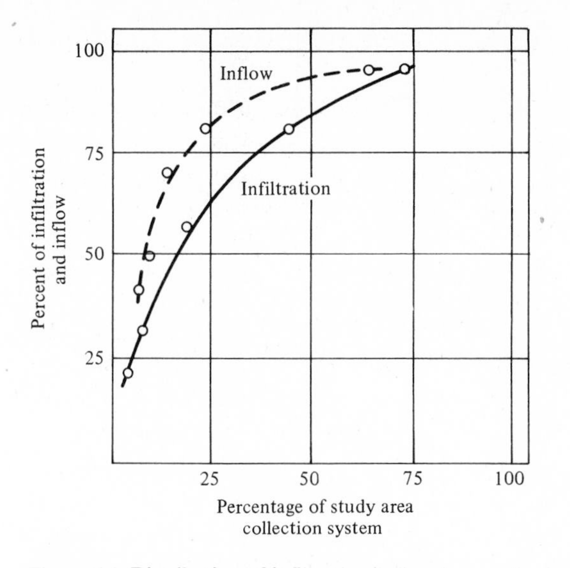

Figure 6-3 Distribution of infiltration/inflow in one district of the Houston, Texas, collection system [1].

Cost-Effectiveness Analysis

The cost-effectiveness analysis can be a simple or complex procedure, depending on the size and complexity of the service (study) area, the nature of the I/I condition, the future system needs, and the extent of the major existing system (transport and treatment) components.

When peak I/I rates are less than 0.15 $m^3/(d)(mm \cdot km)$ [1500 gal/(d)(in · mi)], including service connections, the infiltration/inflow is not usually considered excessive under most recent EPA regulations and a cost-effectiveness analysis is not required.

In a typical I/I analysis, the following information is required to determine the extent of the infiltration/inflow and the evaluation survey program that should be conducted if the infiltration/inflow is excessive:

1. Peak inflow rates by subarea.
2. Average and peak infiltration rates by subarea.
3. Estimates of flows bypassed from system including locations.
4. Projected peak flows tributary to major transport components.
5. Projected average and peak flow tributary to treatment facilities.
6. Capacities of all major existing transport components and treatment facilities.
7. Estimates of I/I reduction levels and costs by subarea.

When this information has been established, the transport and treatment systems are sized to meet future flow conditions at various levels of I/I reduc-

Table 6-1 I/I rates measured in Greenfield, Mass.

Meter	Tributary areas between metering points	Total infiltration to meter, m^3/d[a]	Infiltration between metering points m^3/d	Percent of total infiltration	Characteristics of tributary areas between metering points			Infiltration by various measures			Priority by $m^3/(d)(mm \cdot km)$	Peak inflow to meter, m^3/d	Peak inflow between metering points, m^3/d
					Area, ha	Sewers, $mm \cdot km$[b]	Persons	$m^3/ha \cdot d$ $(4 \div 6)$	$m^3/(d)(mm \cdot km)$ $(4 \div 7)$	$L/capita \cdot d$ $(4 \div 8)$			
(1)	(2)	(3)	(4)	(5)	(6)	(7)	(8)	(9)	(10)	(11)	(12)	(13)	(14)
4	*CRB*-1 2 + 3*A*	833	833	12	118	2,915	1,140	7.1	0.29	731	4	1,022	1,022
3	*CRB*-3*B*	909	76	1	61	1,741	460	14.9	0.04	165	6	833	. . .
7	*GR*-3	492	492	7	42	1,267	330	11.7	0.39	1,491	2	833	833
6	*GR*-4*A*	2,196	2,196	32	136	3,454	3,000	16.1	0.64	732	1	3,975	3,975
5	*GR*-4*B*	4,846	2,650	39	219	7,734	4,800	22.1	0.34	552	3	37,476	33,501
2*A*	*MB*, *GR*-1, 2 + 4*C*	6,474	277	4	148	5,796	2,690	43.7	0.04	84	8	23,470	. . .
2	*GR*-5 + 6	6,625	151	2	128	3,843	1,550	51.8	0.04	97	7	17,034	. . .
1*A*	*GR*-7 + 8	6,625	. . .	. . .	86	2,939	1,710	. . .	. . .	. . .	. . .	15,142	. . .
1	*DR*	189	189	3	27	1,063	230	7.0	0.18	822	5	1,401	1,401
WPCP[c]	Project area summary	6,814	6,814	100	965	30,743	15,910	7.1	0.22	428	. . .	17,413	37,854

[a]During gaging period April 18 to May 5, 1976; 90 to 100 percent of early morning flow.

[b]Including estimated lengths of building connections. The numerical value for each area is obtained by (1) multiplying the sewer diameters, in (mm) by the lengths (km) of sewers of corresponding diameters, and then (2) adding all of the products calculated in step 1.

[c]Water pollution control plant.

Note:
$m^3/d \times 2.6417 \times 10^{-4}$ = Mgal/d
$ha \times 2.4711$ = acre
$mm \times 0.03937$ = in
$km \times 0.6214$ = mi
$m^3/ha \cdot d \times 106.9064$ = $gal/acre \cdot d$
$m^3/(d)(mm \cdot km) \times 10{,}000$ = $gal/(d)(in \cdot m)$
$L \times 0.2642$ = gal

tion. To determine whether the flows are excessive, the capital and operating costs of the facilities are estimated for each I/I reduction level and compared with the costs of eliminating the infiltration/inflow. Several methods of conducting cost-effectiveness analyses are available and well documented in guidelines published by the EPA [2]. A detailed cost-effectiveness analysis is illustrated in Example 6-1.

Example 6-1 Cost-effectiveness analysis for a small municipal wastewater collection system A small municipal wastewater collection system serving about 20,000 people, or about half the city's population, is at capacity, and I/I problems (including one known overflow) have occurred during wet weather. By the year 2005, the population of the city served by the collection system is estimated to be about 33,000 people. The existing secondary treatment facility does not operate effectively during wet weather. It is designed to treat an average flow of 10,600 m^3/d (2.8 Mgal/d) and a peak flow of 30,000 m^3/d (8.0 Mgal/d).

During 1979, the city conducted its own gaging program and determined the levels of infiltration/inflow in the collection system. A schematic configuration of the system and the necessary information to conduct the cost-effectiveness analysis are presented in Fig. 6-4 and Tables 6-2 and 6-3, respectively.

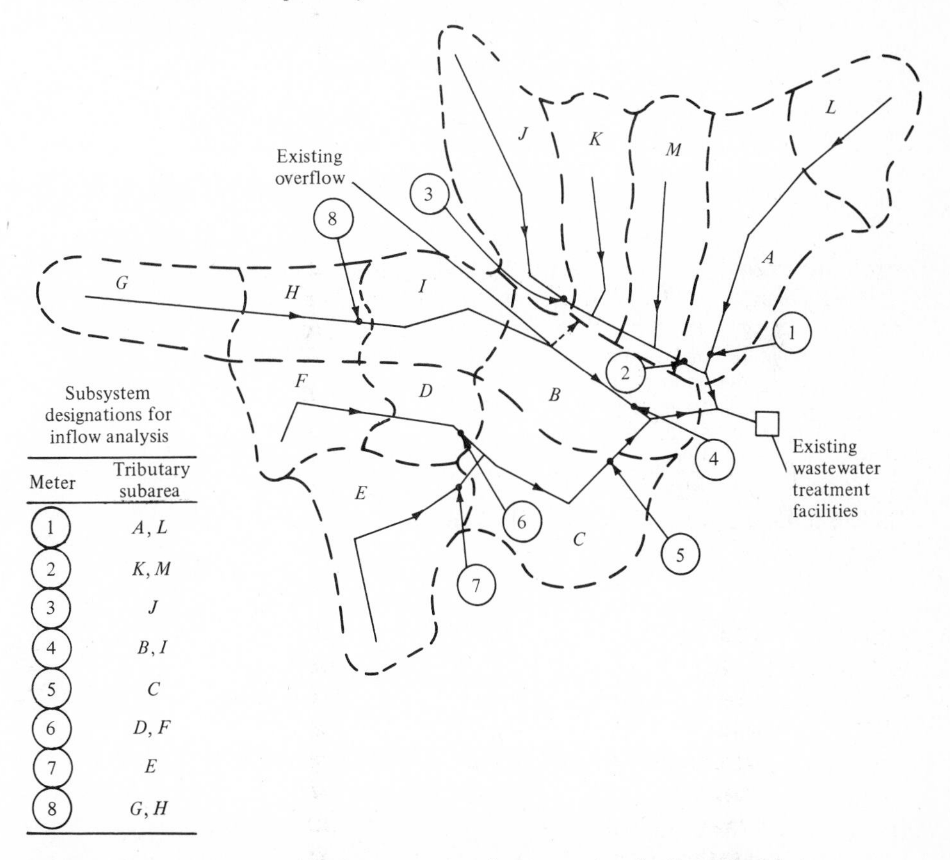

Meter	Tributary subarea
1	*A, L*
2	*K, M*
3	*J*
4	*B, I*
5	*C*
6	*D, F*
7	*E*
8	*G, H*

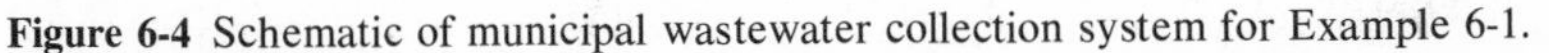

Figure 6-4 Schematic of municipal wastewater collection system for Example 6-1.

Table 6-2 Existing and projected I/I flow for Example 6-1

		Peak rates		Future population served by sewers	Future flows	
Subarea	Subsystem	Infiltration[a], m³/d	Inflow, m³/d		Average[b], m³/d	Peak[c], m³/d
Infiltration						
A		1,287		2,000	1,514	3,785
B		492		3,000	1,249	9,464
C		1,060		3,500	1,779	15,142
D		2,271		3,000	2,423	5,678
E		1,931		3,000	2,196	4,921
F		76		1,500	492	2,650
G		76		2,000	644	5,110
H		1,363		2,000	1,514	5,300
I		1,817		2,500	1,968	9,842
J		757		4,000	1,741	5,678
K		719		3,000	1,401	5,678
L		530		3,000	1,287	4,164
M		871		500	719	1,136
					18,927	78,548
Inflow						
A, L	1		379		2,801	7,949
K, M	2		3,028		2,120	6,814
J	3		379		1,741	5,678
B, I	4		10,599		3,217	19,306
C	5		10,221		1,779	15,142
D, F	6		1,136		2,915	8,328
E	7		0		2,196	4,921
G, H	8		4,542		2,158	10,410
System total		13,250	30,284		18,927	78,548

[a]Average infiltration = peak infiltration ÷ 1.5.
[b]Includes average infiltration.
[c]Includes peak infiltration and peak inflow.
Note: m³/d × 2.6417 × 10^{-4} = Mgal/d

Table 6-3 Estimated costs for I/I reduction for Example 6-1

Subarea/ subsystem designation	Estimated I/I reduction costs, $		
	Sewer system evaluation survey	Rehabilitation	Total
Infiltration			
A	10,000	23,000	33,000
B	11,000	27,000	38,000
C	26,000	40,000	66,000
D	33,000	55,000	88,000
E	36,000	69,000	105,000

Table 6-3 Estimated costs for I/I reduction for Example 6-1 (Continued)

Subarea/ subsystem designation	Estimated I/I reduction costs, $		
	Sewer system evaluation survey	Rehabilitation	Total
F	0	0	0
G	0	0	0
H	30,000	49,000	79,000
I	33,000	63,000	96,000
J	20,000	36,000	56,000
K	19,000	37,000	56,000
L	14,000	33,000	47,000
M	9,000	22,000	31,000
Inflow			
1 *A*, *L*	15,000	12,000	27,000
2 *K*, *M*	25,000	46,000	71,000
3 *J*	16,000	9,000	25,000
4 *B*, *I*	27,000	32,000	59,000
5 *C*	27,000	31,000	58,000
6 *D*, *F*	14,000	10,000	24,000
7 *E*	20,000	6,000	26,000
8 *G*, *H*	20,000	7,000	27,000

Since the purpose of this example is to illustrate the methodology involved in conducting an I/I cost-effectiveness analysis, supporting data needed for the analysis are given and are not derived. In an actual situation, the given data would have to be derived before the analysis could begin.

SOLUTION

1. Assume various levels of I/I reduction and estimate the corresponding future average and peak flows.
   a. Future flows corresponding to four assumed levels of I/I reduction are reported in Table 6-4.
   b. The flows in Table 6-4 are derived by using the data presented in Table 6-2. For example, the future peak flow with an 80 percent reduction in inflow is determined as follows:

$$\text{Future peak flow with 80\% inflow reduction} = 78{,}548 \text{ m}^3/\text{d} - (0.80)(30{,}284 \text{ m}^3/\text{d}) = 54{,}321 \text{ m}^3/\text{d}$$

The future average flow with a 40 percent reduction in infiltration is determined as follows:

$$\text{Future average flow with 40\% infiltration reduction} = 18{,}927 \text{ m}^3/\text{d} - (0.40)(13{,}250 \text{ m}^3/\text{d} \div 1.5) = 15{,}394 \text{ m}^3/\text{d}$$

Table 6-4 Future system flows based on various levels of I/I reduction for Example 6-1

		Future system wastewater flows, m^3/d	
Level	Description	Average	Peak
1	No I/I reduction	18,927	78,548
2	Remove 50% of inflow	18,927	63,406
3	Remove 80% of inflow	18,927	54,320
4	Remove 90% of inflow and 40% of infiltration	15,394	45,992

Note: $m^3/d \times 2.6417 \times 10^{-4}$ = Mgal/d

Table 6-5 Capital costs of transport facilities for Example 6-1

Level[a]	Transport cost, $
1	2,100,000
2	1,000,000
3	300,000
4	0

[a]See Table 6-4.

Table 6-6 Estimated capital costs for treatment components for flow level 1 for Example 6-1

Component added/modified	Design parameter	Affected by flow: Yes	Affected by flow: No	Capital costs, $
Flow equalization	Flow	X		1,400,000
Influent pumping	Flow	X		200,000
Preliminary treatment	Flow	X		150,000
Primary clarifiers	Flow	X		500,000
Aeration tanks	BOD		X	500,000
Secondary clarifiers	Flow	X		500,000
Chlorination	Flow	X		150,000
Solids handling	Settleable solids		X	1,000,000
Miscellaneous sitework and outside piping	. . .		X	400,000
Total				4,800,000

2. Size any needed transport facilities (relief sewers) and estimate the associated capital costs for transport. The estimated transport costs are shown in Table 6-5.
3. Size the additional treatment components required for handling future flows (level 1), and estimate the capital costs by flow- and nonflow-related components. The estimated costs are shown in Table 6-6.
4. Size and estimate the capital costs for the treatment components at the assumed I/I reduction levels. The estimated costs are presented in Table 6-7.
5. Estimate the annual operating and maintenance costs for the treatment facilities for the assumed I/I reduction levels. The estimated annual operating and maintenance (O & M) costs are listed in Table 6-8.

Table 6-7 Estimated capital costs at various I/I reduction levels for Example 6-1

Component added/modified	I/I reduction costs, $			
	Level 1	Level 2	Level 3	Level 4
Flow equalization	1,400,000	800,000	0	0
Influent pumping	200,000	180,000	160,000	150,000
Preliminary treatment	150,000	140,000	120,000	120,000
Primary clarifiers	500,000	430,000	370,000	370,000
Aeration tanks	500,000	500,000	500,000	500,000
Secondary clarifiers	500,000	430,000	370,000	370,000
Chlorination	150,000	140,000	120,000	120,000
Solids handling	1,000,000	1,000,000	1,000,000	1,000,000
Miscellaneous sitework and outside piping	400,000	400,000	400,000	400,000
Total	4,800,000	4,020,000	3,040,000	3,030,000

Table 6-8 Estimated O & M costs for various I/I reduction levels for Example 6-1

Classification	I/I reduction costs, $			
	Level 1	Level 2	Level 3	Level 4
Labor	120,000	120,000	120,000	120,000
Power				
Pumping	30,000	26,000	20,000	18,000
Other	30,000	30,000	30,000	30,000
Miscellaneous utilities	20,000	20,000	20,000	20,000
Chlorine	40,000	35,000	30,000	28,000
Maintenance and repair	50,000	50,000	50,000	50,000
Sludge processing/disposal	30,000	30,000	30,000	30,000
Total	320,000	311,000	300,000	296,000

Table 6-9 Present worth of capital and O & M costs for Example 6-1

Description of costs	Capital and O & M costs, $[a]			
	Level 1	Level 2	Level 3	Level 4
Transport	2,100,000	1,000,000	300,000	0
Treatment				
Capital	4,800,000	4,020,000	3,040,000	3,030,000
O & M	3,200,000	3,110,000	3,000,000	2,960,000
Total	10,100,000	8,130,000	6,340,000	5,990,000

[a]Based on a design period and discount rate of 20 years and 8 percent, respectively.

6. Summarize all transport and treatment costs for each assumed I/I reduction level on a present-worth basis and plot the resultant four costs to establish a transport- and treatment-cost curve. (The design period and discount rate are 20 yr and 8 percent, respectively.)
   a. the present-worth values of the improvements necessary for each I/I reduction level are shown in Table 6-9.
   b. The transport and treatment curves are plotted in Fig. 6-5 as curve *A*.

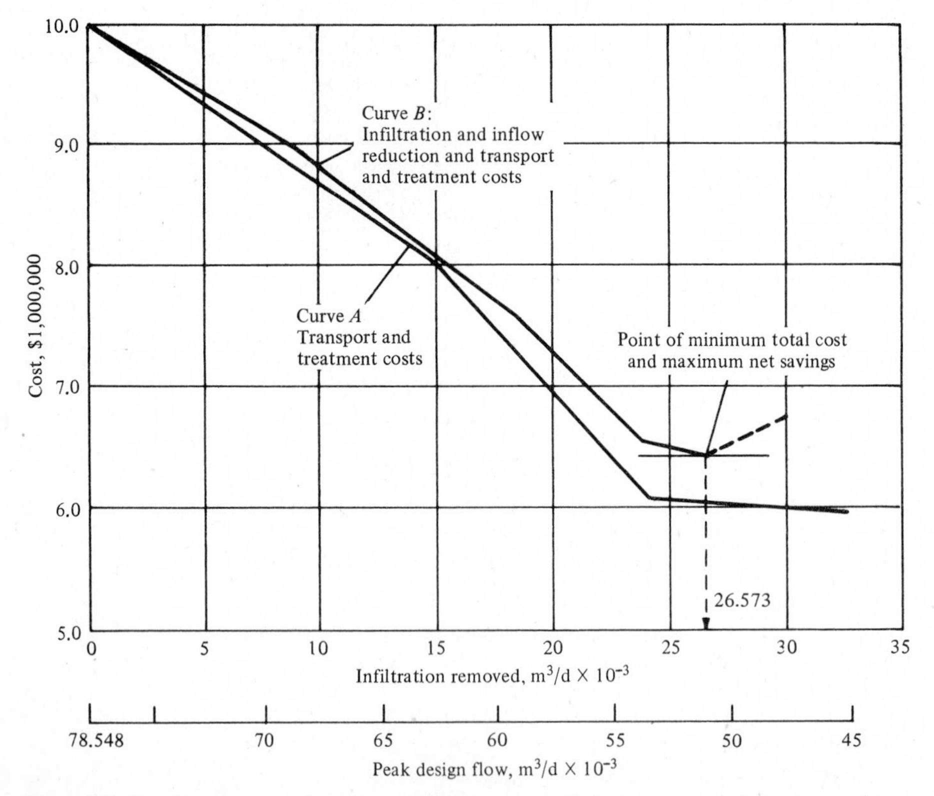

Figure 6-5 Graphic representation of cost-effectiveness analysis for Example 6-1. Note: $m^3/d \times 2.4711 \times 10^{-4}$ = Mgal/d.

7. Assume that 90 percent of the inflow is reduced in the subsystem and that 40 percent of the infiltration is reduced in the subarea, and rank all subareas and subsystems by assigning the highest priorities to those with the lowest I/I reduction cost per cubic meter per day. The required computations are presented in Table 6-10.
8. Prepare a cost-effectiveness summary listing the subareas and subsystems in order of priority along with the associated costs and accumulated savings. The cost-effectiveness summary is given in Table 6-11. For the most part, Table 6-11 is self-explanatory, but the following comments clarify its development and use.
   a. Columns 2 and 3 are data repeated from Table 6-10, cols. 1 and 6, but rearranged in order of priority.
   b. The data in col. 4 are obtained from Table 6-10, col. 3, and are the summation of the I/I reductions.
   c. The data in col. 5 are obtained by subtracting the values in col. 4 from 78,548 m^3/d (the total peak design flow without any reduction in inflow or infiltration).

Table 6-10 Determination of highest priorities for I/I reduction for Example 6-1

Subarea/ subsystem designation	Peak infiltration or inflow, m^3/d		Total cost reduction		Priority ranking by $\$/m^3 \cdot d$
	Total[a]	Reduction	By area, $[b]	Per unit flow, $\$/m^3 \cdot d$	
1	2	3	4	5	6
Infiltration					
A	1,287	515	33,000	64	6
B	492	197	38,000	193	16
C	1,060	424	66,000	156	14
D	2,271	908	88,000	97	10
E	1,931	772	105,000	136	12
F	76	30	0	...	...
G	76	30	0	...	...
H	1,363	545	79,000	145	13
I	1,817	726	96,000	132	11
J	757	303	56,000	185	15
K	719	288	56,000	194	17
L	530	212	47,000	222	18
M	871	348	31,000	89	9
Inflow					
1	379	341	27,000	79	8
2	3,028	2,725	71,000	26	5
3	379	341	25,000	73	7
4	10,599	9,539	59,000	6.2	1
5	10,221	9,199	58,000	6.3	2
6	1,136	1,022	24,000	23	4
7	0	0	0	0	...
8	4,542	4,088	27,000	7	3

[a]From Table 6-3.
[b]From Table 6-4.
Note: $m^3/d \times 2.6417 \times 10^{-4}$ = Mgal/d

Table 6-11 Cost-effective summary for I/I analysis for Example 6-1

Component of I/I reduced	Area designation		Accumulated I/I reduced, m^3/d	Peak design flow after I/I reduction, m^3/d	Accumulated costs, $1,000			
	Priority	Subarea or subsystem			I/I reduction	Transport and treatment	I/I removal and transport and treatment	Net savings
1	2	3	4	5	6	7	8	9
. . .	. . .	. . .	0	78,548	0	10,100	10,100	0
Inflow	1	4	9,539	69,009	59	8,859	8,918	1,182
Inflow	2	5	18,738	59,810	117	7,422	7,539	2,561
Inflow	3	8	22,826	55,722	144	6,616	6,760	3,340
Inflow	4	6	23,848	54,700	168	6,415	6,583	3,517
Inflow	5	2	26,573	51,975	239	6,241	6,480[a]	3,620[b]
Infiltration	6	*A*	27,088	51,460	272	6,220	6,492	3,608
Inflow	7	3	27,429	51,119	297	6,205	6,502	3,598
Inflow	8	1	27,770	50,778	324	6,191	6,515	3,585
Infiltration	9	*M*	28,118	50,430	355	6,177	6,532	3,568
Infiltration	10	*D*	29,026	49,522	443	6,138	6,581	3,519
Infiltration	11	*I*	29,752	48,798	539	6,108	6,647	3,453
Infiltration	12	*E*	30,524	48,024	644	6,075	6,719	3,381
Infiltration	13	*H*	31,069	47,479	723	6,052	6,775	3,325
Infiltration	14	*C*	31,493	47,055	789	6,035	6,824	3,276
Infiltration	15	*J*	31,796	46,752	845	6,022	6,867	3,233
Infiltration	16	*B*	31,993	46,555	883	6,014	6,847	3,253
Infiltration	17	*K*	32,281	46,267	939	6,002	6,941	3,159
Infiltration	18	*L*	32,493	46,055	986	5,993	6,979	3,121

[a]Minimum total cost.
[b]Maximum net savings.
Note: $m^3/d \times 2.6417 \times 10^{-4}$ = Mgal/d

d. The data in col. 6 are the summations of the cost reductions by area given in Table 6-10, col. 4.
e. The cost data in col. 7 are from curve *A* in Fig. 6-5, step 6*b*. (These data can be obtained by direct computation.)
f. The cost data in col. 8 are obtained by summing the corresponding values in cols. 6 and 7.
g. The cost data in col. 9 are obtained by subtracting the corresponding data in col. 8 from 10,100 (cost corresponding to no reduction in inflow or infiltration).

9. Prepare a graphic solution of the cost-effectiveness summary data developed in step 8. The data given in col. 8 of Table 6-11 are plotted in Fig. 6-5 as curve *B*, and the minimum total cost is identified.

Comment The minimum total cost from col. 8 in Table 6-11 represents the most cost-effective level of I/I reduction (26,573 m^3/d) and corresponds to the point of maximum savings in col. 9. It can be concluded from this analysis that (1) inflow in five of the eight subsystems may be excessive and (2) infiltration is not excessive. An evaluation survey should be recommended for the five subsystems to identify inflow; other observations are:

1. Only a small savings in O & M costs can be realized by I/I reduction.
2. Inflow is usually excessive more often than infiltration because of its profound effect on the sizing of flow-related components.
3. Preliminary estimates of the average and peak design flows for the expanded treatment facilities should be 18,927 and 51, 975 m^3/d, respectively, subject to the cost-effectiveness analysis of inflow removal in the evaluation-survey phase.

6-4 SEWER SYSTEM EVALUATION SURVEY (PHASE II)

If it can be shown from an I/I analysis that it may be cost-effective to remove a portion of the flow attibutable to infiltration and inflow, a sewer system evaluation survey (SSES) must be made. The collection system must be surveyed before any repairs are undertaken to verify the extent of the infiltration and inflow and to identify the I/I sources more specifically. Consequently, those portions of the collection system considered the most likely sources of extraneous flows, based on data from the I/I analysis, must be inspected in detail.

The objectives of the evaluation survey are: (1) to establish the type, location, and flowrate of each significant I/I source; (2) to estimate the cost of eliminating or reducing each source; and (3) to identify those sources which are excessive by comparing the costs associated with flow reduction with those for transport and treatment.

Basic Steps in Evaluation Survey

The EPA structured evaluation survey has four basic field-work steps: physical survey, inflow investigations, preparatory cleaning, and internal inspection, followed by the preparation of the cost-effectiveness analysis and report.

Step 1—physical survey of the collection system. The first task in step 1 is to conduct an above-ground survey of the study area primarily to record potential or obvious I/I sources, such as manhole covers in depressed areas, yard and areaway drains outside the limits of the storm-drainage system, or exposed manholes adjacent to watercourses. This survey is followed by a manhole inspection to determine the general condition of the sewers and to select key manholes suitable for flow isolation. Flow measurements and continuous metering, as required, are conducted next.

Step 2—inflow investigations. Interconnection between the sewer and storm-drainage system, whether intended or not, can often be located by introducing smoke into the sewers or by artificially flooding adjacent areas or drainage structures and observing possible transference into the sewers. Various dyes added to the water can also be used for this purpose.

Step 3—preparatory cleaning. If required, the sewer is cleaned before internal inspection of the section. The cleaning ensures that inspection equipment will not be damaged by debris within the pipe. In addition, by removing grit and grease deposits from pipe walls, defects may be detected more easily. Typical sewer-cleaning methods are shown in Fig. 6-6.

Step 4—internal inspection. Sewers are normally inspected with the aid of closed-circuit-television equipment. A small television camera and a high-intensity lamp are mounted on skids and pulled through each length of sewer being inspected (see Fig. 6-7). The television picture is transmitted by cable to an above-ground monitoring station, usually in a truck, where it may be (and should be) videotaped as an inspection record. Photographs may also be taken on the monitor during the inspection (see Fig. 6-8).

Step 5—analysis and report. Following completion of the field work (steps 1 through 4), a report summarizing the results is prepared, and recommendations for a rehabilitation program are made based on a cost-effectiveness analysis. In this analysis, the more detailed costs of transporting and treating the identified extraneous flows are compared with the costs of eliminating the conditions causing them. The procedure followed is essentially the same as that outlined in Example 6-1, but the cost estimates may be more detailed.

Review of Evaluation Survey Methodology

Various aspects of the evaluation survey methodology are examined in the following discussion to provide further insight into the evaluation survey process.

Infiltration survey. In the infiltration survey, the most critical and time-consuming step can be the flow-isolation procedure in the physical survey. The infiltration is isolated by manhole reaches through a systematic series of instantaneous gagings usually made during the early morning hours. The flow isola-

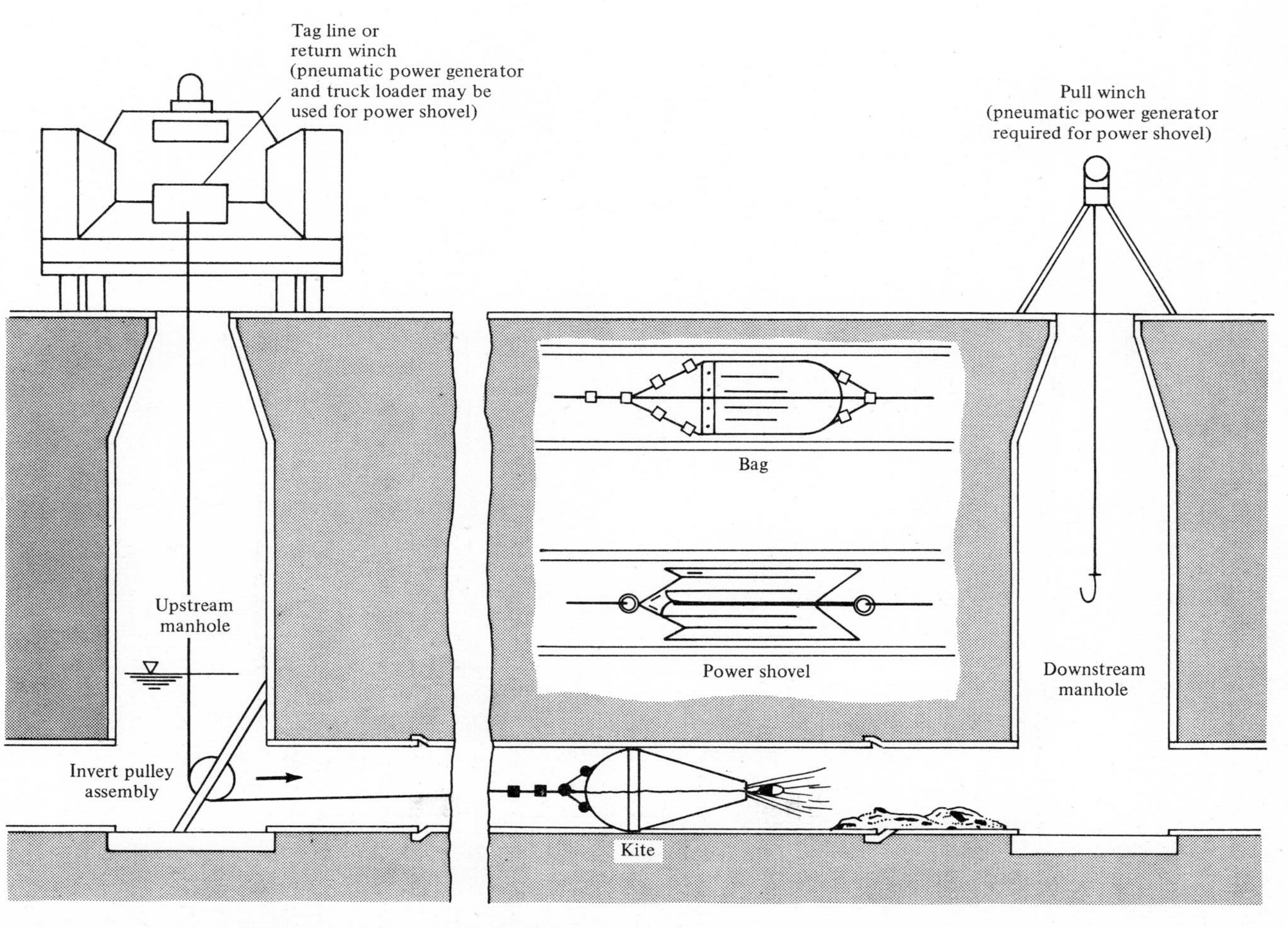

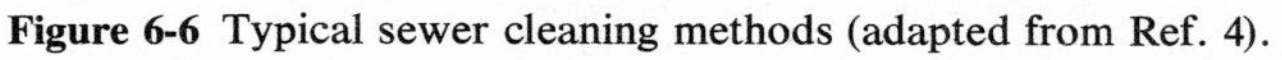

Figure 6-6 Typical sewer cleaning methods (adapted from Ref. 4).

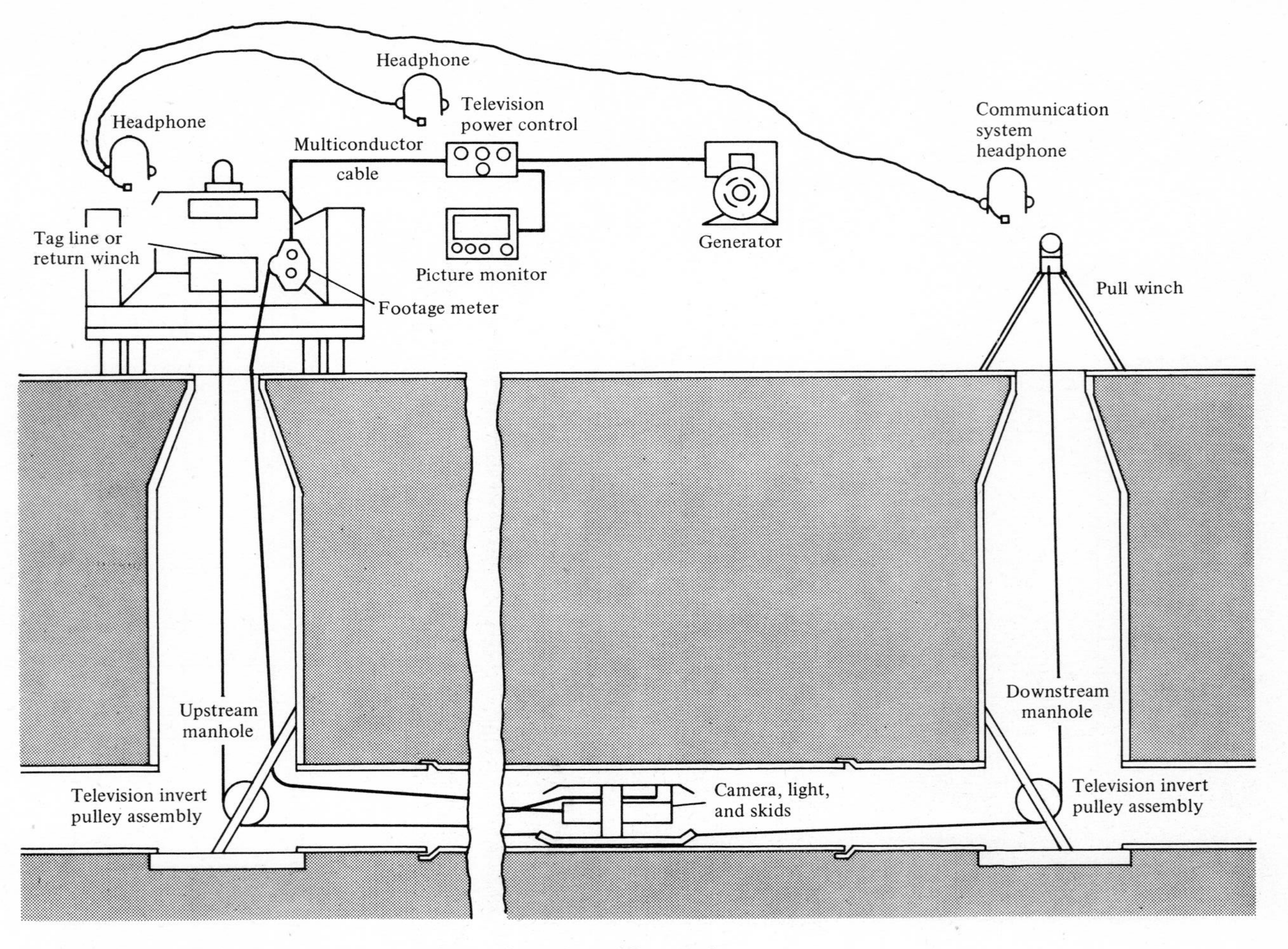

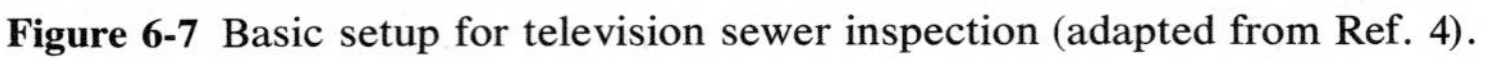
Figure 6-7 Basic setup for television sewer inspection (adapted from Ref. 4).

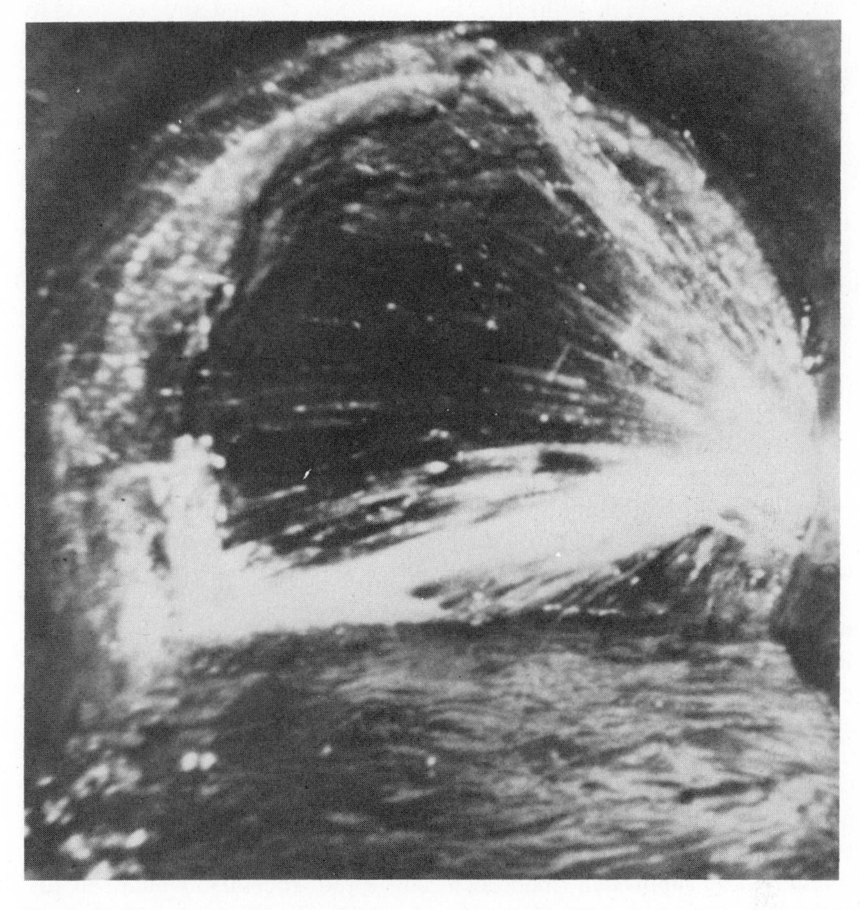

Figure 6-8 Photograph of sewer pipe interior taken during I/I analysis.

tion should be conducted during high-groundwater periods so that the resultant flow measurement approximates peak infiltration rates. The actual gagings should be obtained with portable weirs manufactured and calibrated for this purpose.

Two methods are commonly used for measuring and estimating infiltration rates from manhole to manhole. They are plugging and weiring, and differential measurements. The first method is more reliable and requires that flow in the manhole be plugged while the flow is measured at a manhole downstream. The measurement should be delayed until after the sewer is plugged and the line has drained. The second method, which requires no plugging, involves taking two flow measurements and then subtracting the upstream flowrate from that obtained downstream. The second method is less reliable and should only be used when plugging would be impractical.

Sewer reaches with high infiltration rates are then cleaned and televised to determine the sources of leakage: joints, crushed pipe, leaking service connections, etc. In the past several years, the EPA regulations have been revised to permit "minor rehabilitation" (primarily the sealing of sewers) during the evaluation survey phase.

Inflow survey. Inflow isolation, unlike infiltration, is a more difficult undertaking primarily because inflow is not readily measurable.

The most reliable way of measuring inflow is by using a network of continuous-flow meters that operate before and during a significant storm. The resultant flow hydrographs will provide proof of inflow and permit its quantification, unless the sewer is surcharged. If a substantial portion of the system inflow may come from sources on private property, then a house-to-house survey must be conducted.

The two most widely used techniques for detection of inflow are smoke-testing of sewers and dyed-water flooding and tracing of adjacent drainage structures and areas. Unfortunately, these measures alone cannot always provide the needed information. For example, trapped roof leader connections to the sewer system cannot be found by smoke-testing.

Alternative survey method. Because of the limited amount of data obtained in I/I analyses as now conducted, there is mounting evidence that the scope of an evaluation survey cannot be well defined with those data. Two factors contribute to this problem: (1) The difficulty in making even "order-of-magnitude" cost estimates for I/I reduction and (2) the difficulty in determining the actual level of I/I reduction possible. Both of these factors have a profound effect on the cost-effectiveness analysis, which, in turn, influences the extent of the system included in the evaluation survey.

An alternative approach has been suggested recently which, if implemented, might improve the survey process and, therefore, save time and expense, especially when analyzing a large collection system. The basic steps involved are:

1. Selection of a few small test areas, each with its own homogeneous characteristics (e.g., age or type of sewer pipe, type of developments, geologic conditions, etc).
2. Flow isolation of test areas
3. I/I reduction through rehabilitation
4. Determination of I//I reduction by follow-up flow isolation
5. Establishment of more reliable parameters for (*a*) the percent I/I reduction possible, and (*b*) associated I/I reduction (rehabilitation) costs
6. Preparation of a cost-effectiveness analysis with parameters developed during the I/I analysis to limit the area for the evaluation survey

Although this approach may be more suitable for infiltration surveys, the methodology should be valid for inflow surveys as well.

Cost-Effectiveness Analysis

The cost-effectiveness analysis in the evaluation survey is essentially the same procedure as illustrated in Example 6-1. The only difference is that all cost estimates take into account more detailed information (with positive identification of I/I sources) and are, therefore, more reliable.

6-5 SEWER SYSTEM REHABILITATION (PHASE III)

The reduction of infiltration/inflow through rehabilitation of collection systems is an important aspect of the current I/I program as well as of routine maintenance programs. At present, the equipment and the methods used are in a continual state of development. Therefore, current manufacturers' literature must be consulted, and site visits to ongoing programs are recommended.

I/I Reduction Methods

The principal methods of reducing both infiltration and inflow through rehabilitation are briefly discussed in the following. Additional details may be found in Refs. 3 and 4.

Sealing. When the source of infiltration/inflow is leaking pipe joints and the sewer pipe is otherwise free from defects such as cracks, the joint leakage can be reduced and sometimes eliminated by either of two methods. In the first method—the most popular—a pressure-grouting device (the packer) squeezes a chemical grout into the defective joint and the soil surrounding the pipe (see Figs. 6-9 and 6-10). When the surrounding soil material is not suitable for this process, a second method can be used in which an elastomeric material is injected under pressure into the joint opening. Manhole leakage can also be reduced with these methods.

Sewer lining and replacement. Deteriorated pipes may be either replaced with new pipe or lined by inserting a pipe sleeve, usually of polyethylene plastic. The method chosen depends primarily on the structural condition of the length of pipe. Plastic insertion sleeves or liners are not structurally capable of carrying the same loads as standard sewer pipes and, therefore, should not be used in pipe sections which are structurally unsound. Lining and replacement are both expensive methods of I/I reduction when compared with sealing.

Manhole renovations. In manholes installed at low points in the surrounding terrain, considerable leakage often occurs around exposed surfaces during periods of high rainfall and melting snow. Leakage sources include: holes in the cover; imperfect seating of the cover; and, most frequently, openings in the masonry grade adjustment courses between frame and manhole structure.

Leaks can be corrected either by sealing manhole covers or by raising the manhole. Manhole covers may be sealed by replacing the manhole rim and

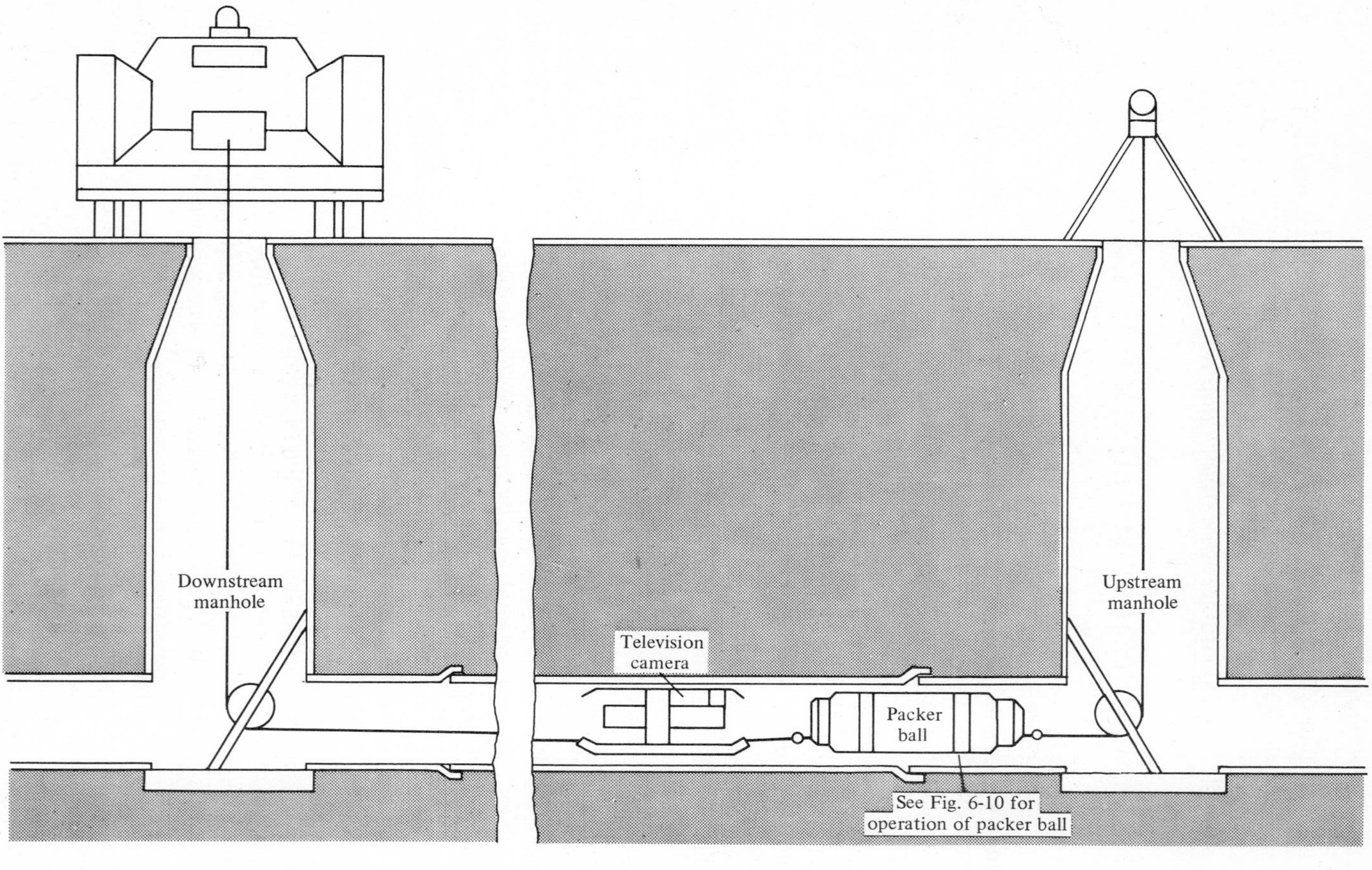

Figure 6-9 Television inspection camera and packer being winched to the nearest leak to be tested or sealed (adapted from Ref. 4).

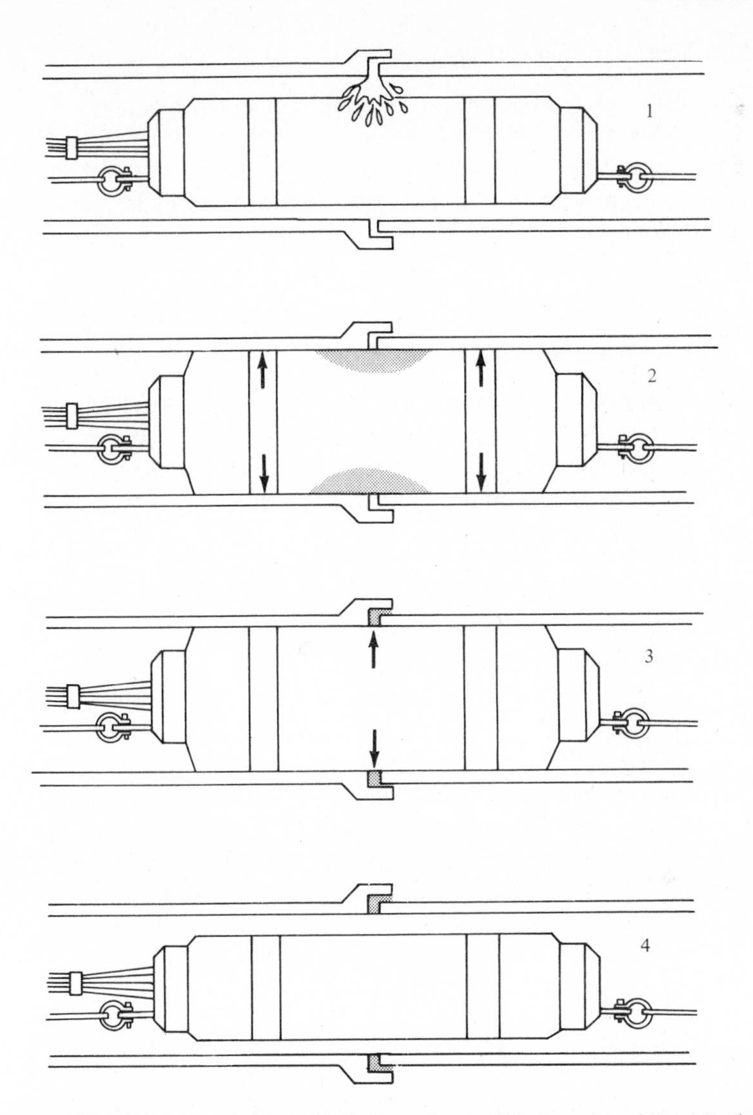

Figure 6-10 Operation of packer ball used to seal leaky sewer joints. Packer ball is positioned (1) and outer edges of ball are inflated to isolate the defective joint. Low-pressure air, which is introduced to the isolated section, confirms the leak, and grout material is inserted (2). The infiltration of the middle section of the packer ball (3) compacts the grout material in the joint or into the surrounding soil to form an effective seal. Packer ball is deflated (4) and moved to the next joint to be sealed [3].

cover with a rim, gasket, and cover set which can be bolted closed. Manholes may be raised by inserting one or more layers (courses) of waterproofed brick masonry between the manhole rim and wall. Leaks through defective manhole covers can be stopped only by replacing the covers.

Removal of Storm-Drainage Connections

The most pronounced inflow sources occur where either catch basins or the adjacent storm-drainage pipes are directly connected to the sanitary sewers. Both sources are eliminated by disconnection and plugging. The cost for this work can vary widely, because flows once transported through such connections must have an alternate route. Catch basins and cross connections must be reconnected to storm-water sewers or drainage ditches of adequate size.

Removal of I/I Sources from Service Connections

There is increasing evidence that a major part of infiltration/inflow can be attributed to the service connection. In public rights of way, service connection sources can usually be traced to broken service pipe, frequently at the junction of the sewer main. Such I/I sources can be eliminated by excavation and repair. On private property, I/I sources include broken service pipes, connections of areaway and foundation drains, sump pumps, and roof drains. These sources can be eliminated by disconnection from the service connection provided the disconnected drainage pipe is connected to storm-water sewers, to dry wells, or to streams or drainage ditches.

I/I Sources on Private Property

Private drainage connections to the sewers (areaway and foundation drains, sump pumps, roof drains), although now illegal in most municipal sewer-use regulations, may have been permissible at one time. Municipal officials are, therefore, often reluctant to institute programs directed at elimination of private I/I contributions, especially since the work is ineligible for federal grants under the current EPA construction-grants program.

Long-Term Effects of I/I Reduction

When I/I sources are removed by such rehabilitation methods as disconnection of drainage connections or replacement/slip-lining of defective sewer pipes, it can be assumed that the sources have been eliminated. However, the same cannot be assumed for the sewer-sealing process.

One particular concern about the lasting effect of the sealing process is groundwater migration. For example, if a leaking sewer main were sealed successfully, the groundwater level near the sealed main could increase and, therefore, cause adjacent service connections or joints to leak more. The answer to this question and others, particularly the level of infiltration reduction possible by sealing, may be forthcoming soon through the follow-up analyses of sewer-sealing programs now in process.

6-6 DESIGN STANDARDS FOR I/I PREVENTION AND CONTROL

The best method for controlling infiltration/inflow into sewer systems is to take measures initially during the planning and construction of new collection systems to prevent its entry. These measures should include:

1. Specifying low allowable leakage limits for pipe acceptance tests and carefully inspecting the quality of the construction work done
2. Requiring appropriate field acceptance testing
3. Specifying pipe products, jointing systems, and appurtenances proved for watertightness
4. Ensuring proper connection practices by local regulations and controls

The last two measures are described briefly in the following. Details on allowable leakage limits and field acceptance tests are provided in Chap. 4.

Pipe Materials, Jointing Systems, and Appurtenances

If the leakage into sewers must be minimized, a wide range of sewer pipe products and jointing systems are now available, which are more watertight than products used in the past.

As noted in Chap. 1, the newer and improved pipe materials, such as plastics, and joints made almost exclusively with gaskets of elastomeric materials practically ensure watertight joints. Significant improvements have also been made to the commonly used precast concrete manhole, including (1) special connecting devices at pipes that are somewhat flexible and usually result in nearly "bottle-tight" joints, and (2) better gasket joints between riser sections of the manhole.

Building Sewers

Building sewers or service connections are frequently the primary contributors of infiltration/inflow. Usually, the portion on public property is installed along with the public sewers and, therefore, can be inspected and tested readily. However, the remainder of the connection is on private property and, if installed without attention to local standards, inspection, and controls, can be a source of extraneous flow to an otherwise tight system.

Standards and inspection. Local jurisdictions should adopt rules governing the installation of connections, including materials and methods to be used. The installation work should be done only by experienced contractors licensed by the local authorities. A permit should be required, and the completed installation should be inspected by the local authority prior to the backfilling of the excavation.

Sewer use regulations. Regulations governing the use of the sewers should also be adopted by the local jurisdiction. The following is typical of regulations with respect to building sewers:

> The applicant for the building sewer permit shall notify the City's Agent when the building sewer is ready for inspection and connection to the public sewer. The connection shall be made under the supervision of the City's Agent or his/her deputy.

DISCUSSION TOPICS AND PROBLEMS

6-1 The diurnal flow patterns for two trunk sewer lines, each serving a residential area, are shown in Fig. 6-11. Which trunk sewer is most affected by infiltration? Explain your answer.

6-2 The typical dry-weather and wet-weather wastewater flows for a community are shown in Table 6-12. Estimate the wet-weather infiltration flowrate.

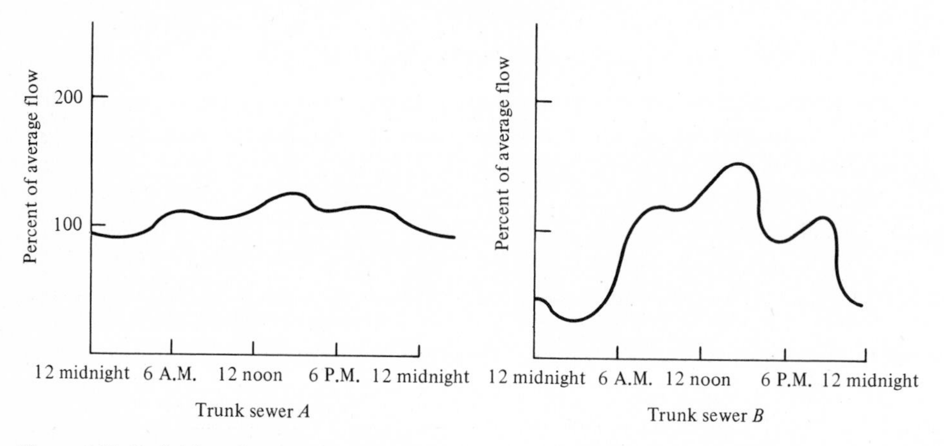

Figure 6-11 Definition sketches for Prob. 6-1.

Table 6-12 Typical dry- and wet-weather wastewater flows for Prob. 6-2

Time	Flow, m^3/s	
	Dry weather	Wet weather
Midnight–1 A.M.	0.21	0.40
1–2	0.18	0.38
2–3	0.17	0.33
3–4	0.23	0.40
4–5	0.40	0.62
5–6	0.62	0.80
6–7	0.81	0.99
7–8	1.05	1.23
8–9	1.20	1.39

Table 6-12 Typical dry- and wet-weather wastewater flows for Prob. 6-2 (Continued)

Time	Flow, m^3/s	
	Dry weather	Wet weather
9–10	1.20	1.39
10–11	1.25	1.46
11–Noon	1.34	1.51
Noon–1 P.M.	1.42	1.61
1–2	1.45	1.67
2–3	1.29	1.47
3–4	1.11	1.30
4–5	1.15	1.37
5–6	1.19	1.39
6–7	1.03	1.21
7–8	0.92	1.10
8–9	0.76	0.91
9–10	0.58	0.75
10–11	0.42	0.61
11–Midnight	0.37	0.55

6-3 Estimate the infiltration and the storm-water inflow for the gaging data reported in Fig. 6-12, which were obtained before, during, and after a rainfall period. The data are from a city located in the western United States that typically receives no rainfall during the summer. The averaged results of a number of measurements of domestic wastewater flows made during the summer are reported in Table 6-13.

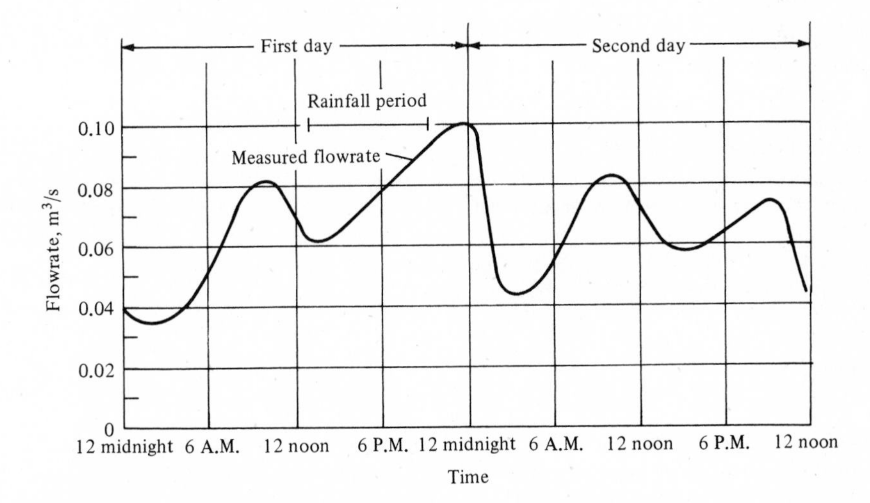

Figure 6-12 Measured wastewater flowrates before, during, and after a rainfall period for Prob. 6-3. Note: $m^3/s \times 22.8245 =$ Mgal/d.

Table 6-13 Averaged domestic flows for Prob. 6-3

Time period	Flow, m^3/s[a]
Midnight–1 A.M.	0.017
1–2	0.015
2–3	0.015
3–4	0.017
4–5	0.021
5–6	0.027
6–7	0.039
7–8	0.048
8–9	0.059
9–10	0.063
10–11	0.063
11–Noon	0.058
Noon–1 P.M.	0.048
1–2	0.039
2–3	0.036
3–4	0.036
4–5	0.037
5–6	0.039
6–7	0.043
7–8	0.047
8–9	0.050
9–10	0.050
10–11	0.037
11–Midnight	0.021

[a]Reported data are the average values for the indicated time period.

6-4 A mountain-resort community has a high visitor population during the summer. Wastewater flows from the community have been monitored late in the summer and late in the winter and have averaged 800 m^3/d and 840 m^3/d, respectively. The population in the summer is normally about 3000, whereas in the winter it is only about 1500. Estimate the winter infiltration flow. Clearly state any assumptions that may be necessary.

6-5 Typical wet-weather wastewater flows for a community and the flows monitored during a rainstorm are shown in Table 6-14. With these data estimate the total volume of inflow resulting from the storm.

6-6 If the average dry-weather wastewater flowrate for the community mentioned in Prob. 6-5 is 7.4 m^3/d, estimate the wet-weather infiltration flowrate.

REFERENCES

1. DeCoite, D. C. W., R. A. Tsugita, and R. Petroff: *Infiltration/Inflow Source Identification by Comprehensive Flow Monitoring,* presented at the 52d Annual Water Pollution Control Federation Conference, Houston, Texas, October 1979.
2. *Handbook for Sewer System Evaluation and Rehabilitation,* U.S Environmental Protection Agency, EPA 430/9-75-021, Washington, D.C., December 1975.

Table 6-14 Typical wet-weather flows for a community and the flows monitored during a rainstorm for Prob. 6-5

	Flows, m^3/s	
Time period	Typical wet weather	Storm related
Midnight–1 A.M.	3.7	...
1–2	3.0	...
2–3	2.7	...
3–4	3.0	...
4–5	3.4	...
5–6	5.5	...
6–7	9.1	...
7–8	9.9	...
8–9	10.2	...
9–10	10.7	...
10–11	11.6	...
11–Noon	12.5	...
Noon–1 P.M.	13.0	...
1–2	10.5	...
2–3	8.5	10.1
3–4	7.5	9.9
4–5	7.4	10.1
5–6	7.7	10.8
6–7	7.5	11.7
7–8	6.5	11.9
8–9	5.6	10.0
9–10	4.9	6.3
10–11	4.6	5.0
11–Midnight	4.2	4.2

3. *Sewer System Evaluation, Rehabilitation, and New Construction: A Manual of Practice,* U.S. Environmental Protection Agency, EPA 600/2-77-017d, December 1977.

4. Kerri, K. D., and J. Brady (eds.): *Operation and Maintenance of Wastewater Collection Systems,* prepared for the U.S. Environmental Protection Agency, Office of Water Program Operations, by Department of Civil Engineering, California State University, Sacramento, Calif., 1976.

CHAPTER

SEVEN

OCCURRENCE, EFFECT, AND CONTROL OF THE BIOLOGICAL TRANSFORMATIONS IN SEWERS

Ideally, a wastewater collection system should perform its function effectively and without creating a nuisance. Unfortunately, in too many locations this has not been the case. Problems which can reduce the performance of sewers include excessive groundwater infiltration and storm-water inflow (discussed in Chap. 6), excessive corrosion in sewers and related appurtenances, and the development of odors. The purpose of this chapter is to discuss (1) some of the major biological transformations, especially the generation of hydrogen sulfide, that can occur in wastewater during transport from its source to the point of treatment, and the implications of these transformations on the design and operation of treatment processes; (2) the corrosion of sewers and the methods to control corrosion; (3) the occurrence of gases in sewers, including those responsible for the development of odors, and their effect and control; and (4) the need for integrated design of collection systems and treatment facilities.

Ultimately, the proper functioning of a wastewater collection system depends on the coordinated efforts of many persons and agencies—the designer, the construction contractor, the owner of the system (usually a municipality) and maintenance personnel, the users (e.g., the residents, commercial centers, and industrial facilities), and the regulatory agencies. The proper management of collection systems is now being recognized as an important aspect of wastewater engineering. The information in this chapter and Chap. 6 is only an introduction to this important subject. Additional details on all aspects of the maintenance of collection systems may be found in Ref. 2.

7-1 BIOLOGICAL TRANSFORMATIONS OCCURRING DURING WASTEWATER TRANSPORT

Two of the most important problems in operating wastewater collection systems are (1) the corrosion of sewers and other facilities and (2) the control of odorous gases in sewers. Both problems are principally related to the generation of hydrogen sulfide.

Generation of Hydrogen Sulfide in Sewers

The hydrogen sulfide found in wastewater—other than that added from industrial sources, infiltrated groundwater, and septage discharged to sewers—results principally from the bacterial reduction of the sulfate ion (SO_4^{2-}) that may be present. Compounds such as sulfite, thiosulfate, free sulfur, and other inorganic sulfur compounds, which occasionally are found in wastewater, can also be reduced to sulfide. In the following, the terminology and equilibrium relationships related to sulfide in sewers are defined; the process microbiology and the processes and rate of sulfide generation, precipitation, oxidation, and emission are described; and the buildup of sulfide in sewers is discussed. References 5 through 8 and 10 should be consulted for additional details.

Definition of terms. The following terms are basic to an understanding of the biological and chemical processes that lead to the generation and buildup of hydrogen sulfide in sewers.

Hydrogen sulfide (H_2S). Hydrogen sulfide is a gas that occurs both in the sewer atmosphere and as a dissolved species in the wastewater. The gas is responsible for the rotten-egg odor in many wastewaters. Hydrogen sulfide can be oxidized by the action of bacteria on exposed sewer surfaces to form sulfuric acid, which leads to significant corrosion problems. The gas is toxic to humans and has caused the death of maintenance workers.

Hydrogen sulfide ion (HS^-). The hydrogen sulfide ion can be formed by the reversible dissociation of dissolved hydrogen sulfide. It is the combination of a positive hydrogen ion and the bivalent, negatively charged sulfide ion.

Sulfide ion (S^{2-}). The sulfide ion can be formed by the second dissociation of aqueous hydrogen sulfide. It can be thought of as elemental sulfur with two additional electrons.

Sulfide. Sulfide is a general term that can refer to any or all species containing the sulfide ion (most frequently the soluble-sulfide species, H_2S, HS^-, and S^{2-}).

Normally, when the concentration of H_2S or HS^- is reported, it is understood that the number given is the concentration of sulfur in the form of H_2S or HS^-. In either case, the actual concentration of each species is found by multiplying the reported value by the ratio of its molecular weight to that of sulfur.

For example, if the concentration of HS^- is reported as 10 mg/L, the actual amount of HS^- is $10 \times (33/32) = 10.3$ mg/L; or, if H_2S is reported as 10 mg/L, the actual amount of H_2S is $10 \times (34/32) = 10.6$ mg/L. Frequently, the use of this convention is denoted by reporting the concentration of the species "as S," but this is not true in general.

There are definite relationships between the concentrations of each of the soluble-sulfide species and the pH, as described next in the discussion of soluble-sulfide equilibrium.

Soluble-sulfide equilibrium. In an aqueous environment, it is possible to determine the portion of the total dissolved sulfide represented by each of the soluble-sulfide species after the pH is known. The relative concentrations of the three species, $H_2S(aq)$, HS^-, and S^{2-}, can be found directly from the following equilibrium relationships:

$$H_2S(aq) \rightleftharpoons H^+ + HS^- \qquad K_1 \simeq 10^{-7} \text{ @ } 25^\circ\text{C} \tag{7-1}$$

$$\frac{[H^+][HS^-]}{[H_2S]} = K_1 \tag{7-2}$$

$$HS^- \rightleftharpoons H^+ + S^{2-} \qquad K_2 \simeq 10^{-14} \text{ @ } 25^\circ\text{C} \tag{7-3}$$

$$\frac{[H^+][S^{2-}]}{[HS^-]} = K_2 \tag{7-4}$$

Although it is beyond the scope of this book to present the derivation, Eqs. 7-1 and 7-3 can be manipulated to produce the log distribution diagram shown in Fig. 7-1. As shown, the predominant forms of sulfide for pH values typically found in wastewater (6.5 to 8.0) are $H_2S(aq)$ and HS^-. Determination of the form of the sulfide present as a functon of pH is illustrated in Example 7-1.

Example 7-1 Determination of sulfide equilibrium in wastewater Determine the concentrations of hydrogen sulfide and hydrogen sulfide ion in the wastewater in a sewer flowing full for a pH range varying from 6.5 to 8.0. Assume that the amount of sulfide generated in the slime layer is sufficient to raise the total sulfide concentration in the wastewater to 3.0 mg/L. Because of the metal in the wastewater, 0.4 mg/L of the sulfide will form insoluble metallic sulfide.

SOLUTION

1. Derive the expression for computing the percentage distributions of H_2S and HS^- in the desired pH range.
   a. From Fig. 7-1, it can be seen that, for the pH range of concern, the amount of S^{2-} present will be negligible and the second dissociation of hydrogen sulfide (see Eq. 7-3) need not be considered. Thus the percentage distribution can be determined by using the first dissociation constant for hydrogen sulfide.
   b. The required expression for computing the percentage distribution of H_2S is derived as follows:

$$H_2S, \% \simeq \frac{[H_2S]}{[H_2S] + [HS^-]} 100 \simeq \frac{100}{1 + [HS^-]/[H_2S]}$$

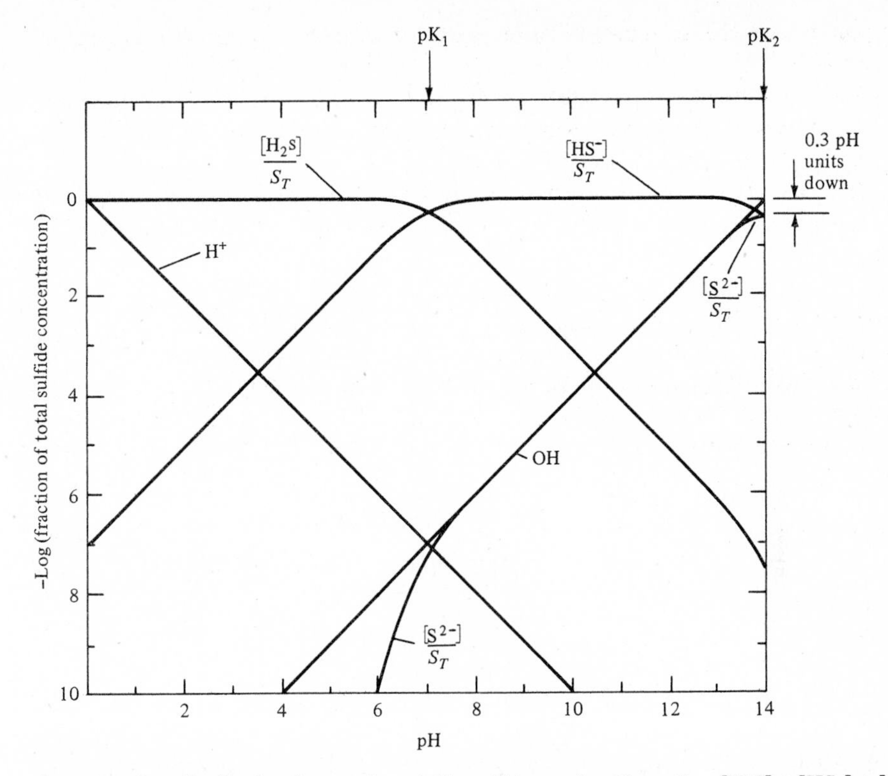

Figure 7-1 Log distribution diagram for soluble-sulfide species. Note: $S_T = [H_2S] + [HS^-] + [S^{2-}]$, all in moles per liter.

where concentrations may be in mol/L or mg/L as *S*. (*Note:* As a general rule, the concentrations in the equilibrium expression are in mol/L. However, because the ratio of concentrations of sulfide species in mol/L is equivalent to the ratio of concentrations in mg/L as *S*, it is permissible to use mg/L in this case.) But from Eq. 7-2, $[HS^-]/[H_2S]$ equals $K_1/[H^+]$ which, when substituted into the foregoing expression, results in

$$H_2S, \% \simeq \frac{100}{1 + K_1/[H^+]}$$

c. It follows directly that the percentage distribution of HS^- is

$$HS^-, \% \simeq 100 - H_2S, \%$$

$$\simeq 100 - \frac{100}{1 + K_1/[H^+]}$$

2. Determine the percentage distributions of H_2S and HS^-.

a. Assume that the value of K_1 in Eq. 7-2 is applicable, and determine the percentage distributions of H_2S and HS^- using the equations derived in steps 2*b* and 2*c*. For a pH value of 6.5, the percentage value of H_2S is

$$H_2S, \% \simeq \frac{100}{1 + 10^{-7}/10^{-6.5}} = 76.0\%$$

The percentage value of HS^- is $100 - 76.0 = 24.0\%$.

b. Determine the percentage distributions of H_2S and HS^- as a function of pH.

pH	H_2S, %	HS^-, %
6.5	76.0	24.0
7.0	50.0	50.0
7.5	24.0	76.0
8.0	9.1	90.0

The percentage distributions of H_2S and HS^- could have been determined from Fig. 7-1 as an alternative to applying the equations derived in steps 2*b* and 2*c*.

3. Determine the concentrations of H_2S and HS^- as a function of pH.
   a. After precipitation, the amount of hydrogen sulfide available is 2.6 mg/L.
   b. The distribution between species as a function of pH is as follows:

	Species, mg/L as *S*	
pH	H_2S	HS^-
6.5	1.98	0.62
7.0	1.30	1.30
7.5	0.62	1.98
8.0	0.24	2.36

Comment The simplified aqueous sulfide system used to illustrate equilibrium calculations in this example is not often encountered in practice. Corrosion and odor problems develop with the transfer of H_2S from the aqueous phase to the gaseous phase. When this mass-transfer process is included, a more complex system must be analyzed.

The transfer of H_2S to the gaseous phase depends on the concentration of H_2S dissolved in the wastewater and, as shown in this example, is therefore affected significantly by the pH. For any given conditions, an equilibrium hydrogen sulfide concentration is approached so that losses from the wastewater stream equal,the supply from the slime layer. This subject is considered further in a subsequent discussion (see "Sulfide buildup in sewers," p. 245).

Process microbiology. The reduction of organic and inorganic sulfur to sulfide can be brought about by a number of microorganisms. For example, many microorganisms can assimilate inorganic sulfur in the form of sulfate, sulfite, and thiosulfate, and reduce it to sulfide in their protoplasm. There it is used to build up sulfur-containing organics, principally proteins and amino acids, i.e., cysteine, methoionine, and cystine. Although these organisms play an important role in nature, they are of no concern in this discussion.

Of particular concern with respect to the generation of hydrogen sulfide in wastewater collection systems are the organisms that reduce sulfate to obtain energy for cell maintenance and growth. Under anaerobic conditions (devoid of oxygen), two genera of obligatory anaerobic bacteria of the species *Desulfovibrio,* commonly called sulfate-reducing bacteria, can convert sulfate to sulfide.

Dv. desulfuricans, Dv. vulgaris, and *Dv. sal xigens* are principal members of the species *Desulfovibrio* associated with this conversion.

The reduction reaction usually is coupled to the oxidation of organic matter and, in special cases, hydrogen. When lactic acid is a source of organic material, the conversion may be represented by the following reaction [3]:

$$\underset{\text{Lactic acid}}{2CH_3CHOHCOOH} + \underset{96}{SO_4^{2-}} \rightarrow 2CH_3COOH + \underset{32}{S^{2-}} + 2H_2O + 2CO_2 \quad (7\text{-}5)$$

In special cases where the organisms involved contain the enzyme *hydrogenase,* the following reaction, which is also important in the corrosion of iron pipes, can occur [3]:

$$4H_2 + \underset{96}{SO_4^{2-}} \rightarrow \underset{32}{S^{2-}} + 4H_2O \quad (7\text{-}6)$$

As shown in both Eqs. 7-5 and 7-6, 96 g of sulfate are required to produce 32 g of sulfide ion.

Because almost all wastewater contains bacteria capable of completing these reactions, as well as organic matter and varying quantities of the sulfate ion, the potential for the production of hydrogen sulfide always exists. To explain more fully why hydrogen sulfide is found in some sewers and not in others, the processes of sulfide generation, precipitation, oxidation, and emission must be understood.

Process and rate of sulfide generation. Because the bacterial reduction of sulfate to sulfide can occur only in an anaerobic environment, it usually occurs in the submerged slime layer that develops on sewer walls. A definition sketch of this slime layer in a sewer is shown in Figs. 7-2 and 7-3. In Fig. 7-2 it is assumed that the wastewater contains some dissolved oxygen (~1 mg/L); in Fig. 7-3 the wastewater is essentially devoid of oxygen.

Typically, the thickness of the slime layer varies from 1 to 1.5 mm (0.04 to 0.06 in), depending on the velocity of flow in the sewer. When the velocity is extremely low, slime layers as thick as 3 mm (0.125 in) or more can develop. If much silt or other abrasive material is in the wastewater, a slime layer may not develop.

The slime layer usually contains a heterogenous population of microorganisms, including filamentous forms, that protrude from the layer. The thickness of the inert anaerobic layer gradually increases, and periodically a portion breaks away from the sewer wall either as a result of the shearing action of moving wastewater or from its own weight. As shown in Figs. 7-2 and 7-3, SO_4^{2-}, organic matter, and nutrients diffuse into the anaerobic layer, and sulfide generated within the anaerobic layer diffuses outward into the moving stream of wastewater.

From studies conducted over a number of years, Pomeroy has suggested the following empirical expression for predicting the rate of sulfide generation

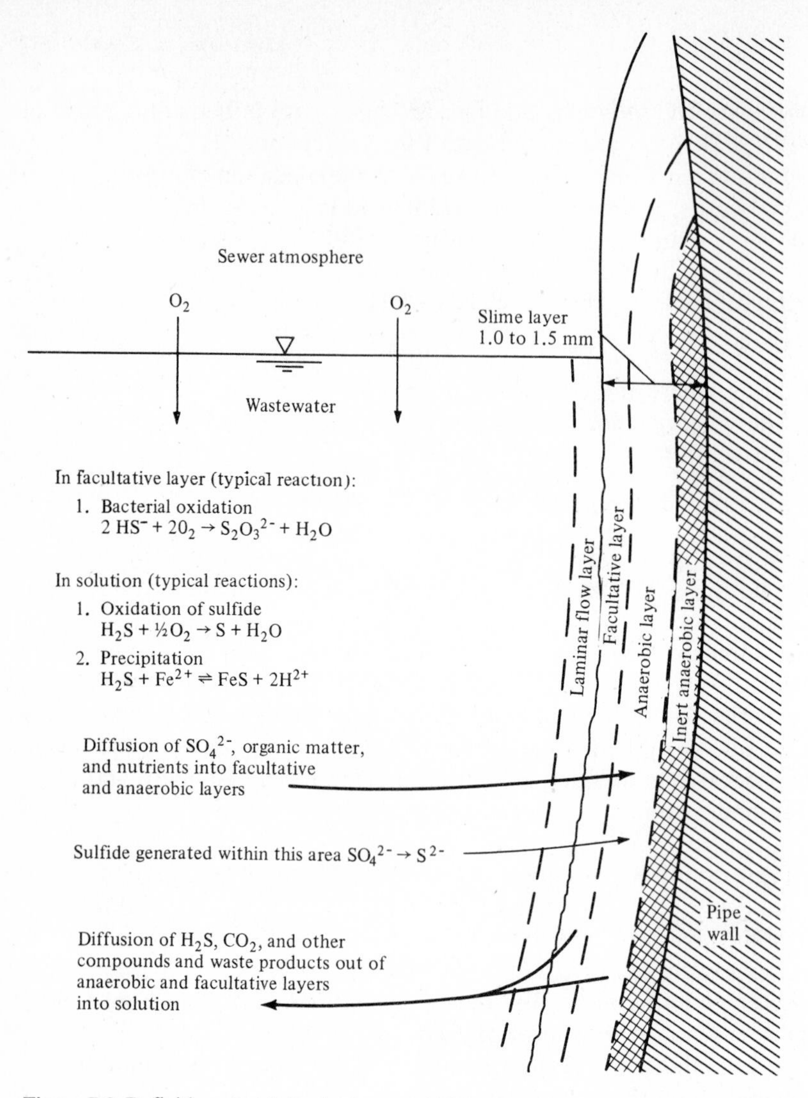

Figure 7-2 Definition sketch for hydrogen sulfide generation in wastewater containing more than 1 mg/L dissolved oxygen (adapted from Ref. 8). Note: mm × 0.03937 = in.

expressed as a flux from the slime layer [5, 6, 7]:

$$\phi_{sg} = M(\text{EBOD}) \tag{7-7}$$

where ϕ_{sg} = rate of sulfide generation expressed as a flux from the slime layer, $g/m^2 \cdot h$

M = observed sulfide flux coefficient, m/h

EBOD = effective BOD_5 (5-day biological oxygen demand at 20°C), g/m^3
= $BOD_5(1.07)^{t-20}$

1.07 = temperature coefficient

t = temperature, °C

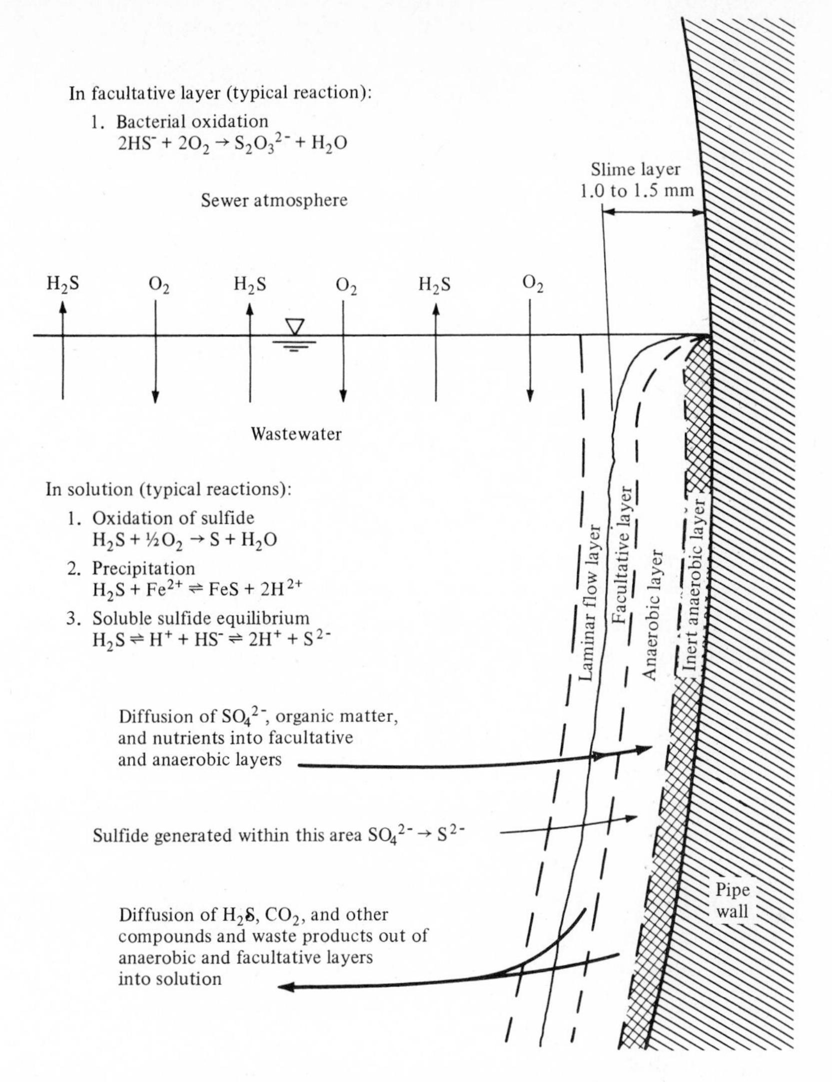

Figure 7-3 Definition sketch for hydrogen sulfide generation in wastewater containing less than 0.1 mg/L dissolved oxygen (adapted from Ref. 8). Note: mm × 0.03937 = in.

The values of M suggested by Pomeroy are 1×10^{-3} m/h for completely full sewers (e.g., force mains) and 0.32×10^{-3} m/h for partly full sewers [7].

These values were determined for sewers of the Los Angeles County Sanitation District in which conditions were favorable for sulfide generation (no dissolved oxygen and a sufficient supply of sulfates). Using these values to determine the rate of sulfide generation in other sewers will not necessarily lead to the actual generation rate; the rates predicted will be approximations of sulfide generation if worst-case conditions develop.

To determine how the sulfide flux from the slime layer affects the sulfide concentration in the wastewater, it is helpful to consider a slug of wastewater moving through a sewer such as the one shown in Fig. 7-4.

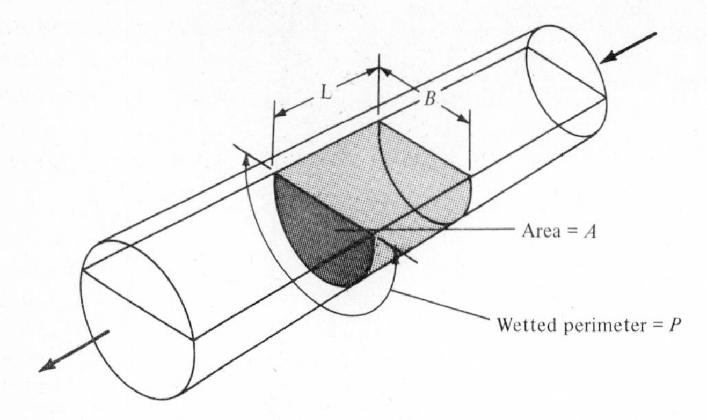

Figure 7-4 Definition sketch for a slug of wastewater moving in a sewer.

The area of the slime layer through which the sulfide flux diffuses into the slug wastewater is $P \times L$, (PL) as shown in Fig. 7-4. Similarly, the volume of liquid into which the sulfide diffuses is given by $A \times L$ (AL). Thus the rate of change of the sulfide concentration in the slug of wastewater caused by sulfide generation in the slime layer can be calculated as follows:

$$r_{sg} = \frac{\phi_{sg} PL}{AL} \qquad \text{(SI units)} \tag{7-8}$$

$$r_{sg} = \frac{3.28\phi_{sg} PL}{AL} \qquad \text{(U.S. customary units)} \tag{7-8a}$$

where r_{sg} = rate of change of sulfide concentration, caused by sulfide generation in the slime layer, $g/m^3 \cdot h$

ϕ_{sg} = rate of sulfide generation expressed as a flux from the slime layer, $g/m^2 \cdot h$

P = wetted perimeter, m (ft)

L = length of slug, m (ft)

A = cross-sectional area, m^2 (ft^2)

Noting that A/P is the hydraulic radius, Eq. 7-8 can be simplified to

$$r_{sg} = \phi_{sg}R^{-1} \qquad \text{(SI units)} \tag{7-9}$$

$$r_{sg} = 3.28\phi_{sg}R^{-1} \qquad \text{(U.S. customary units)} \tag{7-9a}$$

where R = hydraulic radius, m (ft).

For partly filled sewers, it can be assumed that the slime layer is the only source of sulfide generation in the sewer. However, in completely full sewers, more highly reduced conditions can exist because there is no sewer atmosphere to supply oxygen, and sulfide may be generated within the wastewater stream. Pomeroy has developed a correction factor to modify Eq. 7-9 so that the sulfide generated within the wastewater stream is accounted for [7]. The resulting expression for the rate of change of the sulfide concentration caused by sulfide generation in completely full sewers is

$$r_{sg} = \phi_{sg}(1 + 1.57R)R^{-1} \quad \text{(SI units)} \tag{7-10}$$

$$r_{sg} = \phi_{sg}(1 + 0.48R)R^{-1} \quad \text{(U.S. customary units)} \tag{7-10a}$$

where r_{sg} = rate of change of sulfide concentration, caused by sulfide generation in completely full sewers, $g/m^3 \cdot h$

$(1 + 1.57R)$
$(1 + 0.48R)$ = empirical factor for sulfide production in the wastewater stream

R = hydraulic radius, m (ft)

The rate of sulfide generation, which depends on many interrelated factors, is not well quantified theoretically. Among the important factors that affect the rate of generation are:

1. The rate of transfer of SO_4^{2-}, organic matter, and nutrients into the anaerobic slime layer
2. The concentrations of the sulfate ion and the organic matter, usually measured in terms of BOD_5
3. The characteristics (cell age, species distribution, etc.,) and mass of the sulfate-reducing organisms present in the slime layer
4. Environmental conditions, such as pH, the dissolved-oxygen level, and temperature
5. The operational features of the collection system, such as the time of travel, depth of flow, velocity of flow, degree of sewer venting, and the presence of silt deposits

The effects of most of these factors are combined and incorporated into the observed sulfide flux coefficient.

After sulfide has developed in the anaerobic slime layer and has diffused outward into the facultative layer and the wastewater, it may remain in solution, it may be precipitated as metallic sulfide, it may be oxidized, or it may be released as hydrogen sulfide gas to the sewer atmosphere.

Precipitation of metallic sulfides. The reaction of hydrogen sulfide with metallic constituents in the wastewater may be characterized by the following reaction with iron:

$$H_2S + Fe^{2+} \rightarrow FeS + 2H^+ \tag{7-11}$$

Other metals such as zinc, lead, copper, cadmium, and cobalt will also be precipitated. The presence of metals can reduce the concentration of soluble sulfide; in fact, the addition of metals is one of the methods for controlling sulfides in sewers.

Oxidation of sulfide. In the presence of free oxygen (see Fig. 7-2), the hydrogen sulfide ion diffusing outward from the anaerobic layer can be oxidized bacterially in the facultative layer to sulfate and other partially oxidized compounds.

The bacterial oxidation of the hydrogen sulfide ion to thiosulfate can be represented as

$$2O_2 + 2HS^- \rightarrow S_2O_3^{2-} + H_2O \tag{7-12}$$

Hydrogen sulfide diffusing outward into the wastewater can also be oxidized rapidly to elemental sulfur according to the following reaction:

$$H_2S + \tfrac{1}{2}O_2 \rightarrow S + H_2O \tag{7-13}$$

Pomeroy has found that the rate of sulfide oxidation is proportional to the total dissolved-sulfide concentration S_T [$S^{2-} + HS^- + H_2S(aq)$] and the reaeration rate r_r (rate of oxygen transport from the sewer atmosphere into the wastewater stream in $g/m^3 \cdot h$).

The transport of oxygen from the sewer atmosphere into the wastewater stream, expressed as a flux through the exposed water surface, is

$$\phi_O = f_O D_O \tag{7-14}$$

where ϕ_O = flux of oxygen through exposed water surface, $g/m^2 \cdot h$
f_O = oxygen exchange coefficient, m/h
D_O = oxygen deficit in wastewater (the difference between the dissolved-oxygen concentration that would be in equilibrium with the sewer atmosphere and the actual dissolved-oxygen concentration), g/m^3

The oxygen exchange coefficient for flow in sewers has been determined as [8]:

$$f_O = 0.96 C_A \gamma (SV)^{3/8} \quad \text{(SI units)} \tag{7-15}$$

$$f_O = (0.64)(0.96) C_A \gamma (SV)^{3/8} \quad \text{(U.S. customary units)} \tag{7-15a}$$

where f_O = exchange coefficient for oxygen, m/h
0.96 = empirical coefficient applicable to wastewater
C_A = factor accounting for the additional surface area made available from turbulent flow. The term V^2/gd_m in the following expression for C_A is the square of the Froude number used in hydraulic computations.

$$C_A = 1 + \frac{0.17V^2}{gd_m}$$

where 0.17 = empirical constant
V = wastewater velocity, m/s (ft/s)
g = acceleration due to gravity, 9.81 m/s^2 (32.2 ft/s^2)
d_m = mean hydraulic depth, m (ft)
= A/B
A = cross-sectional area of wastewater stream, m^2 (ft^2)
B = width of free surface in sewer, m (ft)
γ = temperature coefficient equal to $(1.07)^{t-20}$
t = temperature, °C
S = slope of energy grade line, m/m (ft/ft)

Thus

$$\phi_O = 0.96 C_A \gamma (SV)^{3/8} D_O \qquad \text{(SI units)} \qquad (7\text{-}16)$$

$$\phi_O = (0.64)(0.96) C_A \gamma (SV)^{3/8} D_O \qquad \text{(U.S. customary units)} \qquad (7\text{-}16a)$$

The reaeration rate may be determined by again considering the slug of wastewater shown in Fig. 7-4. The area through which the oxygen flux occurs is the free-surface area $B \times L$. Thus the reaeration rate is

$$r_r = \frac{\phi_O BL}{AL} \qquad \text{(SI units)} \qquad (7\text{-}17)$$

$$r_r = \frac{3.28 \phi_O BL}{AL} \qquad \text{(U.S. customary units)} \qquad (7\text{-}17a)$$

$$r_r = \phi_O d_m^{-1} \qquad \text{(SI units)} \qquad (7\text{-}18)$$

$$r_r = 3.38 \phi d_m^{-1} \qquad \text{(U.S. customary units)} \qquad (7.18a)$$

where r_r = reaction rate, g/m^3 · h
ϕ_O = flux of oxygen through exposed water surface, g/m^2 · h
B = width of exposed water surface, (ft)
L = length of wastewater slug, m (ft)
A = cross-sectional area of wastewater slug, m^2 m (ft^2)
d_m = mean depth = A/B, m (ft)

Combining Eqs. 7-16 and 7-18 leads to the following relationship for the reaction rate:

$$r_r = 0.96 C_A \gamma (SV)^{3/8} D_O d_m^{-1} \qquad \text{(SI units)} \qquad (7\text{-}19)$$

$$r_r = (2.10)(0.96) C_A \gamma (SV)^{3/8} D_O d_m^{-1} \qquad \text{(U.S. customary units)} \qquad (7\text{-}19a)$$

The rate of oxidation of soluble sulfide, as stated previously, can be proportional to the reaeration rate and the total soluble-sulfide concentration S_T. Thus,

$$r_{so} = K r_r S_T \qquad (7\text{-}20)$$

$$r_{so} = \mathrm{K} 0.96 C_A \gamma (SV)^{3/8} D_O d_m^{-1} S_T \qquad \text{(SI units)} \qquad (7\text{-}21)$$

$$r_{so} = \mathrm{K}(2.10)(0.96) C_A \gamma (SV)^{3/8} D_O d_m^{-1} S_T \qquad \text{(U.S. customary units)} \qquad (7\text{-}21a)$$

where r_{so} = rate of sulfide oxidation, g/m^3 · h
K = proportionality constant
S_T = total concentration of soluble sulfide
$[S^{2-} + HS^- + H_2S(aq)]$, g/m^3 as S

Emission of hydrogen sulfide to the sewer atmosphere. The physical emission of hydrogen sulfide to the sewer atmosphere must occur before odor and corrosion problems can develop. Pomeroy [5, 8] has estimated that the rate at which hydrogen sulfide is released from the wastewater stream is given by

$$\phi_{se} = f_s D_s \tag{7-22}$$

where ϕ_{se} = flux of hydrogen sulfide from the stream surface, $g/m^2 \cdot h$
f_s = exchange coefficient for hydrogen sulfide between an aqueous phase and a gas phase (in this case, between the wastewater and sewer atmosphere, respectively), m/h
D_s = hydrogen sulfide driving force [the difference between the actual concentration of $H_2S(aq)$ and the concentration that would be in equilibrium with $H_2S(g)$ found from Henry's law], g/m^3

The exchange coefficient f_s averaged 0.72 of the oxygen exchange coefficient f_O. Thus,

$$f_s = 0.72 f_O = (0.72)(0.96) C_A \gamma (SV)^{3/8} \quad \text{(SI units)} \tag{7-23}$$

$$f_s = 0.72 f_O = (0.72)(0.64)(0.96) C_A \gamma (SV)^{3/8} \quad \text{(U.S. customary units)} \tag{7-23a}$$

The hydrogen sulfide driving force D_s, a measure of the rate at which a gas will enter or leave the aqueous phase, depends on the relative concentrations of aqueous hydrogen sulfide and hydrogen sulfide in the sewer atmosphere [note that $H_2S(aq)$ depends on the pH value of the wastewater as shown in Example 7-1]. If the concentrations are in equilibrium, there will be no driving force. Frequently, however, the wastewater is supersaturated with hydrogen sulfide, and the driving force is found from

$$D_s = H_2S(aq) - H_2S(aq)\ \text{(saturation)} \tag{7-24}$$

The saturation concentration of aqueous hydrogen sulfide must be determined on the basis of Henry's law: the equilibrium concentration of aqueous hydrogen sulfide is proportional to the partial pressure of hydrogen sulfide gas in the atmosphere, or

$$H_2S(aq)\ \text{(saturation)} = K P_{H_2S} \tag{7-25}$$

where $H_2S(aq)$ (saturation) = saturation concentration of aqueous H_2S, g/m^3
K = Henry's constant
P_{H_2S} = partial pressure of $H_2S(g)$ in the atmosphere, atm

Since the partial pressure of $H_2S(g)$ is directly proportional to the concentration of $H_2S(g)$, Eq. 7-25 may be rewritten as

$$H_2S(aq)\ \text{(saturation)} = K' H_2S(g) \tag{7-26}$$

where K' = constant
$H_2S(g)$ = concentration of H_2S in the atmosphere, g/m^3

Equation 7-27, which is the converse of Eq. 7-26, can be used to determine the concentration of $H_2S(g)$ that would be in equilibrium with a given concentration of $H_2S(aq)$.

$$\frac{1}{K'} H_2S(aq) = H_2S(g)\ \text{(saturation)} \tag{7-27}$$

Dividing Eq. 7-26 by Eq. 7-27 and rearranging leads to

$$H_2S(aq)\text{ (saturation)} = H_2S(aq)\left[\frac{H_2S(g)}{H_2S(g)\text{ (saturation)}}\right] \quad (7\text{-}28)$$

The term in brackets in Eq. 7-28 has been given the symbol q and is referred to as the relative saturation of H_2S in the atmosphere. Substituting q and combining Eqs. 7-24 and 7-28 leads to

$$D_s = (1 - q)H_2S(aq) \quad (7\text{-}29)$$

Typically, q will be between 0.02 and 0.20 [8].

The concentration of $H_2S(aq)$ can be expressed as jS_T, where S_T is the total soluble-sulfide concentration in g/m³ [$H_2S(aq) + HS^- + S^{2-}$ in g/m³ as S] and j is the fraction of S_T represented by $H_2S(aq)$. Thus

$$D_s = (1 - q)jS_T \quad (7\text{-}30)$$

The value of j depends on the pH value and may be determined from Fig. 7-1 or from the equation in Example 7-1, step 2*b*.

Combining Eqs. 7-22, 7-23, and 7-30 results in

$$\phi_{se} = 0.69C_A\gamma(SV)^{3/8}(1 - q)jS_T \quad \text{(SI units)} \quad (7\text{-}31)$$

$$\phi_{se} = 0.44C_A\gamma(SV)^{3/8}(1 - q)jS_T \quad \text{(U.S. customary units)} \quad (7\text{-}31a)$$

It can be shown that the rate of decrease of the hydrogen sulfide concentration caused by losses to the sewer atmosphere is similar to the rate of increase of the dissolved-oxygen concentration caused by reaeration, which was shown to be the flux of oxygen divided by the mean hydraulic depth. The relationship governing the decrease in the hydrogen sulfide concentration is

$$r_{se} = \phi_{se}d_m^{-1} \quad \text{(SI units)} \quad (7\text{-}32)$$

$$r_{se} = 3.28\phi_{se}d_m^{-1} \quad \text{(U.S. customary units)} \quad (7\text{-}32a)$$

where r_{se} = rate of decrease of sulfide concentration in the wastewater caused by emission to the sewer atmosphere, g/m³ · h

d_m = mean hydraulic depth, m (ft)

Thus

$$r_{se} = 0.69C_A\gamma(SV)^{3/8}(1 - q)jS_Td_m^{-1} \quad \text{(SI units)} \quad (7\text{-}33)$$

$$r_{se} = (3.28)(0.44)C_A\gamma(SV)^{3/8}(1 - q)jS_Td_m^{-1} \quad \text{(U.S. customary units)} \quad (7\text{-}33a)$$

Sulfide buildup in sewers. If a sewer is flowing full, essentially all the sulfide generated will remain in the wastewater stream (except that lost by precipitation). However, for partly full sewers the sulfide concentration in the wastewater will ultimately approach a limiting value. When this limiting value is reached, the losses and the inputs to the wastewater stream will be approximately the same.

The limiting value of sulfide can be estimated by equating the rate of generation and the sum of the rates of loss caused by oxidation and emission. (It is assumed that there are no metals present in the wastewater for precipitation of sulfide.)

$$r_{sg} = r_{so} + r_{se} \tag{7-34}$$

Substituting for r_{sg}, r_{so}, and r_{se} in Eq. 7-34 results in the following expression for partly full sewers:

$$M(\text{EBOD})R^{-1} = K(0.96)C_A\gamma(SV)^{3/8}D_O d_m^{-1}S_T + 0.69C_A\gamma(SV)^{3/8}(1-q)jS_T d_m^{-1} \quad \text{(SI units)} \tag{7-35}$$

$$3.28M(\text{EBOD})R^{-1} = K(2.10)(0.96)C_A\gamma(SV)^{3/8}D_O d_m^{-1}S_T + (3.28)(0.44)C_A\gamma(SV)^{3/8}(1-q)jS_T d_m^{-1} \quad \text{(U.S. customary units)} \tag{7-35a}$$

Because of the difficulty involved in specifying some of the terms of the right side of Eq. 7-35, the equation has been modified and simplified by introducing an empirical constant m as follows:

$$M(\text{EBOD})R^{-1} = m(SV)^{3/8}d_m^{-1}S_T \quad \text{(SI units)} \tag{7-36}$$

$$3.28M(\text{EBOD})R^{-1} = 2.10m(SV)^{3/8}d_m^{-1}S_T \quad \text{(U.S. customary units)} \tag{7-36a}$$

where $m = K(0.96)C_A\gamma D_O + 0.69C_A\gamma(1-q)j$

Solving for the limiting value of sulfide $(S_T)_{\text{lim}}$ yields

$$S_T = (S_T)_{\text{lim}} = \frac{M}{m}(\text{EBOD})(SV)^{-3/8}\frac{P}{B} \quad \text{(SI units)} \tag{7-37}$$

$$S_T = (S_T)_{\text{lim}} = 1.56\frac{M}{m}(\text{EBOD})(SV)^{-3/8}\frac{P}{B} \quad \text{(U.S. customary units)} \tag{7-37a}$$

where P = wetted perimeter, m (ft)
B = width of water surface, m (ft)

To determine the rate at which the asymptotic value $(S_T)_{\text{lim}}$ is approached, it is necessary to calculate a mass balance on the soluble sulfide for a slug of wastewater during transport in a sewer. The general statement for the mass balance is

$$\text{Accumulation} = \text{inflow} - \text{outflow} + \text{generation} - \text{utilization} \tag{7-38}$$

where accumulation = rate of change of the total mass of soluble sulfide in the slug of wastewater
inflow = rate of mass transport of soluble sulfide from the slime layer into the slug of wastewater
outflow = rate of mass transport of sulfide in the form of hydrogen sulfide from the slug of wastewater to the sewer atmosphere
generation = rate of production of soluble sulfide in the slug of wastewater

utilization = rate of removal of soluble sulfide in the slug of wastewater (e.g., by oxidation)

For partly full sewers, the generation term can be assumed to be negligible. The terms for outflow and utilization (emission and oxidation) can be combined as was done on the right side of Eq. 7-36. The resulting mass balance for a slug of volume $\mathcal{V}$ is

$$\mathcal{V}\,\frac{dS_T}{dt} = \mathcal{V}[M(\mathrm{EBOD})R^{-1}] - \mathcal{V}[m(SV)^{3/8}S_T d_m^{-1}] \qquad \text{(SI units)} \qquad (7\text{-}39)$$

$$\mathcal{V}\,\frac{ds_T}{dt} = \mathcal{V}[3.28M(\mathrm{EBOD})R^{-1}] - \mathcal{V}[2.10m(SV)^{3/8}S_T d_m^{-1}] \qquad \text{(U.S. customary units)} \qquad (7\text{-}39a)$$

Rearranging leads to

$$\frac{dS_T}{M(\mathrm{EBOD})R^{-1} - m(SV)^{3/8}S_T d_m^{-1}} = dt \qquad \text{(SI units)} \qquad (7\text{-}40)$$

$$\frac{dS_T}{3.28M(\mathrm{EBOD})R^{-1} - 2.10m(SV)^{3/8}S_T d_m^{-1}} = dt \qquad \text{(U.S. customary units)} \qquad (7\text{-}40a)$$

Multiplying both sides of Eq. 7-40 by $m(SV)^{3/8}d_m^{-1}$ and Eq. 7-40a by yields

$$\frac{dS_T}{\left[\dfrac{M(\mathrm{EBOD})R^{-1}}{m(SV)^{3/8}d_m^{-1}}\right] - S_T} = m(SV)^{3/8}d_m^{-1}\,dt \qquad \text{(SI units)} \qquad (7\text{-}41)$$

$$\frac{dS_T}{\left[\dfrac{1.56M(\mathrm{EBOD})R^{-1}}{m(SV)^{3/8}d_m^{-1}}\right] - S_T} = 2.10m(SV)^{3/8}d_m^{-1}d_t \qquad \text{(U.S. customary units)} \qquad (7\text{-}41a)$$

The term in brackets in Eq. 7-41 is recognized to be $(S_T)_{\mathrm{lim}}$ (see Eq. 7-37). Thus,

$$\frac{dS_T}{(S_T)_{\mathrm{lim}} - S_T} = m(SV)^{3/8}d_m^{-1}\,dt \qquad \text{(SI units)} \qquad (7\text{-}42)$$

$$\frac{dS_T}{(S_T)_{\mathrm{lim}} - S_T} = 2.10m(SV)^{3/8}d_m^{-1}\,dt \qquad \text{(U.S. customary units)} \qquad (7\text{-}42a)$$

Integration leads to

$$-ln\,\left|(S_T)_{\mathrm{lim}} - S_T\right| = m(SV)^{3/8}d_m^{-1}t + C \qquad \text{(SI units)} \qquad (7\text{-}43)$$

$$-ln\left|(S_T)_{\mathrm{lim}} - S_T\right| = 2.10m(SV)^{3/8}d_m^{-1}t + C \qquad \text{(U.S. customary units)} \qquad (7\text{-}43a)$$

where C is a constant.

Exponentiating Eq. 7-43 results in

$$|(S_T)_{\lim} - S_T| = Ke \qquad \text{(SI units)} \qquad (7\text{-}44)$$

$$|(S_T)_{\lim} - S_T| = Ke \qquad \text{(U.S. customary units)} \qquad (7\text{-}44a)$$

where $K = e^C$.

At time $t = 0$, $S_T = (S_T)_0$; thus,

$$|(S_T)_{\lim} - S_T| = |(S_T)_{\lim} - (S_T)_0|\, e \qquad \text{(SI units)} \qquad (7\text{-}45)$$

$$|(S_T)_{\lim} - S_T| = |(S_T)_{\lim} - (S_T)_0|\, e \qquad \text{(U.S. customary units)} \qquad (7\text{-}45a)$$

Because the sign of $(S_T)_{\lim} - S_T$ at any time will be the same as it was at time $t = 0$, the absolute values can be dropped, resulting in

$$(S_T)_{\lim} - S_T = [(S_T)_{\lim} - (S_T)_0]e \qquad \text{(SI units)} \qquad (7\text{-}46)$$

$$(S_T)_{\lim} - S_T = [(S_T)_{\lim} - (S_T)_0]e \qquad \text{(U.S. customary units)} \qquad (7\text{-}46a)$$

Thus the value of $(S_T)_{\lim} - S_T$ will decrease exponentially from its initial value to zero. It is therefore possible to define a half-life, which is the time required for $(S_T)_{\lim} - S_T$ to decrease from its value at any given time to one-half that value. At the half-life,

$$e^{-m(SV)^{3/8}} d_m^{-1} t_{1/2} = 0.5 \qquad \text{(SI units)} \qquad (7\text{-}47)$$

$$e^{-2.10m(SV)^{3/8}} d_m^{-1} t_{1/2} = 0.5 \qquad \text{(U.S. customary units)} \qquad (7\text{-}47a)$$

Solving for the half-life yields

$$t_{1/2} = \frac{0.69 d_m}{m(SV)^{3/8}} \qquad \text{(SI units)} \qquad (7\text{-}48)$$

$$t_{1/2} = \frac{0.33 d_m}{m(SV)^{3/8}} \qquad \text{(U.S. customary units)} \qquad (7\text{-}48a)$$

where $t_{1/2}$ = half-life, h
d_m = mean hydraulic depth, m (ft)
m = empirical factor
S = energy grade line, m/m (ft/ft)
V = mean wastewater velocity, m/s (ft/s)

By knowing $(S_T)_{\lim}$, $t_{1/2}$, and the wastewater transit times, the sulfide concentration throughout a test reach can be computed. Values of the constants M and m, needed to calculate $(S_T)_{\lim}$ and $t_{1/2}$, must be determined by trial and error by fitting Eq. 7-46 to test data and adjusting M and m as necessary to obtain good correlation. As stated previously, the worst-case value of M suggested by Pomeroy is 0.32×10^{-3} m/h [7]. Similarly, a conservative value of m for worst-case conditions is 0.64, and for a median rate of sulfide buildup a value of 0.96 might be applicable [7]. Because these values were derived for the specific conditions that existed in the test reaches in Los Angeles County, they do not necessarily apply in every case. However, for most systems these values can be used as first estimates in the trial-and-error computation of M and m.

Table 7-1 has been developed to simplify the determination of sulfide concentrations in a section of sewer in which $(S_T)_{\lim}$ and $t_{1/2}$ are known. The basis for Table 7-1 is Eq. 7-49, which follows directly from Eqs. 7-46 and 7-48.

$$\frac{(S_T)_{\lim} - S_T}{(S_T)_{\lim} - (S_T)_0} = e^{-0.69t/t_{1/2}} \tag{7-49}$$

Other Biological Transformations

In addition to the generation of hydrogen sulfide, the characteristics of the wastewater in a collection system are modified, with the passage of time, by a number of other nonspecific bacterially mediated reactions. That changes occur can be demonstrated by measuring the redox (reduction-oxidation) potential in the wastewater at various points upstream from a treatment facility and some other downstream point.

In an aqueous environment, the redox potential is an approximate measure of the balance between the reducing and oxidizing substances present. It is measured as the electrical potential developed between the standard hydrogen half-cell and the half-cell formed by the solution and a platinum electrode. Because the standard reference hydrogen electrode is difficult to use, the calomel reference electrode is generally substituted. Field measurements relative to the normal hydrogen electrode at pH 7 are usually standardized. They

Table 7-1 Values of $[(S_T)_{\lim} - S_T]/[(S_T)_{\lim} - (S_T)_0]$ for various values of $t/t_{1/2}$ in Eq. 7-49

$t/t_{1/2}$	$\frac{(S_T)_{\lim} - S_T}{(S_T)_{\lim} - (S_T)_0}$
0.1	0.93
0.2	0.87
0.3	0.81
0.4	0.76
0.5	0.71
0.6	0.66
0.8	0.58
1.0	0.50
1.5	0.36
2.0	0.25
2.5	0.18
3.0	0.13
4.0	0.06

are converted to standardized values by using the following expression [3]:

$$E_h = E + r + 60(\text{pH}_m - 7) \qquad (7\text{-}50)$$

where E_h = redox potential referenced to the normal hydrogen electrode at pH 7, mV
E = value of field redox measurement, mV
r = potential of reference electrode, mV
pH_m = value of field pH measurement

A positive potential corresponds to oxidizing conditions and a negative potential corresponds to reducing conditions. Anaerobic conditions are characterized by a negative or small positive redox potential. Typical redox measurements made at various upstream points from a treatment plant are reported in Table 7-2.

The transformations which occur in wastewater depend on whether dissolved oxygen is present. Under aerobic conditions, many of the organic compounds would probably become simpler organic compounds, carbon dioxide, and numerous partially oxidized compounds. The presence of carbon dioxide tends to lower the pH. Under anaerobic conditions, the most likely occurrence is that the complex organic compounds in wastewater become carbon dioxide, methane, simpler organic compounds and acids. Both the carbon dioxide and the acids produced will tend to lower the pH.

Effects of Changes in Wastewater Composition on Treatment

In the following discussion, the emphasis is on the practical implications of hydrogen sulfide generation, and some of the other transformations that can occur in a collection system on the design and operation of wastewater-treatment facilities. The topics covered are: (1) precipitation of trace elements; (2) immediate oxygen demand; and (3) the presence of filamentous microorganisms. The need for integrated design of collection systems and treatment facilities is discussed at the end of this chapter.

Table 7-2 Typical redox measurements upstream from treatment plant

Distance upstream from treatment plant, km	E_h, mV	Dissolved sulfide, g/m³
15.5	+258	0.2
9.7	+122	0.6
8.1	+ 70	1.1
4.8	+ 50	1.2
0.0	+ 48	1.8

[a] Adapted from Ref. 1.
Note: km × 0.6214 = mi

Precipitation of trace elements. Apart from the development of odors and the corrosion of sewers, the presence of hydrogen sulfide can significantly affect the treatability of wastewater because of the precipitation of many of the trace elements necessary for biological growth. As noted earlier, a portion of the hydrogen sulfide diffused into the bulk of the wastewater from the slime layer reacts with various metal ions to form insoluble-sulfide compounds. It has been estimated that 40 to 60 percent of the total sulfide that may be present is in the form of suspended and precipitated sulfide species (FeS, FeS_2, ZnS, CoS, etc.). In fact, it is these precipitated sulfide species that give septic (highly anaerobic) wastewater its dark (in some cases, black) color. The solubility of various metal sulfides is given in Fig. 7-5.

Because many of the inorganic trace elements necessary for biological growth can be precipitated as inorganic sulfides, wastewater containing a considerable amount of hydrogen sulfide may be deficient in trace elements, especially where primary sedimentation is an integral part of the treatment process [11]. Where primary sedimentation is not used, it may be possible for the organisms to extract the necessary trace nutrients from the various precipitated forms.

Immediate oxygen demand. From numerous measurements it has been shown that the immediate oxygen demand of wastewater arriving at a treatment plant (demand occurring within 0.5 to 2.0 h) varies considerably with the distance traveled and the condition of the wastewater. Typical oxygen uptake rates for wastewater samples taken at the upper ends of a collection system vary from 2

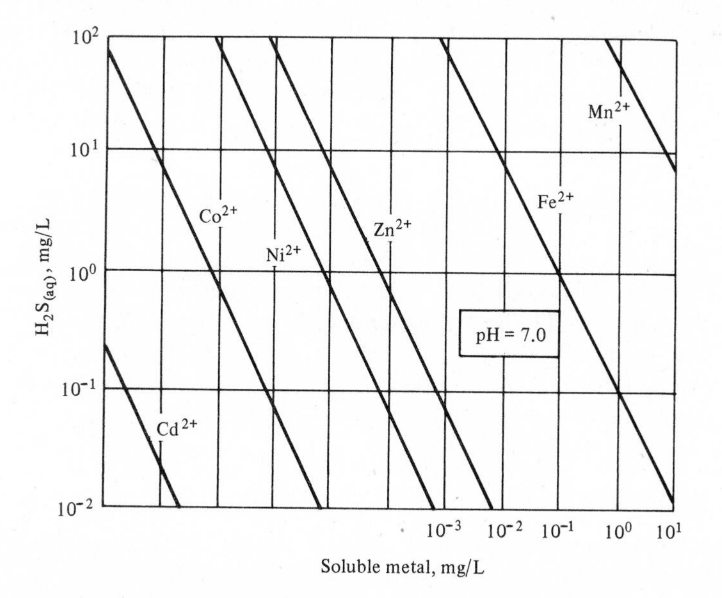

Figure 7-5 Solubility diagram [11].

to 3 $g/m^3 \cdot h$ or $mg/L \cdot h$ [8], whereas uptake rates for wastewater that has traveled a considerable distance have been reported as high as 50 $g/m^3 \cdot d$ or $mg/L \cdot h$.

Of the several possible explanations for these observations, it appears reasonable to suggest the following. Because of the transformations that are occurring in the wastewater as it is transported in the collection system, many of the complex organic compounds originally present have probably been broken down into simpler compounds. In most cases, these compounds can be biologically assimilated and converted more readily in the controlled environment of a treatment plant. This breakdown of organic compounds, and the possibility that many of the facultative organisms sheared from the sewer walls have already developed adaptive enzymes to process the organic compounds found in wastewater, offers a plausible explanation. With respect to the design of aerobic treatment processes, this immediate oxygen demand can have a significant effect on the selection and sizing of aeration equipment.

Growth of filamentous microorganisms. In collection systems where a slime layer has developed, filamentous microorganisms are commonly found in the influent, i.e., in the wastewater entering a treatment plant. In some cases, if an extensive slime layer has developed, the numbers of filamentous organisms can become so significant that they will affect the operation of a treatment process [9].

For example, in the activated-sludge process, these organisms—because of their limited growth requirements—can become the dominant bacterial species on a mass basis in a very short period. If a treatment process is selected with little or no consideration of their presence and ultimate effect, the treatment plant may never function as intended.

7-2 CORROSION CAUSED BY HYDROGEN SULFIDE AND ITS CONTROL

Because of the large capital expenditures made each year to correct the effects of hydrogen sulfide corrosion in sewers, pumping stations, and treatment facilities, it is important to understand how such corrosion occurs and what can be done to control or eliminate it. The following discussion is directed mainly toward corrosion in collection systems, although most of the information is equally applicable to pumping stations and treatment facilities. Corrosion control in pumping stations is also discussed in Chap. 9.

Hydrogen Sulfide Corrosion Process

The hydrogen sulfide corrosion process is illustrated in Fig. 7-6. Although a concrete sewer section is shown, the hydrogen sulfide corrosion process in pumping stations and treatment facilities is essentially the same.

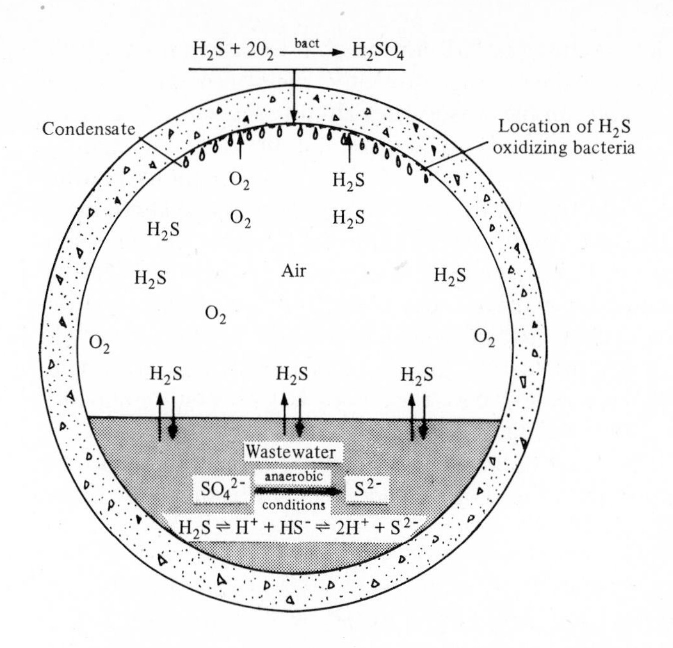

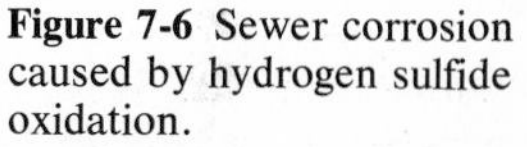
Figure 7-6 Sewer corrosion caused by hydrogen sulfide oxidation.

The form of the sulfide released to the wastewater from the slime layer is controlled by the equilibrium relationships given previously (Eqs. 7-1 and 7-3). As given by these relationships and illustrated in Example 7-1, the amount of dissolved hydrogen sulfide present will increase as the pH decreases (hydrogen-ion concentration increases). From the solution, hydrogen sulfide in the gaseous phase will escape to the sewer atmosphere. The amount present in the atmosphere is directly related to the concentration of aqueous hydrogen sulfide and the value of the relative saturation of the sewer atmosphere q.

After the hydrogen sulfide has escaped into the sewer atmosphere, the next step in the corrosion process is its transfer to the sewer walls above the liquid surface. Because the sewer walls are usually damp with condensate, hydrogen sulfide from the sewer atmosphere will immediately be taken up as it comes in contact with these damp surfaces. The hydrogen sulfide retained in this dampness is then converted to sulfuric acid by aerobic bacteria of the genus *Thiobacillus*.

$$H_2S + 2O_2 \xrightarrow{\text{bacteria}} H_2SO_4 \tag{7-51}$$

Species of *Thiobacillus*, such as *T. concretivorus*, have been reported to remain active in solutions containing up to 7 percent sulfuric acid. However, this reaction will be limited by the moisture and the presence of oxygen.

In the following step of the corrosion process, the sulfuric acid will react with the cement in the concrete pipes. A similar reaction will occur with iron sewer pipes. If the rate of sulfuric acid production is slow, almost all of the acid will react with the cement, producing a pasty mass of material loosely bonded to the inert materials used to manufacture the pipe. If the rate of sulfuric acid production is rapid, much of the acid will not be able to diffuse through the pasty mass. Consequently, it will be carried down the walls of the pipe and into the flowing wastewater stream. In the wastewater stream, the sulfuric acid will react with the alkalinity, and the sulfur will be present in the form of the sulfate ion [6].

Periodically, as the sewer fills, portions of the pasty mass will be sheared off or will fall off the sewer walls because of their mass. This process will repeat itself as the pipe continues to corrode. The pattern of corrosion will vary, depending on the air circulation patterns, the amount of condensate, the rate and amount of hydrogen sulfide produced, and other local factors. Typically, the rate of corrosion will be greatest at the crown of the sewer and near the water surfaces on either side of the pipe (see Fig. 7-7). A photograph of a corroded sewer pipe is shown in Fig. 7-8.

Assuming that all of the hydrogen sulfide released from the wastewater stream is absorbed on the exposed wall of the pipe, the hydrogen sulfide flux to the wall ϕ_{sw} can be determined as follows:

$$\phi_{sw} = \phi_{se} B / P' \tag{7-52}$$

where ϕ_{sw} = flux of H_2S to exposed pipe wall, $g/m^2 \cdot h$

B = wastewater surface width, m (ft)

P' = exposed pipe perimeter, m (ft)

Cement-bonded materials. For cement-bonded materials, the amount of corrosion can be estimated by noting from Eq. 7-53 that 32 g of sulfide is required to

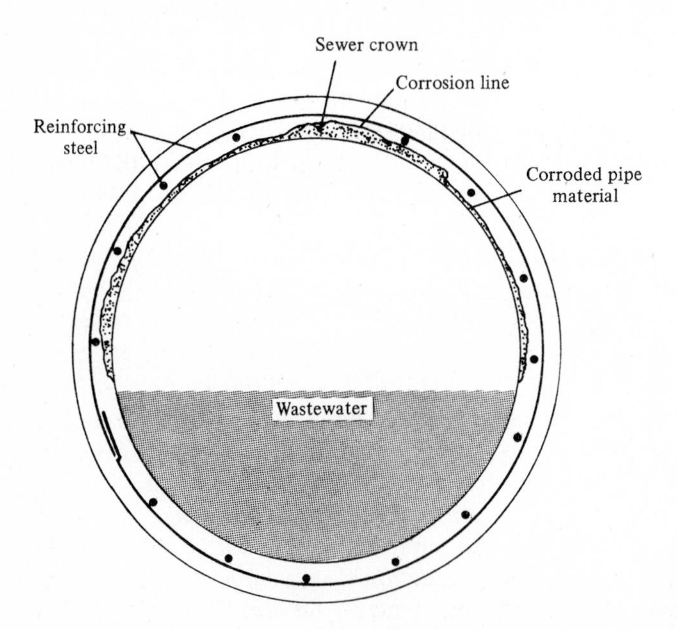

Figure 7-7 Schematic representation of typical corrosion pattern in reinforced-concrete sewer pipe.

Figure 7-8 Sewer pipe corroded by hydrogen sulfide. (*Courtesy E. Rogers*)

produce the acid, which will then react with 100 g of calcium carbonate.

$$H_2S + 20_2 \xrightarrow{\text{bacteria}} H_2SO_4 + CaCO_3 \rightarrow H_2CO_3 + CaSO_4 \tag{7-53}$$

(simplified reaction)

The rate at which the pipe wall is penetrated can be determined by considering a square meter of pipe surface. The rate at which sulfide is absorbed onto this unit area is ϕ_{sw}. Because 32 g of sulfide will ultimately react with 100 g of calcium carbonate, the rate at which the calcium carbonate equivalent of the pipe material in the unit surface area is $\frac{100}{32}\phi_{sw}$ (assuming that all of the acid reacts with the cement-bonded material). The rate at which the actual pipe material is destroyed is calculated by dividing the equivalent calcium carbonate destroyed by the gross alkalinity of the composite cement-bonded material as follows:

The rate of corrosion of pipe material per unit surface area of pipe, $g/m^2 \cdot h$, equals

$$\frac{100}{32}\phi_{sw}\frac{1}{A} \tag{7-54}$$

where A = alkalinity of cement-bonded material expressed as $CaCO_3$ equivalent.

The mass rate of corrosion can be converted to a volume rate of corrosion by dividing the right side of Eq. 7-54 by the specific gravity of the pipe material and the density of water. Accordingly, the volume rate of corrosion of pipe material per unit surface area of pipe, $m^3/m^2 \cdot h$, equals

$$\frac{\frac{100}{32}\phi_{sw}(1/A)}{2.4 \times 10^6 \text{ g/m}^3} \tag{7-55}$$

From an analysis of the units in Eq. 7-55, the rate of penetration into the pipe wall can be expressed in meters per hour, which can be converted to units of millimeters per year and inches per year as follows:

$$c = \frac{\frac{100}{32}\phi_{sw}(1/A)}{2.4 \times 10^6 \text{ g/m}^3}(10^3 \text{ mm/m})(8760 \text{ h/yr}) \tag{7-56}$$

$$c = 11.4\phi_{sw}\frac{1}{A} \qquad \text{(SI units)} \tag{7-57}$$

$$c = 0.45\phi_{sw}\frac{1}{A} \qquad \text{(U.S. customary units)} \tag{7-57a}$$

where c = average rate of penetration, mm/yr (in/yr).

To take into account that all the acid may not react, a factor k is added to Eq. 7-57, resulting in

$$c = 11.4k\phi_{sw}\frac{1}{A} \qquad \text{(SI units)} \tag{7-58}$$

$$c = 0.45\phi_{sw}\frac{1}{A} \qquad \text{(U.S. customary units)} \tag{7-58a}$$

The value of k approaches unity for systems in which acid formation is slow and ranges to 0.3 for systems in which acid formation is fast. If sulfide corrosion has occurred on existing pipes and structures, the actual k value for the system can be determined. By using the field-derived k value, the accuracy of the predicted corrosion rates will be substantially increased. The application of Eqs. 7-56, 7-57, and 7-58 is illustrated in Example 7-2.

The average rate of penetration as defined in Eq. 7-57 can be used in Eq. 7-59 to determine the useful life of a sewer pipe.

$$L = \frac{z}{c} \tag{7-59}$$

where L = useful life of pipe, yr

z = allowable penetration due to corrosion, mm (in)

For reinforced-concrete pipe, the value of z is sometimes chosen as the depth to the reinforcing steel.

Combining Eqs. 7-58 and 7-59 and rearranging yields

$$Az = 11.4k\phi_{sw}L \qquad \text{(SI units)} \tag{7-60}$$

$$Az = 0.45k\phi_{sw}L \qquad \text{(U.S. customary units)} \tag{7-60a}$$

The term Az represents the combined effects of the alkalinity and the allowable penetration and is called the *life factor*. The life factor determines the useful life of a pipeline when k and ϕ_{sw} are given.

When sulfide conditions are anticipated and evaluated in the design of a sewer line, the appropriate value for the life factor to assure performance throughout the desired life (usually 100 yr) should be included in the contract specifications if concrete pipe will be allowed. The pipe manufacturer would have some freedom in the choice of materials which determine the alkalinity and the pipe wall thickness, as long as the life-factor specification could be met.

Ferrous materials. The corrosion of iron and steel materials by sulfuric acid can be estimated in the same way as the corrosion for cement-bonded materials. The same processes are involved: (1) release of hydrogen sulfide, (2) absorption on the exposed wet surfaces, and (3) corrosion reaction. The only difference in the equations for ferrous and cement-bonded materials is the corrosion reaction. By adjusting for specific gravity of iron (7.5) and using the acid-consuming capacity of 1.79 (corresponds to alkalinity), Eq. 7-58 is modified for ferrous materials.

$$c' = 2.04k\phi_{sw} \qquad \text{(SI units)} \tag{7-61}$$

$$c' = 0.08k\phi_{sw} \qquad \text{(U.S. customary units)} \tag{7-61a}$$

where c' = average rate of penetration, mm/yr (in/yr).

Example 7-2 Determination of corrosion rates in an interceptor trunk sewer Estimate the annual rate of corrosion at the midpoint and at the end of a 1200-mm (48-in) trunk sewer that is 10,000 m (32,800 ft) long. The sewer pipe is concrete with a granitic aggregate. The alkalinity of the pipe has been determined to be 0.20. The dissolved-sulfide concentration at the upstream end of the section is negligible and there are no dissolved metals in the wastewater. Assume that conditions favor sulfide production and that the following data apply:

$M = 0.32 \times 10^{-3}$ m/h
$m = 0.64$
BOD_5 = 200 mg/L
T = 20°C
pH = 7.0
S = 0.0008 m/m
n = 0.013
depth of flow = $0.5D$

SOLUTION

1. Determine the velocity of flow with the Manning equation. (R is the same for full pipes and half-full pipes.)

$$V = \frac{1}{n}R^{2/3}S^{1/2}$$

$$= \frac{1}{0.013}\left(\frac{1.2}{4}\right)^{2/3}(0.0008)^{1/2}$$

$$= 0.98 \text{ m/s}$$

2. Determine the limiting sulfide concentration using Eq. 7-35.

$$(S_T)_{\text{lim}} = \frac{M}{m}[\text{EBOD}](SV)^{-3/8}\frac{P}{B}$$

where $M = 0.32 \times 10^{-3}$ m/h
$m = 0.64$
$\text{EBOD} = 200 \text{ mg/L } (1.07)^{20-20}$
$= 200$ mg/L
$S = 0.0008$
$V = 0.98$ m/s
$P = \pi(1.2 \text{ m})/2 = 1.88$
$B = 1.2$ m
$P/B = 1.57$

$$(S_T)_{\text{lim}} = \frac{0.32 \times 10^{-3}}{0.64}(200)(0.0008 \times 0.98)^{-3/8}(1.57)$$

$$= 2.29 \text{ g/m}^3$$

3. Determine the value of $t_{1/2}$ with Eq. 7-48.

$$t_{1/2} = \frac{0.69 d_m}{m(SV)^{3/8}}$$

where

$$d_m = \frac{A}{B} = \frac{\pi(0.6 \text{ m})^2/2}{1.2 \text{ m}} = 0.471 \text{ m}$$

$$t_{1/2} = \frac{(0.69)(0.471)}{(0.64)(0.008 \times 0.98)^{3/8}}$$

$$= 7.4 \text{ h}$$

4. Determine the $t/t_{1/2}$ ratios for the midpoint and end of the interceptor and obtain the corresponding values for $[(S_T)_{\text{lim}} - S_T]/[(S_T)_{\text{lim}} - (S_T)_0]$ from Table 7-1.
   a. For the midpoint of the interceptor:

$$t = \frac{5000 \text{ m}}{0.98 \text{ m/s} \times 3600 \text{ s/h}} = 1.4 \text{ h}$$

$$\frac{t}{t_{1/2}} = 1.4 \text{ h}/7.4 \text{ h} = 0.19$$

$$\frac{(S_T)_{\text{lim}} - S_T}{(S_T)_{\text{lim}} - (S_T)_0} = 0.88$$

   b. For the end of the interceptor:

$$t = 2.8 \text{ h}$$

$$\frac{t}{t_{1/2}} = 2.8 \text{ h}/7.4 \text{ h} = 0.38$$

$$\frac{(S_T)_{\text{lim}} - S_T}{(S_T)_{\text{lim}} - (S_T)_0} = 0.77$$

5. Determine the total soluble sulfide concentration S_T for the midpoint and the end of the interceptor.

a. For the midpoint of the interceptor:

$$\frac{2.29 \text{ g/m}^3 - S_T}{2.29 \text{ g/m}^3 - 0} = 0.88$$

$$S_T = 0.27 \text{ g/m}^3$$

b. For the end of the interceptor:

$$\frac{2.29 \text{ g/m}^3 - S_T}{2.29 \text{ g/m}^3 - 0} = 0.77$$

$$S_T = 0.53 \text{ g/m}^3$$

6. Determine the concentration of H_2S present at a pH value of 7.0.
   a. From Example 7-1, the fraction of the total dissolved sulfide present as H_2S equals 0.500.
   b. The concentration of H_2S is:
      i. For the midpoint of the interceptor,

$$H_2S = 0.500 \ (0.27 \text{ g/m}^3)$$
$$= 0.14 \text{ g/m}^3$$

      ii. For the end point of the interceptor,

$$H_2S = 0.500 \ (0.53 \text{ g/m}^3)$$
$$= 0.27 \text{ g/m}^3$$

7. Determine the flux of H_2S to the exposed pipe wall. Equations 7-31 and 7-52 can be used to obtain the following equation for the flux of H_2S to the exposed pipe wall.

$$\phi_{sw} = 0.69 C_A \gamma (\text{SV})^{3/8} (1 - q) j S_T \text{B/P}'$$

Because the relative saturation of H_2S in the sewer atmosphere q is normally small, it is assumed to be zero. This is a conservative estimate with respect to pipe corrosion.

a. For the midpoint of the interceptor:

$$C_A = 1 + \frac{0.17V^2}{g d_m}$$

$$= 1 + \frac{(0.17)(0.98)^2}{(9.8)(0.471)} = 1.04$$

$\gamma = 1.0$
$S = 0.0008$ m/m
$V = 0.98$ m/s
$q \cong 0$
$jS_T = H_2S = 0.14 \text{ g/m}^3$ [from step 6*b*(i)]
$B = 1.2$ (from step 2)
$P' = P = 1.88$ (from step 2)

$$\phi_{sw} = 0.0044 \text{ g/m}^2 \cdot \text{h}$$

b. For the endpoint of the interceptor:

$C_A = 1.04$
$\gamma = 1.0$
$S = 0.0008$ m/m
$V = 0.98$ m/s
$q \simeq 0$

$$jS_T = H_2S = 0.27 \text{ g/m}^3 \quad \text{[from step 6b(ii)]}$$
$$B = 1.2 \text{ m}$$
$$P' = 1.88 \text{ m}$$

$$\phi_{sw} = 0.0085 \text{ g/m}^2 \cdot \text{h}$$

8. Determine the corrosion in the interceptor using Eq. 7-58. Assume that acid production is slow and k equals 1.0.

$$C = 11.4\phi_{sw}\frac{1}{A}$$

a. For the midpoint of the interceptor:

$$C = \frac{(11.4)(0.0044)}{0.20}$$

$$= 0.25 \text{ mm/yr}$$

b. For the endpoint of the interceptor:

$$C = \frac{(11.4)(0.0085)}{0.20}$$

$$= 0.48 \text{ mm/yr}$$

Comment The values computed for hydrogen sulfide concentration and emission were calculated by assuming there were no junctions, sewer drops at manholes, or other points of increased turbulence in the reach of sewer. Where such points of turbulence occur, there would be a greater emission of hydrogen sulfide from the wastewater and, therefore, more corrosion. Downstream from such points, there would then be less emission of hydrogen sulfide and less corrosion.

The useful life of a concrete pipe can be determined by dividing the available thickness of the concrete used for corrosion protection by the average corrosion rate. With respect to hydrogen sulfide corrosion, the useful life can be extended by increasing either the thickness of the corrosion protection layer or the alkalinity content of the concrete pipe. Higher alkalinities can be obtained by using calcareous aggregate, by blending calcareous and granitic aggregates, or by varying the amount of cement [12].

Control of Hydrogen Sulfide Corrosion

Hydrogen sulfide corrosion can be controlled best by eliminating or limiting the production of hydrogen sulfide. The principal control methods are:

1. Source control of organic and sulfur inputs
2. Aeration
3. Chemical addition, including the use of chlorine, hydrogen peroxide, and other oxidizing agents, sodium nitrate, and toxic substances
4. Periodic cleaning, both mechanical and chemical
5. Ventilation
6. Good design

The most common methods for control in existing systems are aeration, chlorination, and mechanical cleaning. For new systems, the least costly and most effective method is good design. Additional details on the control of hydrogen sulfide corrosion may be found in Refs. 2, 6, 8, and 10.

In collection systems where hydrogen sulfide is not generated, unlined concrete and asbestos-cement pipe may be used. In collection systems that have mild or intermittent corrosive conditions, concrete pipe may be used if the inside concrete cover over the reinforcing steel is increased as a corrosion allowance and if the alkalinity of the pipe material is increased by using a limestone or dolomite aggregate in place of a granitic aggregate. These changes spread the attack of the acid over a much greater mass of material, which should prolong the life of the pipe.

When severe corrosion by hydrogen sulfide is anticipated and cannot be eliminated by the control measures discussed previously, vitrified-clay or plastic pipe should be used, or protective linings of proved performance should be specified. Results with bituminous or coal-tar and epoxy linings have varied; the acid can be diffused through pinholes in the lining and can attack the cement in concrete pipes, thus causing ultimate failure. A plasticized polyvinyl chloride sheet with T-shaped projections on the back that key into the concrete has proved satisfactory for large-diameter concrete pipes.

7-3 OCCURRENCE, EFFECT, AND CONTROL OF ODOROUS AND OTHER SEWER GASES

As wastewater flows in a collection system, the composition of the gases in the sewer atmosphere, above the free liquid surface, is altered from that of normal air by the leakage of natural or manufactured gas, gasoline vapors, carbon monoxide, and industrial gases from commercial and industrial discharges into the sewer. It is also altered by the release of gases produced by the bacterial transformations which can occur on sewer walls, in deposits on the bottoms of sewers, and within the flowing wastewater. Often, the gases released are highly odorous. Hydrogen sulfide, the most important of the sewer gases, has been discussed in detail in the preceding sections. The occurrence of odorous and other sewer gases, some of the effects associated with these gases, and the control of sewer gases are described briefly in this section.

Occurrence of Odorous and Other Sewer Gases

In the early period of development of the wastewater engineering field, it was thought that the various odors in sewers were attributable to a limited list of compounds, including: hydrogen sulfide; organic sulfides; ethyl mercaptan; primary, secondary, and tertiary methylamines; indole and skatole; and ammonia [10]. The structure and offensive odor quality of many of these compounds are reported in Table 7-3. With the advent of modern methods of analysis, more than 50 different gaseous substances have been detected in the sewer atmosphere [10]. Whether all of these substances contribute to the odors in sewers is unknown.

Table 7-3 Major categories of offensive odors in sewers [4]

Compound	Typical formula	Odor quality
Amines	CH_3NH_2, $(CH_3)_3N$	Fishy
Ammonia	NH_3	Ammoniacal
Diamines	$NH_2(CH_2)_4NH_2$, $NH_2(CH_2)_5NH_2$	Decayed flesh
Hydrogen sulfide	H_2S	Rotten eggs
Mercaptans	CH_3SH, $CH_3(CH_2)_3SH$	Skunk
Organic sulfides	$(CH_3)_2S$, CH_3SSCH_3	Rotten cabbage
Skatole	$C_8H_5NHCH_3$	Fecal

Among the gases identified recently, some of those known to be odorous include simple hydrocarbons, paraffinic and aeromatic aldehydes and ketones, carboxylic acids and esters, and terpenoid substances [10]. The presence of this second group of odorous compounds in wastewater can usually be attributed more to commercial and industrial discharges than to the biological transformations in wastewater.

In addition to the odorous gases produced in sewers, other gases are also produced by bacterial conversion reactions. Under aerobic conditions, the input of varying amounts of carbon dioxide and smaller amounts of ammonia would be expected. Under anaerobic conditions, the input of carbon dioxide, methane, and smaller amounts of ammonia would be expected.

Odor characterization and measurement. It has been suggested that four independent factors are required for the complete characterization of an odor: intensity, character, hedonics, and detectability. To date, detectability is the only factor used in the development of statutory regulations for nuisance odors.

Odors can be measured by sensory methods, and specific odorant concentrations can be measured by instrumetnal methods. Under carefully controlled conditions, the sensory (organoleptic) measurement of odors by the human olfactory system can provide meaningful and reliable information. Therefore, the sensory method is now used most often to measure odors emanating from wastewater-treatment facilities.

In the sensory method, human subjects (often a panel of subjects) are exposed to odors that have been diluted with odor-free air, and the number of dilutions required to reduce an odor to its minimum detectable threshold odor concentration (MDTOC) are noted. The detectable odor concentration is reported as the dilutions to the MDTOC. Thus, if four volumes of diluted air must be added to one unit volume of sampled air to reduce the odorant to its MDTOC, the odor concentration would be reported as five dilutions to MDTOC.

In recent years, a number of regulatory agencies have established odor discharge standards on the basis of dilutions to the MDTOC. Fortunately, it appears that the composite odor of sewer gases, which is made up of the odors of the sulfidic gases, amines, aldehydes, and body acids that are present, correlates well with the hydrogen sulfide concentration [10]. Thus the use of hydrogen sulfide appears to be satisfactory for defining the overall odor level in the air contained in collection systems.

Odor levels. Depending on the condition of the wastewater and the location in the collection system, hydrogen sulfide concentrations as high as 500 parts per million (ppm) (by volume) have been noted. Typical values cannot be given because the actual amount of hydrogen sulfide present is highly dependent on the collection system characteristics and the prevailing temperature. For example, because the organisms responsible for the generation of hydrogen sulfide are sensitive to temperature, hydrogen sulfide will not be generated in most collection systems located in cold climates.

Effects of Sewer Gases

One of the consequences of the presence of odorous sewer gases in collection systems, especially hydrogen sulfide, is the potential danger to workers. Some of the effects that exposure to hydrogen sulfide can cause are reported in Table 7-4. The minimum concentration known to have caused death is 300 ppm; 3000 ppm is quickly fatal [1]. Nonodorous gases in collection systems may also be toxic.

Explosions which can result from the ignition of accumulations of various other sewer gases, including methane and many of the odorous gases, are another hazard that must be guarded against in the proper management of collection systems.

Table 7-4 Effects produced by exposure to air containing various hydrogen sulfide concentrations[a]

Exposure time and conditions	Concentration of hydrogen sulfide in sewer atmosphere, ppm by volume	Effects
Prolonged exposure, light work	5–10 (Some persons less)	Little or none.
1 to 2 hr, light work	10–50 (Some persons less)	Slight eye and respiratory irritation, headaches.
6 hr, hard manual labor	About 50	Temporary blindness.
1 hr, hard manual labor	About 100	Maximum limit without serious consequence.

[a] Adapted from Ref. 10.

Control of Sewer Gases

Odorous and other sewer gases may be controlled by designing the collection system properly, by providing ventilation and air relief, and by other methods.

Proper design of collection systems. Sulfide and other odor problems usually develop in collection systems—especially in warm climates—where the velocity of flow is extremely low, where transmission distances or times are long, where the sewers or force mains are filled with wastewater and are in contact with air only intermittently or not at all, and where the collection systems are not self-cleaning or maintained properly. In small sewers where the minimum slope is equal to or greater than 0.006 m/m (ft/ft), there is usually an insufficient buildup of sulfide to cause trouble even in warm climates [8]. In large trunk sewers and intercepting sewers, hydrogen sulfide may be produced, but the rate of buildup is much slower than in smaller sewers. Also, the concentrations seldom reach levels exceeding 1.0 mg/L [8].

The best way to deal with the problem of odorous and other gases is to design these systems so that the gas production is minimized or eliminated. Among the factors that must be considered for odor control when designing sewers are (1) the slopes for both small and large sewers, (2) the sizes of sewers, and (3) the judicious use of points of turbulence. The proper selection of slopes was discussed in Chap. 4. The proper selection of sewer sizes can control the rate of reaeration and the wall area that is wetted. Points of controlled turbulence can be used advantageously to introduce oxygen into the wastewater. Unfortunately, if hydrogen sulfide is present, it will be released to the atmosphere at such points.

Ventilation and air relief. Even if a collection system is designed properly, the ventilation of sewers may be necessary (1) to prevent the accumulation of sewer gases that may be explosive or corrosive; (2) to prevent the occasional buildup of odorous gases; (3) to reduce the accumulation of hydrogen sulfide; and (4) to prevent the creation of pressures above or below atmospheric levels, which may occasionally break household plumbing water seals. Mechanical ventilation with portable blowers should always be provided when workers enter the sewers for inspection and maintenance.

The movement of the gases in the sewer atmosphere is caused by a number of factors. One of these factors is the chimney effect caused by the difference in unit weight between the outside air and the sewer atmosphere and the difference in elevation of the various openings into the sewers. Other factors that affect the movement of gases are the drag of the wastewater flow tending to move the sewer atmosphere and the effect of the wind blowing in through openings, particularly through the outlets of large sewers.

In winter, the gases in the sewer atmosphere move toward the ends of laterals, because the laterals are at higher elevations than the trunk sewers, and the colder outer air enters at the lower openings of the sewers. Practically, however, the wind and the drag on the sewer atmosphere from the downgrade flow of the wastewater have some effect in checking the movement of air up the sewers. The result, at times, is downstream movement of the gases in the sewer atmosphere, especially in the summer when the temperature difference is reversed.

Normally, house sewers connected to vent stacks and unobstructed outlets provide all the ventilation necessary in sanitary or combined sewers. Forced draft may be required occasionally, especially if most building connections have traps between the sewer and the plumbing vent. Storm-water sewers may be ventilated through untrapped inlet connections, unobstructed outlets, and perforated manhole covers. It is undesirable to ventilate combined sewers through untrapped inlet connections because objectionable odors could escape to the street surface. Perforated manhole covers should not be used for sanitary sewer ventilation because surface water could enter through the perforations and odors and visible vapors could escape from manholes.

If the sewer outlet is submerged and there are few ventilated house connections, some special provision for ventilation is usually necessary. A ventilating shaft with a cross-sectional area at least one-half that of the sewer and tall enough to extend above nearby roofs should be adequate. Special means of ventilation should always be provided where main traps are installed in house sewers.

A change of sewer section at depressions should be vented upstream of the depression, so that the sewer atmosphere will not be compressed and can move either upstream or downstream. In large sewers having one or more depressed sewers (inverted siphons), the inlet chambers of the depressed sewers should be vented, possibly with exhaust fans, and should be equipped with facilities to deodorize the exhausted gases in the sewer atmosphere.

Other control methods. In some situations, even with proper collection system design and ventilation, odorous gases may still develop and accumulate. In many older collection systems, for example, where modifications to control or eliminate the production of odors cannot be made, it will be necessary to resort to other methods. The principal physical, chemical, and biological methods used in these cases are reported in Table 7-5. The method selected must be determined by a detailed analysis of local conditions.

7-4 NEED FOR INTEGRATED DESIGN

In summary, it is important to remember that, in addition to the purely functional role of transporting wastewater, the design and operation of the collection system can significantly affect the biological transformations which occur within the moving wastewater and on the walls of the sewer. In turn, these transformations can lead to the corrosion of sewers, the development of odors, and the alteration of the composition of the wastewater to be treated. Until now, the change in the composition of wastewater occurring during transport has seldom been considered in the design of wastewater-treatment facilities. In the future, integrated design, in which both the collection system and treatment facilities are considered together, will improve the performance of each.

Table 7-5 Methods to control odorous gases in wastewater collections systems[a]

Method	Description and/or application
Biological methods	
Special biological stripping towers	Specially designed towers can be used to strip odorous compounds from the sewer atmosphere. Typically, the towers are filled with plastic media of various types on which biological growths can be maintained.
Trickling filters or activated sludge aeration tanks	Sewer gases and pumping station gases can be passed through trickling filter beds or injected into activated-sludge aeration tanks to remove odorous compounds.
Chemical methods	
Chemical oxidation	Oxidizing the odor compounds in the wastewater is one of the most common methods used to control odors. Chlorine, ferric chloride, hydrogen peroxide, ozone, and sulfur dioxide are among the many oxidants that have been used. The use of chlorine also limits the development of a slime layer.
Scrubbing with various alkalies	Sewer gases can be passed through solutions of alkalies, such as slaked lime and sodium hydroxide, to remove odors. If the level of carbon dioxide is very high, costs may be prohibitive.
Physical methods	
Adsorption, activated carbon	Sewer gases can be passed through beds of activated carbon to remove odors. Carbon regeneration can be used to reduce costs.
Adsorption on sand or soil	Sewer gases can be passed through sand or soil. Odorous gases from pumping stations are often vented to the surrounding soils or to specially designed beds containing sand or soils. The role of bacteria in the operation of such systems is not well understood at the present time.
Combustion	Sewer gas odors can be eliminated by combustion at temperatures varying from 650 to 760°C (1200 to 1400°F). The required temperature can be reduced by using catalysts. Sewer gases are often combusted with treatment-plant solids.
Masking agents	Perfume scents can be added to the wastewater to overpower or mask objectionable sewer gas odors. In some cases, the odor of the masking agent is worse than the original odor.
Oxygen injection	The injection of oxygen (either air or pure oxygen) into the wastewater to control the development of anaerobic conditions has proved to be effective. In-line injection with diffusers, U-tube aerators and various other techniques have been used.
Scrubbing towers	Sewer gases can be passed through specially designed scrubbing towers to remove odors. Some type of chemical or biological agent is usually used with the tower.

[a]Developed in part from Refs. 1 and 10.

DISCUSSION TOPICS AND PROBLEMS

7-1 A portion of a combined sewer system is normally odorous during the summer, but odors are reduced significantly during and after a summer rainstorm. Explain this phenomenon.

7-2 List and explain briefly the various mechanisms or processes by which the soluble-sulfide concentration in a slug of wastewater can be increased and decreased.

7-3 A 600-mm sewer normally flows one-half full. The dissolved-oxygen concentration in the wastewater is less than 0.1 mg/L and other conditions favor sulfide production. The temperature of the wastewater is 25°C and the BOD_5 is 225 mg/L. How many grams of sulfide are produced daily in a 100-m section of the sewer, assuming conditions remain constant? Assume that $M = 0.32 \times 10^{-3}$ m/h.

7-4 What percentage increase in the rate of sulfide generation in the slime layer can be expected with a temperature change from 17 to 22°C? From 17 to 27°C?

7-5 By what proportion can the useful life of a concrete sewer be increased if calcereous aggregate is used in the concrete instead of granitic aggregate, if all other factors are identical? Assume that the alkalinity of the concrete with granitic aggregate is 0.2 and the alkalinity of the concrete with calcereous aggregate is 0.4.

7-6 A 450-mm sewer pipe with a slope of 0.005 m/m and a Manning's n value of 0.015 is flowing one-half full. The BOD_5 of the wastewater is 300 mg/L, the temperature is 20°C, and all conditions favor sulfide production. Estimate the limiting sulfide concentration in the wastewater. Assume that $M = 0.32 \times 10^{-3}$ m/h and $m = 0.64$. Neglect precipitation of metallic sulfide.

7-7 For the sewer described in Prob. 7-6, determine the half-life for sulfide buildup. Determine the time required for the sulfide concentration to change from one-half to seven-eighths of its limiting value.

7-8 For the sewer described in Prob. 7-6, plot the sulfide concentration as a function of the distance downstream from a point where the concentration is one-fourth the limiting value.

7-9 What would be the rate of corrosion for the sewer described in Prob. 7-6 at the point where the sulfide concentration is one-half the limiting value? Assume that pH = 7.0, $q = 0.1$, and $k = 1.0$, and A (alkalinity) = 0.2.

7-10 If the sewer described in Prob. 7-6 were cast iron, how long would it take for corrosion to penetrate 3 mm into the crown of the pipe at a point where the sulfide concentration has essentially reached its limiting value? Assume that pH = 6.5, $k = 1.0$, and $q = 0.1$.

7-11 Determine the total amount of sulfide (mg/L) that would have to be produced to suppress iron in the ferrous form from 2.0 to 0.1 mg/L at pH values of 5.5, 7.0, and 8.5, and a temperature of 25°C. Calculate the concentration of total dissolved sulfide that will remain in the wastewater and the concentration that will be removed by precipitation. The solubility product for ferrous sulfide is $[Fe^{2+}] \times [S^{2-}]$, both concentrations in mol/L, is 6.3×10^{-18}. Neglect all other processes that would affect the sulfide concentration.

7-12 After reviewing at least five literature references, discuss the importance of ventilation and air relief in sewers. Cite the references consulted.

REFERENCES

1. County Sanitation District No. 4 of Santa Clara County, Specifications Section 20.2 (part III, pp. 19–23), Campbell, Calif.
2. Kerri, K. D., and J. Brady (eds.): *Operation and Maintenance of Wastewater Collection Systems,* prepared for the U.S. Environmental Protection Agency, Office of Water Program Operations, by Department of Civil Engineering, California State University, Sacramento, 1976.
3. Mara, D. D.: *Bacteriology for Sanitary Engineers,* Churchill Livingston, Edinburgh, 1974.
4. Moncrieff, R. W.: *The Chemical Senses,* 3d ed., Leonard Hill, London, 1967.
5. Parkhurst, J. D., R. D. Pomeroy, and J. Livingston: *Sulfide Occurrence and Control in Sewage Collection Systems,* Report to the U.S. Environmental Protection Agency under Research and Development Grant No. 11010 ENX, 1973.
6. Pomeroy, R. D.: Sanitary Sewer Design for Hydrogen Sulfide Control, *Public Works,* vol. 101, no. 10, 1970.

7. Pomeroy, R. D., and J. D. Parkhurst: The Forecasting of Sulfide Buildup Rates in Sewers, *Progress in Water Technology,* vol. 9, Pergamon Press, New York, 1977.
8. *Process Design Manual for Sulfide Control in Sanitary Sewerage Systems,* U.S. Environmental Protection Agency, Technology Transfer, Washington, D.C., 1974.
9. Tchobanoglous, O.: *Review of Proposed Modification of Wildcat Hill Wastewater Treatment Facility,* report prepared for the city of Flagstaff, Ariz., Davis, Calif., 1975.
10. Thistlethwayte, D. K. B. (ed): *Control of Sulphides in Sewerage Systems,* Butterworth, Melbourne, Australia, 1972, and Ann Arbor Science Publishers, Ann Arbor, Mich., 1972.
11. Wood, D. K., and G. Tchobanoglous: Trace Elements in Biological Waste Treatment, *J. Water Pollut. Control Fed.,* vol. 47, no. 7, 1975.
12. Warren, G., and G. Tchobanoglous: *A Study of the Use of Concrete Pipe for Trunk Sewers in the City of Delano, California,* report prepared for the California Precast Concrete Pipe Association, by Department of Civil Engineering, University of California, Davis, Calif., 1976.

CHAPTER

EIGHT

PUMPS AND PUMP SYSTEMS

This chapter is an introduction to the study of pumps and pump systems. Because the scope of the subject matter is so broad, the contents of this chapter are limited to that information with which the sanitary engineer should be most familiar. This chapter includes an introduction to pump analysis, the types and operating characteristics of pumps and pump drive units, the fundamentals of pump selection, and the analysis of pump systems. The design of pumping stations is discussed in Chap. 9.

Reference books (such as Refs. 1, 3, and 5 through 8), journal articles (such as Refs. 10 and 13), and other current publications should be used to complement the information presented in this chapter. A brief descriptive summary of equipment currently available in the United States and the names and addresses of the manufacturers are contained in Ref. 4. In practice, the engineer must also be familiar with the catalogs and data of leading pump manufacturers, because engineering design problems cannot be solved without reference to such information.

8-1 INTRODUCTION TO PUMP ANALYSIS

The reason for pumping water or wastewater is to transport it from one location to another, usually from a lower to a higher elevation. In this section, the principal concepts of pump analysis are introduced and some of the terms describing pumps and pumping stations are defined. The concepts discussed include (1) capacity, (2) head, and (3) efficiency and power input. Pump and system head-capacity curves are also introduced; they are discussed in detail in Secs. 8-3 and 8-6.

Capacity

The capacity (flowrate) of a pump is the volume of liquid pumped per unit of time, which usually is measured in liters per second or cubic meters per second (gallons per minute or million gallons per day).

Head

The term *head* is the elevation of a free surface of water above or below a reference datum. For instance, if a small, open-ended tube were run vertically upward from a pipe under pressure, the head would be the distance from the centerline of the pipe to the free water surface in the vertical tube.

In pump systems, the head refers to both pumps and pump systems having one or more pumps and the corresponding piping system. The height to which a pump can raise a liquid is the pump head and is measured in meters (feet) of the flowing liquid. The head required to overcome the losses in a pipe system at a given flowrate is the system head.

Terms applied specifically to the analysis of pumps and pump systems include (1) static suction head, (2) static discharge head, (3) static head, (4) friction head, (5) velocity head, (6) minor head loss, and (7) total dynamic head, which is defined in terms of the other head terms. Each of these terms is described in the following and is illustrated graphically in Fig. 8-1. All the terms are expressed in meters (feet) of water.

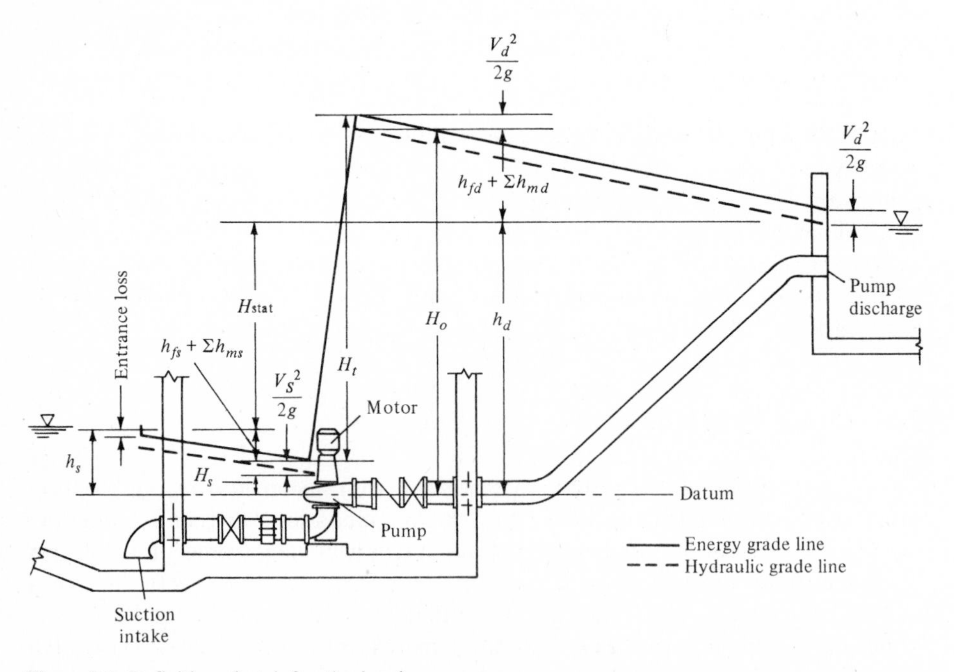

Figure 8-1 Definition sketch for the head on a pump.

Static suction head. The static suction head h_s is the difference in elevation between the suction liquid level and the centerline of the pump impeller. If the suction liquid level is below the centerline of the pump impeller, it is a static suction lift. Pumps handling untreated wastewater are usually installed with a small static suction head (flooded suction), as shown in Fig. 8-1, to avoid having to install a priming system that could be fouled and made inoperable by the solids in untreated wastewater.

Static discharge head. The static discharge head h_d is the difference in elevation between the discharge liquid level and the centerline of the pump impeller.

Static head. Static head H_{stat} is the difference in elevation between the static discharge and static suction liquid levels $(h_d - h_s)$.

Friction head. The head of water that must be supplied to overcome the frictional loss caused by the flow of fluid through the piping system is the friction head. The frictional head loss in the suction (h_{fs}) and discharge (h_{fd}) piping system may be computed with the Darcy-Weisbach or Hazen-Williams equations (see Chap. 2).

Velocity head. The velocity head is the kinetic energy contained in the liquid being pumped at any point in the system and is given by

$$\text{Velocity head} = \frac{V^2}{2g}$$

where V = velocity of fluid, m (ft)
g = acceleration due to gravity 9.81 m/s² (32.2 ft/s²)
In determining the head at any point in a piping system, the velocity head must be added to the gage reading.

Minor head loss. The head of water that must be supplied to overcome the loss of head through fittings and valves is the minor head loss. Minor losses in the suction (h_{ms}) and discharge (h_{md}) piping system are usually estimated as fractions of the velocity head by using the following expression:

$$h_m = K\frac{V^2}{2g}$$

where h_m = minor head loss, m (ft)
K = head loss coefficient
Typical values of K for various pipeline fittings and appurtenances are presented in Appendix C. Standard textbooks and reference works on hydraulics should be consulted for other values.

Total dynamic head. The total dynamic head H_t is the head against which the pump must work when water or wastewater is being pumped. The total dynamic head on a pump, commonly abbreviated TDH, can be determined by considering the static suction and discharge heads, the frictional head losses, the velocity heads, and the minor head losses. The expression for determining the total dynamic head on a pump is given in Eq. 8-1 (see Fig. 8-1).

$$H_t = H_D - H_S + \frac{V_d^2}{2g} - \frac{V_s^2}{2g} \tag{8-1}$$

$$H_D = h_d + h_{fd} + \Sigma h_{md} \tag{8-2}$$

$$H_S = h_s - h_{fs} - \Sigma h_{ms} - \frac{V_s^2}{2g} \tag{8-3}$$

where H_t = total dynamic head, m (ft)
H_D (H_S) = discharge (suction) head measured at discharge (suction) nozzle of pump referenced to the centerline of the pump impeller, m (ft)
V_d (V_s) = velocity in discharge (suction) nozzle, m/s (ft/s)
g = acceleration due to gravity, 9.81 m/s^2 (32.2 ft/s^2)
h_d (h_s) = static discharge (suction) head, m (ft)
h_{fd} (h_{fs}) = frictional head loss in discharge (suction) piping, m (ft)
h_{md} (h_{ms}) = minor fitting and valve losses in discharge (suction) piping system, m (ft). Entrance loss is included in computing the minor losses in the suction piping.

As noted previously, the reference datum for writing Eq. 8-1 is taken as the elevation of the centerline of the pump impeller. In accordance with the standards of the Hydraulic Institute [11], distances (heads) above datum are considered positive; distances below datum are considered negative.

In terms of the static head, Eq. 8-1 can be written as

$$H_t = H_{stat} + h_{fs} + \Sigma h_{ms} + h_{fd} + \Sigma h_{md} + \frac{V_d^2}{2g} \tag{8-4}$$

where H_t = total dynamic head, m (ft)
H_{stat} = total static head, m (ft)
$= h_d - h_s$

In Eq. 8-4, the energy in the velocity head $V_d^2/2g$ is usually considered to be lost at the outlet of the piping system. In practice, this loss of energy is taken as being equivalent to the exit loss and is included as a minor loss.

The energy (Bernoulli's) equation can also be applied to determine the total dynamic head on the pump. The energy equation written between the suction and discharge nozzle of the pump is

$$H_t = \frac{P_d}{\gamma} + \frac{V_d^2}{2g} + z_d - \frac{P_s}{\gamma} + \frac{V_s^2}{2g} + z_s \tag{8-5}$$

where H_t = total dynamic head, m (ft)
P_d (P_s) = discharge (suction) gage pressure, kN/m^2 (lb$_f$/ft^2)
γ = specific weight of water, N/m^3 (lb$_f$/ft^3)
V_d (V_s) = velocity in discharge (suction) nozzle, m/s (ft/s)

g = acceleration due to gravity, 9.81 m/s² (32.2 ft/s²)

z_d (z_s) = elevation of zero of discharge (suction) gage above datum, m (ft)

Head losses within the pump are incorporated in the total dynamic head term in Eq. 8-5.

Pump Efficiency and Power Input

Pump performance is measured in terms of the capacity that a pump can discharge against a given head and at a given efficiency. The pump capacity is a function of the design. Information on the design is furnished by the pump manufacturer in a series of curves for a given pump. Pump efficiency E_p—the ratio of the useful output power of the pump to the input power to the pump—is given by

$$E_p = \frac{\text{pump output}}{P_i} = \frac{\gamma Q H_t}{P_i} \quad \text{(SI units)} \qquad (8\text{-}6)$$

$$E_p = \frac{\text{pump output}}{\text{bhp}} = \frac{\gamma Q H_t}{\text{bhp} \times 550} \quad \text{(U.S. customary units)} \qquad (8\text{-}6a)$$

where E_p = pump efficiency, dimensionless

P_i = power input, kW, kN · m/s

γ = specific weight of water, kN/m³ (lb/ft³)

Q = capacity, m³/s (ft³/s)

H_t = total dynamic head, m (ft)

bhp = brake horsepower

550 = conversion factor for horsepower to ft · lb_f/s

Pump efficiencies usually range from 60 to 85 percent. The energy losses in a pump may be classified as volumetric, mechanical, and hydraulic. Volumetric losses occur because the small clearances necessary between the pump casing and the rotating element can leak. Mechanical losses are caused by mechanical friction in the stuffing boxes and bearings, by internal disk friction, and by fluid shear. Frictional and eddy losses within the flow passages account for the hydraulic losses [3, 14].

Pump Head-Capacity Curve

The head that a pump can produce at various flowrates and constant speed is established in pump tests conducted by the pump manufacturer. The pump head equals the difference between the energy head at the discharge and suction nozzles as given by the energy (Bernoulli's) equation (Eq. 8-5).

During testing, the capacity of the pump is varied by throttling a valve in the discharge pipe and the corresponding head is measured. The results of these tests are plotted to obtain a head-capacity curve for the pump at the given speed (see Fig. 8-2). Simultaneously, the efficiency and the power input are measured, and these values are also plotted in the same diagram. Together these curves are known as pump characteristic curves.

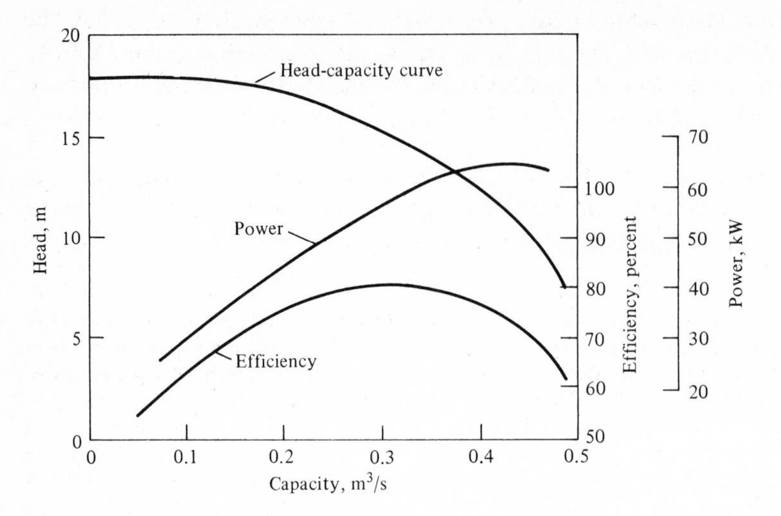

Figure 8-2 Typical pump characteristic curves for a radial-flow centrifugal pump.

System Head-Capacity Curve

To determine the head required of a pump, or group of pumps, that would discharge at various flowrates into a given piping system, a system head-capacity curve is prepared (see Fig. 8-3). This curve is a graphic representation of the system head and is developed by plotting the total dynamic head (static lift plus kinetic energy losses) over a range of flows from zero to the maximum expected value with the use of Eq. 8-4.

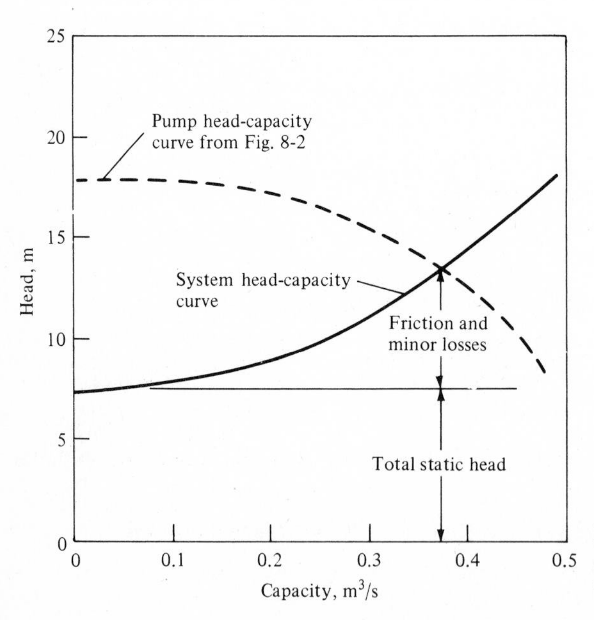

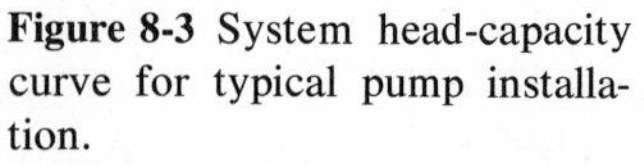

Figure 8-3 System head-capacity curve for typical pump installation.

If the pump head-capacity curve from Fig. 8-2 were plotted in Fig. 8-3, the intersection of the pump head-capacity curve and the system head-capacity curve represents the head and capacity that the pump will produce if operated in the given piping system. This point is also known as the *pump operating point*.

The application of Eqs. 8-1, 8-2, and 8-3 is illustrated in Example 8-1. The development of system head-capacity curves for pumping installations, using these equations and Eq. 8-5, is discussed in Secs. 8-5 and 8-6.

Example 8-1 Determination of head and energy requirements for pump setup A wastewater pump is discharging at a rate of 0.5 m^3/s (7925 gal/min). It has a discharge nozzle diameter of 350 mm (14 in) and a suction nozzle diameter of 400 mm (16 in). The reading on the discharge gage located at the pump centerline is 125 kN/m^2 (18 lb_f/in^2). The value on the suction gage located 0.6 m (2 ft) below the pump centerline is 10 kN/m^2 (1.5 lb_f/in^2). Determine (1) the head on the pump with Bernoulli's equation, (2) the power input required by the pump, and (3) the power input to the motor. Assume that the pump efficiency is 82 percent and that the motor efficiency is 91 percent.

SOLUTION

1. Determine the head on the pump with Bernoulli's equation and the pump centerline as the reference datum.

   a. Bernoulli's equation (Eq. 8-5) for this situation is

$$H = \frac{P_d}{\gamma} + \frac{V_d^2}{2g} + z_d - \frac{P_s}{\gamma} - \frac{V_s^2}{2g} - z_s$$

   b. Determine the values of the individual terms in Bernoulli's equation:

$$\frac{P_d}{\gamma} = \frac{125{,}000 \text{ N/m}^2}{9810 \text{ N/m}^3} = 12.74 \text{ m}$$

$$V_d = \frac{Q_d}{A_d} = \frac{0.5 \text{ m}^3/\text{s}}{(\pi/4)(0.35 \text{ m})^2} = 5.2 \text{ m/s}$$

$$\frac{V_d^2}{2g} = \frac{(5.2 \text{ m/s})^2}{2(9.81 \text{ m/s}^2)} = 1.38 \text{ m}$$

$$z_d = 0 \quad \text{(gage located at pump centerline)}$$

$$\frac{P_s}{\gamma} = \frac{10{,}000 \text{ N/m}^2}{9810 \text{ N/m}^3} = 1.02 \text{ m}$$

$$V_s = \frac{0.5 \text{ m}^3/\text{s}}{(\pi/4)(0.40 \text{ m})^2} = 3.98 \text{ m/s}$$

$$\frac{V_s^2}{2g} = \frac{(3.98 \text{ m/s})^2}{2(9.81 \text{ m/s}^2)} = 0.81 \text{ m}$$

$$z_s = -0.6 \text{ m}$$

   c. To determine the head, substitute the individual terms computed in step 1*b* in Bernoulli's equation:

$$H = 12.74 \text{ m} + 1.38 \text{ m} + 0 - 1.02 \text{ m} - 0.81 \text{ m} - (-0.6 \text{ m})$$

$$= 12.89 \text{ m (42.3 ft)}$$

2. Determine the power input required by the pump (P_i) with the use of Eq. 8-6.

$$p_i = \frac{(9.810\ \text{kN/m}^3)(0.5\ \text{m}^3/\text{s})(12.89\ \text{m})}{0.82}$$

$$= 77.1\ \text{kW}\ (103.4\ \text{hp})$$

(*Note:* kW × 1.3410 = hp.)

3. Determine the required power input to the motor (P_m).

$$P_m = \frac{77.1}{E_m} = \frac{77.1}{0.91}$$

$$= 84.7\ \text{kW}\ (113.4\ \text{hp})$$

8-2 PUMPS

The pumps commonly used in the wastewater field (centrifugal, screw, and positive-displacement pumps), as well as some special pumps that have found application, are discussed in this section. Pump construction is also briefly discussed.

Classification of Pumps

According to the Hydraulic Institute, all pumps may be classified as kinetic-energy pumps or positive-displacement pumps [11]. The pumps related to each class are presented graphically in Fig. 8-4. As shown, centrifugal pumps, the most commonly used pump in the wastewater engineering field, are classified as kinetic-energy pumps. The three types of centrifugal pumps are radial-flow, mixed-flow, and axial-flow. In general, radial-flow and mixed-flow pumps are used to pump wastewater and storm water. Axial-flow pumps may be used for pumping treatment-plant effluent or storm drainage unmixed with wastewater.

Screw pumps, which are becoming more popular, are classified as positive-displacement pumps. In addition, a variety of both positive-displacement and other kinetic-energy pumps are used in specialized applications in the wastewater engineering field.

Centrifugal Pumps

Centrifugal pumps are classified as radial-flow, mixed-flow, and axial-flow. Before discussing these types of pumps individually, it will be helpful to understand the general characteristics of centrifugal pumps.

Pump characteristics. Every centrifugal pump consists of two principal parts: a rotating element called the *impeller,* which forces the liquid being pumped into a rotary motion, and the pump casing, which is designed to direct the liquid to the impeller and lead it away. As the impeller rotates, the liquid leaves the

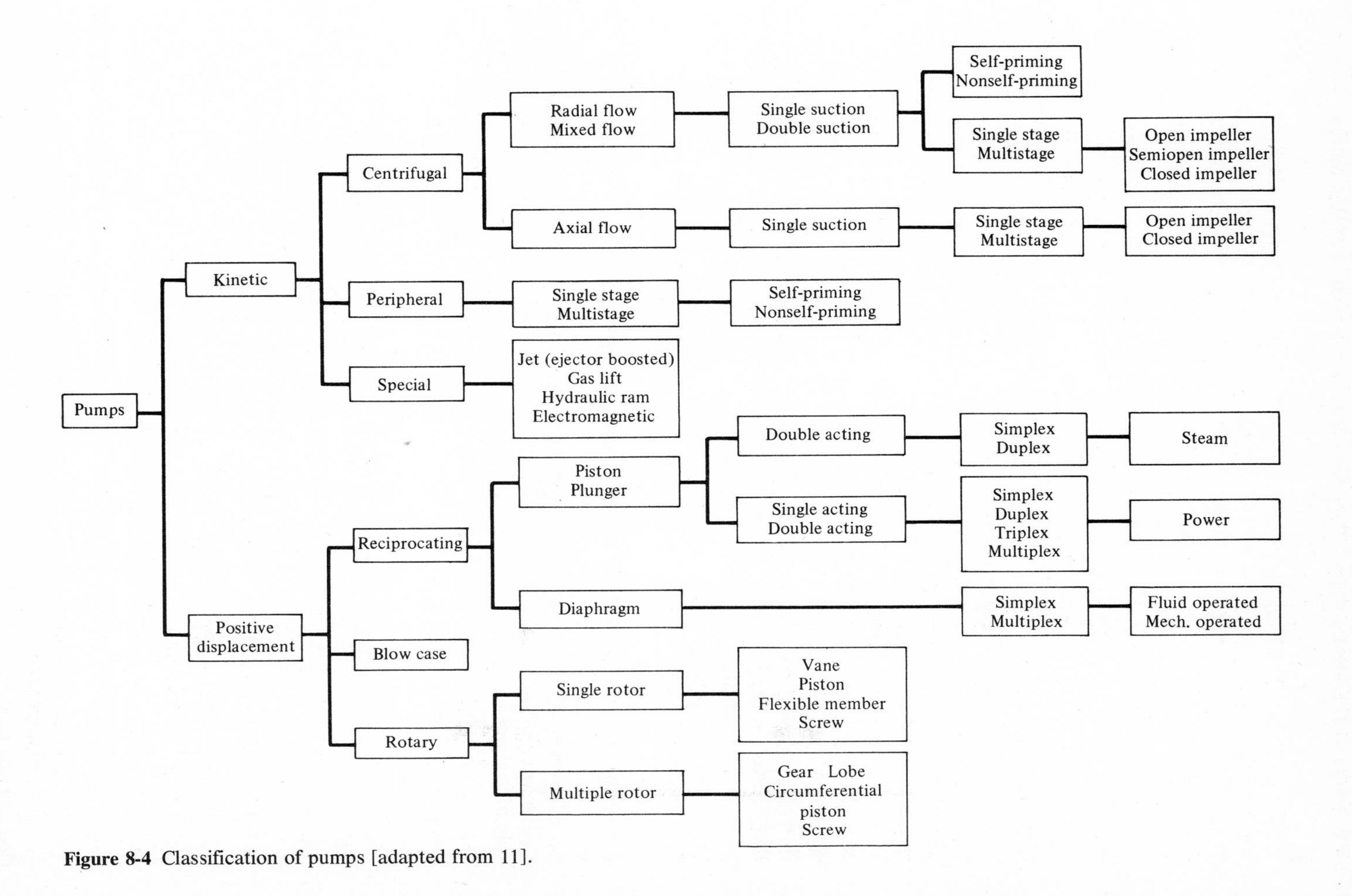

Figure 8-4 Classification of pumps [adapted from 11].

impeller with a higher pressure and velocity than when it entered. The exit velocity of the fluid as it leaves the impeller tip is partially converted to pressure in the pump casing before it leaves the pump in the discharge nozzle. The conversion of velocity to pressure occurs in the pump casing. The pump casing can be of two types—volute or diffusion.

In a volute casing, the size of the channel surrounding the impeller increases gradually to the size of the pump discharge nozzle (see Fig. 8-5), and most of the conversion of velocity to pressure occurs in the conical discharge nozzle. In a diffusion casing, the impeller discharges into a channel provided with guide vanes (see Fig. 8-5). The conversion of velocity to pressure occurs within the vane passages. Pumps with diffusion casings were once known as turbine pumps. They are seldom used for pumping of untreated wastewater, but are often used in high-pressure applications with potable water or treated effluent in wastewater-treatment plants.

The shape of the impeller and the pump casing vary with the type of centrifugal pump. Photographs of the principal types of pump impellers are shown in Fig. 8-6. In a radial-flow pump, the liquid enters the impeller axially through the suction nozzle and is discharged radially into the volute casing. Radial-flow impellers with single- and double-suction inlets (water enters axially from both ends) are shown in Fig. 8-6*a* and 8-6*b*, respectively. In mixed-flow pumps, the liquid enters the impeller axially and is discharged in an intermediate direction somewhere between the radial and axial directions; a mixed-flow impeller is shown in Fig. 8-6*c*. In an axial-flow pump, the liquid enters and leaves the impeller axially; an axial-flow impeller is shown in Fig. 8-6*d*.

Centrifugal pumps are often classified according to a type number known as the *specific speed,* which varies with the shape of the impeller. Typical specific-speed values for the various types of centrifugal pumps are presented in Fig. 8-7. The derivation of specific speed and its use are described in Sec. 8-3.

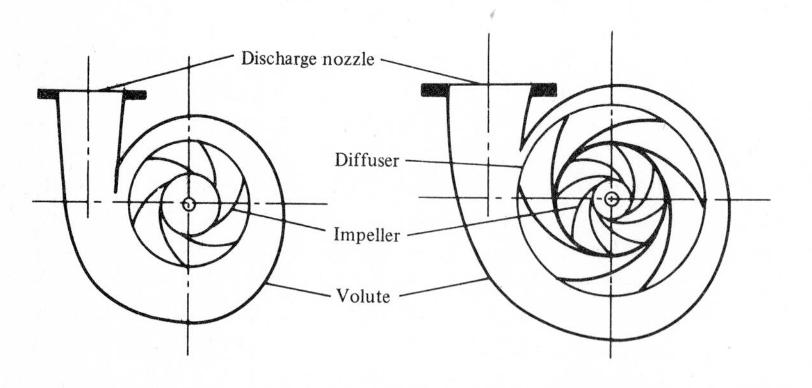

Figure 8-5 Pumping casings [12].

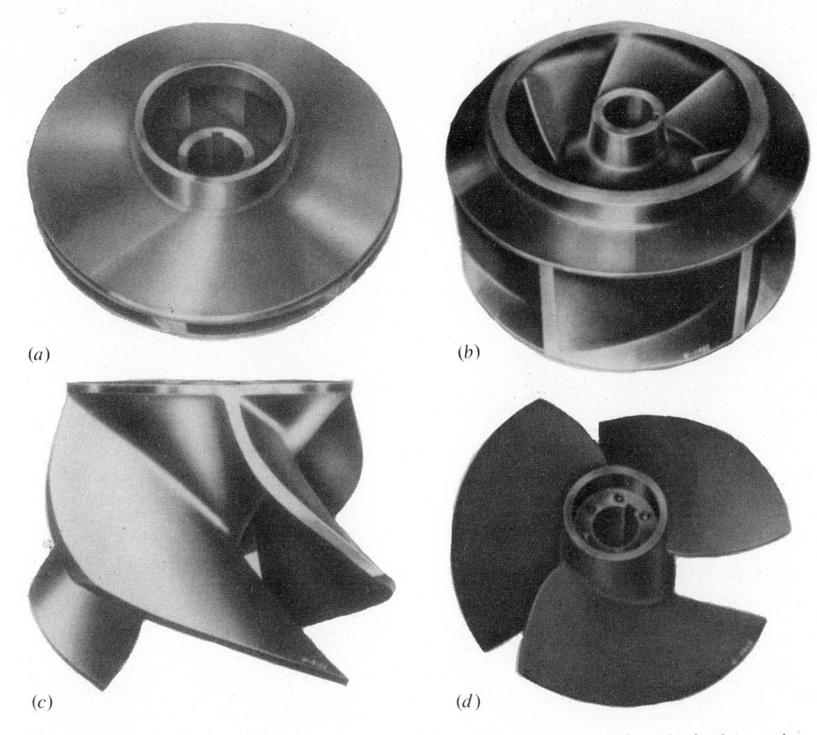

Figure 8-6 Typical impellers used in centrifugal pumps. (*a*) Closed single-suction, (*b*) Closed Francis-vane double-suction, (*c*) Open mixed-flow, (*d*) Axial-flow (propeller). (*Courtesy Worthington Pump, Inc.*)

Radial-flow pumps. The impellers used in radial-flow pumps are classified are single-suction or double-suction impellers. They can also be classified by the shape and the form of their vanes, which may be straight or of double curvature, such as the Francis-type impeller. The radial-flow pumps, including those with Francis-type impellers, have specific speeds varying from 10 to 80 (500 to 4000 in U.S. customary units) (see Fig. 8-7).

Because rags and trash in wastewater (even though screened) would quickly clog the small passages in typical clearwater radial-flow pumps, the pumps used for untreated wastewater are generally the single-end suction, volute type, fitted with nonclog impellers, as shown in Fig. 8-8. Double-suction pumps, where the flow enters the impeller axially from both sides of the impeller and the shaft passes through both suction passages, are not normally used to pump wastewater because rags would tend to wrap around the shaft and clog the pumps.

The pump shafts may be horizontal or vertical. However, vertical pumps are preferred because of space limitations. A typical installation of large, vertical wastewater pumps is shown in Fig. 8-9.

Nonclog pumps have open passages and a minimum number of vanes (not exceeding two in the smaller sizes and limited to three, or at the most four, in the larger sizes) (see Fig. 8-8). Impellers are almost universally enclosed.

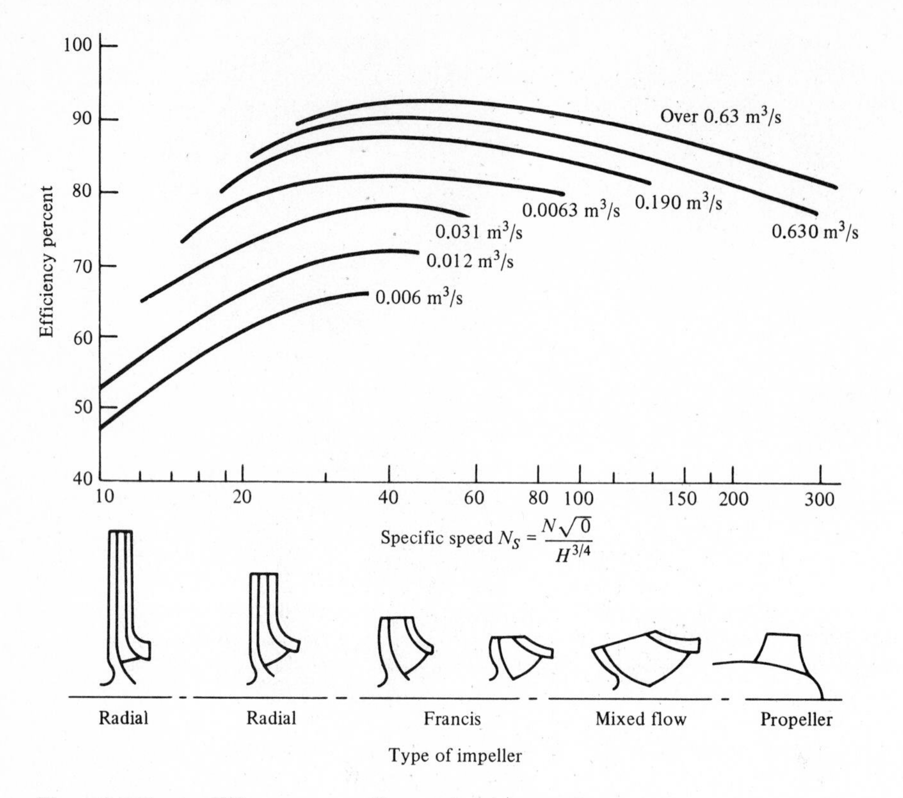

Figure 8-7 Pump efficiency vs. specific speed and pump size.

Wastewater pumps must be able to pass solids that enter the collection system. Because a 70-mm (2.5-in)-diameter solid can pass through most domestic toilets, it is common practice to require that pumps be able to discharge a 75-mm (3-in) solid. Most 100-mm (4-in) pumps—pumps with a 400-mm discharge opening—normally should be able to pass 75-mm (3-in)-diameter solids, and 200-mm (8-in) pumps should be able to pass 100-mm (4-in)-diameter solids, etc. The size of the solid that a pump can discharge becomes correspondingly larger as the pump size increases, up to 200-mm (8-in) or larger solids for 900-mm (36-in) pumps, depending on the design. Nonclog pumps smaller than 100 mm (4 in) should not be used in municipal pumping stations for handling untreated wastewater.

Mixed-flow pumps. The impellers used for mixed-flow pumps may be installed either in volute-type casings, which are designated *mixed-flow volute pumps,* or in diffusion-type casings similar to propeller pumps, which are designated *mixed-flow propeller pumps.* Francis-type and mixed-flow impellers may be available for the same casing design; the Francis-type impellers are designed for heads greater than 30 m (100 ft).

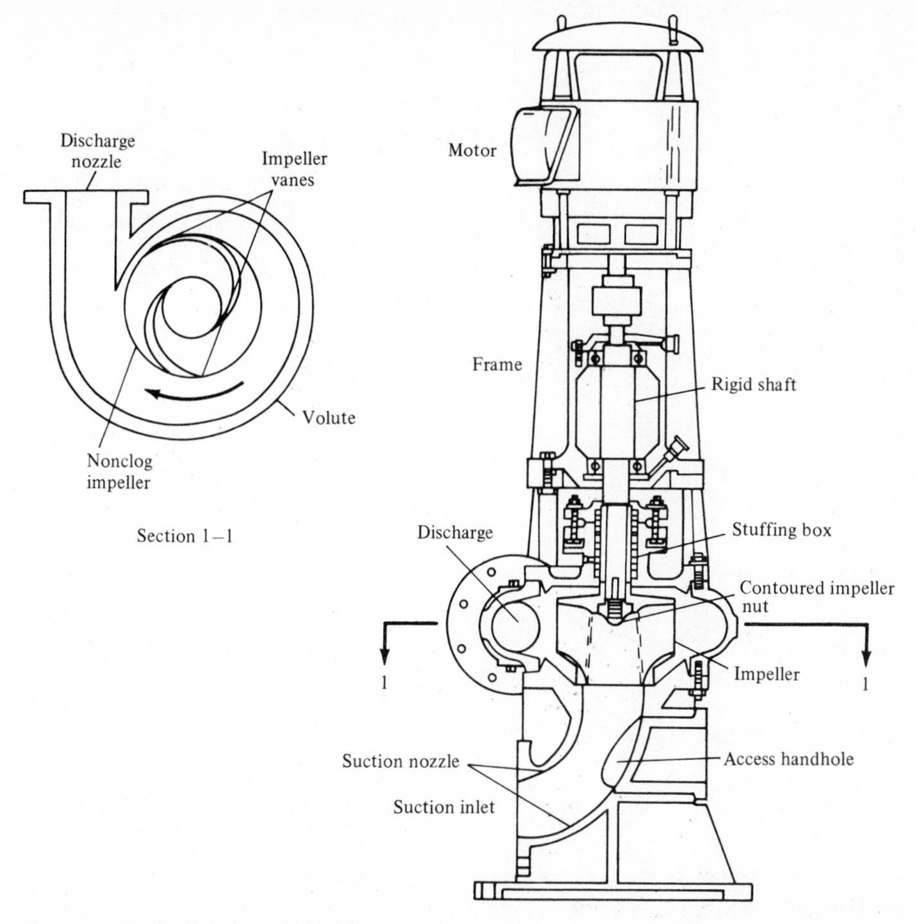

Figure 8-8 Typical vertical radial-flow wastewater pump.

The specific speed of mixed-flow pumps varies from about 80 to 200 (4000 to 10,000 in U.S. customary units). As the specific speed increases from 80 to 120 to 200, the pump characteristics of a mixed-flow pump become more like those of an axial-flow pump. The shape of the impeller and the design of the pump also vary.

Mixed-flow volute pumps are suitable for pumping untreated wastewater and storm water, especially in the specific-speed range between 80 and 120 (see Fig. 8-10). They are available in sizes of 200 mm (8 in) and larger and for heads of up to 15 to 18 m (50 to 60 ft). They operate at higher speeds than the radial-flow nonclog pumps, are usually of lighter construction, and, where applicable, cost less than corresponding nonclog pumps. The size of the solid that the mixed-flow volute pump can pass is much smaller than can be handled by a nonclog pump of the same size, but the 200-mm (8-in) mixed-flow pump can pass a 75-mm (3-in)-diameter solid. Impellers may be either open or enclosed, but enclosed is preferred.

Figure 8-9 Typical large-capacity centrifugal pumps used for pumping wastewater.

Axial-flow pumps. Axial-flow pumps have a multiple-bladed screw rotor, or propeller, placed in a casing, with fixed guide vanes before and after the impeller. These pumps have a specific speed exceeding 200 (10,000 in U.S. customary units). The action of the pump is similar to a ship propeller as it draws water through the outlet guide vanes (see Fig. 8-11).

Axial-flow pumps are used where a large quantity of wastewater must be pumped against a low head and for such services as pumping treatment-plant effluent or storm water. Axial-flow pumps are less expensive than radial- or mixed-flow pumps. They should not be used to transport untreated wastewater, sludge, or unscreened storm water because rags could become lodged and build up on the guide vanes.

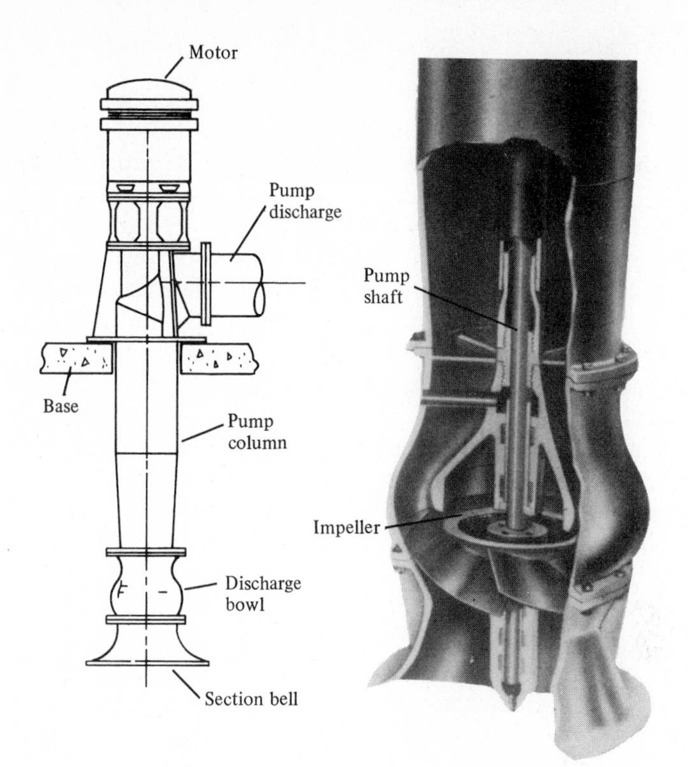

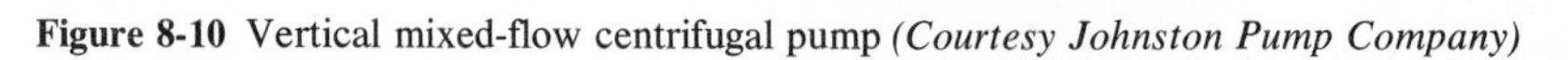

Figure 8-10 Vertical mixed-flow centrifugal pump *(Courtesy Johnston Pump Company)*

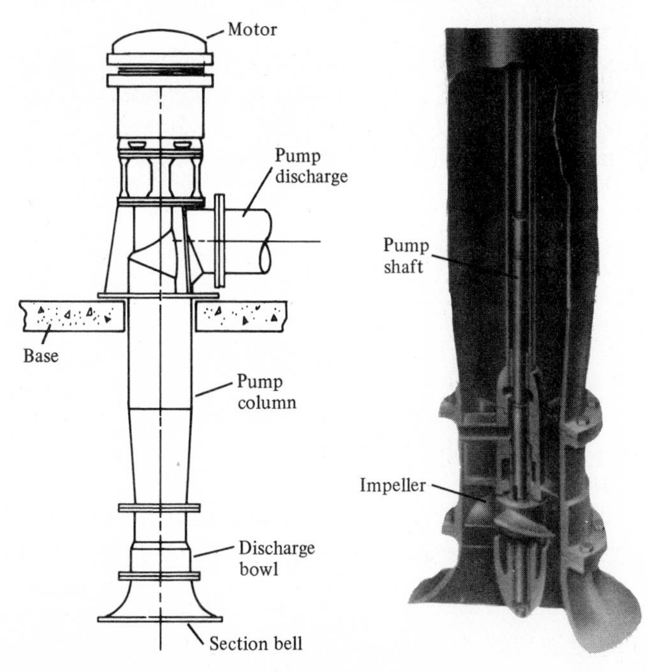

Figure 8-11 Axial-flow centrifugal pump *(Courtesy Johnston Pump Company)*

Screw Pumps

The screw pump, classified as a *positive-displacement pump,* is probably the world's oldest type of pump, but it has only recently received general acceptance in the United States for pumping wastewater. It is based on the Archimedes screw principle in which a revolving shaft fitted with one, two, or three helical blades rotates in an inclined trough and pushes the wastewater up the trough (see Fig. 8-12). The screw pump has two distinct advantages over centrifugal pumps when pumping wastewater: (1) it can pump large solids without clogging, and (2) it operates at a constant speed over a wide range of flows with relatively good efficiencies.

Screw pumps are available in sizes ranging from 0.3 to 3 m (1 to 10 ft) in screw diameter and capacities ranging from 0.01 to 3.2 m^3/s (150 to 50,000 gal/min). Larger pumps are available from some manufacturers. The angle of inclination has been standardized at either 30 or 38°. A pump installed at a 30° angle will pump more than one installed at a 38° angle, but it will take up more space. The total head for a screw pump is limited to a lift of about 9 m (30 ft). This limit is imposed by structural requirements of the screw.

The pumps are usually driven by a constant-speed motor and gear reducer at 30 to 50 rev/min. Normal efficiencies are about 85 percent at maximum capacity and 65 percent at 25 percent of maximum capacity. The pump capacity depends on the depth of the liquid entering the screw; the lower the level, the less the capacity.

Because of its nonclog characteristics and its variable-capacity pumping, the screw pump could be useful in several wastewater pumping applications, including (1) low-lift pumping of untreated wastewater, (2) storm-water pumping, (3) return-sludge pumping, and (4) effluent pumping. Another potential use is at a treatment plant when it is expanded to secondary or advanced waste treatment. A low-lift pump is often required to provide only enough head for

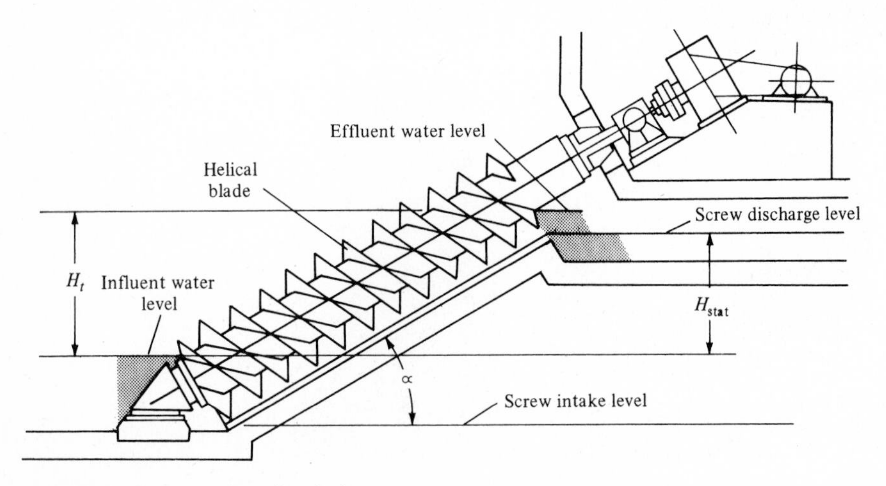

Figure 8-12 Cross section of typical screw pump.

the hydraulic losses through the new process so that an existing outfall can be used.

Most screw pumps are installed outdoors, and only the drive unit is housed. The pumps should be installed in individual inlet channels, with slide gates, so that the inlet to the pump can be isolated and dewatered to allow the

Figure 8-12 Photograph of typical screw pump.

bottom (submerged) bearing to be maintained. The discharge point of the screw is usually above the high-water level in the discharge channel so that outlet check valves and slide gates are not required. Although the screw pump operates at a low speed, the screw trough should be protected for safety. Preferably, the trough should be covered with grating or checkered plate, or, as a minimum precaution, it should be isolated by handrails.

Other Pumps for Wastewater

Other pumps used for various wastewater applications include pneumatic ejectors, bladeless pumps, and air-lift and jet pumps.

Pneumatic ejectors. Pneumatic ejectors are installed where the initial flows are small and the anticipated future flows will not exceed the capacity of the unit. They are used in small installations because they do not clog easily.

A pneumatic wastewater ejector with its associated equipment and controls is shown in Fig. 8-13. Wastewater enters and begins to fill the ejector receiver.

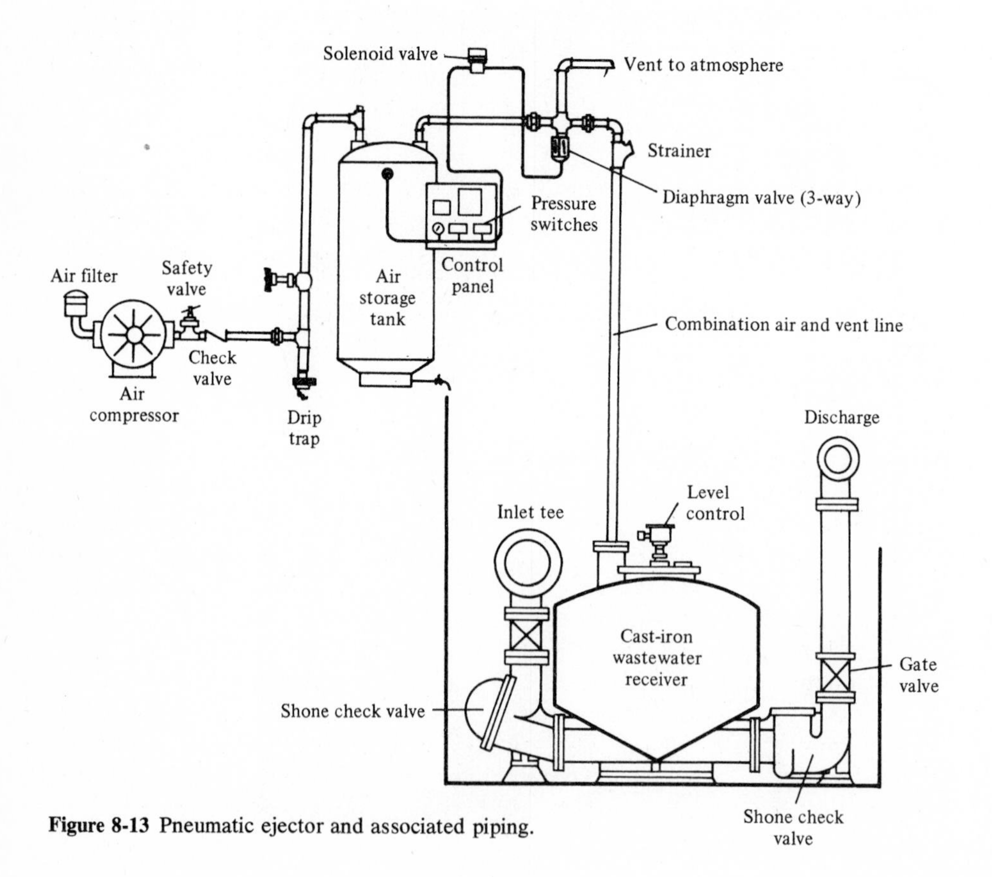

Figure 8-13 Pneumatic ejector and associated piping.

When the receiver is filled, a three-way valve is actuated to close the vent and admit compressed air to the pot, which forces the wastewater into the discharge main. When the wastewater reaches the minimum level, the position of the three-way valve is reversed to shut off the flow of compressed air and open the vent. Venting the air within the pot permits wastewater to flow again into the receiver. The compressed air may be supplied directly by the air compressors or from an air-storage tank which is maintained at the necessary pressure by the air compressors. If an air-storage tank is used, the air compressors can be smaller, and the power of the installed motor can be reduced by one-half.

Pneumatic ejectors are available in capacities varying from 2 to 38 L/s (30 to 600 gal/min) when used singly (simplex operation). Under the usual design conditions, the ejectors operate on a 1-min cycle, filling for 30 s and discharging for 30 s. Consequently, the friction-head losses of the discharge force main should be computed on the basis of 200 percent of the design flowrate.

Wastewater ejectors in municipal installations should be used in pairs operating in alternating cycles (duplex operation). Large units should be alternated in operation. If the incoming and outgoing sewers have sufficient capacity, smaller units may be operated as two separate units, in which case the controls are much simpler.

Pneumatic ejectors are economically feasible for flows of up to 20 L/s (300 gal/min), beyond which point power costs become excessive. For larger flows, nonclog centrifugal wastewater pumps are recommended.

Bladeless pumps. Bladeless pumps are essentially volute-type centrifugal pumps fitted with a special bladeless or single-passage impeller. The capacity of the bladeless pump is about one-half that of the conventional nonclog pump. These pumps have demonstrated superior nonclogging performance and are particularly suitable for small flows. They are available in sizes of up to 125 mm (5 in).

Air-lift and jet pumps. An air-lift pump has no moving parts and is, therefore, practically noncloggable. Compressed air is introduced into the pump at the bottom of the up-draft tube, as shown in Fig. 8-14. Because the density of the air and water mixture is less than that of the surrounding water, the mixture is lifted to a higher elevation by atmospheric pressure. The air-lift pump is limited by available compressed-air pressure with heads normally ranging from 1 to 1.5 m (3 to 5 ft). These pumps have limited use because their efficiency is usually about 30 percent.

Jet pumps are used occasionally in wastewater-treatment plants for such services as priming centrifugal pumps or sump pumps for clearwater sumps. In these pumps, a fluid (gas or liquid) is passed through a Venturi section; the reduced pressure at the throat of the Venturi creates a vacuum to draw water (or air) into the pressure fluid flow. Jet pumps are also known as ejectors.

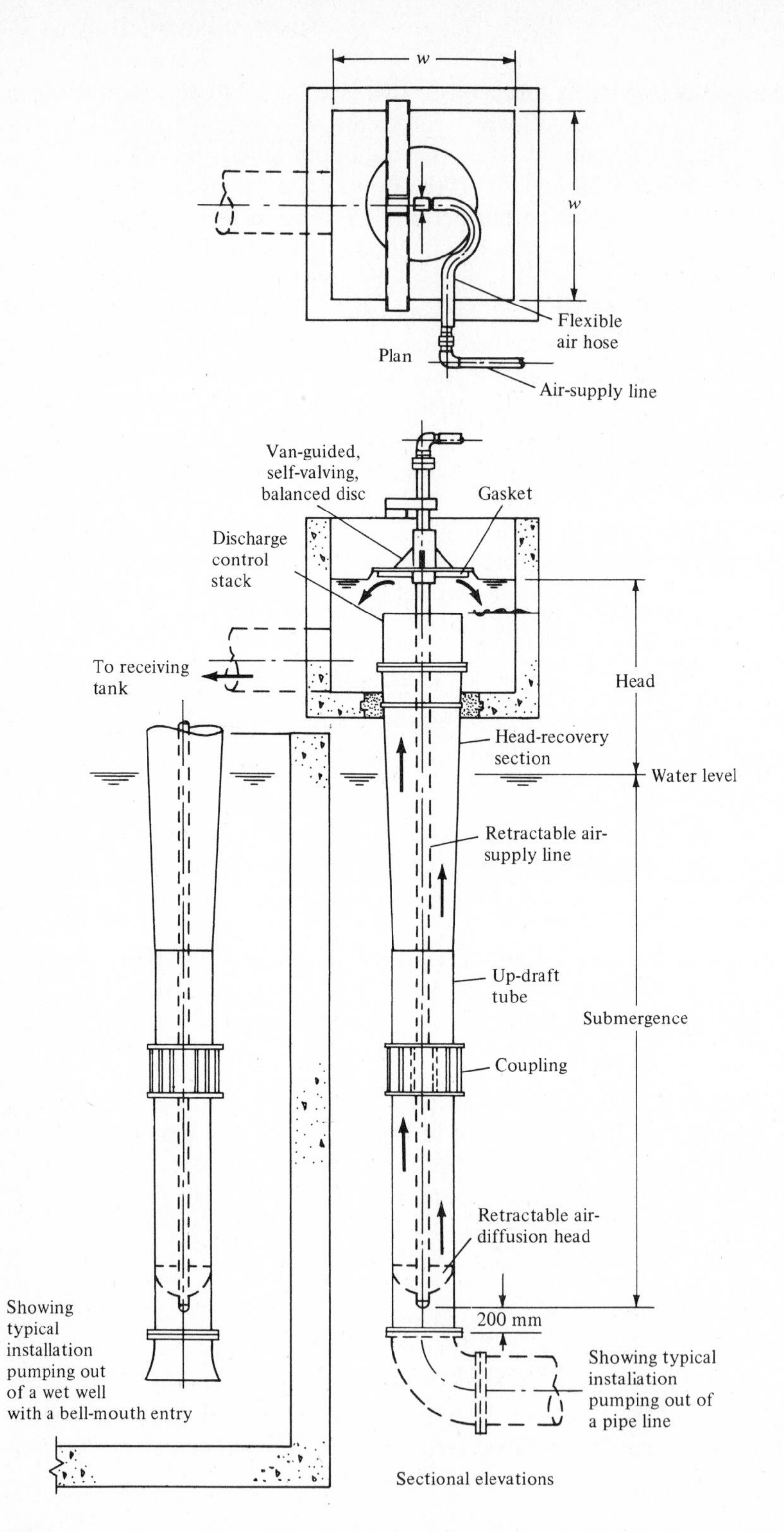
w
w
Flexible air hose
Plan
Air-supply line
Van-guided, self-valving, balanced disc
Gasket
Discharge control stack
To receiving tank
Head
Head-recovery section
Water level
Retractable air-supply line
Up-draft tube
Submergence
Coupling
Retractable air-diffusion head
200 mm
Showing typical installation pumping out of a wet well with a bell-mouth entry
Showing typical instaliation pumping out of a pipe line
Sectional elevations

Pumps for Scum, Grit, and Sludge

Pumps that are used to pump scum, grit, and sludge are torque-flow, plunger, and rotary pumps. These pumps are not customarily used in pumping wastewater.

Torque-flow pumps. Torque-flow pumps, developed by Wemco (trade name) for handling solid materials, are now available from several manufacturers. They have a recessed impeller in the side of the case entirely out of the flow stream. A pumping vortex is set up by viscous drag.

Torque-flow pumps have been installed in many wastewater-treatment plants for pumping grit, sludge, scum, and untreated wastewater, and they have seldom become clogged in locations where the ordinary nonclog pumps clogged repeatedly. Because of their high cost and low efficiency, they are most useful in pumping sludge. Torque-flow pumps cast in Ni-hard material, which has high abrasion resistance, are available for pumping grit and ash slurries.

Plunger and rotary pumps. Plunger-type reciprocating-power pumps are commonly used in wastewater-treatment plants for transferring sludge from primary settling tanks to digestion tanks and from one digester to another in wastewater treatment plants. The progressing-cavity rotary pump has been used for handling concentrated sludge. Rotary gear-type pumps are installed in the lubrication systems of treatment-plant equipment, such as engines and blowers.

Pump Construction

Wastewater pumps are constructed of cast iron with bronze or stainless-steel trim and with either cast-iron or bronze impellers. If the wastewater contains grit, impellers made of bronze, cast steel, or stainless steel last longer. Pump shafts should be high-grade forged steel and protected by renewable bronze or stainless-steel sleeves where the shaft passes through the stuffing box. Stuffing boxes normally have packing or mechanical seals.

Most larger pumps have bronze wearing rings at the suction side of the impeller. Stainless-steel rings can be obtained if specified. Pumps smaller than 250 mm (10 in) usually have no wearing rings or wearing rings with the casing only. The bearings for vertical dry-pit pumps are usually the antifriction type located in the main frame above the impeller with oil-lubricated sleeve bearings.

If packing is provided, it is lubricated with grease or clean water. Grease is suitable for low-head pumps if the abrasive materials in the liquid being pumped are not a problem. If the wastewater contains grit and clean water is available,

Figure 8-14 Typical air-lift pump.

the packing should be lubricated with water at a pressure of 35 to 70 kN/m^2 (5 to 10 lbf/in^2) above the pump discharge pressure. Cross connections between sealing-water and potable-water systems are not permitted. Cross connections are eliminated by using pumped-seal-water systems or, where permitted by local authorities, by installing backflow-prevention devices.

The seal-water system consists of an air-break tank and sealing-water pumps. Water enters through the air break by a float-controlled valve to the tank, and the seal-water pumps take suction from the tank. Backflow-prevention devices consist of two pressure-reducing valves in series that always maintain seal-water pressure below potable-water system pressure. A differential-pressure relief valve relieves the pressure between the two reducing valves if the intermediate pressure approaches the supply pressure.

Wastewater pumps, especially in the smaller sizes, now have mechanical seals. These seals should be of the double mechanical seal type that require a clear fluid, at a slightly higher pressure than the discharge pressure, and should be injected between the two seal faces. The quantity of the liquid that escapes between the seal faces is small. In most cases, continuous flow of the sealing liquid is not required, but a vent connection must be installed at the high point to ensure that the space between the two faces is filled completely.

8-3 PUMP OPERATING CHARACTERISTICS

The operating characteristics of pumps depend on their size, speed, and design. Pumps of similar size and design are produced by many manufacturers, but they vary somewhat because of the design modifications made by each manufacturer. Operating characteristics for various types of wastewater pumps are reported in Table 8-1. The basic relationships that can be used to characterize and analyze pump performance of centrifugal pumps under varying conditions are discussed in detail in this section. Standard pump references should be consulted for the operating characteristics of screw pumps, other pumps for wastewater, and pumps for scum, grit, and sludge.

Pump Characteristic Curves

Pump manufacturers provide information on the performance of their pumps in the form of characteristic curves, commonly called *pump curves*. In most pump curves, the total dynamic head H_t, in meters (feet), the efficiency E in percent, and the power input P in kilowatts (horsepower) are plotted as ordinates against the capacity (flowrate) Q in cubic meters per second (gallons per minute or million gallons per day) as the abscissa. The general shape of these curves varies with the specific speed (see discussion of "Specific Speed," p. 297, in this section).

Characteristic curves for typical radial-flow, mixed-flow volute, mixed-flow propeller, and axial-flow centrifugal pumps are shown in Fig. 8-15. The

Table 8-1 Operating characteristics of various types of wastewater pumps

Characteristics	Centrifugal pumps: Radial-flow	Mixed-flow	Axial-flow	Screw pumps
1. Flow	Even	Even	Even	Even
2. Effect of increasing head on:				
(*a*) Capacity	Decrease	Decrease	Decrease	[a]
(*b*) Power required	Decrease	Small decrease to large increase	Large increase	
3. Effect of decreasing head on:				
(*a*) Capacity	Increase	Increase	Increase	[b]
(*b*) Power required	Increase	Slight increase to decrease	Decrease	[b]
4. Effect of closing discharge valve on:				
(*a*) Pressure	Up to 30% increase	Considerable increase	Large increase	Not applicable
(*b*) Power required	Decrease 50–60%	10% decrease to 80% increase	Increase 80–150%	Not applicable

[a]Head on screw pump can be increased only by lowering suction level, which will decrease capacity and power required.

[b]Head on screw pump can only be decreased by raising suction level and flooding screws. Capacity will remain constant and power required will decrease.

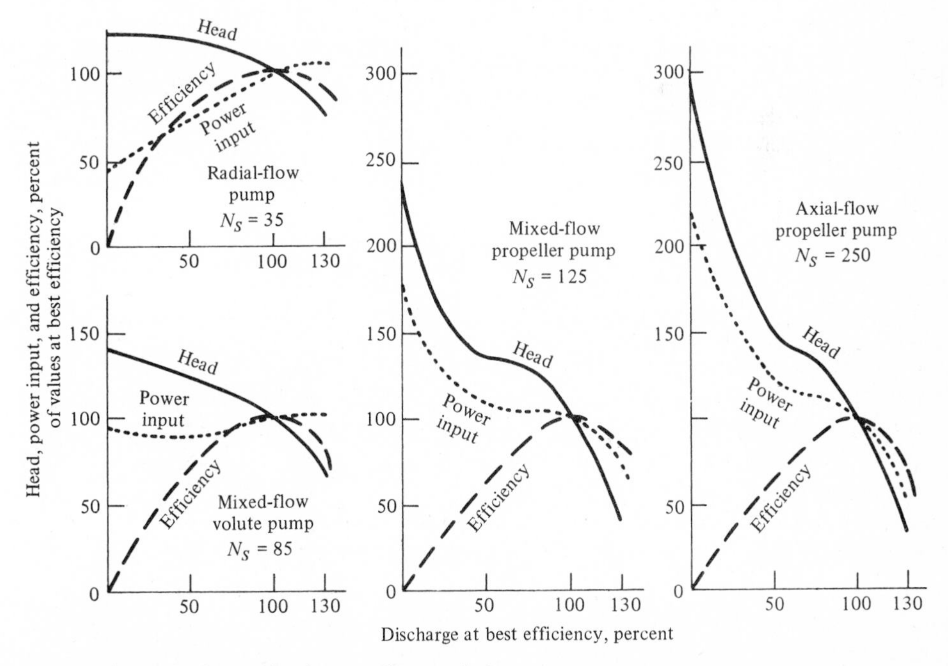

Figure 8-15 Typical centrifugal-pump characteristic curves.

variables have been plotted as a percentage of their values at the best efficiency point (bep). The characteristic curves of a bladeless pump are compared with those of a conventional nonclog pump in Fig. 8-16.

Pump Operating Range

Like most mechanical equipment, a pump operates best at is bep. At the bep, the radial loads on the bearings are at a minimum because the unbalanced radial load on the impeller is at a minimum. These radial loads increase greatly as the pump operating point moves away from the bep, either toward shutoff or toward runout. When the pump discharge rate increases beyond the rate at the bep, the absolute pressure required to prevent cavitation increases so that, in addition to radial load problems, cavitation is a potential problem (see "Cavitation," p. 299, in this section). When the pump discharge rate decreases toward shutoff head (head at zero flow), recirculation of the pumped fluid within the impeller becomes a problem. This recirculation causes vibration and hydraulic losses in the pump and may result in cavitation.

Because of the reasons cited above, it is good practice to limit the operating range of pump operation between 60 and 120 percent of the bep. These ranges may be extended, especially at lower rotational speeds, but care should be taken when operating outside this range. The operating range, or operating envelope, is bounded by the pump head-capacity curves at the high and low rotational speeds and by the head-capacity values corresponding to 60 and 120 percent of the bep. The pump operating envelope for the pump curve shown in Fig. 8-17 is illustrated in Fig. 8-18. The curve is calculated for a pump with a 35-cm (4-in)-diameter impeller.

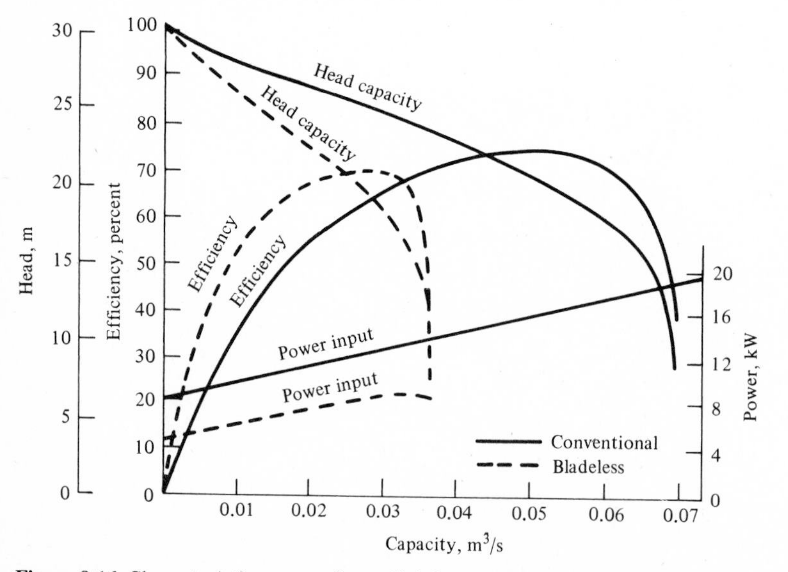

Figure 8-16 Characteristic curves for radial-flow and bladeless-type pumps to handle solids of comparable size.

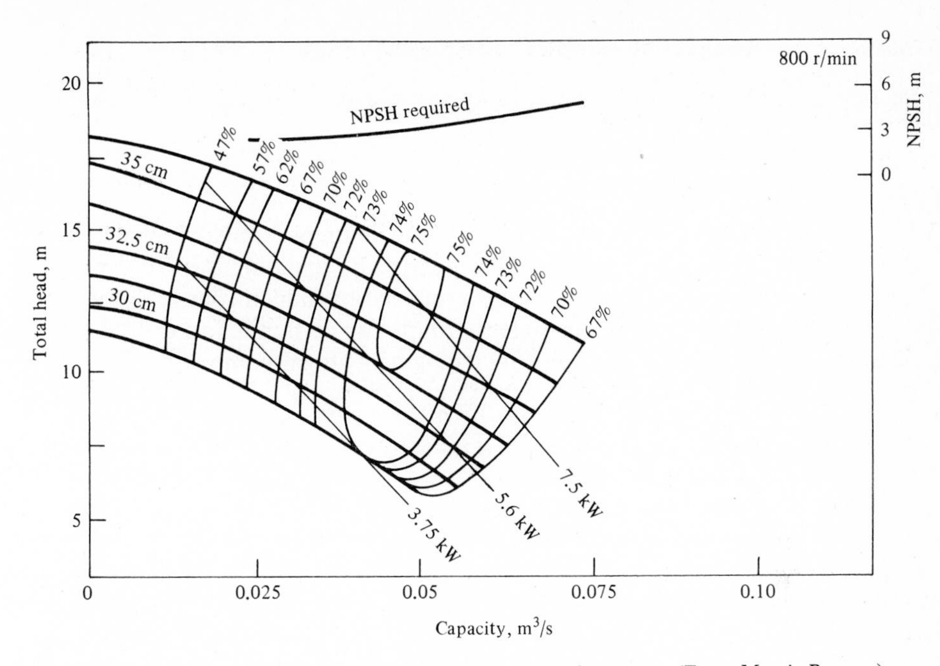

Figure 8-17 Typical manufacturers' curves for 15-cm nonclog pump. (*From Morris Pumps.*)

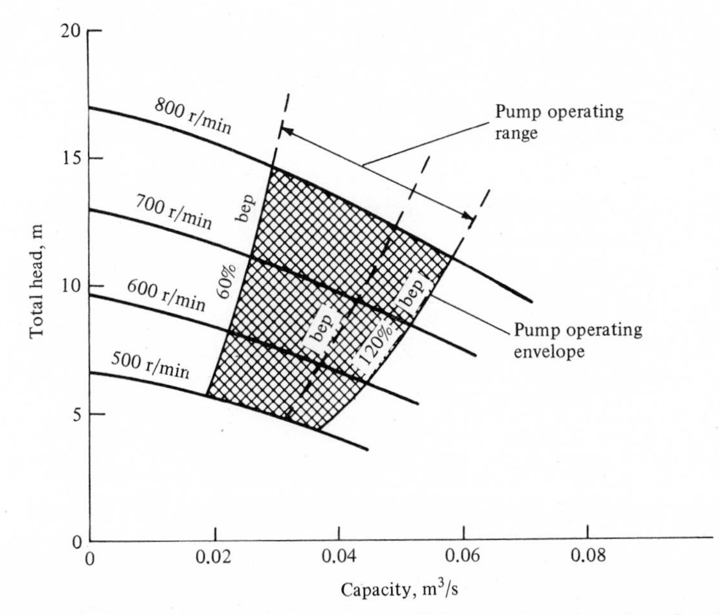

Figure 8-18 Pump operating range and variable-speed envelope (based on 35-mm impeller diameter).

Characteristic Relationships for Centrifugal Pumps

The relationships described in the following are used to predict the performance of centrifugal pumps at rotational speeds other than those for which pump characteristic curves (pump curves) were developed.

Flow, head, and power coefficients. In centrifugal pumps, similar flow patterns occur in a series of geometrically similar pumps. By applying the principles of dimensional analysis and the procedure proposed by Buckingham [2], the following three independent dimensionless groups can be derived to describe the operation of rotodynamic machines including centrifugal pumps [9].

$$C_Q = \frac{Q}{ND^3} \tag{8-7}$$

$$C_H = \frac{H}{N^2D^2} \tag{8-8}$$

$$C_P = \frac{P}{N^3D^5} \tag{8-9}$$

where C_Q = flow coefficient
Q = capacity
N = speed, rev/min
D = impeller diameter
C_H = head coefficient
H = head
C_P = power coefficient
P = power input

The operating points at which similar flow patterns occur are called *corresponding points,* and Eqs. 8-7, 8-8, and 8-9 apply only to corresponding points. However, every point on a pump head-capacity curve corresponds to a point on the head-capacity curve of a geometrically similar pump operating at the same speed or a different speed.

Affinity laws. For the same pump operating at a different speed, the diameter does not change and the following relationships can be derived from Eqs. 8-7 through 8-9.

$$\frac{Q_1}{Q_2} = \frac{N_1}{N_2} \tag{8-10}$$

$$\frac{H_1}{H_2} = \frac{N_1^2}{N_2^2} \tag{8-11}$$

$$\frac{P_1}{P_2} = \frac{N_1^3}{N_2^3} \tag{8-12}$$

These relationships, known collectively as the *affinity laws*, are used to determine the effect of changes in speed on the capacity, head, and power of a pump.

The effect of changes in speed on the pump characteristic curves is obtained by plotting new curves with the use of the affinity laws. The new operating point, the intersection of the pump and system head-capacity curves, will be given by the intersection of the new pump head-capacity curve with the system head-capacity curve, and not by application of the affinity laws to the original operating point only. The use of these relationships is illustrated in Examples 8-6 and 8-7 in Sec. 8-6.

Example 8-2 Determination of pump operating points at different speeds A pump has the characteristics listed in the following table when operated at 1170 rev/min. Develop head-capacity curves for a pump operated at 870 and 705 rev/min, and determine the points on the new curves corresponding to $Q = 0.44$ m³/s (10.0 Mgal/d) on the original curve.

Capacity, m³/s	Head, m	Efficiency, %
0.0	40.0	. . .
0.1	39.0	76.5
0.2	36.6	83.0
0.3	34.4	85.0
0.4	30.5	82.6
0.5	23.0	74.4

SOLUTION

1. Plot the head-capacity curves when the pump is operating at 1170, 870, and 705 rev/min.
   a. The plotting values for the reduced speeds are determined with the affinity laws (Eqs. 8-10 and 8-11), written as follows:

$$Q_2 = Q_1 \frac{N_2}{N_1}$$

$$H_2 = H_1 \left(\frac{N_2}{N_1} \right)^2$$

   b. The corresponding values are listed in the following table and plotted in Fig. 8-19:

1170 rev/min		870 rev/min		705 rev/min	
Capacity, m³/s	Head, m	Capacity, m³/s	Head, m	Capacity, m³/s	Head, m
0.0	40.0	0.000	22.1	0.000	14.5
0.1	39.0	0.074	21.6	0.060	14.2
0.2	36.0	0.149	20.2	0.121	13.3
0.3	34.4	0.223	19.0	0.181	12.5
0.4	30.5	0.297	16.9	0.241	11.1
0.5	23.0	0.372	12.7	0.301	8.4

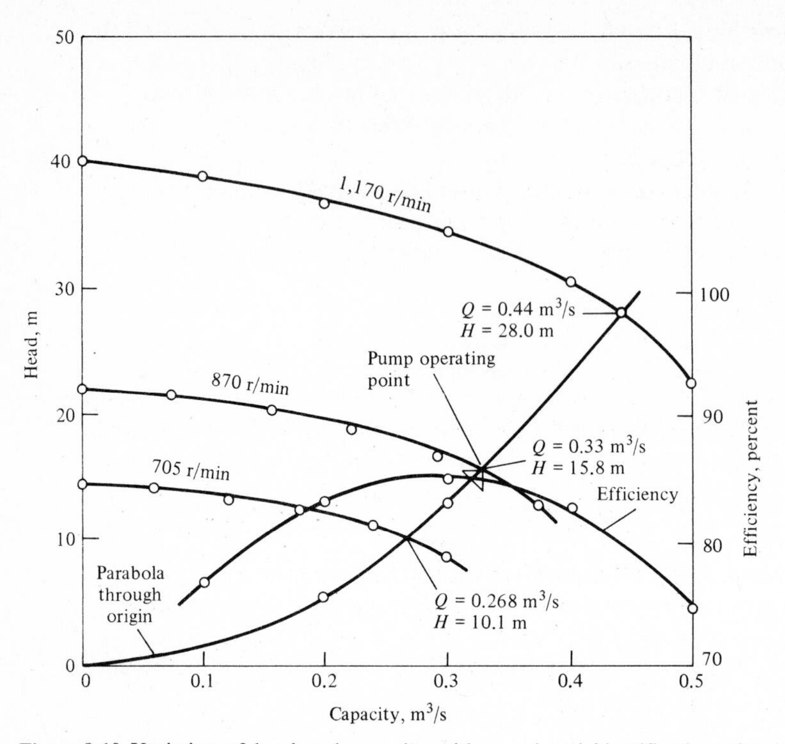

Figure 8-19 Variation of head and capacity with speed and identification of points with same efficiency and specific speed.

2. Determine the capacity and head at the other two speeds corresponding to $Q = 0.44\ m^3/s$ on the original curve.
   a. When the discharge is $0.44\ m^3/s$ and the pump is operating at 1170 rev/min, the head is 28.0 m.
   b. The corresponding head and capacity at 870 and 705 rev/min can be found using the following procedure:
      i. First, eliminate N_1 and N_2 from Eqs. 8-10 and 8-11 to obtain a parabola passing through the origin.

$$\frac{H_2}{H_1} = \frac{Q_2^2}{Q_1^2}$$

or

$$H_2 = \frac{H_1}{Q_1^2} Q_2^2 = kQ_2^2$$

      ii. Second, find the constant k for the curve passing through the points $Q = 0.44\ m^3/s$ and $H = 28.0$ m. This can be determined as follows:

$$k = \frac{28.0\ \text{m}}{(0.44\ \text{m}^3/\text{s})^2} = 144.6\ \text{s}^2/\text{m}^5$$

      iii. Third, determine the two other points on the parabola by using the above k value and plot the data in Fig. 8-19.

$Q = 0.44\ m^3/s \qquad H = 28.0\ m$

$Q = 0.3\ m^3/s \qquad H = 13.0\ m$

$Q = 0.2\ m^3/s \qquad H = 5.8\ m$

iv. Finally, the corresponding points on the reduced-speed head-capacity curves are found at the intersections of the two curves and the parabola through the origin. The corresponding values are as follows:

At 870 rev/min,

$Q = 0.33\ m^3/s \qquad H = 15.8\ m$

At 705 rev/min,

$Q = 0.268\ m^3/s \qquad H = 10.1\ m$

Comment The points determined in step 2*b*(iii) have the same efficiency and specific speed (discussed in the section following this example) as the corresponding point on the original curve.

Specific speed. For a geometrically similar series of pumps operating under similar conditions, the diameter term in Eqs. 8-7 and 8-8 can be eliminated. If the first term is raised to the one-half power, and the second term is raised to the three-fourths power, and if the first term is divided by the second, the relationship obtained is defined as the *specific speed.*

$$N_s = \frac{C_Q^{1/2}}{C_H^{1/2}} = \frac{(Q/ND^3)^{1/2}}{(H/N^2D^2)^{3/4}} = \frac{NQ^{1/2}}{H^{3/4}} \tag{8-13}$$

where N_s = specific speed
N = speed, rev/min
Q = capacity, m^3/s (gal/min)
H = head, m (ft)

To obtain the specific speed based on U.S. customary units of head and capacity, multiply the specific speed based on metric units of head and capacity by 52.

For any pump operating at any given speed, Q and H are taken at the point of maximum efficiency. When using Eq. 8-13 for pumps having double-suction impellers, one-half of the discharge is used, unless otherwise noted. For multistage pumps, the head is the head per stage. Use of the specific-speed parameter is illustrated in Example 8-3.

The computed value of specific speed has no usable physical meaning, except as a type number, but it is extremely useful because it is constant for all similar pumps and does not change with speed for the same pump. Since specific speed for a given pump is independent of both physical size and speed, it depends only on shape and is also sometimes considered a shape factor.

The shape of some of the impellers used in wastewater pumps is shown in Fig. 8-8. The variations in maximum efficiency to be expected with variations in size (capacity) and design (specific speed) are shown in Fig. 8-7. The progressive changes in impeller shape as the specific speed increases are shown along the bottom of Fig. 8-7.

Pump design characteristics, cavitation parameters, and abnormal operation under transient conditions can be correlated satisfactorily with the specific speed. Further consideration of the specific-speed equation reveals the following:

1. If larger units of the same type are selected for about the same head, the operating speed must be reduced.
2. If units of higher specific speed are selected for the same head and capacity, they will operate at a higher speed; hence the complete unit, including the driver, should be less expensive.

It thus becomes obvious why large propeller-type pumps are used in irrigation practice where low-lift high-capacity service is required.

Example 8-3 Use of specific-speed relationships A flow of 0.20 m^3/s (3200 gal/min) must be pumped against a total head of 16 m (52 ft). What type of pump should be selected and at what speed should the pump operate for best practical efficiency?

SOLUTION

1. Select the type of pump. Referring to Fig. 8-7, the best efficiency for a pump at a flow of 0.2 m^3/s is obtained by a pump with a specific speed of about 47 (extrapolate between the 0.190 and 0.63 m^3/s curves). The expected best efficiency is about 87 percent. A pump with a Francis or mixed-flow impeller should be selected.
2. Determine the pump operating speed.
   a. Write the specific-speed formula (Eq. 8-13) and substitute the known quantities to find the operating speed:

$$N_s = \frac{NQ^{1/2}}{H^{3/4}} = 47 = \frac{(N\ \text{rev/min})(0.2\ \text{m}^3/\text{s})^{1/2}}{(16\ \text{m})^{3/4}}$$

$$N_s = 47 = 0.0559N\ \text{rev/min}$$

$$N = \frac{47}{0.0559} = 840\ \text{rev/min}$$

   To meet the above conditions, the pump should operate at 840 rev/min. This speed would be possible if the pump were driven by a variable-speed drive.

   b. From a practical standpoint, if the unit is to be driven by an electric motor, the speed selected should be 870 rev/min for an induction motor (see Table 8-2) and the actual specific speed would be:

$$N_s = \frac{(870\ \text{rev/min})(0.2\ \text{m}^3/\text{s})^{1/2}}{(16\ \text{m})^{3/4}} = 48.6\ \text{rev/min}$$

Changes in impeller diameters. To cover a wide range of flows economically with a minimum number of pump sizes and impeller designs, manufacturers customarily offer a range of impeller diameters for each size casing (see Fig. 8-15). In general, these impellers have identical inlets and only the outside diameter is changed, usually by machining down the diameter. The following relationships for determining the effect of changes in the diameter of the impeller hold approximately, but with less accuracy than the affinity laws [12].

$$\frac{Q_1}{Q_2} = \frac{D_1}{D_2} \tag{8-14}$$

$$\frac{H_1}{H_2} = \frac{D_1^2}{D_2^2} \tag{8-15}$$

$$\frac{P_1}{P_2} = \frac{D_1^3}{D_2^3} \tag{8-16}$$

In some cases, two or more impeller designs may be available, each in a range of sizes, for the same casing. Because these impellers are not geometrically similar, the affinity laws do not hold.

Cavitation

When pumps operate at high speeds and at a capacity greater than the bep, pump cavitation is a potential danger: cavitation reduces pump capacity and efficiency and can damage the pump. It occurs in pumps when the absolute pressure of the inlet drops below the vapor pressure of the fluid being pumped. Under this condition, vapor bubbles form at the inlet and, when the vapor bubbles are carried into a zone of higher pressure, they collapse abruptly and the surrounding fluid rushes to fill the void with such force that a hammering action occurs. The high localized stresses that result from the hammering action can pit the pump impeller.

When determining if cavitation will be a problem, two different net positive suction heads (NPSH or h_{sv}) are used. The NPSH available ($NPSH_A$) is the NPSH available in the system at the eye of the impeller. The NPSH required ($NPSH_R$) is the NPSH required at the pump to prevent cavitation in the pump. The $NPSH_A$ is the total absolute suction head, as given by Eq. 8-4, above the vapor pressure of the water, expressed in meters (feet). Cavitation occurs when the $NPSH_A$ is less than the $NPSH_R$. The $NPSH_A$ is found by adding the term $P_{atm}/\gamma - P_{vapor}/\gamma$ to the right-hand side of Eq. 8-3, or to the energy (Bernoulli's) equation applied to the suction side of the pump. Thus,

$$NPSH_A = h_s - h_{fs} - \Sigma h_{ms} - \frac{V_s^2}{2g} + \frac{P_{atm}}{\gamma} - \frac{P_{vapor}}{\gamma} \tag{8-17}$$

$$NPSH_A = \frac{P_s}{\gamma} + \frac{V_s^2}{2g} + z_s + \frac{P_{atm}}{\gamma} - \frac{P_{vapor}}{\gamma} \tag{8-18}$$

where $NPSH_A$ = available net positive suction head, m (ft)
P_{atm} = atmospheric pressure, N/m^2 (lb_f/ft^2)
P_{vapor} = absolute vapor pressure of water, N/m^2 (lb_f/ft^2)
γ = specific weight of water, N/m^3 (lb_f/ft^2)

In the computation of $NPSH_A$, including the velocity head, $V_s^2/2g$ at the suction nozzle is somewhat illogical because it is not a pressure available to prevent vaporization of the liquid. However, in practice this term cancels out because it is also included in the NPSH required by the pump.

The NPSH required by the pump is determined by tests of geometrically similar pumps operated at constant speed and rated capacity but with varying suction heads. The onset of cavitation is indicated by a drop in efficiency as the head is reduced.

Cavitation constant. The ratio of the $NPSH_R$ to the total dynamic head is known as *Thoma's cavitation constant* σ (sigma).

$$\sigma = \frac{NPSH_R}{H_t} = \text{constant} \tag{8-19}$$

where $NPSH_R = h_{sv}$ = net positive suction head, m (ft)
H_t = total dynamic head, m (ft)

The cavitation constant is used for geometrically similar pumps operating at corresponding points on their head-capacity curves and, unless otherwise indicated, is understood to apply only to the bep. The application of Eq. 8-19 is illustrated in Example 8-4.

Because specific speed is an indication of pump shape, it is not surprising that it has been possible to correlate σ, and therefore NPSH, with the specific speed. The following formula has been developed for single-suction pumps [12].

$$\sigma = \frac{KN_s^{4/3}}{10^6} \tag{8-20}$$

where K = 1210 when N_s is in SI units (m^3/s and m)
K = 6.3 when N_s is in U.S. customary units (gal/min and ft)

The results obtained with this formula closely agree with the values of σ used by the Hydraulic Institute to determine recommended limiting values of suction head based on the NPSH requirement [11].

Cavitation at operating point. The NPSH is not often a problem when heads are 18 m (60 ft) or less, but it should be checked when heads are over 18 m and when the pump is operating under a suction lift or far out on its curve. If requested, pump manufacturers will plot the required NPSH on the pump characteristic curves. Preliminary estimates may be obtained from data in Refs. 3, 11, and 12.

If the pump operates at low head at a capacity considerably greater than the capacity at the bep, the following equation holds approximately:

$$\frac{NPSH_R \text{ at operating point}}{NPSH_R \text{ at bep}} = \left(\frac{Q \text{ at operating point}}{Q \text{ at bep}}\right)^b \tag{8-21}$$

where the exponent b varies from 1.25 to 3.0, depending on the design of the impeller. In most wastewater pumps, b will lie between 1.8 and 2.8. The $NPSH_R$ at the bep increases with the specific speed of the pump. For high-head pumps, it may be necessary either to limit the speed to obtain the adequate NPSH at the operating point or to lower the elevation of the station to provide more $NPSH_A$.

Example 8-4 Determination of allowable suction head for pump Determine the maximum allowable suction head for a pump with a cavitation constant equal to 0.3 when the total dynamic head is 40 m (131 ft). Assume that the pump is to operate at sea level and a temperature of 25°C (77°F).

SOLUTION

1. Note that Eq. 8-19 may be written as follows by setting the required $NPSH_R$ equal to the available $NPSH_A$

$$\sigma = \frac{h_s - h_{fs} - \Sigma h_{ms} - V_s^2/2g + P_{atm}/\gamma - P_{vapor}/\gamma}{H_t}$$

But, from Eq. 8-3, H_s can be substituted for the first four terms in the numerator. The resulting equation is

$$\sigma = \frac{H_s + P_{atm}/\gamma + P_{vapor}/\gamma}{H_t}$$

Then

$$H_s = \sigma H_t - \frac{P_{atm}/\gamma + P_{vapor}/\gamma}{H_t}$$

$$H_s = \sigma H_t - P_{atm}/\gamma + P_{vapor}/\gamma$$

2. Substitute known quantities and solve for H_s.

$$\sigma = 0.3$$

$$H_t = 40 \text{ m}$$

$$P_{atm} = 101.3 \text{ kN/m}^2 \text{ at } 25°\text{C}$$

$$P_{vapor} = 3.17 \text{ kN/m}^2 \text{ at } 25°\text{C}$$

$$\gamma = 9.78 \text{ kN/m}^3 \text{ at } 25°\text{C}$$

$$H_s = 0.3(40 \text{ m}) - \frac{101.3 \text{ kN/m}^2}{9.78 \text{ kN/m}^3} + \frac{3.17 \text{ kN/m}^2}{9.78 \text{ kN/m}^3}$$

$$= 12.0 \text{ m} - 10.0 \text{ m} = 2.0 \text{ m}$$

Comment Because the value of H_s is positive, the head at the inlet nozzle referenced to the centerline of the pump impeller must be 2 m. Thus the centerline of the impeller must be at least 2 m below the water level in the sump.

8-4 PUMP DRIVE UNITS

The most commonly used drives for wastewater pumps are direct-connected electric motors. Units driven by internal-combustion engines are sometimes installed to ensure that the pumps can operate during electric-power outages or where wastewater gas is available for fuel.

Electric Motors

Constant-speed pumps may be driven by squirrel-cage induction motors, wound-rotor induction motors, or synchronous motors (see Fig. 8-20). Squirrel-cage induction motors and synchronous motors operate at a constant speed, but wound-rotor induction motors can operate at different speeds by varying the resistance of the rotor or secondary circuit.

Squirrel-cage motors will normally be selected for constant-speed pumps because of their simplicity, reliability, and economy. Synchronous motors, however, may be more economical for large, slow-speed drives. Wound-rotor motors normally are not used for constant-speed drives, but may be used in special situations where low-inrush starting-current characteristics are required, such as operating in a system powered by an engine-generator set.

(a)

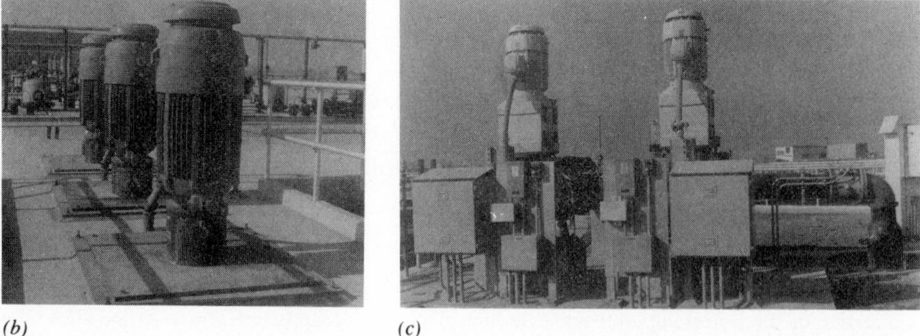

(b) *(c)*

Figure 8-20 Typical electric motor drives. (*a*) 950-kW horizontal with variable-speed drive, (*b*) 40-kW vertical, (*c*) 30-kW vertical.

The synchronous speed of an electric motor is given by the equation:

$$\text{Synchronous speed, rev/min} = \frac{120 \times \text{frequency}}{\text{no. of poles}} \tag{8-22}$$

Synchronous motors must operate at this synchronous speed, but the speed of squirrel-cage induction motors and the full speed of wound-rotor motors is 2 to 3 percent less than synchronous speed because of slip.

Constant- and multiple-speed drives. Typical speeds available for pumps driven by constant-speed motors operating on 60-cycle alternating current are given in Table 8-2. Slight variations in the full speed of induction motors depends on the motor design and the load torque requirements.

Speeds above 1770 rev/min are not used for wastewater pumps; they should only be selected when high system heads require high speeds. However, speeds of 1770 and 3550 rev/min are used at treatment plants for clearwater auxiliary equipment, such as effluent flushing water, process water, and chlorine injection water pumps. Most speeds will be in the mid-range of 1170 to 505 rev/min, depending on pump size and head. Speeds below 440 rev/min are uncommon and normally will be found only in very-large-capacity pumps.

In choosing between two applicable pumps operating at different speeds, the lower speed is preferred for pumping wastewater containing grit to reduce pump wear.

Table 8-2 Operating speeds of constant-speed motors on 60-cycle alternating current[a]

	Motor speed, rev/min	
Poles	Synchronous	Induction
2	[b]	3550
4	1,800	1770
6	1,200	1170
8	900	870
10	720	705
12	600	585
14	514	500
16	450	435
18	400	390
20	350	350
22	327	318
24	300	290
26	277	268
28	257	249

[a]For discussion on motor speeds, see section on "Electric Motors," p. 302.

[b]Not applicable.

Multiple-speed operation can be obtained with squirrel-cage or wound-rotor motors. For squirrel-cage motors, the choice of speeds is restricted to two or more of the speeds listed in Table 8-2. If the lower speed of a two-speed motor is one-half the higher speed, a single winding motor can be used. If the lower speed is not one-half, a two-speed motor with two windings is required.

When operating a pump at two constant speeds, the advantage of the squirrel-cage (or synchronous motor) is that the motor operates at maximum efficiency at both speeds. However, the wound-rotor motor operates at maximum efficiency only at full speed. When operated at less than full speed, the efficiency of the pump motor declines considerably, as explained in the following.

Variable-speed drives. If the operating conditions vary in pumping stations, variable-speed operation of the pumps may be desirable. Variable-speed (stepless) drives have been possible for many years by using liquid resistors for wound-rotor motor controls, magnetic (eddy current) drives, or fluid couplings. Recently, viscous shear couplings have become available. All of these variable-speed drives have one thing in common: the slip losses are converted into heat, resulting in a loss of efficiency.

Slip losses can be computed from the following formula:

$$\text{Slip loss, \%} = \text{load kW}\left(\frac{s}{1 - s}\right)100 \tag{8-23}$$

where

$$s = \frac{\text{full speed} - \text{load speed}}{\text{full speed}}$$

The efficiency of the drive equals the power output divided by the power input, or

$$\text{Drive efficiency, \%} = \frac{\text{power output}}{\text{power input}}100 \tag{8-24}$$

$$= \frac{1 - s}{1}$$

and the overall drive efficiency becomes

$$\text{Overall drive efficiency, \%} = \text{motor efficiency} \times \frac{1 - s}{1} \tag{8-25}$$

Slip losses reach a maximum when the pump operates at two-thirds speed. At this point, the slip losses amount to about 33 percent of the actual-load power requirement or, for variable-torque loads, about 15 percent of the full-load power. Special motors and special drives incorporating additional rotating machines to recover the slip losses are available for large power drives, but in general have not been warranted for wastewater pump drives.

The recent development of solid-waste electronic controls has led to the introduction of newer methods of variable-speed control for both squirrel-cage and wound-rotor motors, as described below. These drives effectively limit the current inrush and provide a soft start of the unit.

1. *Variable voltage*. The reduction of the primary voltage to a special-design, high-slip, squirrel-cage induction motor by saturable reactors or silicon-controlled rectifiers provides variable-speed control. However, this type of control is inefficient, and the losses are rejected in the form of head generated in the motor, causing the motor to run hot. Therefore, this type of speed control is not recommended for municipal service.
2. *Wound-rotor motor, solid-state control.* By including saturable-core reactors as well as resistors in the secondary circuit of a wound-rotor motor and control by silicon-controlled rectifiers, stepless-speed control can be obtained without using the tanks, piping, and heat exchangers required for liquid resistors. Slip losses as described above in (1) still occur. Many wastewater pump drives of this type have been installed.
3. *Variable-frequency drive*. A variable-speed drive is highly efficient when the frequency of the current supplied to a squirrel-cage motor is varied. The alternating current is rectified to direct current and converted back to alternating current at the frequency required to produce the desired speed. A speed range of 3 to 1 is possible, and efficiencies of 95 percent of the usual motor efficiency have been reported. These drives are normally available in sizes of up to 185 kW (250 hp), and are available in larger sizes as specially designed units.
4. *Wound-rotor, regenerative secondary control.* A wound-rotor, variable-speed pump drive, in which the power in the secondary circuit is rectified by solid-state components to direct current, is converted back to alternating current of the same voltage and frequency as the power supply and fed into the primary motor. This type of drive is available for 20-kW (25-hp) motors and larger.

The variable-frequency drive and the wound-rotor motor with regenerative secondary controls are considerably more efficient than the other types of variable-speed drives because the slip losses are either eliminated or recaptured in the form of useful work. The cost of these units is higher than for other types of variable-speed drives, but with the cost of power increasing, these units are becoming more economical.

In general, variable-speed drives are more expensive than constant-speed motors. They are also less efficient and require more maintenance. Since the power required by the pump varies as the speed cubed, the actual power loss is much less than might be supposed by a consideration of the efficiency alone. The loss is not usually excessive when it is considered that with variable-speed pumps, the wet wells for pumping stations are fewer starts and stops of the pump units.

Internal-Combustion Engines

In large pumping stations, internal-combustion engines are used as a source of standby power for driving the pumps and the critical electrical controls if the power fails. These engines are occasionally used to drive pumps at remote sites where electricity is not available or unreliable and at treatment plants where sludge gas is available as a fuel.

Internal-combustion engines usually drive generators so that power is not only available for the pump but also for the auxiliary equipment and the control system. The power generated by these engines can also be used in any of the available pumps instead of being connected to a single pump. Diesel engines or spark-ignited engines firing either natural or propane gas are commonly used for this service. Gasoline engines are occasionally installed, but are not common because of the problems with fuel storage. Internal-combustion engines that drive pumps at remote stations probably will be either diesel engines or spark-ignited gas engines.

Dual-fuel engines. At treatment plants where sludge gas is available, either dual-fuel diesel engines or spark-ignited gas engines can be used. The waste heat available in the jacket water and, in some cases, the exhaust would be reclaimed for heating the digestion tank or the building. Dual-fuel diesels fire a mixture of diesel oil and gas. The ratio of oil to gas can be varied, but a minimum of about 10 percent of diesel oil is required to ignite the gaseous fuel. Spark-ignited engines can fire straight sludge gas. These engines would normally be supplied with dual carburators and a separate source of alternate fuel, such as natural or propane gas, to provide power when the sludge gas is not available.

Direct and gear drives. Although horizontal pumps may be directly driven by engines, the pumps are normally driven through a gear drive to allow both the pump and engine to operate at their optimum speed. The most common arrangement is a horizontal engine driving a vertical pump through a right-angle gear drive.

If the engine drive provides standby power, a combination gear box is installed with an electric motor mounted on the top of the right-angle gear drive and is directly connected to the pump shaft (see Fig. 8-21). The engine is connected to the horizontal shaft of the right-angle gear. A clutch or disconnect coupling is used to disengage the right-angle gear when the motor drives the pump. When the engine drives the pump, the clutch is manually engaged and the motor on the pump shaft rotates like a windmill. If the drive must operate automatically, a clutch is required on the shaft between the engine and the right-angle gear drive so that the engine shaft will not rotate when the motor is driving the pump.

Figure 8-21 Pump with electric motor and diesel engine drive.

8-5 PUMP SELECTION

In selecting equipment for a pumping station, many different and often conflicting aspects of the overall pumping system must be considered. Factors which must be evaluated include (1) design flowrates and flow ranges, (2) location of the pumping station, (3) force main design, and (4) system head-capacity characteristics. When these factors are evaluated properly, the number and sizes of the pumps, the type of drive, and the optimum size of force main can be selected.

Determination of Flowrates

Before pump equipment can be selected, it is imperative that the different flows which might be pumped are determined. The projected flows corresponding to the useful life of the equipment are the design flows. A design period of 20 yr is used for most pumps.

Flowrates that should be considered in pump selection include the design peak flow, the initial and design average flows, and the initial minimum flow. The pumps must be capable of discharging the design peak flow delivered to the pumping station by the collection system. The initial and design average flowrates are important so that equipment can be selected to operate as efficiently as possible at the average flowrates. The initial minimum flowrates are important in sizing the force main so that solids that settle out at low velocities do not plug the force main.

The determination of average and peak flows in sanitary sewers was discussed in Chap. 3. Initial minimum flows to be pumped are calculated from the initial average flows and may be approximated by using the multipliers in Table 8-3.

The older collection systems in many cities include combined systems that collect wastewater and storm water in the same sewers and interceptors. New combined sewer systems are rarely permitted in the United States now, but the existing systems will remain in operation for many years to come. In these systems, the dry-weather flows will approximate the flows from a separate sanitary system, but the storm-water peak flows will greatly exceed those of a separate system.

After the initial and design flowrates are determined, the next important decision is whether to (1) install equipment large enough to handle the entire range of initial and design flows or (2) make suitable provisions for increasing the capacity of the pumping station at some time in the future. The capacity of a

Table 8-3 Factors used for estimating minimum flow

Average flow, m^3/s	Minimum flow factor
0.05	0.25
0.5	0.35
2.5	0.45
5.0	0.50

pumping station is often increased by installing larger impellers, larger pumps, additional units, or, in some cases, motors with larger capacities and higher speeds.

Location of the Pumping Station

The location of a pumping station will affect the selection of equipment, especially the type of drive unit. The remote station serving only a small tributary area may have two identical units, either one capable of handling the design peak flow. A constant-speed motor would normally drive the pumps, so that the station could alternate between no discharge and discharge at the peak rate. If the drainage area is expected to increase substantially in the future, the station may need additional space for one or two more pumps.

However, at a larger station where the entire wastewater flow or major portion thereof is to be pumped to the wastewater-treatment plant, the pumps should be designed, so far as is practicable, to operate continuously. The rate of outflow should change gradually in small increments as the inflow to the station varies, so that the wastewater-treatment plant may be operated at maximum efficiency. This operating mode requires at least one pump with a capacity about equal to or slightly less than the initial minimum flow. In addition to this requirement, provision must be made for the design peak flow. To allow for maintenance and repairs, the design should allow this flow to be handled with the largest pumping unit out of service.

If a large station must operate continuously, variable-speed pumps or a combination of variable-speed pumps and constant-speed pumps are usually selected. It may also be possible to use constant-speed pumps or two-speed pumps, but their use will produce rapid, although small, changes in flow to the treatment plant that could upset the treatment process.

Pumping stations serving combined sewer systems may have two sets of pumps: (1) a set to pump the dry-weather sanitary flow to a treatment plant or to an interceptor leading to a treatment plant and (2) a set of large storm-water pumps to pump combined municipal wastewater and storm water during heavy rains to a storm-water disposal system. The storm-water pumps and their discharge conduit should have ample capacity to prevent street and basement flooding. Many of these pumping stations were built for combined sewer systems, discharging storm water directly to watercourses without treatment. From recent investigations it has been found that such practices result in substantial pollution, especially since the first flush of storm water is heavily laden with solids.

Force Mains

A force main is a sewer designed to receive the wastewater discharged from a pumping station and to convey it under pressure to the point of discharge, which may be a gravity sewer, a storage tank, or a treatment plant. The major considerations in force main sizing are velocity and friction loss. Unfortunately, these two concerns conflict. The velocity should be high enough to maintain carrying velocities in the force main at minimum force main flows. However, high velocities cause high friction losses and higher heads against which the pump must pump. In general, a minimum velocity of 0.6 m/s (2 ft/s) will maintain solids in wastewater suspension, and 1.0 m/s (3.3 ft/s) will resuspend solids that have settled out in the force main. A more detailed discussion on force mains is provided in Chap. 9.

Development of System Head-Capacity Curve

After the system flowrates have been determined, the force main tentatively sized, and the static head defined, the system head-capacity curve can be developed. This curve is required to determine the rating of the pumps. As noted in Sec. 8-1, the system head-capacity curve represents the total dynamic head against which the pumps will be required to operate under various flow conditions. It consists of the static head plus the friction and minor head losses in the piping system plotted against flow (see Fig. 8-3).

In the design of wastewater pumping stations, the static head is the difference in elevation between the free water surface in the wet well and the free water surface at the point of discharge. The water level in the wet well will vary between the control settings calling for pumps to start and stop, and the discharge water surface may also vary. Therefore, the static head will vary, resulting in a family of parallel system head-capacity curves. The maximum system head-capacity curve corresponds to the lowest wet-well level, and the minimum system head-capacity curve corresponds to the higher wet-well level.

To compute the friction and minor head losses in the piping system, the sizes of pipes and fittings must first be assumed. The number of pumping units should be estimated (at least tentatively), and pipe sizes should usually be selected to give velocities between 1 and 2.5 m/s (3.3 and 8 ft/s). The highest velocities are used with the largest units. Velocities ranging from 1.0 to 2 m/s (3.3 to 6.5 ft/s) are desirable but not always economical.

In addition to its use in determining the rating of the pumps, the shape of the system head-capacity curve can often be a guide in assessing the number of pumps required and the type of drive to be used. For example, a single variable-speed (or two-speed) pump is more suitable for a system with a small static head but a large friction head. At reduced speed, the pump head is reduced in proportion to the square of the reduction in speed, but the system head also is reduced in the same proportion. Therefore, if the pump is selected to operate near its bep at high speeds, it will operate near its bep at low speeds. By comparison, a system with a large static head and a small friction head is more suitable for several pumps operating in parallel to discharge the design flow. A small reduction in pump speed will reduce the pump head to below the static head of the system.

8-6 ANALYSIS OF PUMP SYSTEMS

System analysis for a pumping station is conducted to select the most suitable pumping units and to define their operating points. It involves calculating the system head-capacity curves and the use of these curves with the head-capacity curves of available pumps. Two types of pump systems are discussed and illustrated with examples in the following discussion: single-pump and multiple-pump systems. With the cost of energy increasing so rapidly, it is also imperative that energy-efficient pump systems be designed. This subject is considered in homework problems 8-17 through 8-19.

Single-Pump Operation

The pump characteristic curves illustrate the relationship between head, capacity, efficiency, and brake horsepower over a wide range of possible operating conditions, but they do not indicate at which point on the curves the pump will operate. The operating point is found by plotting the pump head-capacity curve on the system head-capacity curve. The pump will operate where the two curves cross.

If too conservative a friction factor is used in computing the system head-capacity curve, the pump may operate farther out on its head-capacity curve than intended. In extreme cases, this may result in a considerable loss of efficiency, an overloaded motor, and possible cavitation. These conditions can be anticipated and guarded against by plotting system head-capacity curves using friction factors for new pipe in addition to the system head-capacity curves based on design friction factors (old pipe). The bep should lie close to the design operating point and within the family of possible system curves. Development of the system head-capacity curve and determination of the operating point for a single pump are illustrated in Example 8-5.

Example 8-5 Development of system head-capacity curve and determination of operating point for single-pump operation A collection system should pump about 0.35 m^3/s (5500 gal/min) through the piping system shown in Fig. 8-1. Assume that the suction piping is 500 mm (20 in) in diameter and 4 m (13 ft) long, and that the discharge pipe is 450 mm (18 in) in diameter and 770 m (2526 ft) long; both are made of cast iron. The static suction head is 1 m (3 ft) and the static discharge head is 21 m (69 ft). Develop a system head-capacity curve for flowrates from 0.0 to 0.5 m^3/s (8000 gal/min). Although two 45° bends are shown in Fig. 8-1, assume that there are five bends in the discharge piping system. A minor loss coefficient of 0.2 can be used for the bellmouth entrance. Minor loss coefficients for the other appurtenances can be found in Appendix C.

If a pump with a 350-mm (14-in) suction and discharge nozzle, operating at 1150 rev/min, has the characteristics given in the following, determine the capacity that the pump will have against the system head-capacity curve. Also determine the operating head and efficiency. If 0.2 m^3/s (3200 gal/min) is to be pumped at reduced speed, determine the new operating speed and efficiency. The pump characteristics are as follows:

Capacity, m^2/s	Head, m	Efficiency, %
0.0	40.0	. . .
0.1	39.0	. . .
0.15	. . .	77.0
0.2	36.6	80.6
0.25	. . .	83.4
0.3	32.5	84.6
0.35	. . .	84.6
0.4	23.0	82.6
0.45	13.5	75.0

SOLUTION

1. Develop and plot the system head-capacity curve. The total dynamic head at various flowrates is determined by using Eq. 8-4 in which the loss of velocity head is considered a minor loss.

$$H_t = H_{\text{stat}} + h_{fs} + \Sigma h_{ms} + h_{fd} + \Sigma h_{md}$$

a. The static head is:

$$H_{\text{stat}} = h_d - h_s = 21 \text{ m} - 1 \text{ m}$$

$$= 20 \text{ m (66 ft)}$$

b. The velocities in the suction and discharge pipes and at the pump nozzles for the desired flow, 0.35 m^3/s, are as follows:

$$V_s = \frac{0.35 \text{ m}^3/\text{s}}{(\pi/4)(0.50 \text{ m})^2} = 1.78 \text{ m/s}$$

$$V_d = \frac{0.35 \text{ m}^3/\text{s}}{(\pi/4)(0.45 \text{ m})^2} = 2.20 \text{ m/s}$$

$$V_{\text{pump nozzle}} = \frac{0.35 \text{ m}^3/\text{s}}{(\pi/4)(0.35 \text{ m})^2}$$

$$= 3.64 \text{ m/s}$$

c. Compute the losses in the suction piping for a flow of 0.35 m^3/s:
i. Friction losses h_{fs}, using the Darcy-Weisbach equation, are:

$$\frac{V_s^2}{2g} = \frac{(1.78 \text{ m/s})^2}{2 \times 9.81 \text{ m/s}^2} = 0.161 \text{ m}$$

$f = 0.017$ (from Fig. 2-11 in Chap. 2, for cast iron pipe, complete turbulence)

$$h_{fs} = f\frac{L}{D} \times \frac{V_s^2}{2g} = 0.017 \frac{4 \text{ m}}{0.50 \text{ m}} 0.16 \text{ m}$$

$$= 0.022 \text{ m}$$

ii. Minor losses Σh_{ms} are:

$$\begin{aligned}
\text{Entrance loss (bellmouth)} &= 0.2\frac{V_s^2}{2g} \\
\text{Elbow loss} &= 0.2\frac{V_s^2}{2g} \\
\text{Gate valve (wide open)} &= 0.07\frac{V_s^2}{2g} \\
\hline
& 0.47\frac{V_s^2}{2g} = 0.47 \times 0.161 \text{ m} \\
& = 0.076 \text{ m}
\end{aligned}$$

Loss in 500- to 350-mm eccentric reducer (used to connect suction piping to pump suction nozzle):

$$0.04\frac{V_p^2}{2g} = 0.04\frac{(3.64 \text{ m/s})^2}{2 \times 9.81 \text{ m/s}^2}$$

$$= 0.04 \times 0.68 \text{ m} = 0.027 \text{ m}$$

Total minor suction losses, Σh_{ms}:

$$0.076 \text{ m} + 0.027 \text{ m} = 0.103 \text{ m}$$

iii. Total suction losses:

$$0.022 \text{ m} + 0.103 \text{ m} = 0.125 \text{ m}$$

d. Compute the losses in the discharge piping for a flow of 0.35 m^3/s.

i. Friction losses h_{fd}, using the Darcy-Weisbach equation, are:

$$\frac{V_d^2}{2g} = \frac{(2.20 \text{ m/s})^2}{2 \times 9.81 \text{ m/s}^2} = 0.247 \text{ m (0.81 ft)}$$

$f = 0.018$ (from Fig. 2-11 in Chap. 2, for cast iron pipe, complete turbulence)

$$h_{fd} = f\frac{L}{D} \times \frac{V_d^2}{2g} = 0.018\frac{770 \text{ m}}{0.45 \text{ m}} 0.247 \text{ m}$$

$$= 7.608 \text{ m}$$

ii. Minor losses, Σh_{md}:

$$\begin{aligned}
\text{Swing check valve (wide open)} &= 2.5\frac{V_d^2}{2g} \\
\text{Gate valve (wide open)} &= 0.07\frac{V_d^2}{2g} \\
\text{Bends (5 at 0.2 each)} &= 1.0\frac{V_d^2}{2g}
\end{aligned}$$

$$\text{Exit loss} = 1.0\frac{V_d^2}{2g}$$

$$4.57\frac{V_d^2}{2g} = 4.57 \times 0.247\,\text{m}$$

$$= 1.129\ \text{m}$$

Loss in 350- to 450-mm increaser (used to connect pump discharge nozzle to discharge pipe):

$$\frac{0.4(V_1 - V_2)^2}{2g} = \frac{(0.4)(3.64\ \text{m/s} - 2.20\ \text{m/s})^2}{2 \times 9.81\ \text{m/s}^2}$$

$$= 0.042\ \text{m}$$

Total minor discharge losses:

$$\Sigma h_{md} = 1.171\ \text{m}$$

iii. Total discharge losses are:

$$7.608\ \text{m} + 1.171\ \text{m} = 8.779\ \text{m}$$

e. Total losses in suction and discharge are:

$$0.125\ \text{m} + 8.779\ \text{m} = 8.904\ \text{m}$$

f. Compute the total head at various flowrates, assuming that the losses vary as the square of the flowrate (friction factor f assumed constant).

Q, m^3/s	$\frac{Q}{0.35}$	$\left(\frac{Q}{0.35}\right)^2$	$\Sigma(h_f + h_m)$	H_{stat}	H_t
0.35	1.0	1.0	8.904	20.0	28.9
0	0	0	0	20.0	20.0
0.10	0.286	0.082	0.730	20.0	20.7
0.20	0.571	0.327	2.912	20.0	22.9
0.30	0.857	0.735	6.544	20.0	26.5
0.40	1.143	1.306	11.629	20.0	31.6
0.50	1.429	2.041	18.173	20.0	38.2

g. Plot the computed values of total head versus the corresponding flowrate (see Fig. 8-22):

2. Determine the flowrate, head, and efficiency at which the given pump will operate against the system head-capacity curve developed in step 1.
   a. Using the given data, plot the pump head-capacity curve and efficiency curve as shown in Fig. 8-22.
   b. The point of intersection of the pump head-capacity curve and the system head-capacity curve is the operating point for the pump. At this point the following data are obtained:

$$\text{Flowrate } Q = 0.35\ \text{m}^3/\text{s}$$

$$\text{Head } H = 29\ \text{m}$$

$$\text{Efficiency } E_p = 84.6\,\%$$

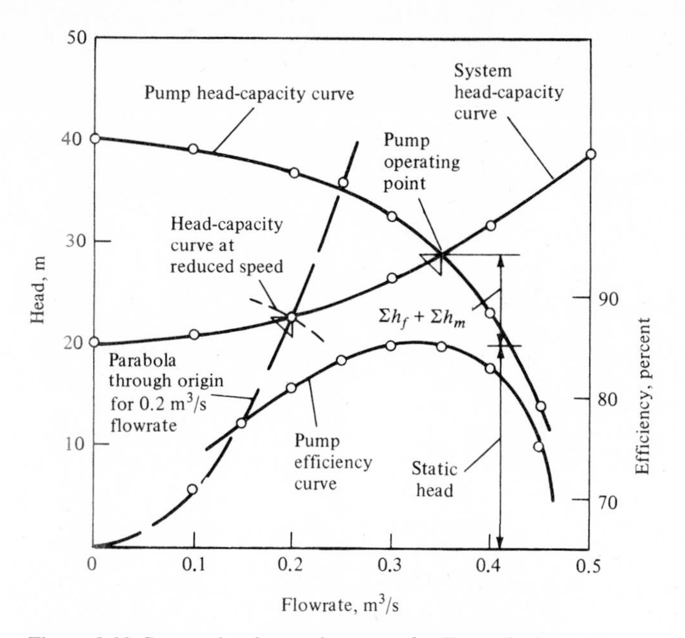

Figure 8-22 System head-capacity curve for Example 8-5.

3. Determine the rotational speed, head, and efficiency when the pump is delivering 0.2 m³/s against the computed system head-capacity curve.
   a. From the system head-capacity curve at a flow of 0.2 m³/s, the new head H equals 23.0 m.
   b. The new speed is given by the application of the affinity laws (Eqs. 8-10 and 8-11):

$$\frac{Q_1}{Q_2} = \frac{N_1}{N_2} \qquad \frac{H_1}{H_2} = \frac{N_1^2}{N_2^2}$$

However, these laws hold only for corresponding points, and the corresponding point on the original pump head-capacity curve is not known. It can be found by the following procedure: eliminate N_1 and N_2 from the foregoing equations, obtaining

$$\frac{H_2}{H_1} = \frac{Q_2^2}{Q_1^2}$$

or

$$H_2 = \frac{H_1}{Q_1^2} Q_2^2 = kQ_2^2$$

This describes a parabola through the origin and the new operating point and is the locus of corresponding points at different speeds.

i. Determine the value of the constant k:

$$k = \frac{23.0 \text{ m}}{(0.2 \text{ m}^3/\text{s})^2} = 575 \frac{s^2}{m^5}$$

ii. Determine at least two other points on the parabola, using the equation $H_2 = 575Q_2^2$, when

$$Q_2 = 0.1\ \text{m}^3/\text{s} \qquad H_2 = 5.75\ \text{m}$$

$$Q_2 = 0.15\ \text{m}^3/\text{s} \qquad H_2 = 12.9\ \text{m}$$

$$Q_2 = 0.25\ \text{m}^3/\text{s} \qquad H_2 = 35.9\ \text{m}$$

iii. Plot the parabola as shown in Fig. 8-22, and determine the coordinates of its intersection with the head-capacity curve at 1150 rev/min. These are found to be

$$Q = 0.248\ \text{m}^3/\text{s} \qquad H = 35.4\ \text{m}$$

This is the point on the original pump head-capacity curve corresponding to the new operating point at reduced speed.

iv. The reduced speed is found by applying the affinity laws:

$$N_2 = N_1 \frac{Q_2}{Q_1} = 1150\ \text{rev/min} \times \frac{0.2\ \text{m}^3/\text{s}}{0.248\ \text{m}^3/\text{s}}$$

$$= 927\ \text{rev/min}$$

or

$$N_2 = N_1 \sqrt{\frac{H_2}{H_1}} = 1150\ \text{rev/min} \sqrt{\frac{23\ \text{m}}{35.4\ \text{m}}}$$

$$= 927\ \text{rev/min}$$

c. The efficiency at the new operating point is assumed to equal the efficiency at the corresponding point on the given pump head-capacity curve. Thus the efficiency will be about 83 percent (see Fig. 8-22).

Multiple-Pump Operation

In the wastewater field, the most common type of station has two or more pumps operating in parallel. Nevertheless, situations will be encountered where pumps operate in series. Each of these situations is described in the following discussion.

Parallel operation. In pumping stations where two or more pumps may operate either individually or in parallel and discharge into the same header and force main, an alternative computation method for determining the pump operating point is recommended:

1. The friction losses in the suction and discharge piping of individual pumps are omitted from the system head-capacity curve.
2. Instead, these losses are subtracted from the head-capacity curves of the individual pumps to obtain modified pump head-capacity curves, which represent the head-capacity capability of the pump and its individual valves and piping combines (see Fig. 8-23).
3. When two or more pumps operate in parallel, the combined pump head-capacity curve is found by adding the capacities of the modified curves at

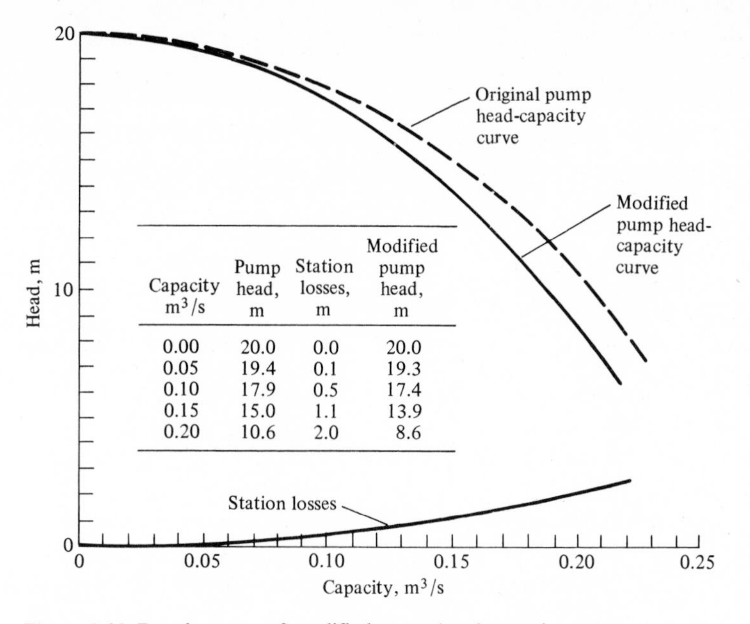

Capacity m^3/s	Pump head, m	Station losses, m	Modified pump head, m
0.00	20.0	0.0	20.0
0.05	19.4	0.1	19.3
0.10	17.9	0.5	17.4
0.15	15.0	1.1	13.9
0.20	10.6	2.0	8.6

Figure 8-23 Development of modified pump head-capacity curve.

the same head (see Fig. 8-24*a*). The point of intersection of the combined curve with the system head-capacity curve gives the total capacity of the combination of pumps and the modified head at which each operates. By entering the modified curves of each pump at this head, determinations can be made of the capacity contributed by each pump, the efficiency of each pump, and the brake horsepower required under these conditions. To find the total head at which each individual pump will operate, one must proceed vertically at constant capacity from its modified pump head-capacity curve to its actual head-capacity curve. The pump specifications or purchase order must be drawn so that the pump will produce this head. Each pump can operate at several points on the head-capacity curve, with the head increasing and the discharge decreasing as more pumps go into operation. An effort should be made to limit these operating points to a range of flows between 60 and 120 percent of the bep.

The process involved in selecting pumps operating in parallel is demonstrated in the following two examples. The selection of constant-speed and two-speed pumps is described in Example 8-6, and the selection of variable-speed and constant-speed pumps is illustrated in Example 8-7.

Series operation. Often one or more booster pumps may be installed in the suction line or the force main leading from a pumping station to meet specific site conditions. Pumps installed in series with existing pumps are used to increase the capacity of the pumping station and the discharge of the pumps.

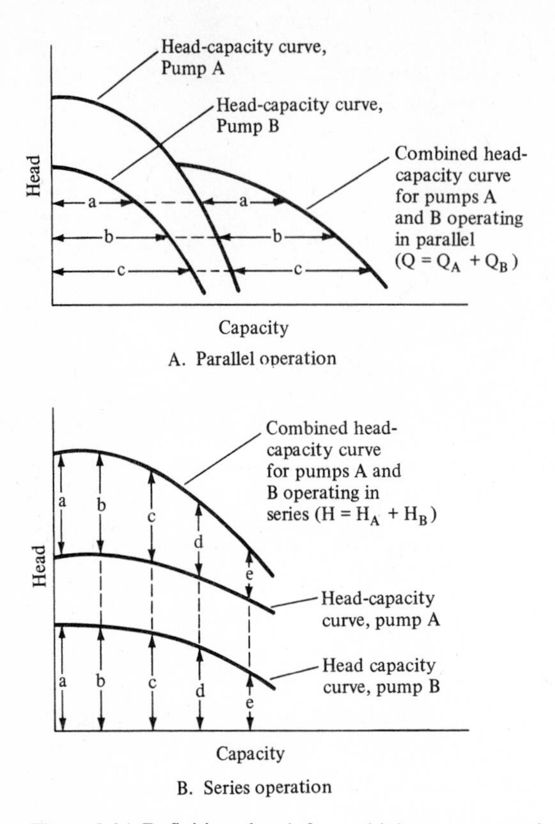

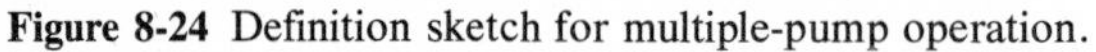

Figure 8-24 Definition sketch for multiple-pump operation.

When two or more pumps operate in series, the combined head-capacity curve is found by adding the head of each pump at the same capacity. This procedure is illustrated in Fig. 8-24*b*. When a booster pump is added to a force main fed by parallel pumps, the combined head-capacity curve is found by adding the head of the booster pump to the modified head of the parallel pumps at a given capacity.

Example 8-6 Selection of constant-speed and two-speed pumps A pumping station is to be designed for a collection system that receives flow from a partially developed tributary area. Wastewater from the pumping station is to be discharged to a large interceptor. Select a pump system that will meet the requirements for both a 10-yr (interim) flow and 20-yr (design) flow by using the data specified in the following.

Ten years in the future (interim design), the estimated average flowrate is 0.044 m^3/s (700 gal/min) and the peak flowrate is 0.095 m^3/s (1500 gal/min). The estimated average and peak design flowrates 20 yr in the future are 0.075 m^3/s (1200 gal/min) and 0.15 m^3/s (2400 gal/min), respectively.

The force main is a long pipe 300 mm (12 in) in diameter. At the end of the 20-yr design period, the friction loss in the force main at the peak flowrate (0.15 m^3/s) is expected to be 15 m (49 ft). The system static head from the high wet-well level to the discharge elevation is

7 m (23 ft), and the difference between the high and low levels in the pumping station wet well is 1 m (3.3 ft). The pumping station losses will be limited to 1.3 m (4.3 ft) at the pump rating point by correctly sizing the pump suction and discharge piping.

SOLUTION (Preliminary Analysis)

1. Plot the system head-capacity curve. Because there is only a 1-m difference between the high and low wet-well levels, and the peak flow will be pumped at the high wet-well level, only the system head-capacity curve, assuming a static head of 7 m, need be plotted.
   a. The system head-capacity curve is plotted by using the following relationship (see Fig. 8-25):

$$H_t = 7.0\text{ m} + 15\text{ m}\left(\frac{Q\text{ m}^3/\text{s}}{0.15\text{ m}^3/\text{s}}\right)^2$$

   The curve plotted reflects the friction that will occur in the force main at the end of the 20-yr design period.

   b. Because the frictional loss will be less when the force main is new, another curve should be drawn to reflect that situation. If it is assumed that the Hazen-Williams C value for the force main is 130 when new and 80 at the end of the 20-yr design period, the corresponding values for the total dynamic head for the new curve would be 62

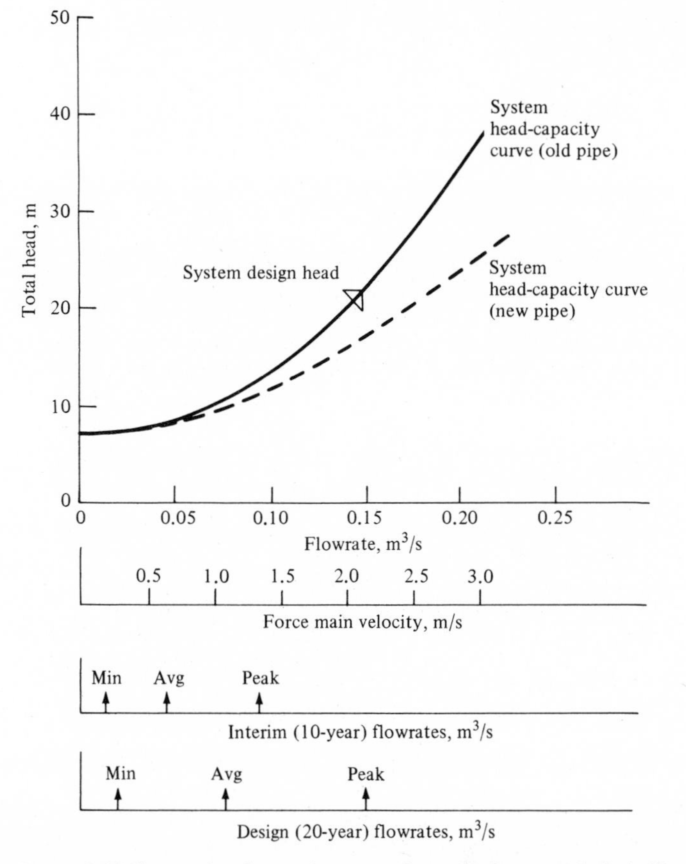

Figure 8-25 System head-capacity curve for preliminary analysis in Example 8-6.

percent of the values computed in step 1*a*. The system head-capacity curve when the force main is new is plotted by using the following relationship (see dashed line Fig. 8-25):

$$H_t = 7.0\text{ m} + (0.62)(15\text{ m})\left(\frac{Q\text{ m}^3/\text{s}}{0.15\text{ m}^3/\text{s}}\right)^2$$

2. Using the following relationship, plot the force main velocities in the system head-capacity curve diagram (see Fig. 8-25).

$$V\text{ m/s} = \frac{Q\text{ m}^3/\text{s}}{A\text{ m}^2} = \frac{Q\text{ m}^3/\text{s}}{0.071\text{ m}^2}$$

3. Plot the minimum, average, and peak flowrates for both the 10-yr interim and the 20-yr design period on the system head-capacity curve diagram (see Fig. 8-25). The interim and future average and peak flowrates were given in the problem statement. The interim minimum flow is developed with data from Table 8-3. Because both of the average flows are close to 0.05 m³/s, a factor of 0.25 is used.

$$\text{Interim design minimum flow} = 0.25(0.044\text{ m}^3/\text{s}) = 0.011\text{ m}^3/\text{s}$$

$$\text{Future design minimum flow} = 0.25(0.075\text{ m}^3/\text{s}) = 0.019\text{ m}^3/\text{s}$$

4. Analyze the system head-capacity curve and develop an approach to solving this problem.
   a. The following conclusions can be made:

      i. The system head-capacity has a low static head and a high friction head.
      ii. Both the interim and future minimum flows are below the minimum acceptable force main velocities of 0.6 m/s.
      iii. The velocity at interim average flow is about 0.6 m/s, but the interim peak flow velocity is about 1.35 m/s, which is sufficient to resuspend solids which may settle out at 0.6 m/s or when there is no flow in the force main.
      iv. There is a sizable difference between the interim peak flow and the future design peak flow. A pump installed for the future design flow will be oversized for the interim peak flow.
      v. Because this station is remote and is not the major contributor to a treatment plant, variable-speed pumping is not a requirement. Constant-speed or two-speed pumps may be used. However, the pumps installed in the station must be capable of discharging the future design peak flow with the largest unit out of service. Because it is not known if the area will be developed fully, the interim design flow should be pumped efficiently.

   b. On the basis of the foregoing analysis, two alternative designs will be evaluated.

      i. *Alternative* A. One operating pump, plus a standby unit, each sized to pump the future design flow. To pump the interim design flows efficiently, two-speed pumps should be used.
      ii. *Alternative* B. Two operating pumps, plus a standby, each sized to pump one-half the future design peak flow. These pumps may be constant-speed or two-speed pumps.

SOLUTION—Alternative A

1. Determine the pump rating point.
   a. The flowrate at the operating point is

$$Q = 0.15\text{ m}^3/\text{s}$$

b. The head at the operating point is

$$H = 22 \text{ m} + \text{station losses} = 22 \text{ m} + 1.3 = 23.3 \text{ m}$$

2. Select a pump from a manufacturer's catalog.
   a. A pump with a relatively high speed must be selected, about 1170 rev/min (see Table 8-2), because of the relatively high head. A pump is found that has the following characteristics when operating at 1170 rev/min:

Operating point	Capacity, m^3/s	Head, m	Efficiency, %
Shutoff	0.000	33.5	. . .
	0.060	29.0	. . .
60% bep	0.100	26.8	68
Rating	0.150	23.3	76
Bep	0.180	21.2	78
120% bep	0.215	18.2	76
Runout	0.228	16.4	70

3. Develop the modified pump head-capacity curve. The modified pump-head capacity curve is developed by subtracting the station losses from the pump head-capacity curve developed at various flows.

   Station losses at the rating point (0.15 m^3/s) equal 1.3 m. Station losses at other flows are proportional to the square of the ratio of the flows; they can be computed with the following relationship.

$$\text{Station losses at } Q = 1.3 \text{ m}\left(\frac{Q}{0.15 \text{ m}^3/\text{s}}\right)^2$$

The necessary computations are summarized below.

	Capacity, m^3/s						
Item	0.0	0.6	0.1	0.15	0.18	0.215	0.228
Pump head, m	33.5	29.0	26.8	23.3	21.2	18.2	16.4
Station losses, m	0	0.2	0.6	1.3	1.9	2.7	3.0
Modified pump head, m	33.5	28.8	26.6	22	19.3	15.5	13.4

4. Plot the original and modified pump head-capacity curves on the system head-capacity curve (see Fig. 8-26).

   Reviewing Fig. 8-26, it can be seen that the pump will operate at the rating point on the system head-capacity curve for old pipe and at 0.177 m^3/s at 21.4 m on the system head-capacity curve for new pipe.
5. Determine the power required by the pump and the power consumed for one day's operation.
   a. During interim operation, the pump will operate at 0.177 m^3/s at 21.4 m and 78 percent efficiency. The required power for these conditions is determined by using Eq. 8-6:

$$P_i = \frac{\gamma Q H}{E_p}$$

$$P_i = \frac{(9.81 \text{ kN/m}^3)(0.177 \text{ m}^3/\text{s})(21.4 \text{ m})}{0.78} = 47.6 \text{ kW}$$

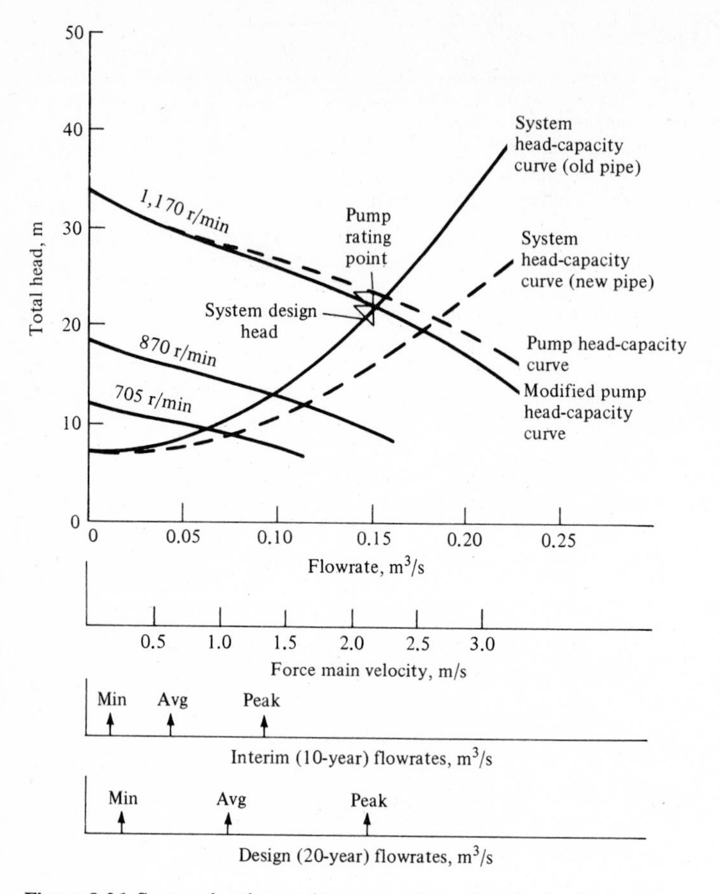

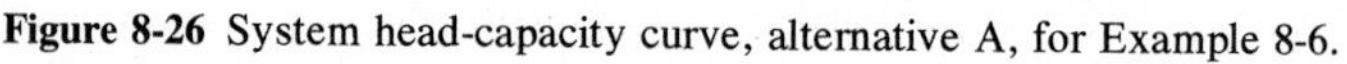

Figure 8-26 System head-capacity curve, alternative A, for Example 8-6.

Motor input, assuming 90 percent motor efficiency, is

$$P_m = \frac{47.6 \text{ kW}}{0.9} = 53 \text{ kW}$$

b. Find the hours per day that the pump will operate and determine the power consumed:

$$\text{Operating hours} = (24 \text{ h/d})\frac{0.044 \text{ m}^3/\text{s}}{0.177 \text{ m}^3/\text{s}} = 5.97 \text{ h/d}$$

$$\text{Power consumed} = (5.97 \text{ h/d})(53 \text{ kW}) = 316 \text{ kWh/d}$$

6. From the system head-capacity curve it can be seen that the pump is considerably oversized for the interim design flows. Therefore, a two-speed pump must be used to pump interim flows efficiently.

   Using the affinity laws and the modified pump head-capacity curve, check pump operation at 870 rev/min and 705 rev/min (see Table 8-2). The required computations are given below.

	1170 rev/min		870 rev/min		705 rev/min	
Operating point	Capacity, m^3/s	Head, m	Capacity, m^3/s^a	Head, m^b	Capacity, m^3/s^a	Head, m^b
Shutoff	0.000	33.5	0.000	18.5	0.000	12.2
	0.060	28.8	0.045	15.9	0.036	10.5
60% bep	0.100	26.2	0.074	14.5	0.060	9.5
Rating point	0.150	22.0	0.110	12.2	0.090	8.0
Bep	0.180	19.3	0.134	10.7	0.109	7.0
120% bep	0.215	15.5	0.160	8.6	...	...
Runout	0.228	13.4	...	...	...	...

[a] $Q_1/Q_2 = N_1/N_2$.
[b] $H_1/H_2 = N_1^2/N_1^2$.

Note: The affinity laws can be applied to modified pump head-capacity curves because both the head developed by the pump due to speed change and the head losses in the suction and discharge piping vary as the square of the flow.

7. Plot the reduced-speed modified curves on the system head-capacity curve (see Fig. 8-26) and analyze the resulting pump system curve. The following observations can be made.
   a. At 870 rev/min, the pump will operate at 0.097 m^3/s at 13.9 m (13.3 m system head plus 0.6 m station losses) on the system head-capacity curve for old pipe. Pump efficiency is about 73 percent (extrapolated from 1170 rev/min curve for corresponding points). This operating point is essentially the same as the interim design peak flow of 0.095 m^3/s. The force main velocity is about 1.3 m/s, which is sufficient to resuspend solids that settle when the pump is not operating.
   b. At 870 rev/min, the pump will operate at 0.11 m^3/s at 12.9 m and 76 percent on the system head-capacity curve for new pipe.
   c. At 705 rev/min, the pump will operate at 0.06 m^3/s at 9.2 m and 71 percent efficiency on the system head-capacity curve. The operating point is at 60 percent of the bep so that it is within the allowable envelope of pump operation. Force main velocity at this speed is about 0.85 m/s. Heavier solids may settle out at this velocity. However, since this is the low speed and the pumps will be operated periodically at a higher speed, this is not a problem.
8. Determine the power required by the pump motor. Assume that the new pipe system head-capacity curve applies.
   a. At 870 rev/min, assuming 88 percent motor efficiency,

$$P_m = \frac{(9.81)(0.11)z)12.9)}{(0.76)(0.\ 88)} = 20.8\ \text{kW}$$

   b. At 705 rev/min, assuming 86 percent motor efficiency,

$$P_m = \frac{(9.81)(0.07)(0.2)}{(0.68)(0.86)} = 10.8\ \text{kW}$$

9. Check the power consumed by the pump. Two subalternates must be considered:

Alternative A-1. Install 1170/870 rev/min motor to provide for both interim and future design flows.

Alternative A-2. Install 870/705-rev/min motor initially to provide efficient pumping of interim flows and replace the motor when required to provide for future design flows.

a. *Alternative* A-1. All interim flows would be pumped at 870 rev/min.

$$\text{Operating hours} = (24\ \text{h/d})\frac{0.044}{0.11} = 9.6\ \text{h/ d}$$

$$\text{Power consumed} = (9.6)(20.8) = 200\ \text{kWh/d}$$

b. *Alternative* A-2. Assume that 80 percent of the flow is pumped at 705 rev/min and 20 percent is pumped at 870 rev/min.

$$\text{Total flow pumped} = (0.044\ \text{m}^3/\text{s})(86{,}400\ \text{s/d}) = 3800\ \text{m}^3/\text{d}$$

i. At low speed,

$$\text{Operating hours} = \frac{(3800\ \text{m}^3/\text{d})(0.8)}{(0.07)(60)(60)} = 12\ \text{h/d}$$

$$\text{Power consumed} = (12)(10.8) = 130\ \text{kWh/d}$$

ii. At high speed,

$$\text{Operating hours} = \frac{3800 - 3040}{(0.11)(3600)} = 1.9\ \text{h/d}$$

$$\text{Power consumed} = (1.9)(20.8) = 40\ \text{kWh/d}$$

Total power consumption for alternative A-2 is 130 + 40 = 170 kW.

SOLUTION—Alternative B

1. Determine the pump rating point.
   a. The flowrate at the operating point for each pump is

$$Q = 0.075\ \text{m}^3/\text{s}\frac{0.15\ \text{m}^3/\text{s}}{2}$$

   b. The head at the operating point is

$$H = 23.3\ \text{m}$$

2. Select a pump from a manufacturer's catalog. A pump is found with the following characteristics when operating at 1170 rev/min:

Operating point	Capacity, m³/s	Head, m	Efficiency, %
Shutoff	0.000	33.5	. . .
	0.025	29.0	. . .
60% bep	0.050	26.5	68
Rating point	0.075	23.3	75
Bep	0.085	21.8	76
120% bep	0.100	19.2	72
Runout	0.108	17.4	69

3. Develop the modified pump head-capacity curve. Station losses at the rating point equal 1.3 m. The necessary computations are summarized below.

Item	Capacity, m^3/s						
	0	0.025	0.05	0.075	0.085	0.10	0.10
Pump head, m	33.5	29.0	26.5	23.3	21.8	19.2	17.4
Station losses, m[a]	0.0	0.2	0.6	1.3	1.7	2.3	2.7
Modified pump head, m	33.5	28.8	25.9	22.0	20.1	16.9	14.7

[a] See step 3 under "Solution—Alternative A" for computational procedure.

Note: The pump requires 10 m NPSH at runout. The impeller selected is in the middle of the available diameters so that the pump capacity can either be increased by selecting a larger impeller or decreased by trimming the impeller.

4. Plot the pump head-capacity curve and the modified pump head-capacity curve on the system head-capacity curve (see Fig. 8-27).
5. Plot the combined curve for two pumps operating in parallel on the system head-capacity curve (see Fig. 8-27). This curve is developed by noting the flow at a given head on the modified pump head-capacity curve and doubling the flow at the head for two pumps operating in parallel.

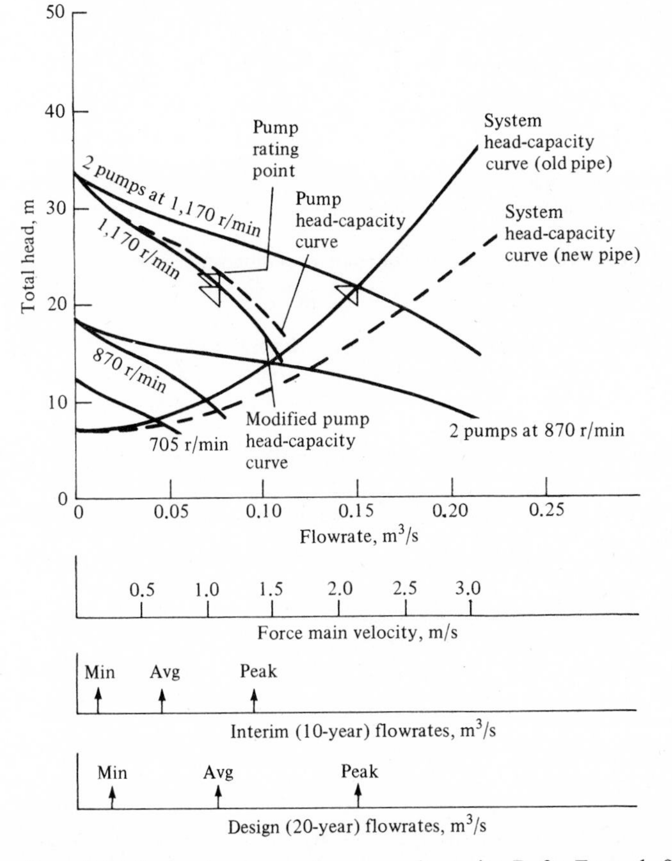

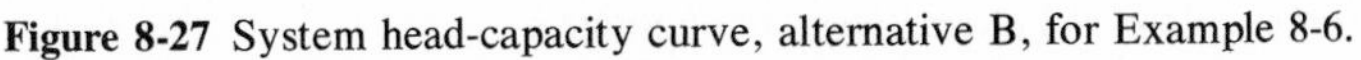

Figure 8-27 System head-capacity curve, alternative B, for Example 8-6.

6. To operate two pumps at low flows, the speed must be reduced. Using the affinity laws and the modified pump head-capacity curve, check pump operation at 870 and 705 rev/min. The modified pump head-capacity curves are computed below.

	1170 rev/min		870 rev/min		705 rev/min	
Operating point	Capacity, m^3/s	Head, m	Capacity, m^3/s	Head, m	Capacity, m^3/s	Head, m
Shutoff	0.000	33.5	0.000	18.5	0.000	12.2
	0.025	29.0	0.019	15.9	0.015	10.5
60% bep	0.05	26.5	0.037	14.3	0.030	9.4
Rating point	0.075	23.3	0.056	12.2	0.045	8.0
Bep	0.085	21.8	0.063	11.0	0.051	7.2
120% bep	0.100	19.2	0.074	9.3	...	...
Runout	0.108	17.4	0.080	8.1	...	...

7. Plot the reduced-speed modified curves on the system head-capacity curve (see Fig. 8-27) and analyze the resulting pump system. The following conclusions can be made.
   a. If 705 rev/min were selected, the sequence of pump operation would be:

   Step 1—one pump, low speed (0.046 m^3/s at 7.9 m)
   Step 2—two pumps, low speed (0.06 m^3/s at 8.3 m) (not shown in Fig. 8-27)
   Step 3—two pumps, high speed (0.175 m^3/s at 19.5 m)

   Because there is too large a jump between step 2 and step 3, step 3 would be required to discharge interim peak flows. Therefore, operation at 705 rev/min is not a good selection and this alternative can be rejected.

   b. If 870 rev/min were selected, the sequence of pump operation when operating against the new pipe system curve would be:

   Step 1—one pump, low speed (0.075 m^3/s at 9.0 m, 72 percent efficiency)
   Step 2—two pumps, low speed (0.125 m^3/s at 13.5 m, 74.5 percent efficiency)
   Step 3—two pumps, high speed (0.175 m^3/s at 19.5 m, 76 percent efficiency)

8. Check the power rating of the pump and the power consumed when the pump is operating against the system head-capacity curve for new pipe.
   a. The power required by a pump operating during step 1 is 12.1 kW, assuming 86 percent motor efficiency.
   b. The power required for two pumps (step 2) is 21.7 kW, again assuming 86 percent motor efficiency.
   c. The power required for two pumps (step 3) operating at high speed, for the new pipe system curve is 55.6 kW, assuming 88 percent motor efficiency.
9. Check power consumption for the interim design flows. Since the discharge at low speed of this alternative is slightly more than for alternative A-1 (0.075 m^3/s versus 0.7 m^3/s), assume that 85 percent of the flow is pumped at low speed and 15 percent at high speed. On the basis of this assumption, the power required at low speed is 148.8 kWh/d and 31.2 kWh/d at high speed, for a total of 180 kWh/d.

COST COMPARISON OF ALTERNATIVES

1. Compare the power costs of the alternatives.

Alternative	A-1	A-2	B
Power, kWh/d	200	170	180

a. The savings of alternative A-2 over A-1 is 30 kWh/d, or $438/yr, at an assumed power cost of $0.04/kWh. The savings over the 10-yr interim period, based on 8 percent interest compounded annually, is approximately $6400. The replacement cost of a motor if the 20-yr design flows occur must be balanced against this savings.

b. The savings of alternative A-2 over alternative B is 10 kWh/d, $146 per year, and $2110 over the interim design period.

FINAL PUMP SELECTION Either alternative A or B is an acceptable solution. The selection of alternative A-1, A-2, or B should be based on economic and other considerations.

1. Some of the possible advantages of alternative A over alternative B are: potentially a smaller building can be used since only two pumps are required instead of three; lower maintenance costs on two pumps versus three; and a simpler control system.
2. The advantage of alternative B over alternative A-2 is that the installed equipment is capable of discharging the 20-yr design flows, whereas a pump motor may have to be replaced in alternative A-2 to meet future conditions. The future design flows, however, may not occur.
3. To illustrate the effect that the system head-capacity curve has on pump selection, assume that the static head in this example is three-quarters of the total head at the rating point. Assume also that there is a corresponding decrease in friction loss, such as might occur with a shorter length of force main. The system head-capacity curves would then be as shown in Figs. 8-28 and 8-29. As shown in Figure 8-28, the pump will discharge the design

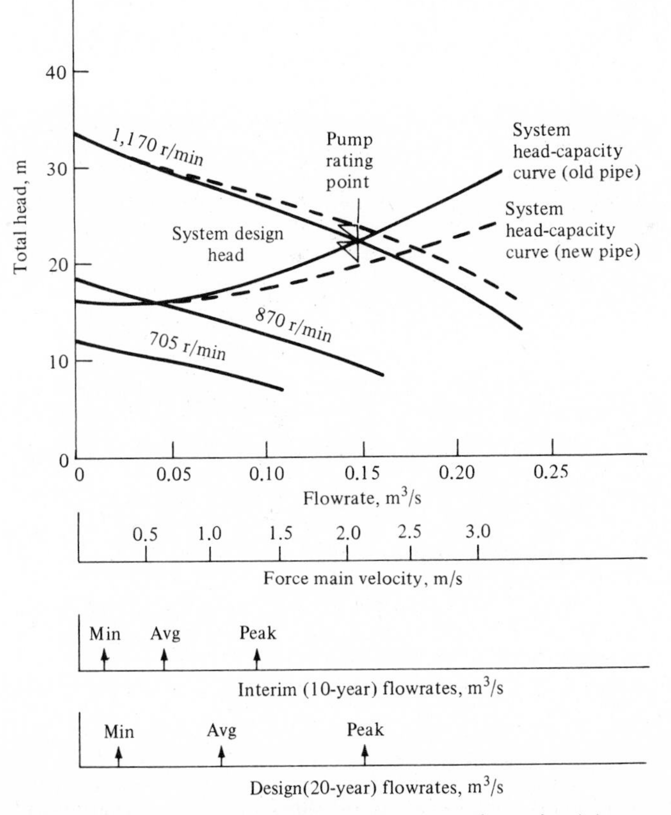

Figure 8-28 Revised system head-capacity curve, alternative A (one-pump operation), for Example 8-6.

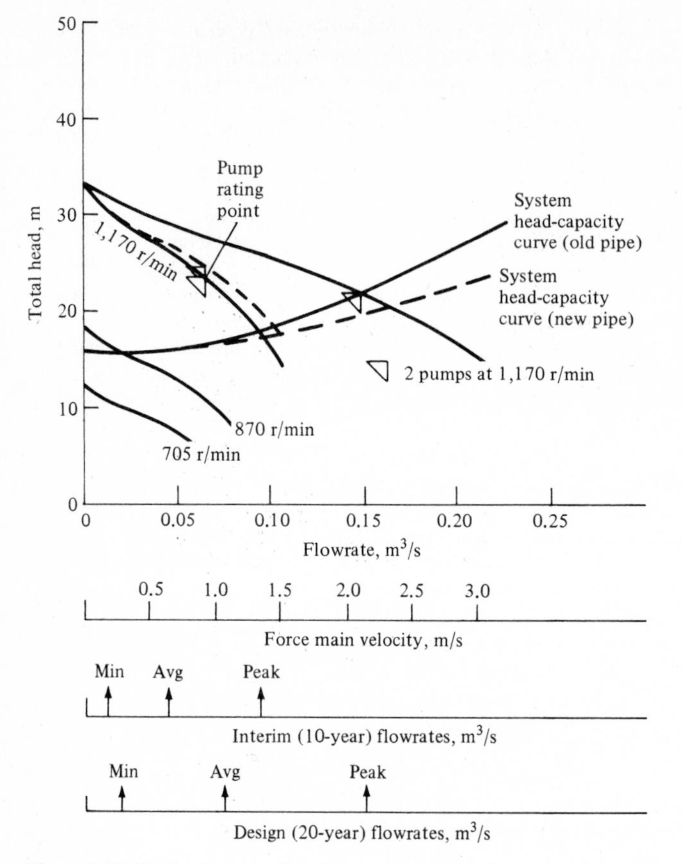

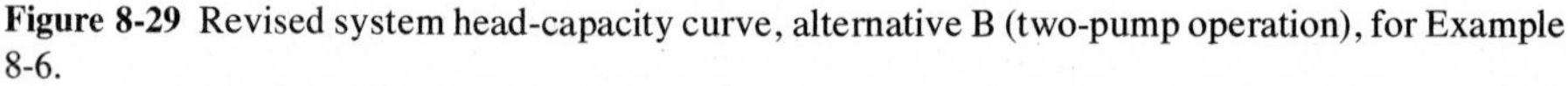

Figure 8-29 Revised system head-capacity curve, alternative B (two-pump operation), for Example 8-6.

quantity at 1170 rev/min, but will discharge only 0.037 m^3/s at 870 rev/min. The 705-rev/min speed cannot be used because the system static head is greater than the pump shutoff head. Because the pump operates near shutoff head at 870 rev/min (0.037 m^3/s is 33 percent of the bep), the pump should not be operated continuously at low speed. With this alternative, the interim flow is pumped inefficiently and should, therefore, not be used.

As shown in Fig. 8-29, one pump at high speed will discharge about 0.095 m^3/s at 17.3 m total head and two pumps in parallel would discharge the design year peak flow. As noted above for alternative A, the pumps in alternative B should not be operated at a reduced speed because they would then operate near shutoff head.

Of the two, alternative B, with single-speed pumps, would be the preferred solution for a system with a high static head. Also, only two pumps (one operating plus a standby) need to be installed initially to pump the interim flows. A third pump could be installed in the future when the flows to the station exceed the interim design flows.

PUMP SPECIFICATIONS When specifying the pumps, enough information must be provided so that potential manufacturers can correctly size the pumps they propose to supply. All pumps can be sized to meet one rating point, but pumps from different manufacturers cannot be expected to meet the requirements of two rating points. In this example, the rating point would correspond to the future design peak flow discharge. Other operating characteristics to be specified would be the requirements to prevent cavitation and operation at low speed. With the use of the foregoing information, the following pump specifications could be written:

1. Alternative A rating points:

	Capacity, m^3/s	Total head, m	Remarks
Rating point	0.15	23.3	At maximum speed, 1170 rev/min
Alternate rating point	≥0.06	9.7	Low-speed operation

2. Alternative B rating points:

	Capacity, m^3/s	Total head, m	Remarks
Rating point	0.075	23.3	At maximum speed, 1170 rev/min
Runout	. . .	. . .	Successful high-speed operation at a total head of 17 m
Alternate rating point	≥0.068	10.6	Low-speed operation

Example 8-7 Selection of variable- and constant-speed pumps Select two or more pumps for installation in a proposed large pumping station that is to pump to a treatment plant through a 1-m (42-in)-diameter force main. The initial average flow is 1.0 m^3/s (22.8 Mgal/d). The initial minimum flow is 40 percent of average and the initial peak flow is two times average flow. Estimated future flows are expected to increase by 50 percent over initial flows. The initial design capacity of the station is to be 2.5 m^3/s.

The friction loss in the force main at the future design peak flow is expected to be 6 m (19.5 ft). Initial friction losses will be 60 percent of the design losses because new pipe is smoother and has a higher Hazen-Williams C value. Station losses are assumed to be 1.3 m (4.25 ft) at the pump rating point (design conditions). The static head is 14 m (46 ft) at the high wet-well level and 16 m (52 ft) at the low wet-well level. The pumps are to be controlled automatically or manually between these levels.

SOLUTION

1. Tabulate design flow conditions in m^3/s.

Flowrate	Minimum	Average	Peak
Initial	0.40	1.0	2.0
Future	0.60	1.5	3.0
Initial peak	. . .	. . .	2.5

2. Plot the system head-capacity curve and plot the various flowrates and force main velocities on the system head-capacity curve (see Fig. 8-30). The system head-capacity curve should also include the minimum system head-capacity curve with new pipe to ensure that there will not be a problem with runout.

   The system head-capacity curve is plotted by using the following relationship:

   a. Future design conditions for old pipe:

$$H_t = \text{static head} + 6\text{ m}\left(\frac{Q}{3\text{ m}^3/\text{s}}\right)^2$$

   b. Initial design conditions for new pipe:

$$H_t = \text{static head} + (0.6)(6\text{ m})\left(\frac{Q}{3\text{ m}^3/\text{s}}\right)^2$$

3. Analyze the system head-capacity curve. The following conclusions can be made.

   a. The system head-capacity curve is basically flat with large static head and a small friction head.

   b. The initial minimum flow will be pumped by a single pump operating at the low-water level in the wet well (highest static lift). Under these conditions, the pump must be capable of operating at 0.4 m^3/s at a head of 16.1 m plus station losses.

   c. The initial design peak of 2.5 m^3/s will be pumped by one or more pumps operating at the high-water level in the wet well (minimum system head-capacity curve). Therefore, the design head corresponding to this flow will be 18.2 m plus the station loss of 1.3 m for a total head of 19.5 m.

4. Selection of pumps.

   a. Because the pumps must operate continuously, variable- and constant-speed pumps must be used. Under this condition, the lead pump at minimum speed must discharge not more than one-half of the full-speed capacity. This provides a system where (1) the lead variable-speed pump operates at 50 to 100 percent of capacity, and (2) the two variable-speed pumps operate at 100 (50 + 50) to 200 percent capacity (based on the capacity of a single pump). A constant-speed pump can then operate at 100 percent capacity so that the three pumps operating together can produce 200 (50 + 50 + 100) to 300 percent of the capacity of a single pump.

   b. The initial design peak flow may be pumped by two or three pumps operating in parallel. It is best to use the minimum number of pumps possible, but each pump must be capable of operating at reduced speed at the initial minimum flow.

      i. The rating point for two pumps is

$$\frac{2.5\text{ m}^3/\text{s}}{2} = 1.25\text{ m}^3/\text{s at } 19.5\text{ m}$$

      ii. The rating point for three pumps is

$$\frac{2.5\text{ m}^3/\text{s}}{3} = 0.83\text{ m}^3/\text{s at } 19.5\text{ m}$$

   c. As shown in Fig. 8-7, the maximum efficiency for pumps having a capacity over 0.6 m^3/s is about 90 percent and occurs at a specific speed of about 60, which is in the Francis or mixed-flow impeller range.

   d. To provide a pump that will have the maximum turndown from full speed to minimum speed, the pump should be selected so that the rating point is to the right of the bep on the pump head-capacity curve.

   e. Taking into consideration the foregoing factors, pumps with the following characteristics were selected.

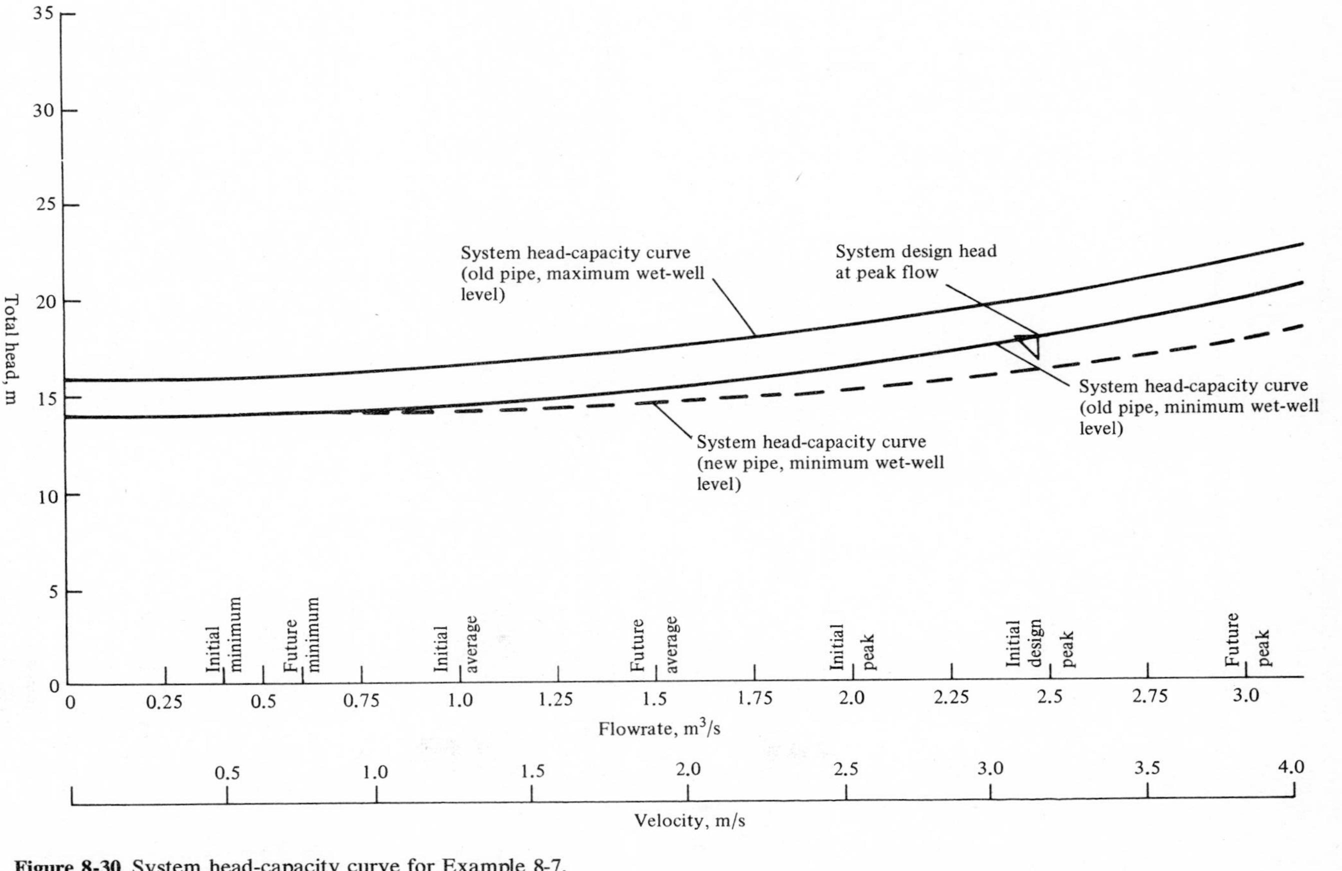

Figure 8-30 System head-capacity curve for Example 8-7.

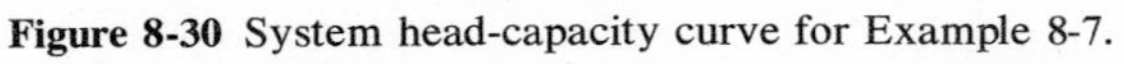

i. For two pumps plus standby pump [600-mm (24-in) mixed-flow pump (585 rev/min)]:

Operating point	Capacity, m^3/s	Head, m	Efficiency, %
Shutoff	0	31.1	...
	0.5	28.2	...
60% bep	0.72	26.5	70
Bep	1.2	20.5	85
Rating point	1.25	19.5	84
120% bep	1.43	15.8	79

Note: 1. NPSH required at 120% bep = 6.7 m.
2. Impeller is mid-sized so capacity can be increased or decreased.

ii. For three pumps plus standby pump [500-mm (20-in) mixed-flow pump (705 rev/min)]:

Operating point	Capacity, m^3/s	Head, m	Efficiency, %
Shutoff	0	26.9	...
	0.25	23.5	...
60% bep	0.46	22.6	76
Bep	0.78	20.1	86.5
Rating point	0.83	19.5	86
120% bep	0.96	16.5	81
Runout	1.0	14.9	77

Note: 1. NPSH required at 120% bep = 6.7 m.
2. Impeller diameter at rating point = 560 mm. Maximum impeller in volute = 590 mm.

5. Develop the modified pump head-capacity curves; station losses at the rating point equal 1.3 m.

a. For a 600-mm (24-in) mixed-flow pump:

	Capacity, m^3/s					
Item	0	0.5	0.72	1.2	1.25	1.43
Pump head, m	31.1	28.2	26.5	20.5	19.5	15.8
Station losses, m	0	0.2	0.4	1.2	1.3	1.7
Modified pump head, m	31.1	28.0	26.1	19.3	18.2	14.1

b. For a 500-mm (20-in) mixed-flow pump:

	Capacity, m^3/s						
Item	0	0.25	0.46	0.78	0.83	0.96	1.0
Pump head, m	26.4	23.5	22.6	20.1	19.5	16.5	14.9
Station losses, m	0	0.1	0.4	1.1	1.3	1.7	1.9
Modified pump head, m	26.4	23.4	22.2	19.0	18.2	14.8	13.0

6. Develop the pump characteristics at reduced speeds and plot the pump envelopes.
   a. Use the affinity laws to determine operation at variable speeds. The following curves are developed with the modified pump head-capacity curve.
      i. For a 600-mm (24-in) mixed-flow pump:

	585 rev/min[a]		500 rev/min		450 rev/min	
Operating point	Capacity, m/s	Head, m	Capacity, m^3/s	Head, m	Capacity, m^3/s	Head, m
Shutoff	0	31.1	0	22.7	0	12.4
	0.5	28.0	0.43	20.4	0.38	16.6
60% bep	0.72	26.1	0.62	19.0	0.55	15.4
Bep	1.2	19.3	1.03	14.1	0.92	11.7
Rating point	1.25	18.2	1.07	13.3	0.96	10.7
120% bep	1.43	14.1	1.22	10.3	1.1	8.3

[a]Modified pump head (see step 5).

      ii. For a 500-mm (20-in) mixed-flow pump:

	705 rev/min[a]		650 rev/min		600 rev/min	
Operating point	Capacity, m^3/s	Head, m	Capacity, m^3/s	Head, m	Capacity, m^3/s	Head, m
Shutoff	0	26.4	0	22.4	0	19.1
	0.25	23.4	0.23	19.9	0.21	17.0
60% bep	0.46	22.2	0.43	18.9	0.39	16.1
Bep	0.78	19.0	0.72	16.1	0.66	13.7
Rating point	0.83	18.2	0.77	15.5	0.71	13.2
120% bep	0.96	14.8	0.88	12.6	0.82	10.7
Runout	1.0	13.0	0.92	11.0	0.85	9.4

[a]Modified pump head (see step 5).

   b. Plot the operating data determined in step 6*a* on a graph to obtain an operating envelope (see Fig. 8-31).

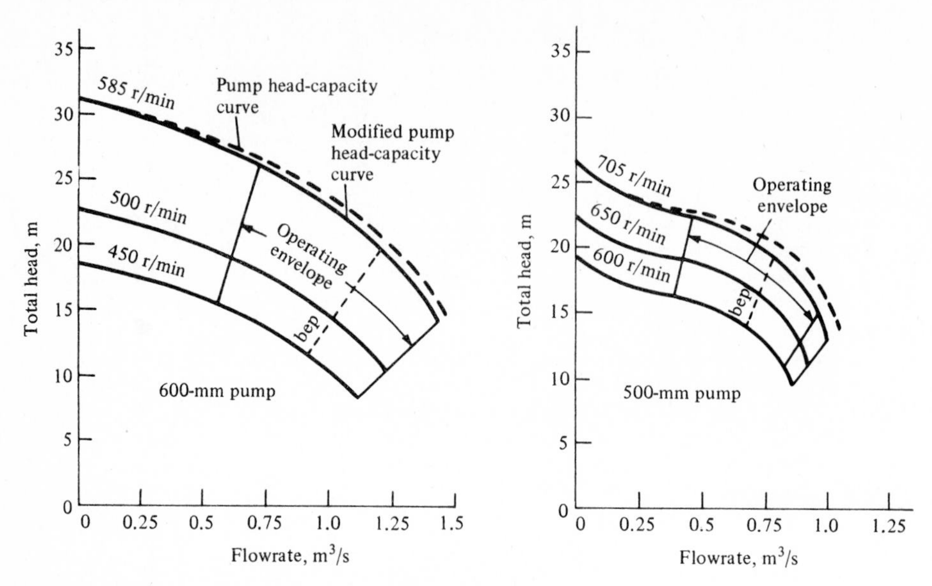

Figure 8-31 Pump operating envelope for Example 8-7.

The operating envelope is described by the head-capacity values for 60 and 120 percent of the bep. The operating envelope is bounded by the pump head-capacity curves at the high and low rotational speeds and by the pump head-capacity values corresponding to 60 and 120 percent of the bep.

7. Compare the pump operating envelopes plotted in Fig. 8-31 to the system head-capacity curve plotted in Fig. 8-30, and select the pump arrangement to be used. (*Note:* This comparison is best accomplished if the pump curve envelopes are drawn on tracing paper that can be superimposed over the system head-capacity curve.)
   a. From Fig. 8-31, it can be concluded that the 600-mm pump, operating at low speed, cannot produce the minimum flow of 0.4 m^3/s at 16.1-m system head and stay within the operating envelope. This pump has a minimum flow of approximately 0.57 m^3/s at 16.1 m head. Therefore, this alternative is rejected.
   b. From Fig. 8-31 it can be concluded that the 500-m pump can operate at the minimum flow. Therefore, design the station using three operating pumps plus one standby pump. Three pumps should be variable speed and one pump should be constant speed. The lead and first follow pump on line are to be variable-speed pumps, the second follow pump should be a constant-speed pump. The third variable-speed pump should be the standby.
   c. When the system flows approach the initial design peak flows, the pumping capacity should be increased to meet the future peak flow requirement by adding a fifth pump. Because variable-speed capability is furnished by the existing pumps, the future pump can be a constant-speed pump.

SUMMARY ANALYSIS The operation of the system is represented graphically in Fig. 8-32. Although the final pump head-capacity curves are shown in Fig. 8-32, in practice they would be added only after the following analyses were completed.

1. Stage I: Operation of lead pump.
   a. The speed of the pump would vary in response to changes in the wet-well level to discharge all station flows between the initial minimum flow of 0.4 m^3/s and a flow of about 0.96 m^3/s.

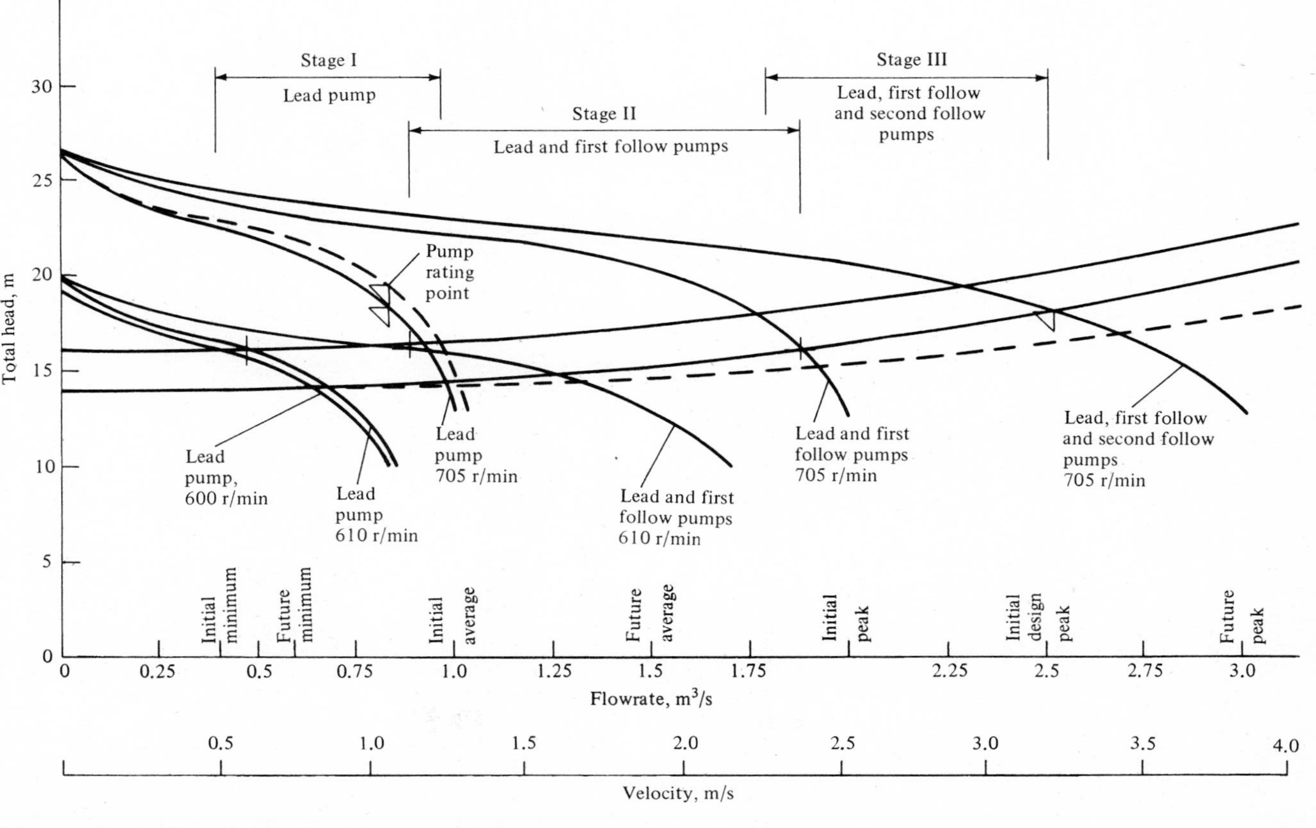

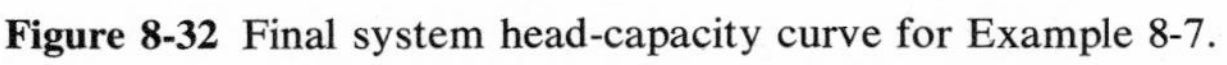

Figure 8-32 Final system head-capacity curve for Example 8-7.

b. At the minimum flow of 0.4 m^3/s, the system would be operating at a low wet-well level, or on the maximum system head-capacity curve. From Fig. 8-32 it can be seen that the pump would be operating at 0.4 m^3/s at a head of 16.1 m. Figure 8-31 shows that the pump must operate at about 600 rev/min to meet these conditions.

c. The pump controls would be adjusted to permit the lead pump to operate to slightly less than 600 rev/min. The controls would also start the first follow pump when the lead pump reached its maximum speed.

2. Stage II: Operation of lead and first follow pumps.

a. The lead and first follow pumps would operate in parallel and their speed would vary simultaneously to discharge all flows between 0.88 m^3/s and 1.88 m^3/s. The minimum flow for this step is selected to provide a small overlap between one- and two-pump operation to prevent frequent starts of the first follow pump.

b. Again, the minimum flow of this step would occur when the wet-well level is low. Figure 8-32 shows that the system head at 0.88 m^3/s (or 0.44 m^3/s from each pump) is 16.3 m. As shown in Fig. 8-31, the pump speed, when operating at 0.44 m^3/s and a head of 16.3 m, is about 610 rev/min. Therefore, the controls would be adjusted to shut down the first follow pump when the pump speed is reduced to 610 rev/min.

c. The controls would also be adjusted to start the second follow pump when both the lead and first follow pumps reach full speed.

3. Stage III: Operation of lead, first follow, and second follow pumps.

a. The lead, first follow, and second follow pumps would discharge all flows between 1.78 m^3/s and the initial design peak flow of 2.5 m^3/s. Again, there is an overlap between the minimum flow of this stage and the maximum flow of stage II to prevent short cycles of the second follow pump.

b. The second follow pump is the constant-speed pump. Therefore, station flow is varied by a change in the speed of the lead and first follow pumps.

Figure 8-32 shows that the system head at a flow of 1.78 m^3/s and low wet-well level is about 17.7 m. As shown in Fig. 8-31, the constant-speed pump at a head of 17.7 m would discharge 0.86 m^3/s. The two variable-speed pumps must, therefore, discharge 0.92 m^3/s ($1.78 - 0.86 = 0.92$), or 0.46 m^3/s each. As shown in Fig. 8-31, the pumps would operate at about 630 rev/min to discharge 0.46 m^3/s at a head of 17.7 m.

c. The controls would be adjusted to shut down the second follow pump when the variable-speed pumps drop below 630 rev/min. If all three pumps, operating at maximum speed, could not discharge the station flow, the controls would sound an alarm and later start the standby pump.

PUMP SPECIFICATIONS When specifying these pumps, sufficient information must be given by the designer to define the required pump operation. For this problem, information is required on the following items.

1. Pump rating point. The pump rating point is normally chosen as the pump operating point to discharge the design peak flow to the station. Also, the maximum pump speed should be included. In this case, the pump rating point is

$$\text{Rating point} = 0.83\ \text{m}^3/\text{s at } 19.5\ \text{m}$$

$$\text{Maximum speed} = 705\ \text{rev/min}$$

2. Requirements to prevent cavitation. Information must be provided so that the supplier will furnish a pump that will not cavitate under the worst-system operating condition. In this example, the worst condition occurs when one pump is operating at full speed against the system head-capacity curve for new pipe. From Fig. 8-32 it can be seen that the minimum head is about 16.0 m (14.3 m system head plus 1.7 m pumping station losses).

To select the pump, the manufacturer must know the NPSH available in the system as well as the NPSH required by the pump. Assume that the suction static head is 3 m, the pipe friction and minor losses are 0.08 m, and the temperature of the wastewater is 25°C (vapor pressure at 25°C = 3.17 kN/m²). The available NPSH is determined with Eq. 8-17 and by neglecting the velocity head term.

$$NPSH_A = h_s - h_{fs} - \Sigma h_{ms} + \frac{P_{atm}}{\gamma} - \frac{P_{vapor}}{\gamma}$$

$$= 3\text{ m} - 0.08\text{ m} + \frac{101.3\text{ kN/m}^2}{9.78\text{ kN/m}^3} - \frac{3.17\text{ kN/m}^2}{9.78\text{ kN/m}^3} = 13.1\text{ m (43 ft)}$$

Information on the requirements to prevent cavitation can be provided by a statement similar to the following:

> The pump shall operate successfully at full speed and without cavitation or vibration to a total head of 16 m when operating with 13.1-m available NPSH.

3. Minimum operating speed. The requirement for the initial minimum flow should be specified as an alternative rating point. Unlike two-speed pumps where the speed ratio is fixed, the operating speed of variable-speed pumps can be adjusted to meet the specified conditions exactly. For this problem, the reduced speed is:

   Pump rating point at initial minimum flow = 0.4 m³/s at 16.0 m

   Reduced speed = 606 rev/min

DISCUSSION TOPICS AND PROBLEMS

8-1 A wastewater pump has a 300-mm discharge and a 350-mm suction. The reading on the discharge gage located at the pump centerline is 130 kPa (kN/m²). The reading on the suction gage located 0.5 m below the pump centerline is 15 kPa (kN/m²). If the total head on the pump is 15 m, determine (1) the pump discharge and (2) the energy input to the motor, assuming a pump efficiency of 82 percent and a motor efficiency of 91 percent.

8-2 Solve Prob. 8-1, but assume that the total head is 12 m and the reading on the discharge gage is 100 kPa (kN/m²).

8-3 A centrifugal pump with an impeller diameter of 0.2 m delivers 0.02 m³/s against a head of 18 m at a power input of 4 kW when operating at 1170 r/min. If it is assumed that the efficiency remains the same, determine the (1) head, (2) discharge, and (3) power input for a geometrically similar pump with an impeller diameter of 0.25 m operating at 870 r/min.

8-4 A mixed-flow volute pump is to operate at a head of 5 m and discharge 0.17 m³/s. It is to be driven by a direct-coupled squirrel-cage induction motor operating on 60-cycle (60-Hz) current. If the specific speed is not to exceed 115, what should be the operating speed? What pump efficiency could be expected, and how much power will be required?

8-5 If the ratio of heads for a prototype and a model pump is 4:1, what is the power ratio if the scale ratio of the impeller diameter is 5:1?

8-6 A radial flow centrifugal pump with a cavitation constant of 0.25 must develop a head of 12 m. Determine the maximum permissible suction lift on the pump at sea level if the temperature of the wastewater is 25°C.

8-7 A wastewater pump is designed to operate at 705 r/min, pumping 1.0 m³/s against a head of 30 m. It will operate at, or close to, the best efficiency point with flooded suction (water at top of pump volute). The barometric pressure is 750 mm, the temperature of the wastewater is 20°C, and the losses in the suction pipe amount to 0.3 m. Determine both the available and required net positive suction head. If this pump is operated at reduced head to deliver 1.3 m^3/s, can it be expected to cavitate?

8-8 A mixed flow pump is operating at 350 r/min. If the head on the pump is 3 m and the cavitation constant is 0.5, determine the rate of discharge.

8-9 A centrifugal pump is to be used to pump treated effluent a distance of 300 m through a pipe with a diameter of 0.3 m. If the static lift is 5.0 m and the friction factor f for the pipe is 0.025, determine the flowrate, head, and efficiency when using a pump with the following characteristics. Neglect minor losses.

Flowrate, m^3/s	Head, m	Efficiency, %
0.00	15.0	. . .
0.05	14.7	67
0.10	12.8	79
0.14	. . .	85
0.15	9.3	. . .
0.16	. . .	84
0.18	. . .	78
0.20	4.0	65

8-10 Using the data given in Prob. 8-9, determine the head and discharge if two identical pumps were (1) operated in parallel and (2) operated in series.

8-11 Develop a relationship that can be used to estimate the efficiency of two or more pumps operating in (1) parallel and (2) series.

8-12 Using the relationships developed in Prob. 8-11, estimate the efficiency of two pumps operating in (1) parallel and (2) series using the following information.

1. Parallel operation

	Pump	
Item	1	2
H, m	15	15
Q, m^3/s	0.26	0.33
Ep, %	75	80

2. Series operation

	Pump	
Item	1	2
H, m	15.5	21.2
Q, m^3/s	0.25	0.25
Ep, %	75	65

8-13 The following data are for a 0.1-m pump operating at 1770 r/min. What would be the head and power required for a 0.2-m geometrically similar pump discharging 0.5 m^3/s when operating at 1170 r/min?

Flowrate, m^3/s	Head, m	Efficiency, %
0.00	16.0	...
0.05	15.5	67
0.10	13.7	79
0.14	...	85
0.15	10.5	...
0.16	...	84
0.18	...	78
0.20	5.5	65

8-14 A variable-speed pump delivers 0.315 m^3/s at a speed of 1150 r/min against a system head of 27 m, of which 22.5 m is static head and 4.5 m is friction head. The pump characteristics are as follows.

Capacity, m^3/s	Head, m	Efficiency, %
0	34.0	0
0.10	33.8	57
0.15	33.4	68
0.20	32.2	79
0.25	30.5	86
0.30	28.0	88
0.35	24.5	87
0.40	21.0	82

Determine the speeds necessary to pump 0.150 and 0.225 m^3/s against the system curve, and determine the heads, efficiencies, and brake horsepower required at all these operating points. Assume that the friction losses vary as the square of the flow.

8-15 A variable-speed pump is to be used to pump treated wastewater through a force main. When the flowrate is 0.15 m^3/s, the total system head is equal to 10 m; the corresponding station losses are equal to 1.13 m. The system static head is 5 m. The characteristics of the variable speed pump when operating at 870 r/min are as follows.

Capacity, m^3/s	Head, m
0.0	20.0
0.05	19.4
0.10	17.9
0.15	15.0
0.20	10.6

Develop the modified pump head-capacity curve and determine the head and flowrate when the pump is operating at 870 r/min. What would be the flowrate if two identical pumps were operated in parallel?

8-16 Using the data of Prob. 8-15, determine the operating speed of the second variable-speed pump so that the combined output of the two pumps operating in parallel is equal to 0.2 m^3/s. What is the head and discharge for each pump? Assume the first variable-speed pump will operate at 870 r/min.

8-17 Determine the power requirements for a constant-speed and variable-speed pump operating against a constant head of 12.5 m and a discharge of 0.05 m^3/s. Assume that the pump characteristics for both pumps are the same as those given in Fig. 8-17. The constant-speed pump is to operate at 800 r/min. The variable-speed pump is to operate over the range from 800 to 300 r/min. How much power is lost (wasted) using the constant-speed pump in this application? At what rotational speed should the variable-speed pump be operating under these conditions?

8-18 The friction loss in a force main is estimated to be 10 m when the flowrate is equal to 0.15 m^3/s. The system static head is 5 m. If, after construction, the actual friction loss is 6 m when the flowrate is 0.15 m^3/s, what is the annual cost of the energy wasted if a pump with the characteristics given in Prob. 8-15 is used to pump at a flowrate of 0.14 m^3/s? What is the cost of the energy that would have been wasted in the original design? Assume that the pump efficiency is 76 percent, the motor efficiency is 96 percent, and the cost of energy is $0.06/kWh.

8-19 Using the pump curves shown in Fig. 8-33, determine which pump would be most energy efficient if the average wastewater flow is 0.35 m^3/s and the peak flow is 0.56 m^3/s. The coordinates of the system head-capacity curve are as follows.

H, m	Q, m^3/s
35	0.0
40	0.225
45	0.375
50	0.450
55	0.525
60	0.560

REFERENCES

1. Addison, H.: *Centrifugal and Other Rotodynamic Pumps,* 3d ed., Chapman & Hall, London, 1966.
2. Buckingham E.: Model Experiments and the Form of Empirical Equations, *Trans. ASME,* vol. 37, pp 263–296, 1915.
3. Church, A. H.: *Centrifugal Pumps and Blowers,* Wiley, New York, 1944.
4. *Environmental Wastes Control Manual and Catalog File,* 1976 ed., chap. 2, "Public Works," 1976.
5. Hicks, T. G., and T. W. Edwards: *Pump Application Engineering,* McGraw-Hill, New York, 1971.
6. Karassik, I. J., and R. Carter: *Centrifugal Pumps,* F. W. Dodge Corporation, New York, 1960.
7. Karassik, I. J., et al.: *Pump Handbook,* McGraw-Hill, New York, 1977.
8. Kristal, F. A., and F. A. Annet: *Pumps,* 2d ed., McGraw-Hill, New York, 1953.
9. Norrie, D. H.: *An Introduction to Incompressible Flow Machines,* Edward Arnold Publishers Ltd., London, 1963.
10. Potthoff, E. O.: Motor Drives for Sewage Pumping, *Wastes Eng.,* vol. 23, no. 6, 1952.
11. *Standards of Hydraulic Institute,* 13th ed., New York, 1975.
12. Stepanoff, A. J.: *Centrifugal and Axial Flow Pumps,* 2d ed., Wiley, New York, 1957.
13. Stratton, C. H.: Raw Sewage Pumps, *Sewage and Industrial Wastes,* vol. 26, no. 12, 1954.
14. Webber, N. B.: *Fluid Mechanics for Civil Engineers,* SI ed., Chapman & Hall, London, 1971.

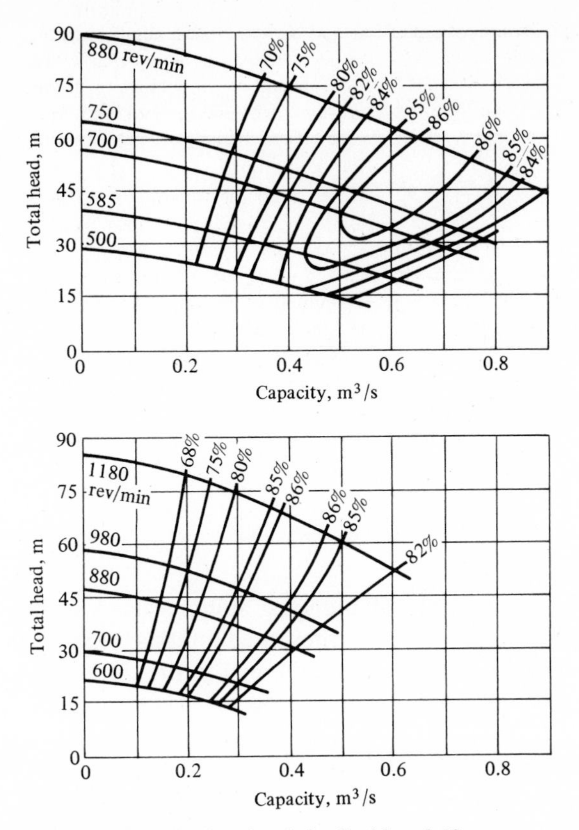

Figure 8-33 Definition sketch for Problem 8-19.

CHAPTER

NINE

PUMPING STATIONS

This chapter is an introduction to the study of pumping stations used with wastewater collection systems. The information presented in this chapter applies to stations at a wastewater-treatment plant or at a remote site. Many of the concepts presented also apply to pumping stations used for water supplies and other similar purposes.

The contents of this chapter are limited to that information with which the sanitary engineer should be most familiar. The major topics covered are (1) the types of pumping stations that are commonly used, (2) the more important details of pumping station design, (3) the design of force mains, and (4) the theory, analysis, and control of waterhammer.

To gain additional information about the design of pumping stations, the engineer should take every opportunity to inspect existing pumping stations and to study the plans and specifications for existing or proposed pumping stations.

9-1 TYPES OF PUMPING STATIONS

Pumping stations are often required for the pumping of

1. Untreated domestic wastewater
2. Storm-water runoff
3. Industrial wastewater
4. Combined domestic wastewater and storm-water runoff
5. Sludge at a wastewater-treatment plant
6. Treated domestic wastewater
7. Circulating water systems at treatment plants

Apart from pumping facilities required at wastewater-treatment plants, the principal conditions and factors necessitating the use of pumping stations in the wastewater collection system are as follows:

1. The elevation of the area or district to be serviced is too low to be drained by gravity to existing or proposed trunk sewers.
2. Service is required for areas outside the natural drainage area but within the service district.
3. Omission of pumping, although possible, would involve excessive construction costs because of the deep excavations required for the installation of a trunk sewer to drain the area.

Modern pumping stations are usually automatic in routine operation. The smaller stations are normally unattended and require little attention other than a daily check on the proper functioning and lubrication of the equipment. The larger stations, especially those serving major areas of large cities, are usually attended stations, but operating staffs are small. In most cases, one or two persons per shift are sufficient for operation.

Classification of Pumping Stations

Pumping stations have been classified in various ways, none of which is entirely satisfactory. Some of the common methods of classifying stations are as follows:

1. Capacity (cubic meters per second, cubic meters per day, or liters per second) (gallons per minute or million gallons per day)
2. Source of energy (electricity, diesel, etc.)
3. Method of construction (conventional custom built, factory-assembled, etc.)
4. Specific function, or purpose

A classification of pumping stations by capacity and the normal method of construction for that capacity is presented in Table 9-1. As shown, there is a considerable overlap in the capacity range between the factor-assembled pumping stations, commonly called *package pumping stations,* and the conventional custom-built pumping stations.

Factory-assembled pumping stations are shipped from the factory in modules with all the equipment and components mounted and connected permanently within the module. They are currently available with three types of pumping equipment: pneumatic ejectors, wet-pit submersible pumps, and dry-pit pumps. Pneumatic ejectors are normally used to pump low flows because centrifugal pumps, sized to discharge a 75-mm (3-in) solid, should not be operated below 0.006 m^3/s (100 gal/min). Small submersible wet-pit pumps, designed so that they can be removed for maintenance without disturbing the

Table 9-1 Classification of pumping station by capacity and method of construction

Class/type	Capacity range	
	m^3/s	gal/min
Pneumatic ejector	<0.02	<300
Factor assembled		
Wet pit	0.006 – 0.03	100 – 500
Dry pit	0.006 – >0.1	100 – >1,600
Conventional		
Small	0.2 – 0.09	300 – 1,400
Intermediate	0.06 – 0.65	1,000 – 10,000
Large	>0.65	>15 Mgal/d[a]

[a]Large stations are rated in Mgal/d (million gallons per day).

Note: $m^3/s \times 15{,}850.3 = \text{gal/min}$

$m^3/s \times 22.8245 = \text{Mgal/d}$

discharge piping, are also used for low flows. Both the ejector and the submersible pumps are available in factory-assembled stations or can be installed in conventional stations. The capacity of factory-assembled dry-pit pumping stations has increased greatly in recent years. Stations are now furnished on special order with capacities exceeding 0.3 m^3/s (5000 gal/min).

The capacity of conventional pumping stations ranges from about 0.02 to more than 0.65 m^3/s (from 300 to 10,000 gal/min). They are used where (1) local conditions make a factory-assembled station impractical, and (2) the quantity of flow or its variation, or both, exceeds the capacity of the available factory-assembled station. Although the term "conventional" is used to describe these stations, each one is a special design matched to the local conditions.

General Features of Pumping Stations

The main purpose of a pumping station structure is to pump water, and the station houses the pumps and the auxiliary equipment required by the pumps Therefore, the design features of pumping stations vary with the capacity of the station and the method of construction. Typical schematic views of a modern conventional wastewater pumping station are shown in Fig. 9-1. A typical factory-assembled station is shown in Fig. 9-2. The general features of conventional and factory-assembled pumping stations are summarized in Table 9-2. Details on the design of conventional stations are discussed in Sec. 9-2; package stations are described in more detail in Sec. 9-3.

9-2 DESIGN OF CONVENTIONAL PUMPING STATIONS

Conventional stations are required for larger flows or where the wastewater must be screened and, in some cases, comminuted to protect the pumps. Unlike factory-assembled pumping stations where the site is adapted to the stan-

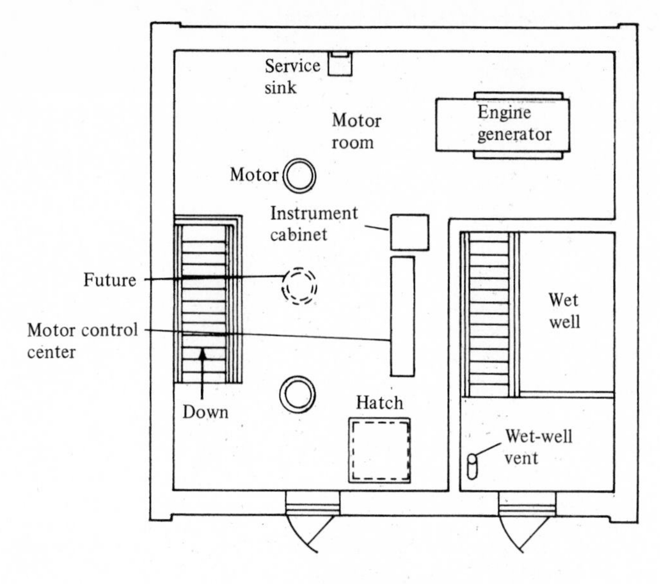

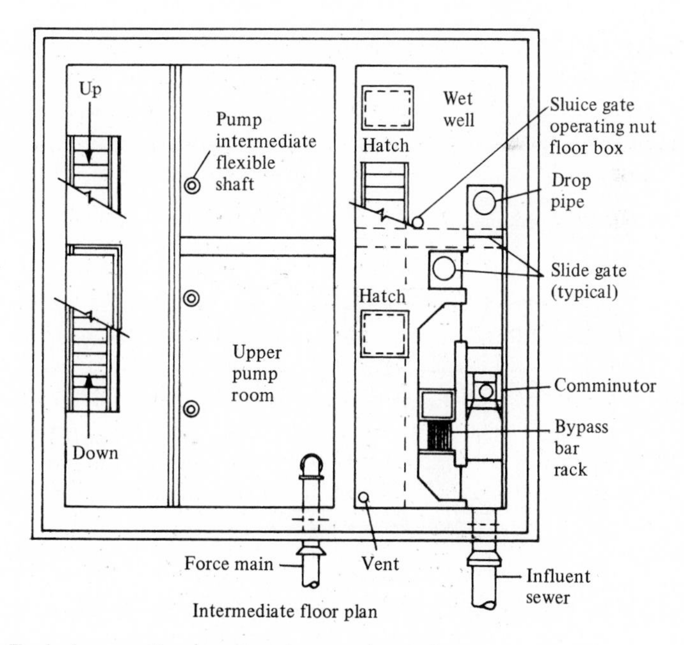

Figure 9-1 Typical conventional wastewater pumping station.

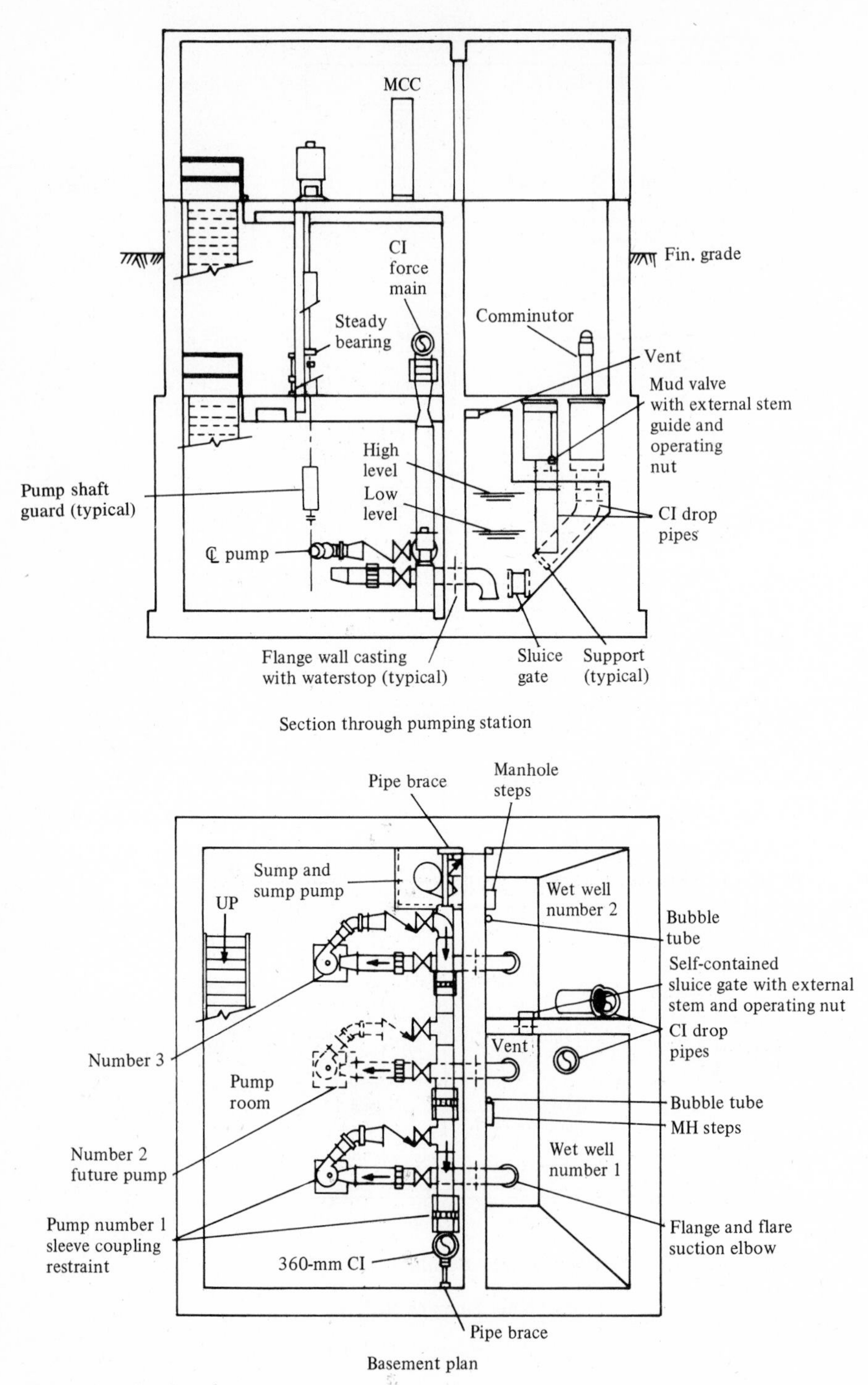

Figure 9-1 *Continued.*

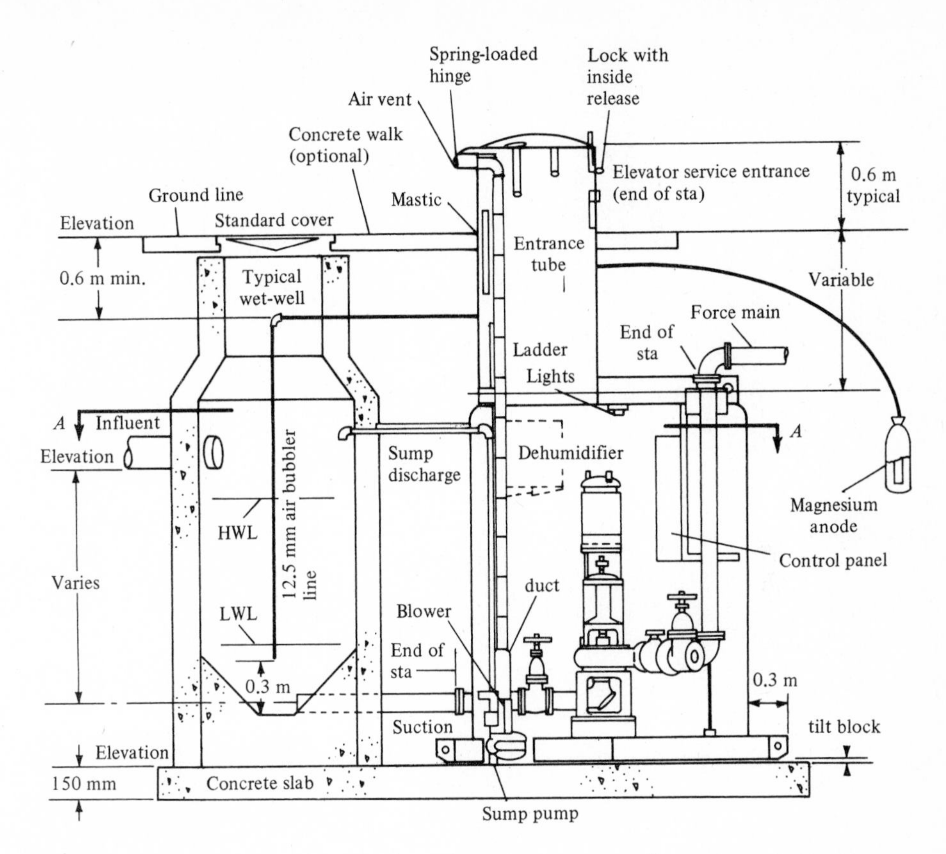

Section $A - A$

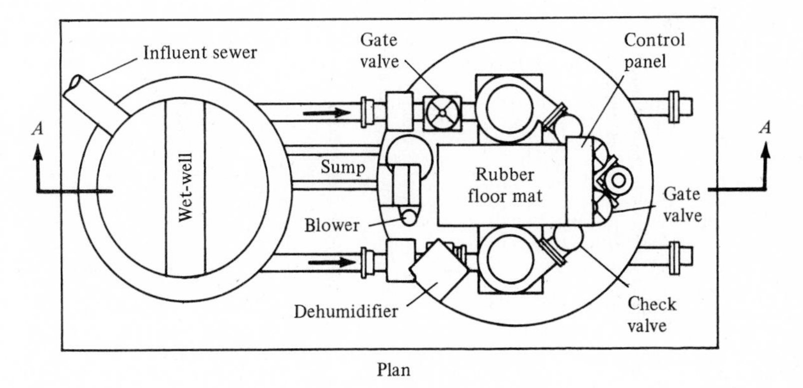

Plan

Figure 9-2 Factory-assembled pumping station with two pumps.

Table 9-2 General features of conventional and factory-assembled pumping stations

Item	Common function	Conventional[a]	Factory assembled[b]
Construction		Reinforced concrete substructure; superstructure may be of masonry, reinforced concrete, or wood or metal panels.	Steel, fiberglass.
Wet well	Used to receive wastewater from the collection system and to store it before it is pumped.	Pump protection equipment, including bar racks and comminutors, is often installed in the wet well of large stations. Entrance to the wet well must be directly from outdoors. Access should be by stairs.	Concrete manholes are often used as the wet well in small stations.
Dry well	Used to house pumps.	Pumps motors and control panels are located on an intermediate floor of the dry well or on the ground floor of the station.	Pump motors are usually located in the dry well along with the control panel. Dehumidifier usually provided to control corrosion.
Pumps	Set in the bottom of the dry well with the top of the pump casing below the low-water level in the wet well.		
Suction and discharge piping	Suction piping is used to connect the wet well to the pump. Discharge piping is used to connect the pump discharge to the station force main. Valves are usually located in the suction and discharge piping system so that the pumps can be isolated for maintenance and cleaning.		
Pumping station instrumentation	Includes automatic and manual controls for pumps, high and low wet-well level alarms, and flow metering.	Motor control panel is located on the ground floor in large stations.	Control panel is set within the dry well.
Electrical equipment	Electrical motors are the most commonly used method of powering the station pumps.	Pump motors are located on an intermediate floor of the dry well or on the ground floor of the station. Dual-fuel engines are sometimes used in large stations to drive the pumps.	Pump motors are usually coupled directly to the pump and are located in the dry well.

Power source for pumping station		To make station operation reliable, power must be available from two sources. This requirement can be met by providing two electric feed lines or one feed line and one or more engine-generator sets in the station.	A single power source is generally used, but an engine-generator can be provided as an auxiliary power source.
Heating and ventilation	Depending on temperature conditions, the wet well may need to be heated to prevent condensation and freezing. In cold climates, heating is required in the dry well to prevent freezing. Both the dry well and the wet well require ventilation to purge these areas of any dangerous fumes.		
Plumbing		Plumbing system in large stations normally consists of a sump pump that is used to drain the dry well.	
Miscellaneous		Access openings or hatches must be located to permit the removal of pumps, motors, and other equipment. To assist in the maintenance and removal of equipment, lifting hooks or, preferably, monorails with trolleys should be located over floor hatches and each piece of large equipment. Stations sometimes house other facilities, such as toilets and workshops, and have storage space. In some stations, especially in warm climates, septic wastewater and hydrogen sulfide can cause odor and corrosion problems. In these cases, chlorination or other chemical treatment facilities can be located in the station to eliminate these problems.	

[a]See Fig. 9-1.
[b]See Fig. 9-2.

dard factory-assembled plant design, conventional stations are custom designed for a particular location. To provide an introduction to the design of conventional pumping stations, the subjects identified in Table 9-2 are discussed in detail in this section.

Pumping Station Construction

Pumping station substructures should be made of reinforced concrete. The exterior walls below grade and wet-well walls below the maximum high wet-well level should be coated with tar to prevent leakage. Superstructures should blend in with the surroundings and should be of fireproof construction. Windows are normally omitted at unattended stations to reduce risk of vandalism. The ground floor of the station must be set above the flood plain of the surrounding area to eliminate the possibility of flooding the station. In both the substructure and the superstructure, the wet and dry wells must be isolated from each other. This requires that all walls between the wet and dry wells be made vapor-tight and that all pipe and conduit penetrations between them be caulked gas-tight.

Rectangular or square stations are more common because the rectangular shape provides more usable space and can be divided more easily into wet and dry wells. However, for deep stations, circular stations may be considered; a circular structure withstands structural loadings better than a rectangular structure. Additional details on the structural design of pumping stations may be found in Ref. 3.

Pumping stations must have facilities for the servicing of equipment and the removal and replacement of equipment in the building. In large stations, overhead bridge cranes or monorails are provided for equipment handling. In smaller stations, lifting hooks are located over larger items of equipment. In addition, doors must be made large enough for the removal of equipment, and floor openings or floor hatches must be provided for removing equipment from lower floors.

Wet Wells

Wet wells are required in pumping stations to store the wastewater before it is pumped. To protect the pumps from clogging, bar racks, screens, or comminutors are also used with the wet well. The storage volume needed depends on the type of pump operation, constant-speed drive or variable-speed drive. If constant-speed operation is selected, the volume must be adequate to prevent the short cycling of the pumps, i.e., frequent starting and stopping.

Equally important functions of the wet well are to provide sufficient submergence of the pump suction inlet to prevent vortexing and to make the transition of flow from the sewer to the pump suction pipe as smooth as possible. Many pump problems have resulted from improperly designed wet wells. Prerotation and other hydraulic turbulence that affect the pump suction head and pump performance are the principal problems.

Wet-well design considerations. Wastewater enters the pumping station through the wet well. Because the sewer gas and the volatile and flammable material carried with the wastewater are vented to the atmosphere in the wet well, many explosions have resulted from careless design and operation of the wet well. To minimize such occurrences, all equipment and electrical work in the wet well must be of explosion-proof and spark-proof construction.

It is good practice, and most regulatory authorities require, that the wet well be divided into two or more sections so that a portion of the station may be taken out of service for inspection and cleaning. Because of the size of the wet well, turbulent flow is not always present and grit and heavier wastewater solids settle out. For this reason, the bottom of the wet well should slope toward the flanged inlet.

Each section of the wet well should have an individual inlet and slide gate, or sluice gates should be arranged to divert the flow from that section when it is taken out of service. The sections, however, must be interconnected by sluice gates so that the total storage capacity of the station is available to prevent short-cycling of the pumps. If a floor covers the wet well, access must be provided to each section.

A small station provided with a comminutor and two wet wells is shown in Fig. 9-2. The station houses two pumps with space for a future pump. The future pump is located so that a dead spot is not created in the wet well. The operating pump is at the opposite end of the wet well from the inlet and the future pump is located next to the inlet. If the positions had been reversed, a dead spot would have been created at the end of the wet well and wastewater solids could settle out, causing odors and other problems.

The wet well of a large pumping station with four operating pumps and space for a future pump is illustrated in Fig. 9-3. As shown, the station is divided into three wet wells with a separate mechanically cleaned bar rack for each section of the wet well.

The shape of the wet well is important to minimize the deposition of solids. The floor is level from the wall to a point 0.3 to 0.4 m (12 to 18 in) beyond the outermost edge of the suction bell and should then slope upward to the opposite wall, as shown in Fig. 9-3. A slope of 1:1 or greater is recommended and is required by most state regulatory agencies [4].

Wet-well volume. The required wet-well volume depends on the method of pump operation. If the pumps are driven by a variable-speed drive that varies the pumping rate to match the inflow to the station, the storage volume required in the wet well is small. The storage required in this case should be sufficient to allow time for the change in capacity when a pump is started or stopped before the next start or stop point is reached. This time is normally less than 1 min.

Constant-speed or multiple-speed pumps need a larger storage volume to prevent short-cycling of the pump and motor. For squirrel-cage induction motors that operate between 15 and 75 kW (20 and 100 hp), the time between starts of the motor should not be less than 15 min. For motors over 75 kW but

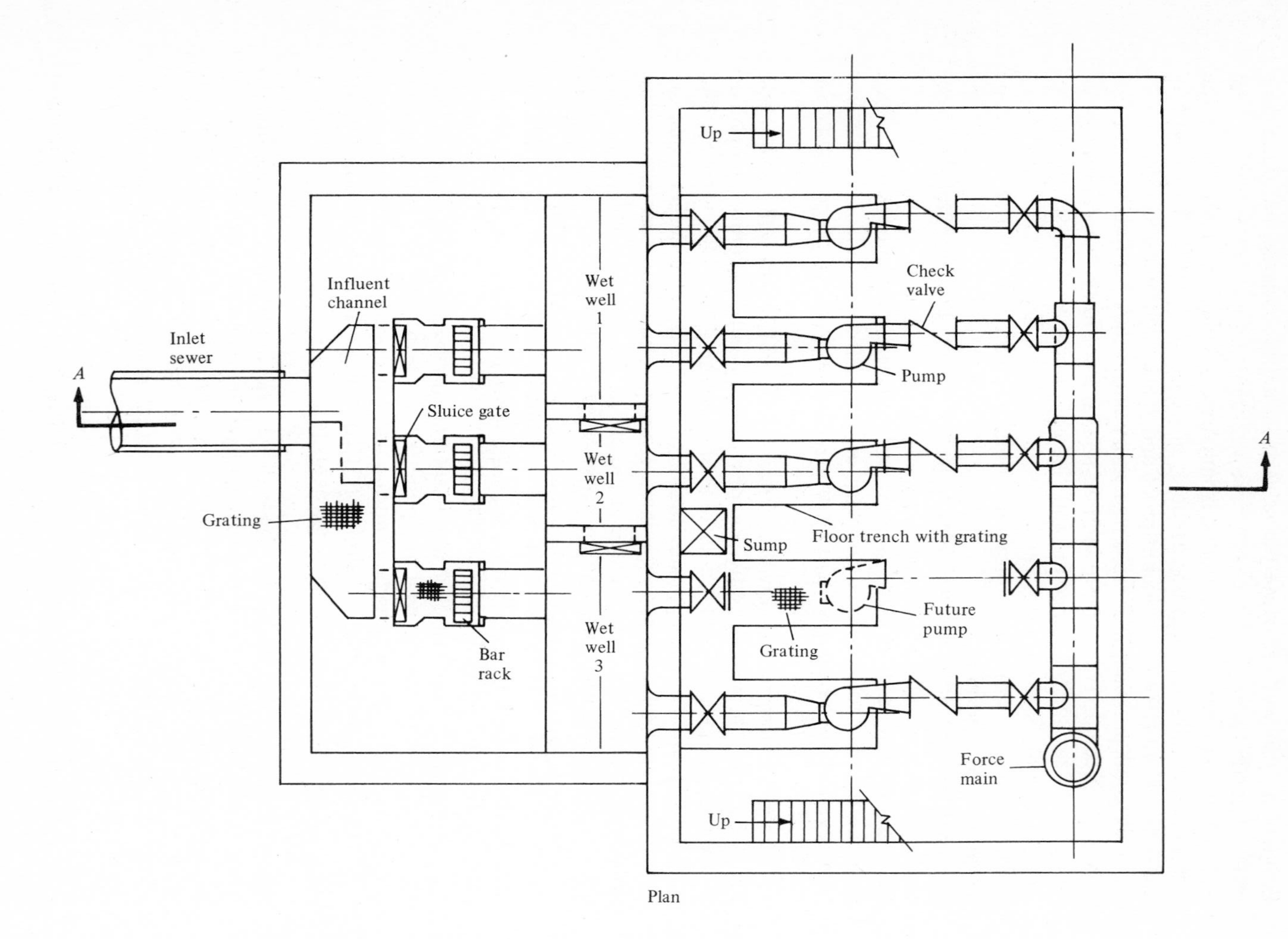
Up
Influent channel
Wet well 1
Check valve
Inlet sewer
A
Pump
Sluice gate
Wet well 2
A
Grating
Sump
Floor trench with grating
Future pump
Wet well 3
Bar rack
Grating
Force main
Up
Plan

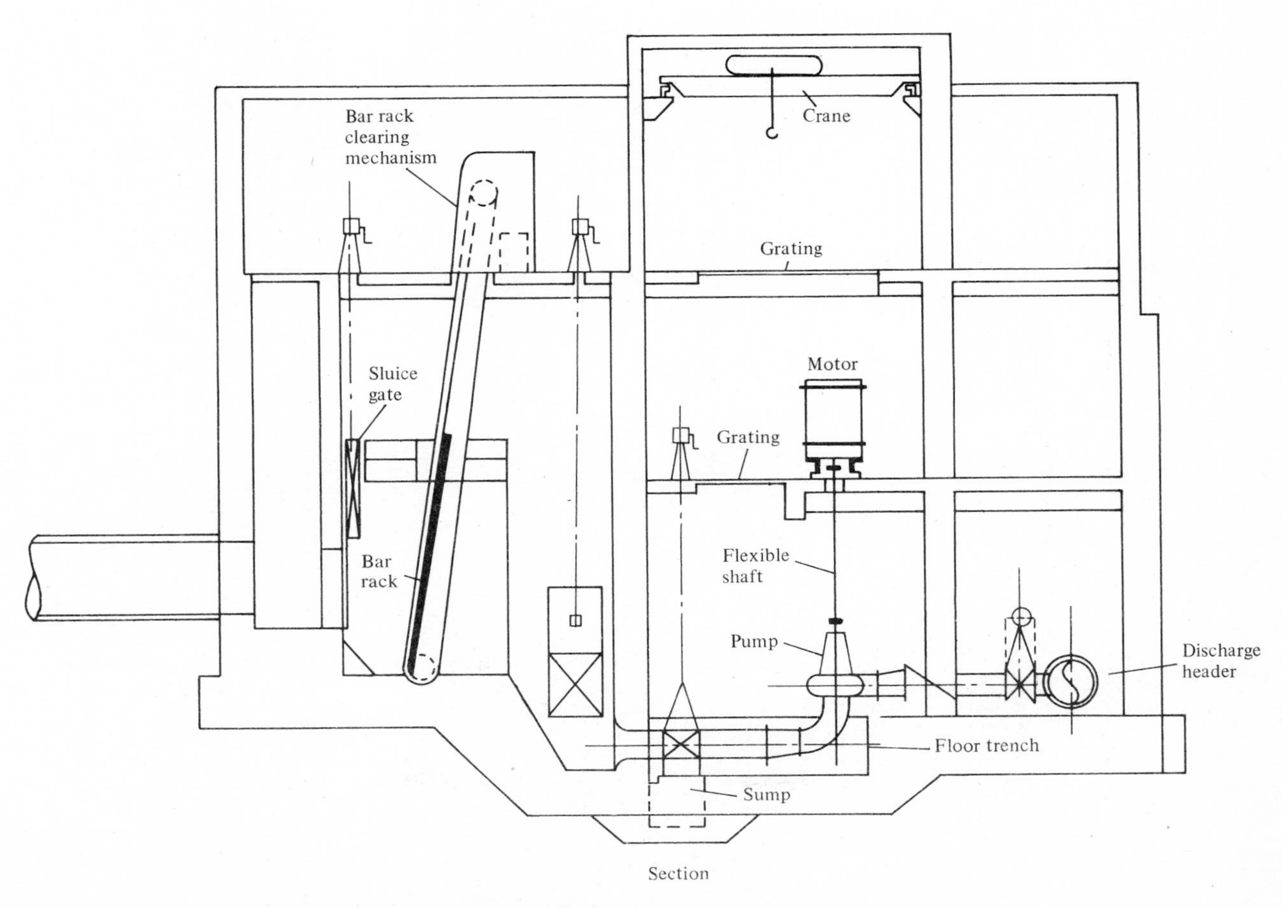

Figure 9-3 Typical large conventional wastewater pumping station.

less than 200 kW (250 hp), the time between starts should not be less than 20 to 30 min. The manufacturer should be contacted for allowable time between starts for motors larger than 200 kW. The allowable time between starts for motors smaller than 15 kW can be reduced to 10 min, but 15 min is recommended.

The time between starts is a function of the pumping rate and the quantity of flow entering the station. For a multiple-speed pump, the pumping rate is the difference in flow between the two steps.

The volume of the wet well between start and stop elevations for a single pump or a single-speed control step for multiple-speed operation is given by the following equation [9].

$$V = \frac{\theta q}{4} \tag{9-1}$$

where V = required capacity, m³ (gal)

θ = minimum time in minutes of one pumping cycle (time between successive starts or changes in speed of a pump operating over the control range)

q = pump capacity, m³/min (gal/min), or increment in pumping capacity where one pump is already operating and a second pump is started, or where pump speed is increased

This equation is derived in Example 9-1. The minimum cycle time for single-pump operation occurs when the inflow is exactly half the pump capacity. Under this condition, the on and off times are equal. The pump is on a longer time and off a shorter time for larger inflows and vice versa for smaller inflows; in both cases, the cycle time is greater.

Example 9-1 Derivation of formula for determining wet-well volume Derive the expression for determining the volume of the wet well required between start and stop elevations, as given in Eq. 9-1. Assume that pump operation is intermittent.

SOLUTION

1. Derive an expression for the time required for one complete cycle, using the definitions given for Eq. 9-1 and the following definition for i.

   i = rate of inflow for single-pump operation or, for multiple-step operation, rate of inflow in excess of previous step pumping rate. For example, one pump discharges 0.05 m³/s and two pumps discharge 0.08 m³/s. For an inflow of 0.07 m³/s, $i = 0.07 - 0.05 = 0.02$ m³/s

   a. The time required to fill the wet well when the pump is not operating:

   $$t_f = \frac{V}{i}$$

   b. The time to empty the wet well when the pump is operating:

   $$t_e = \frac{V}{q - i}$$

c. The total time for one complete pump cycle:

$$\theta = t_f + t_e = \frac{\forall}{i} + \frac{\forall}{q - i}$$

2. Determine the value of i that will make θ a minimum.
   a. Clear fractions in the expression derived for:

$$\theta(iq - i^2) = +(q - i) + \forall_i = \forall q$$

$$\frac{\forall}{\theta} = i - \frac{i^2}{q}$$

b. To find the value of i that will make θ a minimum and the term $\forall/\theta$ a maximum, differentiate the term $\forall/\theta$ with respect to i and set the resulting derivative equal to zero:

$$\frac{d(\forall/\theta)}{di} = 1 - \frac{2i}{q} = 0$$

or

$$q = 2i$$

c. To make sure the value of $\forall/\theta$ is a maximum, find the second derivative:

$$\frac{d^2(\forall/\theta)}{di^2} = -\frac{2}{q}$$

Since the second derivative is negative, the term $\forall/\theta$ is a maximum. Therefore, when $q = 2i$, $\forall/\theta$ is a maximum, or for any preselected value of θ, the maximum required wet-well volume occurs when $i = q/2$.

3. Solve for the wet-well volume. Substitute $i = q/2$ in the expression for θ determined in step 1.

$$\theta = \frac{\forall}{q/2} + \frac{\forall}{q - q/2} \qquad \text{or} \qquad \frac{\theta q}{2} = 2\forall$$

and

$$\forall = \frac{\theta q}{4}$$

Comment As noted in step 1, the equation given in step 3 can be used to find the cycle time, or the difference between start and stop levels, for any number of constant-speed pumps or speed steps of multiple-speed pumps.

Wet-well modifications. If the volume computed requires an unusually large wet well in a small station containing two identical pumps, one of which is a standby, the wet-well volume can be reduced by half by installing an automatic alternator in the pump control circuit. An alternator, which starts and runs the pumps alternately, has the effect of reducing the value of θ by half for a single pump and motor.

Most state regulatory agencies now include maximum retention time in the wet-well design criteria to minimize the potential for the development of septic conditions and the resultant odors. A maximum retention time of 10 min at average design flowrates is often quoted. Unfortunately, this requirement may conflict with the need for adequate volume to prevent short-cycling of the pumps. In these cases, multiple pumps or multiple-speed pumps should be considered to reduce the incremental change in the pumping rate and, therefore, the required volume. Also, odors can be minimized if the lowest liquid level in the well is set above the sloping portion of the wet well. This can be accomplished by making this level the stop point for the lead pump in the sequence.

The more common problem is obtaining sufficient wet-well volume at a reasonable cost. In larger stations served by large sewers, considerable effective volume can be added by using the available storage in the incoming sewers. If the pump start elevation in the wet well is below the invert of the sewers, no storage is available. However, if the pump start elevation is above the invert, backwater curves can be computed to obtain the effective volume in the sewers between the various control settings. This storage often amounts to over 50 percent of the total volume. This system is most commonly used in stations that have mechanically cleaned screens.

When using the storage in the incoming sewers, care should be taken to ensure that adequate velocities are maintained in the sewers and through the screens. The use of sewer storage is not common in small stations having comminutors because the storage available in the smaller sewers is small and the comminutors could be flooded.

Wet-Well Appurtenances

All pumps, regardless of size, can clog on rags and other material commonly found in wastewater. The larger the pump, the larger the solid the pump can discharge, but all pumps can become plugged by rags. Rags tend to wrap around pump parts and, eventually, build up until the pump is clogged. To protect pumps from clogging, equipment is installed in the wet well of all but the smallest stations to screen or cut up rags and other material, commonly called *screenings*. The most frequently used devices are bar racks and comminutors.

Bar racks. A bar rack is a device composed of parallel bars that is used to remove, by straining, large objects from the wastewater as it passes through the bar rack. The spacing of the bars varies from about 25 to 150 mm (1 to 6 in), depending on the degree of protection required. Bar racks at pumping stations usually have mechanical cleaning devices, although hand-cleaned bar racks are often used in emergency bypass channels when the main screening device is out of service.

Known as screenings, the objects removed from the wastewater must be disposed of. The screenings can be removed from the station to some disposal point, or they can be ground up and returned to the wastewater flow.

Comminutors. A comminutor is a mechanical device that strains the wastewater as it passes through the unit, and then cuts up the screenings into pieces small enough so that they can pass through the strainer and the pump without clogging.

Selection of screening devices. In selecting screening devices, the following should be considered:

1. Rags that are comminuted tend to agglomerate when agitated into a rag ball that looks like an oily waste rag. This is not a problem at a remote pumping station where only pump protection is required. However, at a treatment plant, the rag ball could disrupt downstream treatment processes.
2. In bar racks, the smaller the opening between bars, the greater the quantity of screenings that will be removed. Therefore, the bar spacing should be small enough to protect the pump but as large as feasible to minimize the amount of screenings that must be removed. If the only criterion is to protect the pump, a suggested distance between bars is one-third the size of the maximum solid that the pump can discharge.
3. The smallest clear opening normally used is 100 mm (1 in). Problems can be expected with a bar spacing less than 75 mm (0.75 in) because putrescible wastewater solids will often be removed with the screenings, causing odor problems.

Dry Wells

Located adjacent to the wet well in conventional pumping stations, the dry well is used to house the pumps and the related suction and discharge piping and valves. A gutter should be located in the dry well along the wall separating the wet and dry wells to convey seepage, pump drainage, and floor-washdown water to a sump. The pump room floor should slope to this gutter, and the gutter should slope to the sump with a minimum pitch of 10 mm/m (0.125 in/ft). Typical examples of dry wells are shown in Figs. 9-1 and 9-3.

Stairs should be installed in all but the smallest stations. Stairs should be galvanized steel or aluminum of satisfactory rigidity. Ship ladders should only be installed when local building codes allow them and conventional stairs cannot fit into the station. Vertical ladders and circular stairs should not be used.

In deep pumping stations, an intermediate floor is often located between the ground floor and the bottom of the dry well (see Figs. 9-1 and 9-3). Motors used to drive the pumps are usually located on the intermediate floor. In shallow stations, motors are placed on the ground floor. In laying out the dry well, care should be taken to ensure that access openings or hatches are provided on the intermediate and ground floors to permit the removal of motors, pumps, piping, and other equipment. To make the maintenance and repair of pumps more convenient, adequate clearances must be provided around each pump.

Pump Settings

The pumps in conventional stations are normally vertical-shaft, single-suction units, installed in a dry well, with motors mounted on a floor above the pump and driving the pump through a vertical flexible shaft, as shown in Fig. 9-1. The pumps should be set so that the high point of the casing is below the minimum level of the wastewater in the wet well. This setting ensures that air cannot enter the pump through the packing when the pump is not operating and that the pump is full when started automatically.

Pumps should be lined up and equally spaced for an orderly arrangement, and there must be ample clear space for access and maintenance. A minimum clear space between pumps of 1 to 1.3 m (3 to 4 ft) is recommended for small pumps and space equal to the width of the volute for larger pumps.

Cleanouts are required on the pump volute and in the suction elbow. The inside surface of the cleanout cover plate should be contoured to the curve of the volute or suction elbow.

The high and low points of the volute should have vent and drain connections with not less than 75-mm (0.75 in)-diameter vent and drain valves. The casing drain and the drain from the stuffing box area should be piped to the floor drainage gutter.

Since most pump maintenance involves the rotating parts of the pump (impeller, shaft, shaft sleeve, and bearings), the pump should be accessible. Pumps are more accessible if the motor is on an upper floor and driving the pump through a flexible universal-joint shaft. With this arrangement, the pump can be disassembled by removing only the flexible shaft. This does not disturb the alignment between the pump and motor. In direct-connected pumps where the motor is mounted on a frame, supported by the pump, the motor must be removed before the pump can be disassembled.

Suction and Discharge Piping

The velocity of the wastewater at the pump suction and discharge nozzles ranges from 3 to 4.25 m/s (10 to 14 ft/s). If the velocity is more or less than this, a better pump could probably be selected. Pumps with higher discharge velocities are necessary for heads of 30 m (100 ft) or more. It is recommended that the suction piping be one or two sizes larger than the suction nozzle and that the discharge piping be at least one size larger than the discharge nozzle. Most wastewater pumps have suction and discharge nozzles of the same size, but occasionally the suction nozzle may be one size larger.

Suction piping. Desirable velocities in pump suction piping are 1.2 to 1.8 m/s (4 to 6 ft/s). An eccentric decreaser with the flat side uppermost is located ahead of the pump suction nozzle (see Fig. 9-4). Vertical pumps normally have a suction elbow. If the suction elbow is not supplied as part of the pump, a reducing elbow, preferably with a long radius, should be installed under the

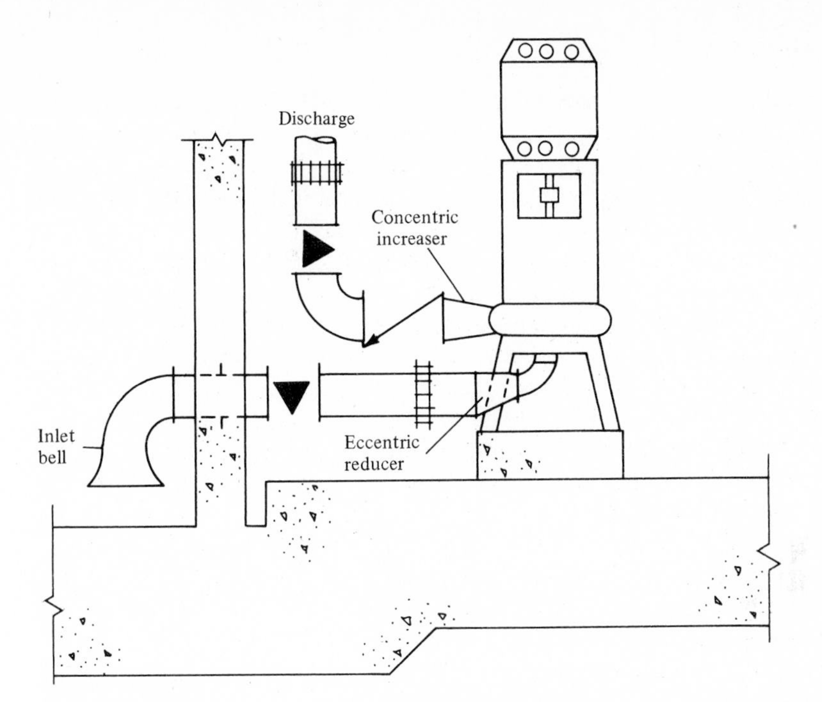

Figure 9-4 Typical wastewater pump with concentric increaser and eccentric reducer.

pump. A gate valve should be installed on the wet-well wall casting, and a flexible coupling should be installed between the pump and the suction gate valve. This arrangement permits the pump to be opened without flooding the pump room. The preferred gate valve is the solid-wedge, outside-screw-and-yoke type.

The end of the suction pipe in the wet well has either (1) a 90 or 45° flange and flare elbow, or (2) a 90 or 45° flanged elbow and a flange and flare fitting, as shown in Fig. 9-5. If D is the diameter of the flared inlet, the lip of the flare at mid-height should not be less than $\frac{1}{3}D$ nor more than $\frac{1}{2}D$ above the floor. An inlet flush with the wall is also sometimes used.

Adequate submergence must always be provided to prevent air from being drawn into the pump suction by a vortex when the system is operating at low wet-well levels. The submergence required above a flared inlet is a function of the inlet velocity, and the submergence required for various inlet velocities is presented in Table 9-3.

Discharge piping. Desirable velocities in the discharge pipe at maximum pump discharge range from 1.8 to 2.4 m/s (6 to 8 ft/s). A concentric increaser, shown in Fig. 9-4, should be provided on the pump discharge, followed by a check valve and a gate valve. Preferably, the gate valve should be the solid-wedge, outside-screw-and-yoke type. The check valve should be one of the following

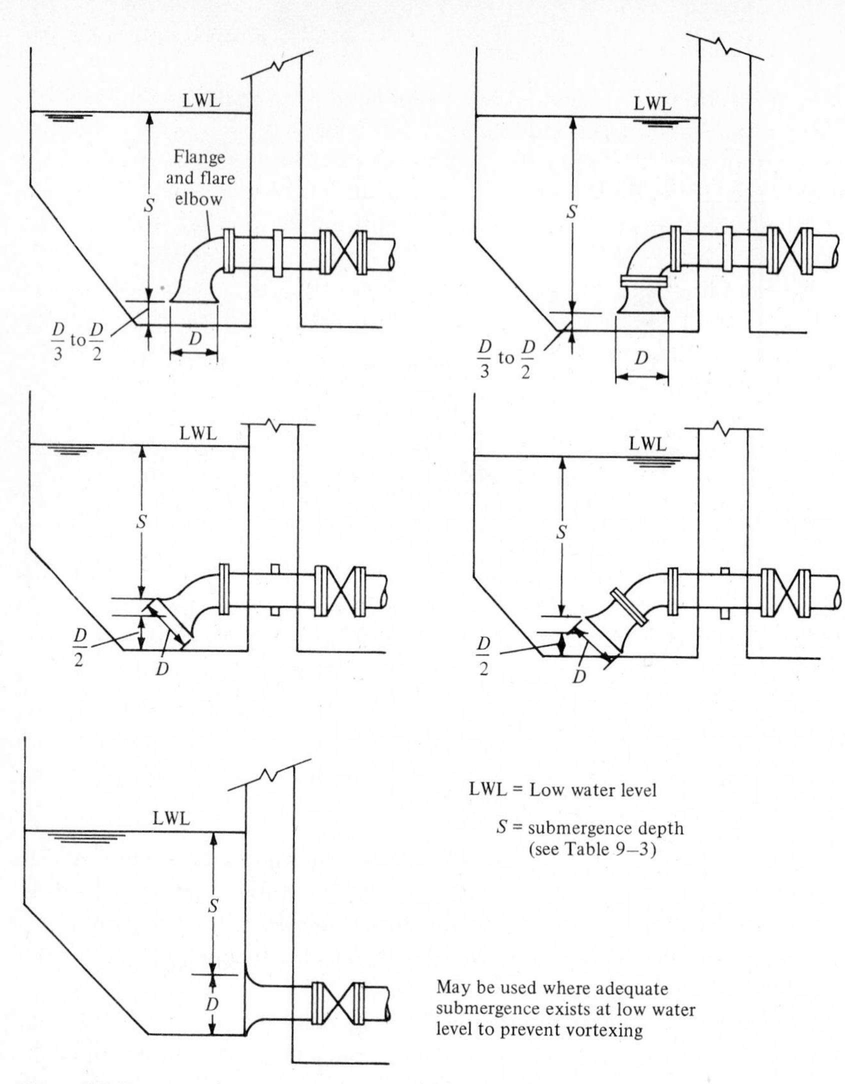

Figure 9-5 Pump suction connections to wet well.

Table 9-3 Submergence depth required to prevent vortexing in pump suction connection

Velocity at diameter D		Required submergence depth, S	
m/s	(ft/s)	m	(ft)
0.6	(2)	0.3	(1)
1.0	(3.3)	0.6	(2)
1.5	(5)	1.0	(3.4)
1.8	(6)	1.4	(4.5)
2.1	(7)	1.7	(5.7)
2.4	(8)	2.15	(7.1)
2.7	(9)	2.6	(8.5)

Note: m × 3.2808 = ft

types: swing check valve, cone valve, tilting disk check valve, or butterfly valve. These four valve types are described in Table 9-4.

The pump discharge should connect to the station discharge header on a horizontal plane. If the discharge is connected to the header vertically, solids being pumped could settle out in an idle pump discharge riser and clog the discharge.

In large lift stations at wastewater-treatment plants, where the elevation of the water surface in the discharge channel does not vary greatly, the pumps may be provided with individual, unvalved discharge pipes discharging over a siphon. The invert of the siphon would be 1 m or more above the maximum water surface in the discharge channel and the high point of the siphon would have an automatically controlled vent valve. When the pump is operating, the vent valve would close to establish the siphon. When the pump stops, the vent valve would open to break the siphon and prevent reverse flow through the pump.

An alternative arrangement would involve discharge over weirs, with the weir crest set above the maximum water level in the discharge conduit. Large low-head pumps sometimes discharge through a flap valve or backwater gate into individual chambers with provisions for stop logs (used to control the depth of the chamber). The arrangements also apply to large storm-water pumping stations.

Pumping Station Instrumentation

Pumping station instrumentation includes the automatic controls used to sequence the operation of the pumps, the manual controls used for the same purpose, and the alarms used to signal operational problems. All of these must be housed in a specially designed control panel whose features vary with the type of pumping station.

Automatic controls. Automatic control of pumps in wastewater pumping stations is almost always based on the liquid level in the wet well. The heart of any pump control system is the method for measuring the liquid level in the wet well. Common controls include floats, electrodes, bubble tubes, sonic meters, and capacitance tubes. Pressure diaphragm sensors have also been used. Each of these controls is described in Table 9-5.

Manual controls. In addition to automatic control, the pumps must be capable of being controlled manually for operation during emergencies, when the automatic controls are inoperative, and for maintenance. In addition to constant-speed operation, manual-speed control should be furnished for variable-speed control. The manual control should bypass the low-water cutoff but not the low-water alarm.

Alarms. Alarms should be included in the control system. In unattended stations, a common alarm should be transmitted by telemetering or other means to

Table 9-4 Types of check valves

Type of valve	Description	Installation
Swing check valve	Available in sizes up to 0.75 m (30 in). Used on all systems except those that require a different valve for waterhammer control or a size over 0.75 m. Should have an outside weight and level with the weight adjusted to assist closing.	Should be installed horizontally. If installed vertically, debris can settle out on top of the clapper when the pump is not operating. This debris would enter the bonnet of the valve when the pump starts to operate and would prevent the valve from opening fully.
Cone valve	Used on systems requiring valves larger than 0.75 m in size or if timed opening and closing are required to control waterhammer. Operated by a hydraulic cylinder. The use of cone valves for waterhammer control is discussed further in Secs. 9-4 and 9-5.	The valves and the hydraulic control system are expensive and are only installed where absolutely necessary. In the open position, the cone valve has a smooth pipe passageway; therefore, to reduce costs, the cone valve is normally the same size as the pump nozzle, or even smaller if space is available for both a decreaser and an increaser.
Tilting-disk check valve	Available in sizes up to 1.8 m (72 in). Used only if the right size of swing check valve is not available. Has a shaft that passes through the flow passage of the body.	Rags and other debris can wrap around the shaft and prevent the valve from operating properly. Should be installed only when a large size is necessary and when the screenings are removed by mechanically cleaned screens with openings no larger than 25 mm (1 in) or have been comminuted.
Butterfly valve	Like the tilting-disk valve, operated by a hydraulic cylinder and has a shaft passing through the flow passage of the body. Used only in large sizes for waterhammer control.	Should be installed only when the screenings have been removed or comminuted.

Note: mm $\times$ 0.03937 = in.

Table 9-5 Devices for measuring the liquid level in wet wells

Type of measuring device	Description	Installation
Float	Used in simple systems that require only on-off control. One type of float is a mercury switch which is inside a weighted float and suspended above the wet well. As the liquid level rises to the float level, the float tips on its side, closing (or opening) the mercury switch.	The float switches available can operate many different control actions. Examples include (1) a mercury-type float which is often installed in factory-assembled pumping stations or to control sump pumps, and (2) a float, located in a pipe or cage, which can be attached to a spring-loaded drum by a cable or tape; this device indicates the wet-well level, but is seldom used because the float action can become sluggish as grease builds up in the float tube.
Electrode	A control system using electrodes consists of a series of electrode probes mounted at different elevations in the wet well. When the liquid level rises to the electrode, an electric circuit is energized.	Often placed in pneumatic-ejector stations where the compressed air used during the ejector cycle "blows off" the electrode, but electrodes are seldom installed in pumping stations. Grease and other material in wastewater can coat the electrodes in wet wells and the electrodes require frequent cleaning.
Bubble tube	Probably the most commonly used device for measuring the wet-well level. A small quantity of compressed air is fed into an open-ended pipe that is submerged in the wet well. The back pressure in the pipe depends on the depth of the liquid over the open end of the pipe. This pressure is used to indicate the liquid depth and to control pumps by pressure switches.	Because the differential in back pressure in the bubble tube is small, a level transmitter is often used to magnify the differential pressure between the low and high levels in the wet well. This system has low maintenance requirements; the compressed air keeps the bubble tube free of dirt and grease. If the tube does become clogged, it can be cleaned out by isolating the tube from the control elements and blowing the tube out with high-pressure air.
Sonic meter	Used to measure the distance from the meter to the liquid surface in the wet well.	The location of the meter is important because the meter beam takes the shape of a cone with about 10° angle at the meter. Obstructions within this cone, such as walls, etc., can give false readings. The meter should also be isolated from stray electrical or acoustical signals.
Capacitance tube	A tube suspended in a wet well; its capacitance as measured by an electronic circuit, is proportional to the length of the tube submerged in the liquid. The output is converted to a signal that can be used for wet-well level indication and control.	Foreign matter can build up on the sensing tube. Some tubes automatically compensate for grease buildup, but are more expensive. To date, capacitance tubes are not in common use.

Table 9-5 Devices for measuring the liquid level in wet wells

Type of measuring device	Description	Installation
Pressure diaphragm sensors	Pressure diaphragm box-type sensors operate on a simple principle: the diaphragm box is fixed at a location that becomes the reference point for the measurement. As the liquid rises above the diaphragm, the pressure on the diaphragm compresses the air trapped in a closed tubing system connected to a pressure (sensing) element, thereby providing a continuous level detecting system.	Grease can build up on the diaphragm. The diaphragm must be taken out to remove the grease and then replaced at the (submerged) reference point.

a location that will always be attended. Alarms should include the following:

1. High wet-well liquid level
2. Low wet-well liquid level
3. Pump-failure alarm (for variable speed or pumping station with complex controls)

The high-level alarm should be set above the start point of the last pump in the normal sequence, but before the start of the standby pump. This setting will indicate a failure of one of the normally operating pumps while the standby pump is still available.

The low-level alarm is set below the shutoff point of the lead pump to indicate a malfunction in the pump control system. An emergency low-level cutoff is normally furnished that shuts the pumps off before they pump the wet well dry. The low-level cutoff is located below the low-level alarm.

Pump-failure alarms are normally furnished for variable-speed pumps to indicate a malfunction in the speed controls. The origin of the signal depends on the type of variable-speed pump installed.

Control panel design. In larger stations, a control panel or a control cabinet should be furnished to centralize the control instrumentation. The following may be included on the face of the panel:

1. Wet-well level indication.
2. System flow indication if flow measurement is provided in the station.
3. Sequence selection switches to select lead, first follow, etc., and standby pumps.
4. Pump operating controls for each pump including:
   (*a*) Hand-off-automatic switch
   (*b*) Pump on-off indicating lights
   (*c*) Variable-speed controls if furnished
      (I) Manual-automatic speed control
      (II) Speed indicator
   (*d*) Pump ammeter or kilowatt meter
5. Alarm annuciator panel to indicate individual alarms and test and alarm-horn silence button.

The number of these or other items on the control panel depends on the complexity of the pump system and the individual requirements of each system.

Flow measurement. Flow measurement is sometimes included in the station instrumentation. The simplest measuring system for constant-speed pumps is the use of running-time meters in the pump motor starters. Since constant-speed pumps discharge at a fairly constant rate when they operate, an approximation of the total volume pumped can be computed if the operating time is known.

In the larger stations, flow measurement is provided by Venturi meters or flow tubes. Venturi meters are rarely used today because space is not available to provide the necessary straight run of pipe upstream of the meter. Flow tubes are most common because flow tubes are now available that require a minimum straight section upstream of the meter.

Selection of Pump Control Points

The selection of pump control points for constant- and multiple-speed pumps and variable-speed pumps is discussed in this section. A control range of at least 1 m (3.3 ft) is desirable between maximum and minimum levels in the wet well. A minimum of 75 mm (3 in) and preferably 150 mm (6 in) should be allowed between control functions (starts and stops of successive pumps). Such a minimum setting is used to allow for hystereses in the control system and to ensure repeatability in the components of the system.

Constant- or multiple-speed pumps. Constant- or multiple-speed pumps are the simplest to control and require only on-off switches to change from one step to the next. An example of how to determine the control points for the on-off switches is provided in Example 9-2.

Example 9-2 Determination of pump controls for multiple-speed pumps Set up a control sequence for alternative B of Example 8-6. The wet-well configuration is similar to that shown in Fig. 9-1, and the elevations of the wet well are shown in Fig. 9-6. Assume that a bubble-tube level-control system is used and that a 0.15-m separation should be maintained between each control function.

SOLUTION

1. Review Example 8-6. In Example 8-6, alternative B involves a pumping system with two operating two-speed pumps plus a standby two-speed pump. The active control level in the wet well is 1 m (3.3 ft). The three stages of operation are as follows:

   Step I —lead pump, low speed
   Step II —lag pump, low speed
   Step III—both pumps high speed

2. Establish the low wet-well level. The top of the pump volute is at elevation 51.4 m and the low wet-well level must be above this. Set the low wet-well level at 51.5 m. The lead pump in step I stops at this elevation.
3. Establish the high wet-well level. The active control band is 1.0. Therefore, the high wet-well level is 52.5 m. Both pumps start at this elevation in step III.
4. Establish the pump start elevations. Because both pumps start at a high wet-well level elevation of 52.5 m, and a 0.15-m separation should be maintained between control functions, the following start elevations are established:

Start lead and first follow pumps (high speed)	52.50 m
Separation	− 0.15 m
Start lead and first follow pumps (low speed)	52.35 m
Separation	− 0.15 m
Start lead pump (low speed)	52.20 m

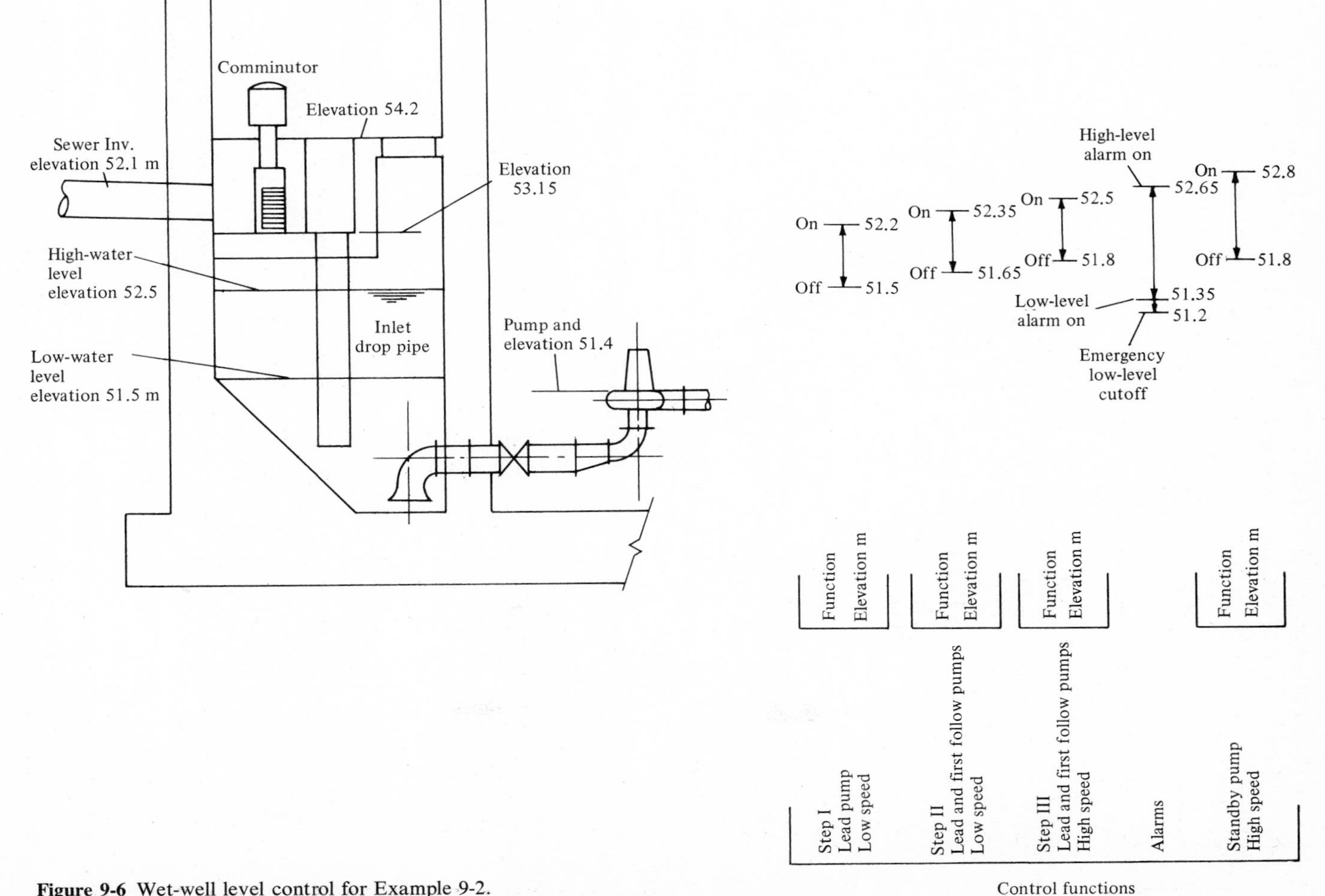

Figure 9-6 Wet-well level control for Example 9-2.

5. Establish the pump stop elevations.

Stop lead pump	51.50 m
Separation	0.15 m
Stop lead pump and first follow pumps (low speed)	51.65 m
Separation	0.15 m
Stop lead pump and first follow pumps (high speed)	51.80 m

6. Establish the alarm and emergency low-level pump cutoff elevations.
   a. The high-level alarm is set above the normal high level and below the start of the standby pump.
   b. The low-level alarm is set below the normal low-water level but above the emergency low-water cutoff.
   c. The emergency low-level cutoff is set to protect the pumps and wet-well equipment. In this example, the level should be set to prevent lowering the wet-well level below the invert of the bar racks.
   d. The alarms and emergency low-level cutoff are set as follows:

High-level-alarm elevation	52.65 m (52.50 m + 0.15 m)
Low-level-alarm elevation	51.35 m (51.50 m − 0.15 m)
Low-level-cutoff elevation	51.2 m (51.35 m − 0.15m)

7. Establish the operating range of the standby pump.
   a. The standby pump only operates at high speed. Therefore, it stops at the same elevation as in step III.
   b. The start elevation of the standby pump is set at 52.80 m (52.65 m + 0.15 m). This elevation is above the high-level alarm so that the operator is alerted when one of the pumps fails.

Variable-speed pumps. Variable-speed pumps require more complex controls than constant- or multiple-speed pumps. The two basic types of variable-speed control are variable-level control and constant-level control. Variable-level control is the simplest. A band is set up in the wet well to produce a signal and control the speed of any variable-speed pump that is operating. The pumps would operate at maximum speed at the high wet-well level and at a selected minimum speed at the low wet-well level. The pumps are started and stopped, independently of the speed control band, by the level switches. An example of variable-level speed control is presented in Example 9-3.

Constant-level speed control is occasionally used when a system requires a narrow control band. In a constant-level system, a level in the wet well is selected and as the liquid level rises above or drops below this setting, a speed-control signal is produced to increase or decrease the speed of the pump. The more the liquid level deviates from the set point, the stronger the signal to change speed. When a pump is operating at full speed and the level continues to rise, a second pump can be started. As the level begins to fall, the second pump can be stopped when the pump is operating at its minimum speed.

Example 9-3 Determination of pump controls for variable- and constant-speed pumps Set up a control system for the pump system in Example 8-7. The wet-well configuration is similar to the wet well shown in Fig. 9-3, and the critical elevations are shown in Fig. 9-7. Assume that a bubble-tube level-control system is used and that a 0.15-m separation should be maintained between each control function.

The storage capacity available in the sewers should be used. At design peak flow, the inlet sewers are to operate at 80 percent of depth. The head loss allowance from the inlet channel through the sluice gate and partially clogged bar rack is 0.2 m (8 in). Maintaining velocities in the inlet sewer and through the screens is not a part of this problem. Assume that the velocity through the bar rack will be controlled by the proper operation of the sluice gates to maintain velocities within allowable limits.

SOLUTION

1. Review Example 8-7. In Example 8-7, the pump system involves three variable-speed pumps and one constant-speed pump that are sized to deliver a design peak flow of 2.5 m^3/s. One of the variable-speed pumps is used for standby service. The three stages of operation are as follows:

   Stage I. One variable-speed pump; stage capacity 0.4 to 0.96 m^3/s; speed range 600 to 700 rev/min.

   Stage II. Two variable-speed pumps; stage capacity 0.88 to 1.88 m^3/s; speed range 610 to 705 rev/min.

   Stage III. Two variable-speed pumps plus one constant-speed pump; stage capacity 1.78 to 2.5 m^3/s; speed range 630 to 705 rev/min.

   The active control band in the wet well is 2 m and the variable-speed pumps range in speed from a low of 600 rev/min to a high of 705 rev/min.
2. Establish the high wet-well level.
   a. Because the high-water level in the inlet sewer is to be at 80 percent depth, the high wet-well level is at elevation 53 m.

Invert elevation of 1.5-m sewer	52.0 m
Depth of flow in sewer (1.5 m × 0.8)	1.2 m
Sewer water-surface elevation	53.2 m
Head loss through gates and bar rack	− 0.2 m
	53.0 m

   b. The second follow pump starts at 53.0 m.
3. Establish the low wet-well level. The active wet-well control band is 2 m. Therefore, the low wet-well level is 51.0 m (53.0 m − 2.0 m).
4. Check the submergence of the pump volute. The centerline of the 500-mm pump is at 50.0 m. Therefore, the top of the volute is 50.0 m + 0.5 m/2 = 50.25 m. This is below the low-water level and is acceptable.
5. Check the submergence of the suction bell. The diameter of the suction bell is about 1 m and the area of the suction is

$$A = \frac{\pi}{4}(1)^2 = 0.78 \text{ m}^2$$

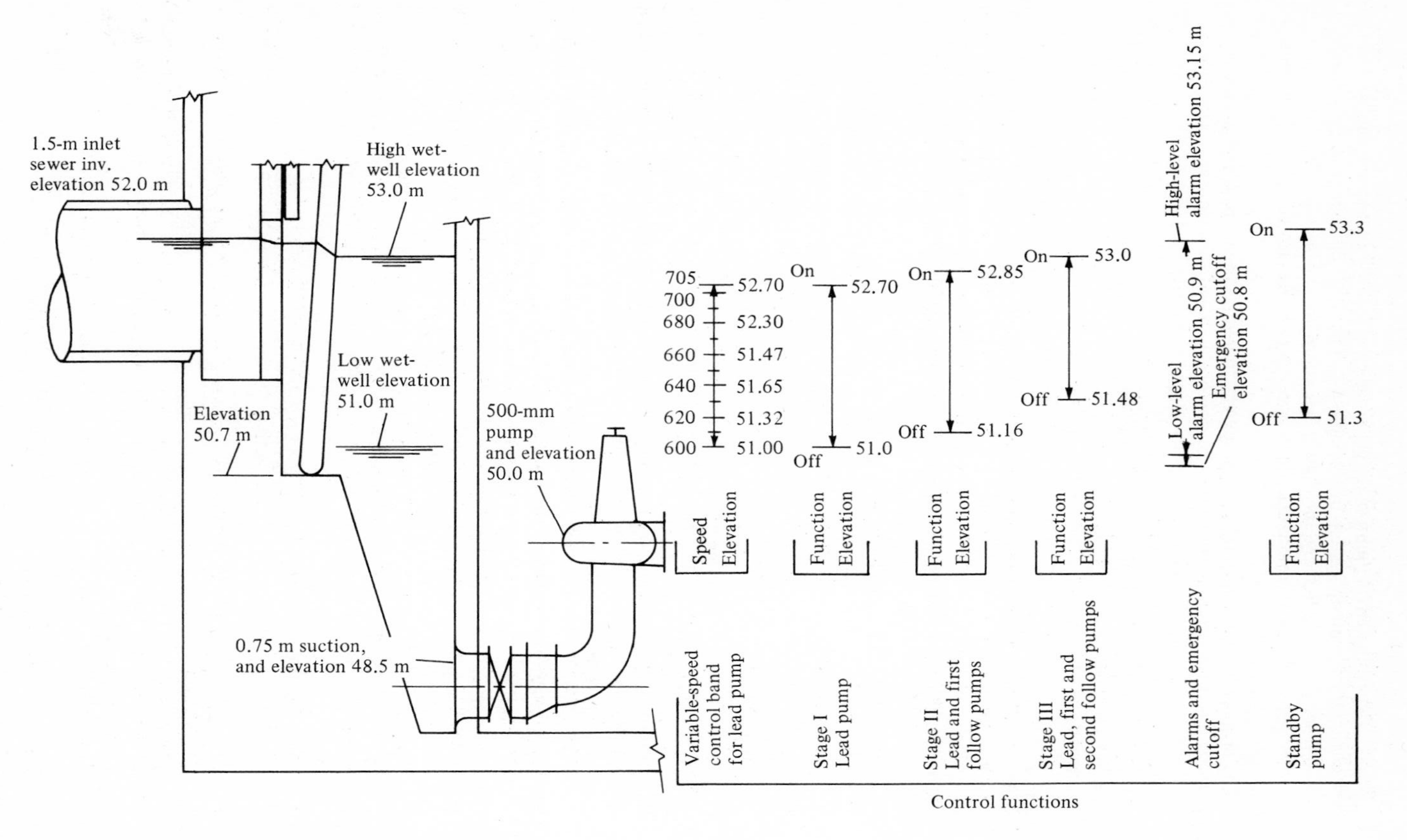

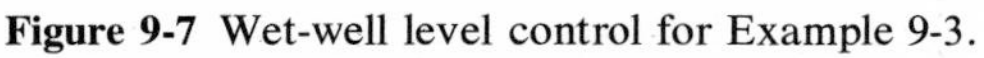

Figure 9-7 Wet-well level control for Example 9-3.

The maximum velocity occurs when the lead pump operates by itself at high speed. From Example 8-7, the flow for this condition is 0.96 m^3/s. Therefore,

$$\text{Velocity} = \frac{Q}{A} = \frac{0.96 \text{ m}^3/\text{s}}{0.78 \text{ m}^3/\text{s}} = 1.23 \text{ m/s}$$

From Table 9-3, the required submergence for a velocity of 1.23 m/s is about 0.8 m. The actual submergence is

Low-water level in wet well	51.0 m
Top of suction bell (48.5 + 0.5)	−49.0 m
Submergence (difference)	2.0 m

Because 2.0 m > 0.8 m, the submergence is adequate.

6. Establish the pump start elevations. The second follow pump starts at the high wet-well elevation of 53.0 m, and a 0.15-m separation should be maintained between control functions. Therefore, the following start elevations are established:

Start second follow pump elevation	53.00 m
Separation	− 0.15 m
Start first follow pump	52.85 m
Separation	− 0.15 m
Start lead pump	52.70 m

7. Establish the speed control band. The lead pump operates at high speed at the start elevation and operates at minimum speed at the low wet-well level. Intermediate speeds are proportional to the wet-well level above the minimum speed. The variable-speed control band is shown in Fig. 9-7.
8. Establish the pump stop elevations. In stage I, the lead pump must operate at a minimum speed of 600 rev/min to discharge the minimum flow of 0.4 m^3/s. Therefore, the lead pump stops at elevation 51.0 m (53.0 m − 2.0 m), at which point the speed is 600 rev/min.

   In stage II, the lead and first follow pumps must operate at a minimum speed of 610 rev/min to discharge the stage minimum flow of 0.88 m^3/s. Therefore, the first follow pump stops at elevation 51.16 m [51.0 m + 51.32 m)/2], at which point the speed is 610 rev/min. However, the lead pump continues to operate.

   In stage III, the lead and first follow pumps must operate at a minimum speed of 630 rev/min to discharge the stage minimum flow of 1.78 m^3/s. Therefore, the second follow pump stops at elevation 51.48 m [(51.32 m − 51.65)/2], at which point the speed is 630 rev/min.
9. Establish the alarm and emergency low-level pump cutoff elevations.
   a. The high-level alarm is set above the normal high level and below the start of the standby pump.
   b. The low-level alarm is set below the normal low-water level, but above the emergency low-water cutoff.
   c. The emergency low-water cutoff is set to protect the pumps and wet-well equipment. In this example, the level is set to prevent lowering the wet-well level below the invert of the bar racks.
   d. The alarms and emergency low-water pump cutoff are set as follows:

High-level-alarm elevation	53.15 m (53.0 m + 0.15 m)
Low-level-alarm elevation	50.90 m
Low-level-cutoff elevation	50.80 m

10. Establish the operating range of the standby pump.
   a. The start elevation of the standby pump is set at elevation 53.3 m (53.15 m + 0.15 m). This is above the high-level alarm so that the operator is alerted when one of the pumps fails.
   b. The stop elevation is set to prevent short-cycling of a pump. The worst condition occurs if one of the variable-speed pumps fails. The maximum flow with one variable-speed pump at 705 rev/min is 0.96 m^3/s. The minimum flow with one variable-speed pump and the constant-speed pump is about 1.3 m^3/s (0.44 m^3/s from the variable-speed pump and 0.92 m^3/s from the constant-speed pump). Therefore, the standby pump operates in place of the constant-speed pump to allow continuous discharge of flows between 0.96 and 1.3 m^3/s. This is accomplished by setting the stop elevation of the standby pump below the stop elevation of the constant-speed pump. In this case, the standby pump stops at elevation 51.3, at which point the speed is about 620 rev/min.

Electrical Equipment

Small-station pumps and equipment will operate at 460 V, three phases, and 60 Hz. Pumps in large stations may operate at 2300 or 4160 V with auxiliary equipment on 460 V. Standby power is now required at all stations.

Critical stations should be supplied by two separate feeders from separate substations of the power company. If one fails, the other can supply power automatically. If two separate feeders are not available, standby power can be supplied by one or more engine-generator sets sized to start and run sufficient pumps to avoid flooding of streets and basements and to prevent overflows. In isolated cases, dual-drive units may be installed so that the pumping equipment may be either motor- or engine-driven. In unattended stations, the transfer of power sources and operation of engines must be completely automatic. Power transformers would normally be installed in an outdoor fenced enclosure or on poles.

The motor starters and control should be located in a factory-assembled, freestanding control center located at ground-floor level in a clean, dry area, as shown in Fig. 9-1. This type of construction is neater, safer, and more satisfactory than the wall-mounted assemblies of individual starters and circuit breakers used in the past. Large stations should have a separate electrical room containing the motor starters, switch gear, meters and instruments, and a control bench board with adequate heating and ventilation.

All electrical equipment and lights in the wet well must be of explosion-proof construction because of the possible danger of explosion of vapors and gases in the incoming wastewater. Adequate lighting and a convenient number of explosion-proof receptacles for power tools should be provided.

Heating and Ventilation

Heating and ventilating systems are now common features of most large pumping stations.

Heating systems. All pumping stations, except those in warm climates, should have automatically controlled heating systems to prevent freezing in cold weather. Comfortable temperatures should be maintained in the dry well of attended stations and lower temperatures may be maintained in unattended stations.

During some temperature conditions, condensation on the walls and fog in the wet well may be severe and, in the coldest climates, icing on the upper-floor level and on the door may occur. In cold climates, thermostatically controlled heaters of explosion-proof construction are recommended in the wet well.

The type of heating system used depends on the size of the station and the heating requirements. In larger stations, gas- or oil-fired hot-water or steam boilers can supply the heat to the unit heaters, radiators, or heating and ventilating units. In smaller stations, dry wells may be heated by gas- or oil-fired warm-air furnaces with a separate heating system provided for the wet well. Electric heat is often used in the smaller stations.

Ventilating systems. Ventilating systems in the wet- and dry-well sides of the station must be entirely separated, and all openings for pipes or electric conduits must be caulked gas tight. Heating and ventilating equipment in the wet well must be of explosion-proof construction.

Ample fresh-air ventilation is essential. Wet wells should have a mechanical supply of fresh air distributed throughout the wet-well operating area with a gravity exhaust system discharging through the roof. The system should maintain a positive pressure in the wet well to minimize the quantity of air brought into the station through the sewer. The wet-well exhaust intake should be close to the inlet sewer.

Intermittent ventilation is usually provided at small unattended stations. Continuous ventilation is provided at large attended stations or where there are mechanically cleaned screens or other expensive equipment that could be damaged by the moist atmosphere found in most wet wells with intermittent ventilation. It is recommended, and most regulatory agencies now require, that the ventilation rate for the wet well, based on the station volume below grade and above the minimum wet-well level, be (1) a minimum of 30 air changes per hour if the fan does not operate continuously, and (2) 12 air changes per hour if the fan operates continuously. The 30 air changes per hour for intermittent operation is required for safety. This ventilation rate quickly flushes out or dilutes the contaminated air and gases that have accumulated in a station.

If the station has intermittent ventilation, it may be advisable to equip the supply fan with a two-speed motor to operate continuously on low speed and on high speed only when the station is inspected. In developed areas, such operation may dilute the odors enough to prevent the discharge of objectionable odors to the atmosphere when the high-speed fan is used. The fan switch should be just inside the entrance door. As an extra precaution, the fan can be interlocked with the light switch.

The dry well should be positively ventilated with either supply or exhaust fans or, for larger systems, with both supply and exhaust fans. It is recommended, and most regulatory agencies require, that the ventilation rate, based on the volume of the dry well below grade, be (1) a minimum of 15 air changes per hour if the fan operates intermittently, and (2) 6 air changes per hour if the fan operates continuously. Attended stations are usually ventilated continuously.

Additional ventilation capacity may be required to remove the heat generated by the pump motors and other electrical equipment. This is especially true of variable-speed drives. In some cases, depending on the type of drive, it may be possible to capture this heat and use it for space heating. In all cases, the ventilation system should be automatically controlled to provide the necessary cooling whenever space temperatures rise above acceptable levels.

Plumbing

A supply of potable water is recommended in pumping stations. Potable water is used for washing and to provide seal water for pumps furnished with water-sealed packing glands. It is recommended that, as a minimum, a service sink be furnished in the station with a small hot-water heater so that the operator may wash after working on equipment. Hose gates should be in both the wet and dry wells for washdown of the floors, and nonfreeze wall hydrants should be provided for watering of grass and shrubs.

Where building codes allow and water pressure is adequate, seal water and wet-well and pump-room hose connections can be supplied from the potable-water system through backflow-prevention devices. These consist of two pressure-reducing valves in series, with a pressure-relief valve between them. If the intermediate pressure approaches the potable-water pressure, the relief valve opens to prevent contaminated water from entering the potable-water system.

If backflow-prevention devices cannot be used, seal water may be furnished by a seal-water system that consists of a tank and two seal-water pumps. The potable water enters the tank through an air break to prevent the possibility of contamination and the level in the tank is controlled by a float valve. The suction intake for seal-water pumps is from this tank. The pumps are often small centrifugal pumps or turbine-type pumps, depending on the flow and pressure required. Two pumps are recommended to provide standby reliability.

Toilets are required in attended stations and may be desirable in remote unattended stations. Toilets and the service sink are on the ground floor so that they may drain directly to the wet well.

Pumping Station Drainage

Floor drainage is discharged from the sump in the dry well to the wet well by automatically controlled sump pumps. Duplex units should be installed to provide reliability, and the controls should include a high-level alarm between the lead and standby start setting to warn of failure of the lead pump.

It is strongly recommended that the sump pumps be the nonclog type and not the drainage type. Large balls of rags that are removed from pumps frequently enter the drainage pump. The discharge of the sump pump should be fitted with two check valves in series and should discharge to the wet well above the highest expected wet-well liquid level. This is done to minimize the possibility of flooding the dry well by reverse flow from the wet well.

9-3 DESIGN OF FACTORY-ASSEMBLED PUMPING STATIONS

Factory-assembled pumping stations are common in collection systems with low flows and where the need to protect the pump from clogging is minimal. Design details for three types of factory-assembled stations—pneumatic ejector, wet pit, and dry pit—are discussed in this section.

Major Physical Features

Factory-assembled stations are shipped from the factory in modules with all the equipment and components mounted and connected permanently within the module. When the station arrives in the field, the modules are interconnected, all external piping and power connections are made, and the station is ready for operation. Because a manhole normally serves as the wet well in factory-assembled stations, the use of screens or comminutors is not feasible.

Most factory-assembled stations are small enough so that the station consists of only two modules: (1) the station module that houses the pumps, controls, and other auxiliary equipment, and (2) the entrance tube that projects above the ground level. To prevent flooding, the top of the entrance tube is located above the maximum flood level of the surrounding area. Large factory-assembled stations may consist of three or more modules. In some of the largest stations, one pump, the motor, and the piping are mounted in individual modules; the controls are mounted in another module; and two entrance tubes are installed.

The entrance tube is normally a minimum of 1 m (3 ft) in diameter, but it must be large enough to allow the largest item of equipment in the station to be removed. The entrance tube also provides the ventilation needed in the station.

The electric control panel is in the pump module, but the electric meter and disconnect switch are on a power pole adjacent to the station. The station receives its power through an above-ground connection on the entrance tube.

Materials of Construction

Factory-assembled pumping stations are most commonly fabricated from steel plates reinforced with steel beams as required. However, they are also available in concrete or fiberglass shells. Both concrete and fiberglass shells resist corrosion more and are recommended where soil conditions are corrosive. In steel stations, sacrificial magnesium anodes are electrically bonded to the shell and buried outside the station to provide cathodic protection for the steel shell. The three types of factory-assembled stations available are described in the following.

Pneumatic-Ejector Station

Pneumatic-ejector stations are used to pump low flows from remote areas. The ejector capacity available usually ranges from 0.005 to 0.0125 m^3/s (75 to 200 gal/min); sizes over 0.02 m^3/s (300 gal/min) are uncommon. A typical pneumatic-ejector station that might be constructed in a remote area is shown in Fig. 9-8.

Station appurtenances. As shown in Fig. 9-8, the duplex station has a common suction line that serves two wastewater receivers. The air system consists of two compressors and an air receiver. Stations are available with individual inlets from the manhole to the wastewater receivers.

The station wet well illustrated in Fig. 9-8 is a conventional concrete manhole and the inlet piping is shown pitching down to the station. The station inlet must be below the bottom of the manhole so that the head is sufficient for the flow to open completely the inlet check valve. The inlet pipe should be the same size from the manhole to the receiver. If a larger pipe were used with a buried pipe reducer, the inlet line could become clogged. Unlike a centrifugal pump, which can put additional pressure on a partially plugged suction line, an ejector must rely on the gravity head to clear a partially clogged line.

Because the most common place where plugging occurs is in the inlet connection in the manhole, a tee is sometimes installed in the inlet line and a second connection is made to the manhole at a higher elevation. If the lower inlet becomes clogged, the upper inlet would still enable the station to operate. This method of connection to the manhole is shown in Fig. 9-8 and is recommended where a single pipe connects the manhole and the station.

Air compressors. Air compressors in ejector stations may discharge the air directly into the wastewater receiver, or the air can be stored in an air receiver until the wastewater must be discharged. The direct-connected compressor requires a larger compressor and motor because it must deliver the full volume of air to fill the receiver during the discharge cycle, whereas the storage system can discharge this quantity over the total fill and discharge cycle. The stored system can also reduce the frequency of starting the compressor motor because air can be stored at a pressure higher than is required to discharge the wastewater.

Both piston-type and rotary sliding-vane air compressors are used in ejector stations. The piston type is preferred, however, because the rotary com-

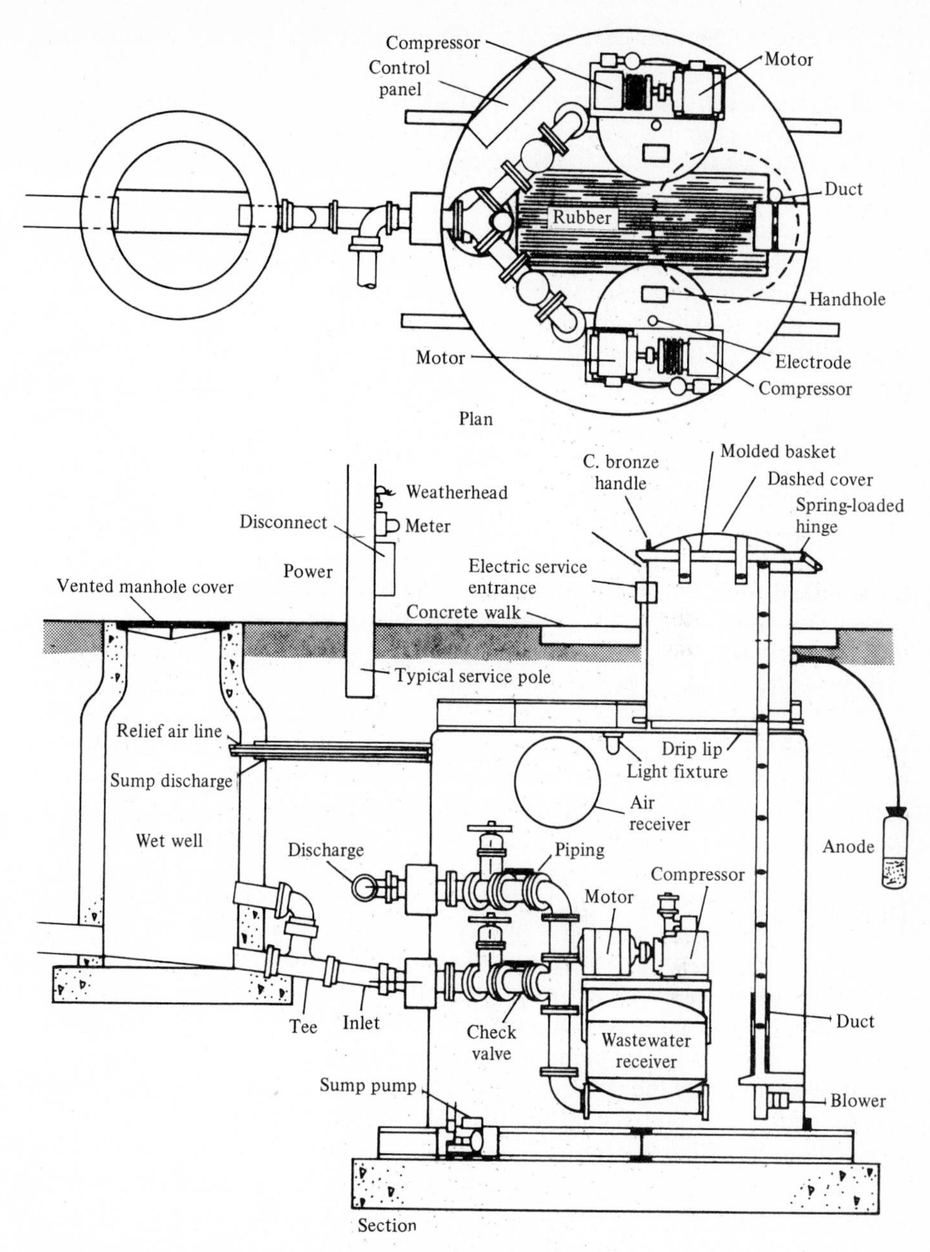

Figure 9-8 Typical duplex pneumatic-ejector pumping station.

pressor has only about a 150-kN/m² (20 lb_f/in^2) discharge pressure and needs to be continuously lubricated with oil.

System controls. Float switches or electrodes are used to control the operation of pneumatic-ejector stations. Both control elements must be readily accessible for cleaning. The controls sense the high-wastewater level in the receiver and start the discharge cycle. The discharge cycle may be ended either by (1) the controls sensing the low-wastewater level in the receiver, or (2) a timer switch that stops the air supply and vents the receiver after the time required to empty the receiver.

Pneumatic-ejector stations can overheat. Unlike a pump system where the power input is transferred to the pump fluid, the heat of compression in a pneumatic ejector is released within the station from the compressor and the air receiver, if used. Therefore, the ventilation fan should be controlled thermostatically to prevent overheating, or the vent fan should be interconnected with the air compressor.

Wet-Pit Station

Nonclog submersible wet-pit pumps have been on the market for many years, but their use has been restricted for municipal service because of maintenance problems. However, the pumps have now been improved so that the pump can be removed from the wet well for servicing without disturbing the discharge piping. The pump slides along and is positioned by fixed guide rails. In the lowered position, the pump discharge engages the discharge pipe. A typical installation is shown in Fig. 9-9.

This type of station is available as a steel factory-assembled station, but the pumps and hardware are normally installed in a conventional masonry manhole. The valves may be installed in the pumping station manhole. However, valve maintenance is simplified if the valves are placed in a separate manhole, as shown in Fig. 9-9.

Dry-Pit Station

A typical dry-pit pumping station with two pumps was illustrated in Fig. 9-2. The wet well is an oversized manhole with a sloped bottom. The low-water elevation is set at an elevation so that no air can enter the suction pipe by vortexing. The top of the pump volute is set below the low-water level of the wet well to eliminate the possibility of air binding the pump. Adequate wet-well storage must be provided between the high and low wet-well levels to prevent frequent starts of the pumps.

Each pump has its individual suction pipe to the wet well. Isolation valves are located both between the pump and the wet well and downstream of the discharge check valve so that the pump may be taken out of service for maintenance. Check valves are usually the spring-loaded type because the station has limited space.

A variation of this design involves the use of self-priming nonclog pumps in place of the volute-type centrifugal pump. This design variation, now approved by several regulatory agencies, can reduce the depth to which the station must be buried. Self-priming pumps have mechanical seals to minimize the leakage of air into the pump when it is not operating, and provisions must be made to vent the air evacuated from the suction pipe during the priming cycle.

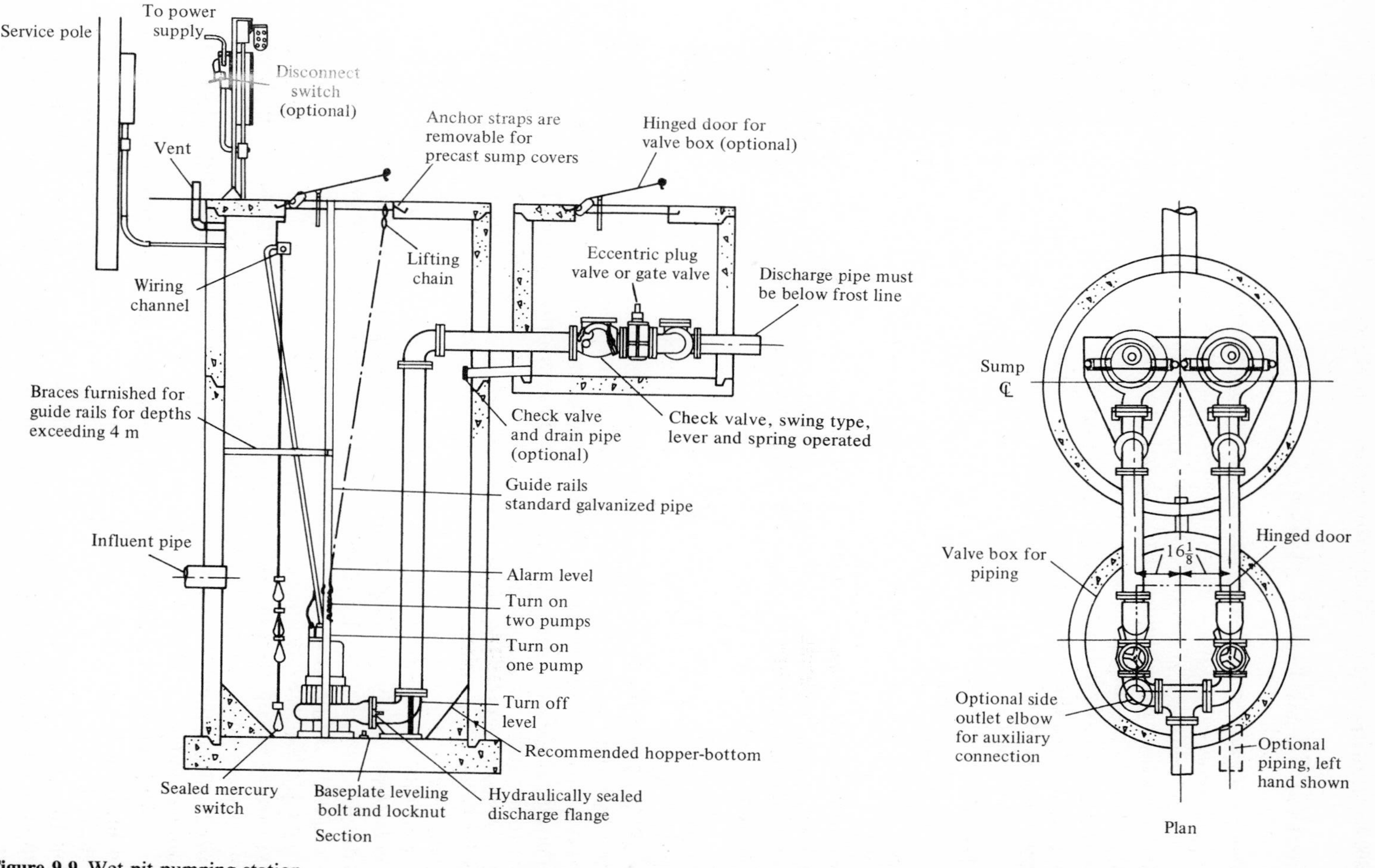

Figure 9-9 Wet-pit pumping station.

9-4 DESIGN OF FORCE MAINS

In a wastewater collection system, a pipeline designed to receive the wastewater discharged from a pumping station and to convey it under pressure to the point of discharge, which may be a gravity sewer, a storage tank, or a treatment plant, is called a *force main*. The internal operating pressure of a force main is usually at a maximum adjacent to the pumping station and decreases to, or nearly to, atmospheric pressure at its point of discharge. A force main is an integral part of the pumping system, and its design is necessarily influenced by the number, size, and types of pumps selected for the pumping station.

Other factors that influence the design of force mains are the minimum velocity requirements of the wastewater and the pipe strengths necessary to withstand the maximum internal pressures, including the transient pressures caused by waterhammer. The external earth and surface loads to be imposed on the pipe should also be considered. In this section, the hydraulic design of force mains and related design considerations, including pipeline appurtenances, are discussed. The theory, analysis, and control of waterhammer are discussed in Sec. 9-5. Details on the structural design of force mains are not covered in this book, but may be found in Refs. 2 and 10 to 12 and in standard structural design texts.

Hydraulic Design of Force Mains

The hydraulic design of a force main for routine operating conditions principally involves determining the size of the force main that will meet velocity and pump operation requirements. The hydraulic performance of a force main can be represented by a system head-capacity curve, the development and use of which have been discussed in Chap. 8.

Determination of force main size. Theoretically, the most economical size of force main should be determined on the basis of the power costs for pumping plus the annual costs for capital investment of the force main and pumping equipment. In practice, the selection of the force main size is frequently governed by the need to maintain (1) a velocity at minimum flow adequate to prevent solids deposition or (2) a velocity capable of resuspending settled solids at least once a day. The selection of pump equipment then depends on whether the pumps can discharge the desired flows at the heads required by the size of the force main.

Often, however, it is not feasible to size long force mains solely on the basis of the velocity at minimum flow. In such cases, the best approach is first to select the most economical size of force main with adequate carrying velocities for the entire range of initial and design flows, and then to select the various pumps. After the initial sizing, it may be found that a larger force main is necessary to reduce the friction losses so that a reasonable selection of pumps can be made. If the initial and design flows differ considerably, it may be necessary to build a smaller force main initially and to install a second one at some later date.

Force mains are generally 200 mm (8 in) or larger in diameter. In some cases, 150-mm (6-in) pipe may be used for small pumping stations with short force mains, and 100-mm (4-in) pipe may be used for small ejector stations.

Energy losses in force mains. As discussed in Chap. 8, the system head-capacity curve is a plot of the total dynamic head (static lift plus the kinetic energy losses) versus the corresponding flowrates. Friction losses in force mains usually are determined by using the Hazen-Williams equation. The following C values are recommended for design conditions.

C = 100 for unlined cast-iron and ductile-iron pipes.
C = 120 for cement-lined cast-iron and ductile-iron pipes, reinforced-concrete pressure pipe; prestressed concrete cyclinder pipe; asbestos-cement pressure pipe; steel pipe, 500 mm (20 in) or larger, with bituminous or cement-mortar lining; and various types of plastic pipe.

Friction losses may also be calculated by using the Darcy-Weisbach equation with the appropriate f values (see Chap. 2).

Minor losses caused by valves and fittings and losses at entrances and exits may be calculated by the use of data presented in Appendix C. These minor losses plus the friction losses are the kinetic-energy losses in the force main.

In many cases, a pump that will operate satisfactorily in accordance with the system head-capacity curve for the design year may cavitate when first placed into operation. Cavitation occurs because the energy losses in the new force main are less than those calculated for the design year. Therefore, system head-capacity curves should be developed for both old pipe (design year) and new pipe (initial year). For new ductile-iron pipe and concrete pipe, a C value of 140 is recommended; for new plastic pipes, C values of 150 or higher should be used.

Force main velocities. Velocity criteria for force mains are derived from observations that solids do not settle out at a velocity of 0.6 m/s (2.0 ft/s) or greater. Solids do settle at lower velocities or when the pump is stopped, and a velocity of 1.1 m/s (3.5 ft/s) or greater is required to resuspend the deposited solids.

For small-size or medium-size pumping stations serving only part of a sewered area where flow may be pumped intermittently at any rate up to the maximum, the desirable force main velocities range from 1.1 to 1.5 m/s (3.5 to 5 ft/s). A small station would have only two pumps, one of which would be a standby, and these would discharge at the maximum rate or not at all. For pumping stations that operate intermittently, solids in the wastewater remaining in the line when the pump stops will settle out. A velocity of 1.1 m/s (3.5 ft/s) is desirable to ensure that these deposited solids are resuspended.

In a small station with two pumps, it should be possible to operate both pumps together, even though only one is needed for design-year conditions. If the flows are too small to warrant a 1.1 m/s (3.5 ft/s) design velocity, pumps can be selected to produce a 1.1 m/s minimum velocity with both pumps operating. In such a station, both pumps should be operated together by manual control once a week for a sufficient length of time to flush out the line.

Larger pumping of this type may have three or four pumps, all of the same size, one of which is a standby. For a station with three pumps, force main velocities of about 0.9 and 1.5 m/s (3.0 and 5.0 ft/s) might be selected with one and two pumps in operation, respectively. With stations having four pumps, force main velocities of about 0.7, 1.2, and 1.7 m/s (2.25, 4.0, and 5.5 ft/s) might be selected with one, two, and three pumps in operation, respectively. These velocities allow for a reduction in pump capacity because of greater friction losses at increased flows.

The pump capacities required to maintain velocities of 0.6 and 1.1 m/s (2.0 and 3.5 ft/s) in 150- to 300-mm (6- to 12-in) force mains are shown in Table 9-6.

Force main design usually becomes more complicated for pumping stations serving all or a major part of a sewered area where it is required to pump continuously at or close to the incoming flowrate. These pumping stations may have several sizes of pumps, some of which may be constant- or multiple-speed units and some variable-speed units. These pumps must operate continuously and must be sized so that, by operating either singly or in combination, they can pump at continuously varying flowrates ranging from initial minimum to design peak.

The range in discharge and velocity to provide for the flowrates shown in Table 9-6 may be on the order of 7 or 8 to 1. If the maximum velocity in the force main is set at 1.8 m/s (6 ft/s), the initial minimum flow would produce a velocity of only 0.22 to 0.26 m/s (0.7 to 0.9 ft/s) in the force main. For continuous pumping, one pump would have to be sized for this flow. Because these are not self-cleaning velocities, there will be some solids deposition, but this can be accepted for the following reasons:

1. At minimum flow, the solids and grit content of wastewater is lowest, and it is the grit that may settle out.
2. At daily peak flows, the pumping rate is 1.5 to 2.0 times the daily average, resulting in velocities that flush out any material which has settled out during minimum flow.
3. The alternative solution of two force mains is more expensive and operationally undesirable, and should be avoided if at all possible.
4. Pumping stations and force mains designed on this basis have worked satisfactorily.

Other Design Considerations

After the size of the force main has been determined, a number of other design details must be resolved to ensure that the force main can operate successfully.

Table 9-6 Pump capacities for minimum force main velocities

Force main diameter, mm	(in.)	Pump capacity, m^3/s (gal/min)			
		$V = 0.6$ m/s	(2.0 ft/s)	$V = 1.1$ m/s	(3.5 ft/s)
150	(6)	0.011	(174)	0.019	(301)
200	(8)	0.020	(317)	0.035	(555)
250	(10)	0.031	(491)	0.054	(856)
300	(12)	0.045	(713)	0.078	(1236)

Note: mm × 0.03937 = in
m^3/s × 15,850.3 = gal/min
m/s × 3.2808 = ft/s

The pipe materials must be selected, the alignment of the force main must be defined, and the need for and design of various force main appurtenances must be determined.

Pipe materials. The materials from which force main pipes are made must be suitable for use in the soil environments in which they will be laid; they must also resist any internal corrosion caused by the character of the wastewater to be conveyed. Some of the more common materials now or once used for force main pipelines are presented in Table 9-7.

Table 9-7 Pipe materials used for force mains

Pipe material	Usual size range, mm (in)	Type of joint[a]	Remarks
Asbestos cement (AC)	100–1050 (4–42)	Sleeve	May be susceptible to deterioration by aggressive soils or waters.
Cast iron (CI)	. . .	Push-on or mechanical joint	Commonly used in past in sizes up to 600 mm (24 in). Has been replaced by ductile-iron pipe.
Ductile iron (DI)	100–1350 (4–54)	Push-on or mechanical joint	May need protection against aggressive soils or waters.
Fiberglass-reinforced plastic (FRP)	600–2000 (24–78)	Push-on or sleeve	Highly resistant to corrosion.
Polyvinyl chloride (PVC)	100–300 (4–12)	Push-on	Highly resistant to corrosion.
Prestressed concrete, steel cylinder type	400–3600 (16–144)	Push-on with steel joint rings	May be susceptible to deterioration by aggressive soils or waters.
Steel	500–3600 (20–144)	Sleeve	Must be lined with bituminous material or cement mortar, and protected on outside against corrosion. Largely replaced by ductile-iron and prestressed concrete pipe.

[a]Bell and spigot (push-on or mechanical joint) unless otherwise stated. All joints have gaskets.

Depth of cover. Force mains are usually laid at a comparatively shallow depth. A minimum depth of cover of 0.9 m (3.0 ft) is recommended to minimize the impact of live loads, but a somewhat greater depth is desirable. The pipes should be laid at such a depth that frost will not endanger them (in cold climates), and that other utilities—future as well as existing—are avoided as much as possible. Public agency codes usually specify the minimum distances between pipelines carrying potable water and force mains (as well as sewers).

Anchorage. Force mains must be anchored to resist the thrusts that develop at angles, bends, branches, and plugs in the pipe. The magnitude of the forces to be resisted can be calculated by well-known formulas found in standard texts on hydraulics and mechanics (see Chap. 2). The required anchorage may be attained by installing restrained pipe joints, concrete thrust blocks, or concrete anchor blocks.

Anchorage design at wastewater force main fittings should be based on pipeline pressures at least 25 percent greater than the maximum pump design shutoff head plus a waterhammer allowance with an appropriate factor of safety. Pipe specifications should be coordinated with the design to avoid the possibility of specifying field test pressures greater than pressures on which anchorage design is based, even though the nearest commercial-pipe pressure rating may be much greater than the design pressure.

For resisting horizontal thrust, the use of flexible self-restraining pipe joints or joints restrained with tie rods and clamps is recommended. Either method can be effective provided the component parts are made of corrosion-resistant materials or are protected from corrosion by a liberal application of asphalt or other acceptable corrosion-retarding coatings. Some examples of restrained joints are shown in Fig. 9-10.

In the past, the most common anchorage for resisting horizontal thrust was a concrete thrust block placed between the fitting to be restrained and the undisturbed trench wall. If this method of anchorage is selected, safe thrust block bearing areas must be determined and indicated in the contract drawings. A typical thrust block is shown in Fig. 9-11.

Since the development of reliable restrained joints the use of thrust blocks has become less common. In fact, the use of concrete thrust blocks in some situations may not be advisable or even possible. Soil bearing conditions may be unreliable because of poor soil conditions, such as filled land or high groundwater. The possibility also exists that the soil bearing capacity may be reduced because of future excavations that may be made adjacent to the force main.

The proper functioning of restrained joints depends on the frictional resistance between the pipeline and the soil, and the proper functioning of a thrust block depends on the bearing pressure capacity of the soil against which it is placed. Therefore, soil information from the project area (preferably from soil boring logs) is necessary, and competent geotechnical advice or review should be obtained whenever pipeline anchorage is required.

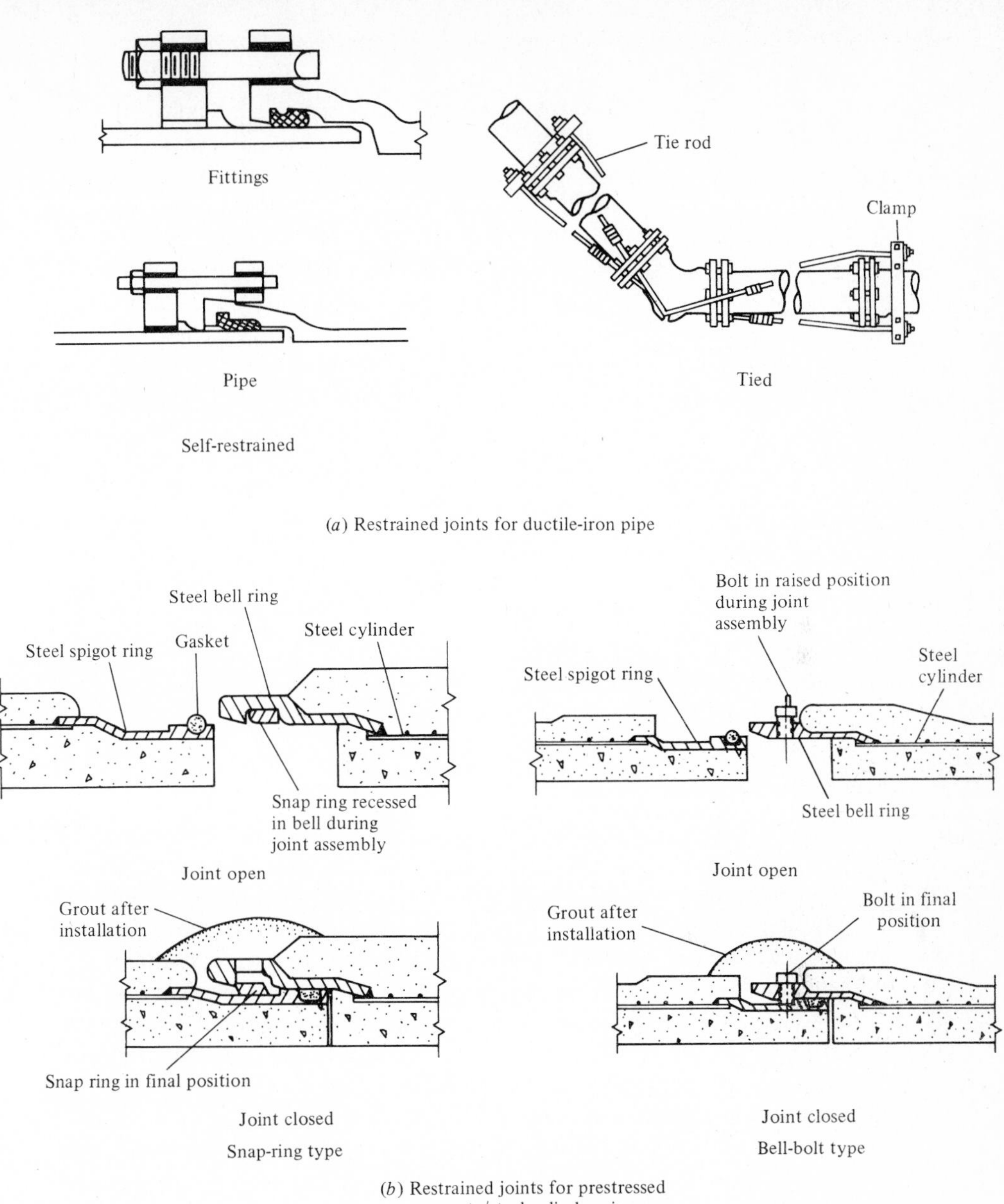

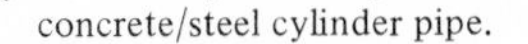

(*b*) Restrained joints for prestressed concrete/steel cylinder pipe.

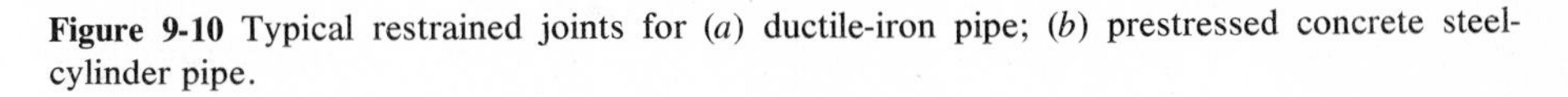

Figure 9-10 Typical restrained joints for (*a*) ductile-iron pipe; (*b*) prestressed concrete steel-cylinder pipe.

Vertical bends in force mains where the thrust is upward should be avoided wherever possible. If this is unavoidable, the fitting must be restrained by tied pipe joints or a concrete anchor block. The anchor block would depend solely on its own weight to resist the thrust. If the block is below the groundwater level, only the buoyant weight of the concrete is effective.

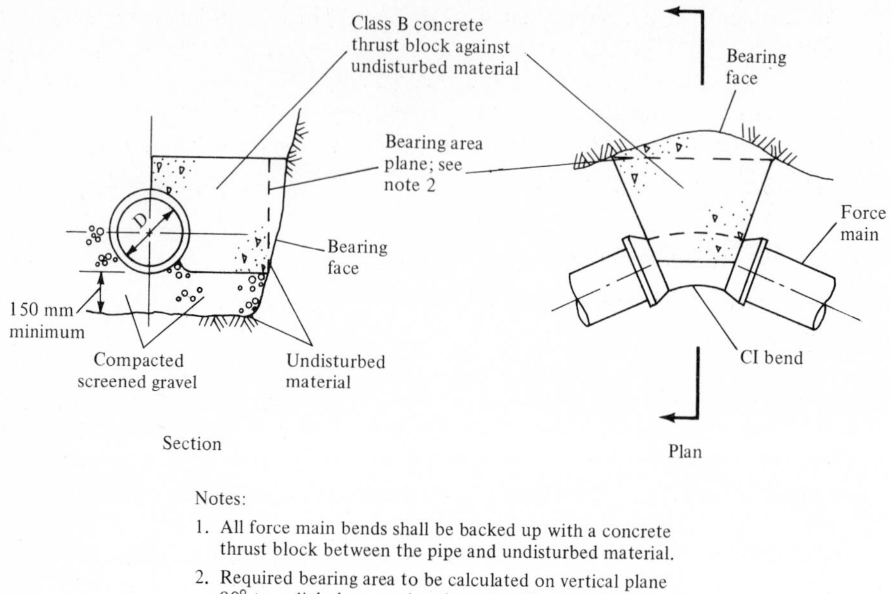

Notes:

1. All force main bends shall be backed up with a concrete thrust block between the pipe and undisturbed material.
2. Required bearing area to be calculated on vertical plane 90° to radial plane passing through midpoint of bend.
3. Minimum bearing area: 0.2 m^2.

No scale

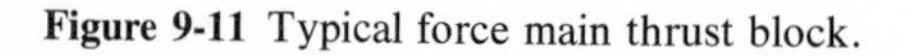
Figure 9-11 Typical force main thrust block.

Reduction of outlet turbulence. Special attention should be given to the horizontal and vertical layout at the outlet of a force main to minimize turbulence and the resulting liberation of hydrogen sulfide, which not only smells bad but also may be oxidized to sulfuric acid. This acid could severely attack the concrete and mortar of the outlet structure and, in some cases, the sewer downstream of the outlet structure. This condition may be intensified by prolonged storage of the wastewater in the wet well and in the force main during hot weather. This phenomenon was discussed in Chap. 7.

One common method of minimizing turbulence is to align the crown of the force main outlet with the crown of the sewer at the receiving manhole. The invert of the manhole would then be sloped smoothly between the force main and sewer inverts. If the velocity in the force main is higher than about 2 m/s (6.5 ft/s), or if more than one force main is involved, it may be necessary to design a special entrance to minimize turbulence. A provision for energy dissipation may also be necessary.

If the gravity sewer is considerably lower than the normal elevation of the force main, and especially if the sewer is embedded in rock, the force main may stop in a manhole at ordinary depth, and the connection at the low-level sewer may be made with a short gravity sewer and a standard drop inlet. If the connection is to be made to an existing manhole or other structure, an inside drop inlet may be used where suitable.

As a further precaution against hydrogen sulfide problems, especially in hot climates, it may be desirable to seal the outlet of the force main pipe. The sewer can be sealed by placing the crown of the force main pipe outlet at the same elevation as the invert of the gravity sewer, with the manhole invert sloping upward from the force main outlet to the sewer. The seal may also be provided with a trap formed with pipe fittings.

Force Main Appurtenances

Common force main appurtenances include blowoffs, access manholes, and air and vacuum valves.

Blowoffs. A blowoff is a controlled outlet on a pipeline, so arranged that it can be used to drain or flush the pipeline. Blowoffs normally are not required on force mains. However, where the force main contains a long depressed section between high points, a blowoff may be desirable in case the force main must be drained and pumped out.

A suitable blowoff might consist of a valved connection in a manhole at the low point, discharging to a manhole or vault that would serve as a wet well for a portable pump. The size of the blowoff should not be less than 150 mm (6 in). It should, if possible, be large enough to provide flushing velocities in the force main.

Air valves. Force mains are usually constructed at approximately uniform depths below the ground surface. This practice often results in high and low points along the course of the pipeline. In such cases, it may be necessary to provide valves for the release of trapped air when putting the line in service (filling) and during regular operation, or for admitting air should it be necessary to drain the force main. The release or admittance of air for filling or draining can be accomplished by use of a manually operated air valve. Specially designed automatic air release valves have been used for bleeding air from high points while the force main is under pressure.

If possible, force mains should be designed without high points, and with the top of the force main below the hydraulic grade line at the minimum pumping rate so that air-relief valves will not be needed. If the elimination of high points is not feasible, a manual air-release valve should be installed at each significant high point where air could become trapped. A high point may be considered significant if it is 0.6 m (2 ft) or more above the minimum hydraulic grade line or, when pumping is intermittent, above the static head line. Air-release valves should not be less than 20 mm (3/4 in) nor more than 50 mm (2 in) in size. Larger sizes should be used with larger-diameter force mains. Preferably, these valves should be located in vaults, but a valve of 40-mm (1 1/2-in) diameter or larger may be buried in the ground and operated with a wrench through a box. An air-release valve may discharge to a sewer manhole, a vented dry well, or other suitable place. Consideration should be given to providing a means of rodding out or back flushing the valve. A typical air valve installation for a force main is shown in Fig. 9-12.

Automatic air-release valves should not be installed if their use can be avoided. From past experience it has been found that automatic air-release valves require frequent maintenance in order for them to function as intended. Inadequate maintenance causes these valves to clog and malfunction, often soon after they are installed. In most cases, manual air valves could be used instead of automatic air valves. For example, if after the force main has been put into service, the need develops for frequent use of a manually operated valve to relieve entrapped air or gas, the valve may be left at a part-open setting for continuous bleeding of air or sewage. As a last resort, an automatic air-release valve may be installed.

Automatic air-release valves, if used, must be specially designed to keep the valve operating mechanism free from contact with sewage to inhibit clogging and resulting malfunction. They must be located in a manhole or vault and protected against freezing. Automatic air valves should be installed on top of the force main with a shutoff valve close to the force main. A 25-m (1-in)

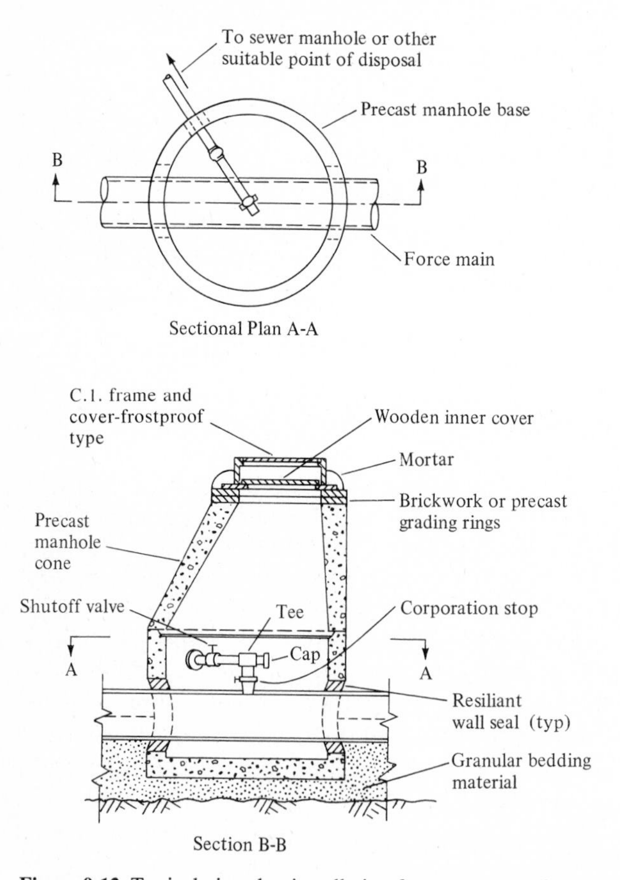

Figure 9-12 Typical air valve installation for wastewater force main.

blowoff valve should be installed either above the shutoff valve or on the air valve body. A back-flushing connection should be provided by the valve manufacturer.

Automatic air and vacuum valves have been used to allow the quick automatic admission of air that might be needed to prevent collapse of a thin-walled pipeline during the fast drainage that would take place through a broken force main, or during water-column separation following a power failure. They also have been used for venting air during the filling of the force main. However, these valves are subject to maintenance problems similar to those of air-release valves. Furthermore, their malfunction could create additional waterhammer problems.

In general, automatic air and vacuum valves should not be used on sewage force mains. Instead, the problem of possible collapse of force main pipes because of internal pressures less than atmospheric should be solved by the use of pipe having walls sufficiently strong to withstand the induced added crushing load.

9-5 WATERHAMMER IN WASTEWATER FORCE MAINS

The rapid changes that can occur in the velocity of flow in force mains and pressure pipe can be caused by pump startup, pump shutdown, or power failure and can result in a considerable change in pressure. The change in pressure can be either positive (above normal) or negative (below normal) and is sometimes accompanied by a hammering-type noise. This transient pressure and flow condition in force mains and pressure pipes is known as *waterhammer*. The most severe waterhammer conditions generally result during pump shutdown or power failure. The theory, analysis, and control of waterhammer are discussed in this section.

Theory of Waterhammer

In the case of rapid pump shutdown or power failure, the flow and velocity in the force main are initially steady. However, when the power supply to the pump motor is cut off, the pump rapidly decelerates from full speed to zero speed, causing a rapid decrease in the pump discharge into the force main. This decrease creates a negative-pressure wave (below normal pressure) that rapidly travels from the pumping station end of the force main to its discharge end. This negative-pressure wave decelerates the flow in the force main in accordance with Newton's second law of motion. When this negative-pressure wave reaches the discharge end of the force main, it is reflected back toward the pumping station as a positive-pressure wave, which further decelerates the flow during its passage.

A cycle of pressure wave travel is completed when the positive-pressure wave reaches the pumping station, where it is again reflected and a second wave of reduced pressure travels up the pipe. This cycle is repeated while the speed and discharge of the pump continue to decrease. In a short period, the pump speed is reduced to the point where it can no longer develop sufficient head to pump against the discharge head. Then, the swing check valve on the pump discharge, normally used to prevent reverse flow through the pump, closes, and this closure isolates the pump from the transients in the force main. However, the transient flows and pressures in the force main continue until all energy is dissipated by pipe frictional resistance. During this sequence of events, the head at the end of the force main remains constant. At intermediate points, the head is determined by the sum of the pressures of the positive and negative waves.

The roundtrip time of travel of the pressure wave from and back to the point of flow change is termed the critical period and is given by

$$T = \frac{2L}{a} \tag{9-2}$$

where T = critical period, s
L = length of force main between point of flow change and point of reflection, m (ft)
a = velocity (celerity) of pressure wave, m/s (ft/s)

The magnitude of the pressure buildup at the point of flow change is related to whether the flow is stopped in a time interval equal to or less than or greater than the critical period. The velocity of the pressure wave and the magnitude of the pressure buildup are discussed next.

Pressure wave velocity. The velocity of a waterhammer pressure wave depends on the physical properties of the fluid and the force main pipe, and the acceleration due to gravity. It can be calculated with the following equation:

$$a = \frac{1440}{\sqrt{1 + C_1(Kd/Ee)}} \qquad \text{(SI units)} \tag{9-3}$$

$$a = \frac{4720}{\sqrt{1 + C_1(Kd/Ee)}} \qquad \text{(U.S. customary units)} \tag{9-3a}$$

where a = velocity of pressure wave, m/s (ft/s)
C_1 = 1, for pipe with expansion joints throughout
= $1 - \mu^2$ for pipes anchored against axial movement (buried force mains, for example)
= $\frac{5}{4} - \mu$, for pipes without expansion joints and anchored at the upstream end
μ = Poisson's ratio
K = bulk modulus of water, taken as 2070 MN/m² (300,000 lb/in²)
d = pipe diameter, mm (in)
E = modulus of elasticity of pipe material, MN/m² (lb/in²)
e = thickness of pipe wall, mm (in)

The modulus of elasticity, Poisson's ratio, the normal ranges of diameter-to-wall thickness ratios, and the normal range of waterhammer wave velocities in force mains made of various pipe materials are presented in Table 9-8. In specific designs, actual waterhammer wave speed should be calculated by Eq. 9-3 and 9-3*a* with the use of the manufacturer's data for thickness and properties of materials in the case of special or nonstandard pipe materials.

Parmakian has computed and plotted values of waterhammer wave speed a for various types of pipe against values of d/e, and he has also presented a method for determining an equivalent thickness of steel pipe for pipes made of composite materials [8]. These may also be used to determine waterhammer wave speeds in force mains of different types of materials as well as of composite materials.

Magnitude of waterhammer pressure buildup. If the velocity at the point of flow change decreases to zero in a time interval equal to or less than T (see Eq. 9-2), the change is said to be "instantaneous." Such a change would be caused by a rapid valve closure. The magnitude of the waterhammer pressure at the location of the valve for an "instantaneous" change in velocity can be derived from the impulse momentum principle [6, 15] and is given by

$$h_{w(\max)} = \frac{aV}{g} \tag{9-4}$$

where $h_{w(\max)}$ = maximum head caused by waterhammer, m (ft)
a = velocity of pressure wave, m/s (ft/s)
V = velocity of fluid in pipeline, m/s (ft/s)
g = acceleration due to gravity, 9.81 m/s^2 (32.2 ft/s^2)

Table 9-8 Typical physical properties and waterhammer wave velocities in force main

			Normal range of values		
Force main pipe material	Modulus of elasticity, E, MN/m^2 × 10^{-3}	Poisson's ratio, μ	Diameter, D, mm	Ratio of pipe diameter to wall thickness, d/e	Waterhammer wave velocity, a, m/s
Asbestos cement (AC)	23.4	0.2	150–900	10–20	820–1040
Cast iron (CI)	53–103	0.3	150–1200	15–50	1000–1250
Ductile iron (DI)	165.5	0.3	150–1350	25–95	940–1280
Steel	200	0.2	600–2100	50–225	820–1200
Prestressed concrete (PC)	200[a]	0.2	400–3600	60–110[b]	1000–1160
Polyvinyl chloride (PVC)	2.8	0.4	150–300	15–25	350–450
Fiberglass reinforced plastic (FRP)	14.5–20.5	0.3	600–2000	70–110	370–470

[a] Since wave velocity a is based on an equivalent steel pipe, the value of E must be that for steel. See Ref. 8.
[b] Ratio of diameter to equivalent steel thickness for composite material.
Note: MN/m^2 × 145.0 = lb/in^2
mm × 0.03937 = in
m/s × 3.2808 = ft/s

If the valve is at the discharge end of the pipeline, the waterhammer pressure is positive and is added to the initial pressure. If the valve is at the beginning of a pipeline as on the discharge of a pump, the waterhammer pressure is negative and is subtracted from the initial pressure. Since the total pressure in the line cannot be less than the vapor pressure, an "instantaneous" valve closure on the discharge of a pump may result in the formation of a vapor cavity instead of reducing the velocity to zero.

In situations where the velocity decreases to zero in a time interval T greater than $2L/a$, which would correspond to a slow valve closure or a short force main, the change is said to be slow. Under these conditions, the full pressure does not develop because it is reduced by the reflected waves.

In most wastewater force mains, heads are small, normally ranging from 7 to 20 m (23 to 65 ft), with velocities at design flow from 1.5 to 2.0 m/s (5 to 6.5 ft/s). With the drop in pressure limited to the pumping head or somewhat more if a negative pressure or a vacuum is produced on power failure, several roundtrips of the pressure wave are required to reduce the velocity in the force main to zero. Each travel of the wave along the pipe in either direction decreases the velocity in the force main by

$$\Delta V = \frac{gH}{a} \tag{9-5}$$

where H is the difference in head at the two ends of the force main plus the friction head at the average velocity during the passage of the wave. During this time, the head at the end of the force main remains constant while the head at the outlet of the swing check valve varies only slightly, rising gradually to the wet-well level.

When the velocity has been reduced to zero, three events occur: (1) the check valve closes; (2) one passage of the positive-pressure wave from the outlet back to the pump raises the pressure to normal and the reverse velocity to ΔV, where $\Delta V = g/a$ times the static head; and (3) the reverse velocity ΔV comes up against a closed valve, which causes a further increase of $\Delta H = (a/g)\Delta V$, resulting in a waterhammer surge pressure to double the static pressure.

Analysis of Waterhammer

In the past, waterhammer analyses of force mains to determine maximum and minimum pressures were carried out by using graphical methods, arithmetic integration, or waterhammer charts [1, 8, 13, 14]. Simplifying assumptions had to be made to keep these graphical and arithmetic analyses manageable. It was sometimes necessary to consider an equivalent force main of uniform diameter and material and, where desirable due to high points in the line, it was impractical to consider the force main as consisting of more than two or three segments, preferably of equal length. Analysis and evaluation of designs to determine their effectiveness were, nevertheless, accurate but time-consuming, and the comparison of alternate methods of control was usually out of the question. One of the advantages of using graphical or arithmetic integration methods was that they provided the analyst with clear pictures of the sequence of events, and familiarity with the graphical method is recommended for this reason.

Widespread accessibility to digital computers in recent years has provided a convenient and cost-effective means to analyze complex waterhammer and hydraulic transient problems rapidly without making some of the simplifying assumptions that have been necessary in the past. For example, in an adequate computer program, the force main can be divided into 10 or more segments of equal length with division points corresponding closely to high points in the profile or changes in diameter or materials of construction. The effect of high points in the profile, the possibility of water-column separation at the high points, and the surge pressures developed on rejoinder of the columns cannot be overemphasized. They can be analyzed effectively and economically by an adequate program.

The shape of the pressure wavefront depends on the ratio of the decelerating torque to the inertia (WR^2) of the pump, liquid, and motor, and the closing characteristics of the discharge valve. These all vary with time and are incorporated in the program as well as the characteristics of any relief valves or suppressors contemplated. As a result, the use of numerical methods and computer programs to analyze various waterhammer conditions in force mains has now become a routine design procedure. Adequate programs make use of the method of characteristics [5, 15, 16], which is based on the nonsteady-state equations of motion and continuity for flow in pipes.

In recent years, many computer programs for waterhammer analysis of force mains have been developed. Some of these programs are available for use through computer service bureaus, and others are available from people who developed them. Manufacturers and equipment vendors in many cases have programs based entirely on the use of their control equipment. Before any computer program is used to analyze wastewater force mains, the engineer must be satisfied that the program is applicable to a particular problem. This is especially so in the case of complicated pumping stations and force mains that require elaborate control systems and safety features. In some special cases, it may be necessary to modify available programs or write new programs. In such cases it is always advisable to use the services of a specialist in waterhammer.

Waterhammer Control

The objective of waterhammer control is to limit pressure changes in the force main within allowable ranges by minimizing the rate of change of velocity. Waterhammer problems in wastewater pumping stations and force mains can be either simple or complex. Consequently, the waterhammer control systems needed to protect pumping stations and force mains range from simple to elaborate. The control systems generally used to limit waterhammer pressures in wastewater pumping stations and force mains are:

1. Swing check valve on pump discharge with outside weight and lever to assist closure.
2. Spring-loaded swing check valve on pump discharge.
3. Swing check valve either spring-loaded or having outside weight and lever together with high-pressure relief valve.
4. Positively controlled valve on pump discharge set to open at a preset pressure during startup and to close at a predetermined rate after power failure.
5. Positively controlled discharge valve, together with one or more bypass surge-relief valves, set to open at preset pressures and close at predetermined rates.
6. Air and vacuum valves at the pumping station and at high points along the force main to limit subatmospheric pressures.

Swing check valves. In simple cases such as small- and medium-size pumping stations with a gradually rising short force main (less than about 500 m) and a small static lift (about 15 to 20 m), the control system may consist of a swing check valve with outside weight and lever on the pump discharge to assist closure of the valve disk when the flow reverses.

Disks of swing check valves without an outside weight and lever may tend to remain open at times when the flow through the valve drops to zero. This allows an unbalanced head to develop on the pipe, which causes the wastewater remaining in the pipe to accelerate in the reverse direction. If the valve disk remains open long enough, the reverse flow would cause the pump and motor to run backward until equilibrium is attained with the pump operating as a turbine at runaway speed. Under these conditions, maximum reverse flow could range from 50 to 110 percent of rated flow; maximum transient reverse speed could range from 125 to 150 percent of rated forward speed; and maximum transient pressure could range from 150 to 175 percent of rated head. A severe waterhammer condition could result if the valve disk were to close suddenly following establishment of high reverse velocity. This sudden closure would result in a sudden increase in pressure head, given by Eq. 9-4.

A large increase in pressure caused by a sudden valve closure might rupture the force main. Also, a sudden closure of the valve disk could cause severe slamming. If this occurs frequently, as is the case with intermittently operated pumps, it could eventually loosen the seals and joints in the pipeline, thereby causing leaks. The use of an outside weight and lever with a swing check valve reduces the chances of the disk remaining open. For this reason, swing check valves with an outside lever and weight are recommended for small- and medium-size pumps. The slamming problem can be eliminated or greatly reduced by adjusting the location of the lever weight. Spring-loaded check valves can also be used to ensure closure of the valve disk and eliminate slamming. Usually, in small stations, check valves as just described are all that is needed to control waterhammer, unless the force main has significant high points.

High-pressure relief valves allow the flow from a force main to reenter the wet well. One or more of these valves may be used with a swing check valve on the pump discharge to control waterhammer pressure. The relief valves generally used in wastewater application are essentially spring- or weight-loaded valves. The disk of this type of valve is set to open when the pressure in the force main exceeds a predetermined pressure. When the force main pressure drops below the set pressure, the valve disk is designed to close slowly. This type of control is used primarily as an added protection to the force main in relatively simple cases. For complicated cases, positively controlled valves should be used for waterhammer control.

Positively controlled valves. Ideally, the force main would exit the pumping station at the same level as the pumps and run level or rise gently at a uniform slope until near the end where it would rise abruptly to the outlet. This layout is seldom possible. For economy, force main profiles conform to existing ground elevations and may contain both high and low points. Furthermore, at many deep stations, individual pump discharges rise vertically to the main header, which then exits the station close to the ground level, or the main header itself rises just inside or outside the station and then runs approximately level near the ground surface.

Even in a small- or medium-size pumping station, severe waterhammer pressures can occur if the force main rises sharply at the pumping station or if the force main profile has intermediate high points. In either case, during the initial downsurge following power failure or pump shutdown, the absolute pressure may drop to nearly complete vacuum equal to vapor pressure, resulting in formation of vapor cavities and water-column separations at the high points. The two portions of the water column on either side of a cavity would then behave independently of each other, coming to rest or slowing down and reversing, while the pressures at the high points would remain at the vapor pressure.

If the pump column continues its forward movement for a few seconds, the cavity may not be large and the columns may come together with very little difference in velocity and inconsequential surge pressures. If the pump column comes to rest quickly, the cavity may grow large enough to cause the other column to accelerate in the reverse direction almost to its initial velocity at the moment the columns rejoin. If any of the reversing water columns should come against a closed valve or a stagnant water column, the resulting pressure increase would be high and is given by Eq. 9-4.

In large stations, check valves should be positively controlled cone valves, plug valves, or butterfly valves. Butterfly valves should not be used unless rags or other solids are removed or comminuted ahead of the pumps as discussed in "Suction and Discharge Piping" in Sec. 9-2.

Normal shutdown will be accomplished by slowly closing the valve while the pump continues to run; when the valve reaches the closed position, it trips a limit switch to stop the motor. On power failure, the power to the pump is cut off with the valve wide open, and hydraulic pressure from a reliable source or a hydraulic accumulator operating through a power cylinder is used to close the valve. Normal closing, emergency closing, and normal opening time of the valve may be different and should be individually adjustable.

The time of closure of these valves is of paramount importance, because they can break the water column at the pump discharge if they close too fast. Water-column separation might result in large pressures in the force main. Consequently, these valves must operate slowly and should be about half closed when the velocity in the force main has dropped to zero and begins to reverse.

In a force main without high points, the wave velocity a may be written as $L/\Delta t$, where $\Delta t = L/a$, the time required for a wave to travel the length of the pipe. If $L/\Delta t$ is substituted for a in Eq. 9-5, the following equation results by rearrangement of terms:

$$\Delta t = \frac{L\,\Delta V}{gh} \qquad \text{or} \qquad t = \frac{LV}{gH_{av}} \tag{9-6}$$

where t = time for velocity to drop to zero
H_{av} = average decelerating head including friction

This equation provides the time t for the velocity in the force main at the pumping station end to drop to zero. A valve closing time of $2t$ should then be used for positively controlled valves. For cone valves, time $2t$ would be from the beginning of unseating to the end of rotation of the plug but not including reseating time. The specifications should call for the operating time to be adjustable at least over the range from t to $4t$.

Alternatively, the time t for the velocity to drop to zero can be determined by a preliminary computer run, assuming swing check valves on the pump discharge. This method is particularly suitable in the case of a high point in the discharge line, resulting in water-column separation, as it will also reveal the time of the rejoinder of the columns and maximum resulting surge pressure. This time would then be used instead of t for selecting the valve closing time for the initial computer runs with the proposed positively controlled valves, and additional computer runs would determine the optimum closure time.

With this scheme, there will be reverse flow through the pump when a power failure occurs. Under these conditions, the pumps could attain a reverse speed of 60 to 125 percent of rated speed, depending on the number of pumps in the station and the number in operation at the time of the power failure. Pumps and motors should be specified to withstand safely a reverse speed of 150 percent of full speed. The maximum surge pressure following power failure should not exceed 175 percent of the rated head, measured above the top of the pump volute. By careful analysis and correct timing, it can usually be limited to 125 percent of the rated head.

Surge-relief valves. In some large pumping stations, the use of positively controlled discharge valves alone may not provide sufficient control to limit waterhammer pressures. For large pumping stations with long force mains, where only the discharge valves are used for waterhammer control, the required closing time for the positively controlled valves might be excessively long, resulting in excessive backflow to the wet well as well as excessive and undesirable reverse rotation of the pump and motor. These conditions could occur with a force main having high points resulting in water-column separation. In such cases, bypass surge-relief valves are used with the positively controlled discharge valves.

When used with bypass surge-relief valves, the discharge valves are designed to close by the time the pump discharge drops to zero. This prevents reverse flow through the pumps. The bypass surge-relief valves are usually positively controlled cone or butterfly valves and are designed to be fully open by the time the discharge valves close. These valves allow the reverse flow in the force main to flow into the wet well at a reduced rate while bypassing the pumps. Usually, more than one bypass surge-relief valve is used to ensure that at least one bypass valve will open if one of the others malfunctions. The bypass surge-relief valves are made to close slowly to keep pressures in the force main within allowable limits.

Air and vacuum valves. Automatic air and vacuum valves have also been used in wastewater force mains to prevent formation of vacuums at the pumping station and at high points in the force main. When the pressure in the force main drops below atmospheric pressure, these valves allow enough air to enter the force main to keep the pressure close to atmospheric. By keeping the pressure in the air cavity close to atmospheric, the velocity with which the water columns on either side of the cavity rejoin is reduced and some air-cushioning effect is obtained. This limits the pressure increase in the force main following collapse of the vapor cavity.

Quick-opening, slow-closing air valves are also available for use at high points on the force main. These allow a small discharge of the force main contents at each operation. Because of their susceptibility to malfunctioning, the use of air and vacuum valves or quick opening–slow closing air valves should be restricted to force mains carrying treated wastewater.

Other control measures. Hydropneumatic tanks have been used to some extent to control waterhammer in small- and medium-size stations. The tank is connected to the main header and the pumps are equipped with swing check valves, preferably with dashpots. On pump shutdown or power failure, the tank supplies water to the force main to limit the downsurge in pressure and prevent water-column separation. After the check valves close, the return-pressure wave forces the wastewater to flow into the tank, where it is effectively cushioned. Several companies supply such tanks and use computer programs for proper sizing of the tanks and other components.

Where the maximum steady-state discharge-head elevation does not exceed the roof of the station, control can sometimes be obtained by extending one or two relief risers extending vertically from the main header to a point above the roof with discharge back to the wet well. The top of the loop is vented to break any siphon action. No valves are required other than the usual check and gate valves on each pump discharge. The system positively limits surge pressures on startup and power failure and has been used successfully for both wastewater pumping stations and low-lift water pumping stations.

It should be recognized that many of the devices used to control waterhammer in water pumping stations, such as surge suppressors and relief valves, are not applicable in wastewater pumping stations because of the solids in wastewater.

In the design of pumping and force main systems, it is essential that all piping and equipment, including joints, be capable of withstanding the expected waterhammer pressures as well as the nominal system pressures. This also applies to the pipe supports, anchors, and bracing both inside the station and outside above or below ground. Buried elbows and fittings should be adequately backed up by concrete or suitably strapped to withstand the expected pressures.

DISCUSSION TOPICS AND PROBLEMS

9-1 A pumping station containing three identical constant-speed pumps, one of which is to be a spare, is to be designed to pump wastewater through a force main 750 m long. The flow varies from a minimum of 0.013 m^3/s to a maximum of 0.110 m^3/s. The incoming sewer is 450 mm in diameter, and its invert is 9 m below the water surface at the point of discharge of the force main. Allow a 0.15-m loss through the bar screens, and a 0.15-m difference in wet-well levels between the start levels for no. 1 and no. 2 pumps in sequence.

Select the size of force main using cement-lined cast-iron pipe (relative roughness equals 0.0015); compute the system curves when no. 1 and no. 2 pumps start; and determine the head, capacity, and efficiency of each pump when it operates alone and when two pumps operate together. Assume that the pumps are radial-flow centrifugal pumps, that the shape of the pump curves will be as given in Fig. 8-15, that the efficiency at the best efficiency point will be 75 percent, and that by proper selection of pump size, speed, and impeller diameter, any desired head and capacity can be obtained. Assume that the losses in the individual suction and discharge pipes of each pump amount to 1 m when the pump capacity is equal to the capacity at the best efficiency point.

9-2 In Prob. 9-1, assume a drawdown of 0.75 m for the control of each pump. What area of wet well is required? State the basis for your answer.

9-3 Two pumps, each with a capacity of 0.065 m^3/s, are installed in a lift station (no force main). The inflow varies from 0.013 to 0.120 m^3/s. The individual pumps are not to start more often than three times per hour. Assume that the difference in water level with one and two pumps in operation is 0.20 m, and that the water depth is 1 m when the first pump in sequence stops. If the area of the wet well is 12 m^2, determine and show on a diagram the float switch settings for each pump.

9-4 Two pumps, each with a capacity of 60 L/s, are installed in a lift station (no force main). The inflow varies from 12 to 150 L/s. The individual pumps are not to start more often than four times per hour. Assume that the difference in water level with one and two pumps in operation is 0.15 m, and a water depth of 1 m when the first pump in sequence stops. If the area of the wet well is 10 m^2, determine and show on a diagram the float switch settings for each pump.

9-5 The liquid volume of an existing concrete sump is 29 m^3 and the cross-sectional area is 15 m^3. If the sump is to be used as a wet-pit lift station (no force main) and three equal-capacity pumps are to be installed, estimate the maximum flowrate that can be pumped from the pump station. Assume that the pumps are not to start more than four times in an hour, the distance between control functions is 0.15 m, and the minimum liquid level that must be maintained in the sump is 0.8 m.

9-6 Visit one or more of the pumping stations in your community or on your school campus. Classify the pump stations according to type and capacity (see Table 9-1). How do the general features of the pumping stations compare to those described in this chapter?

9-7 Lay out a pumping station similar to the ones shown in Figs. 9-1 and 9-3 using the information from Probs. 9-1 and 9-2 or using information supplied by your instructor.

9-8 Derive Eq. (9-4) for the pressure increase in a pipeline due to the instantaneous valve closure from the impulse-momentum equation [see Eq. (2-11)].

REFERENCES

1. Bergeron, Louis: *Waterhammer in Hydraulics and Wave Surges in Electricity,* Wiley, New York, 1961.
2. *Handbook of PVC Pipe,* Uni-Bell Plastic Pipe Association, Dallas, Texas, 1977.
3. Joint Committee of the American Society of Civil Engineers and the Water Pollution Control Federation: *Design and Construction of Sanitary and Storm Sewers,* chap. 9, "Structural Requirements," ASCE Manual and Report 7, New York, 1969.
4. Joint Committee of the American Society of Civil Engineers and the Water Pollution Control Federation: *Design and Construction of Sanitary and Storm Sewers,* chap. 12, "Wastewater and Stormwater Pumping Stations," ASCE Manual and Report 7, New York, 1969.
5. Lister, M.: *The Numerical Solution of Hyperbolic Partial Differential Equations by the Method of Characteristics,* chap. 15, "Mathematical Methods for Digital Computers," vol. I, (eds.) Ralston, A. and H. S. Wilf, Wiley, New York, 1967.
6. Moody, L. F.: Simplified Derivation of Water Hammer Formulas, *Symposium on Water Hammers, ASME,* New York, 1933.
7. Morrison, E. B.: Nomograph for the Design of Thrust Blocks, *Civil Eng.,* vol. 50, no. 6, June 1959.
8. Parmakian, J.: *Water Hammer Analysis,* Prentice-Hall, Englewood Cliffs, N. J., 1955.
9. Seminar Papers on Waste Water Treatment and Disposal, *Pumps, Measuring Devices, Hydraulic Controls,* Boston Society of Civil Engineers, 1961.
10. Standard Pipe Specifications, ANSI A21.1.
11. Standard Pipe Specifications, ANSI A21.50.
12. Standard Pipe Specifications, AWWA C301.
13. Stepanoff, A. J.: *Centrifugal and Axial Flow Pumps,* 2d ed., Wiley, New York, 1956.
14. Stepanoff, A. J.: Elements of Graphical Solution of Water Hammer Problems in Centrifugal Pump Systems, *Trans. ASME,* vol. 71, 1949.
15. Streeter, V. L., and E. B. Wylie: *Hydraulic Transients,* McGraw-Hill, New York, 1967.
16. Wylie, E. B., and V. L. Streeter: *Fluid Transients,* McGraw-Hill, New York, 1978.

APPENDIX

A

CONVERSION FACTORS

Table A-1 Metric conversion factors (SI units to U.S. customary units)

Multiply the SI unit		by	To obtain the U.S. customary unit	
Name	Symbol		Symbol	Name
Acceleration				
meters per second squared	m/s^2	3.2808	ft/s^2	feet per second squared
meters per second squared	m/s^2	39.3701	in/s^2	inches per second squared
Area				
hectare (10,000 m^2)	ha	2.4711	acre	acre
square centimeter	cm^2	0.1550	in^2	square inch
square kilometer	km^2	0.3861	mi^2	square mile
square kilometer	km^2	247.1054	acre	acre
square meter	m^2	10.7639	ft^2	square foot
square meter	m^2	1.1960	yd^2	square yard
Energy				
kilojoule	kJ	0.9478	Btu	British thermal unit
joule	J	2.7778×10^{-7}	kW · h	kilowatt-hour
joule	J	0.7376	ft · lbf	foot-pound (force)
joule	J	1.0000	W · s	watt-second
joule	J	0.2388	cal	calorie
kilojoule	kJ	2.7778×10^{-4}	kW · h	kilowatt-hour
kilojoule	kJ	0.2778	W · h	watt-hour
megajoule	MJ	0.3725	hp · h	horsepower-hour
Force				
newton	N	0.2248	lbf	pound force
Flowrate				
cubic meters per day	m^3/d	264.1720	gal/d	gallons per day
cubic meters per day	m^3/d	2.6417×10^{-4}	Mgal/d	million gallons per day
cubic meters per second	m^3/s	35.3147	ft^3/s	cubic feet per second
cubic meters per second	m^3/s	22.8245	Mgal/d	million gallons per day
cubic meters per second	m^3/s	15,850.3	gal/min	gallons per minute
liters per second	L/s	22,824.5	gal/d	gallons per day
liters per second	L/s	0.0228	Mgal/d	million gallons per day
liters per second	L/s	15.8508	gal/min	gallons per minute

Length				
centimeter	cm	0.3937	in	inch
kilometer	km	0.6214	mi	mile
meter	m	39.3701	in	inch
meter	m	3.2808	ft	foot
meter	m	1.0936	yd	yard
millimeter	mm	0.03937	in	inch
Mass				
gram	g	0.0353	oz	ounce
gram	g	0.0022	lb	pound
kilogram	kg	2.2046	lb	pound
megagram (10^3 kg)	Mg	1.1023	ton	ton (short: 2000 lb)
megagram (10^3 kg)	Mg	0.9842	ton	ton (long: 2240 lb)
Power				
kilowatt	kW	0.9478	Btu/s	British thermal units per second
kilowatt	kW	1.3410	hp	horsepower
watt	W	0.7376	ft/lbf/s	foot-pounds (force) per second
Pressure (force/area)				
pascal (newtons per square meter)	Pa (N/m^2)	1.4504×10^{-4}	lbf/ft^2	pounds (force) per square inch
pascal (newtons per square meter)	Pa (N/m^2)	2.0885×10^{-2}	lbf/ft^2	pounds (force) per square foot
pascal (newtons per square meter)	Pa (N/m^2)	2.9613×10^{-4}	in Hg	inches of mercury (60°F)
pascal (newtons per square meter)	Pa (N/m^2)	4.0187×10^{-3}	in H_2O	inches of water (60°F)
kilopascal (kilonewtons per square meter)	kPa (kN/m^2)	0.1450	lbf/in^2	pounds (force) per square inch
kilopascal (kilonewtons per square meter)	kPa (kN/m^2)	0.0099	atm	atmosphere (standard)
Temperature				
degree Celsius (centigrade)	°C	1.8(°C) + 32	°F	degree Fahrenheit
degree kelvin	K	1.8(K) − 459.67	°F	degree Fahrenheit
Velocity				
kilometers per second	km/s	2.2369	mi/h	miles per hour
meters per second	m/s	3.2808	ft/s	feet per second
Volume				
cubic centimeter	cm^3	0.0610	in^3	cubic inch
cubic meter	m^3	35.3147	ft^3	cubic foot
cubic meter	m^3	1.3079	yd^3	cubic yard
cubic meter	m^3	264.1720	gal	gallon
cubic meter	m^3	8.1071×10^{-4}	acre · ft	acre · foot
liter	L	0.2642	gal	gallon
liter	L	0.0353	ft^3	cubic foot
liter	L	33.8150	oz	ounce (U.S. fluid)

Table A-2 Metric conversion factors (U.S. customary units to SI units)

Multiply the U.S. customary unit		by	To obtain the SI unit	
Name	Symbol		Symbol	Name
Acceleration				
feet per second squared	ft/s²	0.3048[a]	m/s²	meters per second squared
inches per second squared	in/s²	0.0254[a]	m/s²	meters per second squared
Area				
acre	acre	0.4047	ha	hectare
acre	acre	4.0469×10^{-3}	km²	square kilometer
square foot	ft²	9.2903×10^{-2}	m²	square meter
square inch	in²	6.4516[a]	cm²	square centimeter
square mile	mi²	2.5900	km²	square kilometer
square yard	yd²	0.8361	m²	square meter
Energy				
British thermal unit	Btu	1.0551	kJ	kilojoule
foot-pound (force)	ft · lbf	1.3558	J	joule
horsepower-hour	hp · h	2.6845	MJ	megajoule
kilowatt-hour	kW · h	3600[a]	kJ	kilojoule
kilowatt-hour	kW · h	3.600×10^{6}[a]	J	joule
watt-hour	W · h	3.600[a]	kJ	kilojoule
watt-second	W · s	1.000[a]	J	joule
Force				
pound force	lbf	4.4482	N	newton
Flowrate				
cubic feet per second	ft³/s	2.8317×10^{-2}	m³/s	cubic meters per second
gallons per day	gal/d	4.3813×10^{-5}	L/s	liters per second
gallons per day	gal/d	3.7854×10^{-3}	m³/d	cubic meters per day
gallons per minute	gal/min	6.3090×10^{-5}	m³/s	cubic meters per second
gallons per minute	gal/min	6.3090×10^{-2}	L/s	liters per second
million gallons per day	Mgal/d	43.8126	L/s	liters per second
million gallons per day	Mgal/d	3.7854×10^{3}	m³/d	cubic meters per day
million gallons per day	Mgal/d	4.3813×10^{-2}	m³/s	cubic meters per second

Length				
foot	ft	0.3048[a]	m	meter
inch	in	2.54[a]	cm	centimeter
inch	in	0.0254[a]	m	meter
inch	in	25.4[a]	mm	millimeter
mile	mi	1.6093	km	kilometer
yard	yd	0.9144[a]	m	meter
Mass				
ounce	oz	28.3495	g	gram
pound	lb	4.5359×10^2	g	gram
pound	lb	0.4536	kg	kilogram
ton (short: 2000 lb)	ton	0.9072	Mg (metric ton)	megagram (10^3 kilogram)
ton (long: 2240 lb)	ton	1.0160	Mg (metric ton)	megagram (10^3 kilogram)
Power				
British thermal units per second	Btu/s	1.0551	kW	kilowatt
foot-pounds (force) per second	ft · lbf/s	1.3558	W	watt
horsepower	hp	0.7457	kW	kilowatt
Pressure (force/area)				
atmosphere (standard)	atm	1.0133×10^2	kPa (kN/m^2)	kilopascal (kilonewtons per square meter)
inches of mercury (60°F)	in Hg (60°F)	3.3768×10^3	Pa (N/m^2)	pascal (newtons per square meter)
inches of water (60°F)	in H_2O (60°F)	2.4884×10^2	Pa (N/m^2)	pascal (newtons per square meter)
pounds (force) per square foot	lbf/ft^2	47.8803	Pa (N/m^2)	pascal (newtons per square meter)
pounds (force) per square inch	lbf/in^2	6.8948×10^3	Pa (N/m^2)	pascal (newtons per square meter)
pounds (force) per square inch	lbf/in^2	6.8948	kPa (kN/m^2)	kilopascal (kilonewtons per square meter)
Temperature				
degrees Fahrenheit	°F	0.555(°F − 32)	°C	degrees Celsius (centigrade)
degrees Fahrenheit	°F	0.555(°F + 459.67)	K	degrees kelvin
Velocity				
feet per second	ft/s	0.3048[a]	m/s	meters per second
miles per hour	mi/h	4.4704×10^{-1}[a]	m/s	kilometers per second

Table A-2 (continued)

Multiply the U.S. customary unit		by	To obtain the SI unit	
Name	Symbol		Symbol	Name
Volume				
acre-foot	acre-ft	1.2335×10^3	m^3	cubic meter
cubic foot	ft^3	28.3168	L	liter
cubic foot	ft^3	2.8317×10^{-2}	m^3	cubic meter
cubic inch	in^3	16.3871	cm^3	cubic centimeter
cubic yard	yd^3	0.7646	m^3	cubic meter
gallon	gal	3.7854×10^{-3}	m^3	cubic meter
gallon	gal	3.7854	L	liter
ounce (U.S. fluid)	oz (U.S.fluid)	2.9573×10^{-2}	L	liter

[a] Indicates exact conversion.

Table A-3 Conversion factors for commonly used wastewater-treatment plant design parameters

Parameter (in SI units)	To convert, multiply in direction shown by arrows			
	SI units	→	←	U.S. units
Screening				
m^3 screenings/10^3 m^3 wastewater	$m^3/10^3\ m^3$	133.6806	7.4805×10^{-3}	ft^3/Mgal
Grit removal				
Air supply				
m^3 air/m of tank length · min	m^3/m · min	10.7639	0.0929	ft^3/ft · min
Grit removal				
g grit/m^3 wastewater	g/m^3	8.3454	0.1198	lb/Mgal
kg grit/m^3 wastewater	kg/m^3	8345.4	1.1983×10^{-4}	lb/Mgal
Surface overflow rate				
m^3 flow/m^2 surface area · h	m^3/m^2 · h	589.0173	0.0017	gal/ft^2 · d
m^3 flow/m^2 surface area · d	m^3/m^2 · d	24.5424	0.0407	gal/f^2 · d
Volume				
m^3 grit/10^3 m^3 wastewater	$m^3/10^3m^3$	133.6806	7.4805×10^{-3}	ft^3/Mgal
Flow equalization				
Air supply				
m^3 air/m^3 tank volume · min	m^3/m^3 · min	133.6806	7.4805×10^{-3}	$ft^3/10^3$ gal · min
Mixing horsepower				
kW/m^3 tank volume	kW/m^3	5.0763	0.1970	hp/10^3 gal
Sedimentation				
Particle settling rate				
m/h	m/h	3.2808	0.3048	ft/h
m/h	m/h	0.4090	2.4448	gal/ft^2 · min
Sludge scraper speed				
m/h	m/h	0.0547	18.2880	ft/min
Solids loading				
kg solids/m^2 surface area · d	kg/m^2 · d	0.2048	4.8824	lb/ft^2 · d

Table A-3 (continued)

Parameter (in SI units)	SI units	To convert, multiply in direction shown by arrows →	←	U.S. units
Surface overflow rate				
m^3 wastewater/m^2 surface area · d	$m^3/m^2 \cdot d$	24.5424	0.0407	$gal/ft^2 \cdot d$
m^3 wastewater/m^2 surface area · h	$m^3/m^2 \cdot h$	589.0173	0.0017	$gal/ft^2 \cdot d$
Volume of sludge				
m^3 sludge/10^3 m^3 wastewater	$m^3/10^3\ m^3$	133.6806	7.481×10^{-3}	ft^3/Mgal
Weight of dry sludge solids				
g dry solids/m^3 wastewater	g/m^3	8.3454	0.1198	lb/Mgal
Weir overflow rate				
m^3 wastewater/m weir length · d	$m^3/m \cdot d$	80.5196	0.0124	gal/ft · d
Activated sludge				
Aeration device mixing intensity, diffused aeration				
m^3 air/m^3 tank volume · min	$m^3/m^3 \cdot min$	1000.0	0.001	$ft^3/10^3\ ft^3 \cdot min$
Aeration device mixing intensity, mechanical aeration				
kW/10^3 m^3 tank volume	$kW/10^3\ m^3$	0.0380	26.3342	$hp/10^3\ ft^3$
Air flowrate				
m^3 air/h	m^3/h	0.5886	1.6990	ft^3/min
Air requirements, organic removal				
m^3 air/kg BOD_5 removed	m^3/kg	16.0185	0.0624	ft^3/lb
Air requirements, volume of wastewater				
m^3 air/m^3 wastewater	m^3/m^3	0.1337	7.4805	ft^3/gal
Organic load				
kg BOD_5 applied/m^3 aeration-tank volume · d	$kg/m^3 \cdot d$	62.4280	0.0160	$lb/10^3\ ft^3 \cdot d$
Oxygen requirements				
kg O_2/kg BOD_5 applied · d	kg/kg · d	1.0	1.0	lb/lb · d
Oxygen-transfer rate				
kg O_2 transferred/kW · h	kg/kW · h	1.6440	0.6083	lb/hp · h
kg O_2 transferred/m^3 wastewater · h	$kg/m^3 \cdot h$	0.0624	16.0185	$lb/ft^3 \cdot h$

Trickling filters and rotating biological contactors				
Hydraulic load				
m^3 wastewater/m^2 bulk surface area · d	$m^3/m^2 \cdot d$	24.5424	0.0407	$gal/ft^2 \cdot d$
m^3 wastewater/m^2 bulk surface area · h	$m^3/m^2 \cdot h$	589.0173	0.0017	$gal/ft^2 \cdot d$
m^3 wastewater/m^2 bulk surface area · d	$m^3/m^2 \cdot d$	1.0691	0.9354	Mgal/acre · d
L wastewater/m^2 bulk surface area · min	$L/m^2 \cdot min$	35.3420	0.0283	$gal/ft^2 \cdot d$
Organic load				
kg BOD_5/m^3 filter-medium volume · d	$kg/m^3 \cdot d$	62.4280	0.0160	$lb/10^3\ ft^3 \cdot d$
Specific surface loading, hydraulic				
m^3 wastewater/m^2 filter medium surface area · d	$m^3/m^2 \cdot d$	24.5424	0.0407	$gal/ft^2 \cdot d$
m^3 wastewater/m^2 filter medium surface area · d	$m^3/m^2 \cdot d$	0.0170	58.6740	$gal/ft^2 \cdot min$
m^3 wastewater/m^2 filter medium surface area · h	$m^3/m^2 \cdot h$	589.0173	0.0017	$gal/ft^2 \cdot d$
Specific surface loading, organic				
kg BOD_5/m^2 filter medium surface area · d	$kg/m^2 \cdot d$	0.2048	4.8824	$lb/ft^2 \cdot d$
Tank volume				
L/m^2 medium surface area (rotating biological reactor)	L/m^2	2.4542×10^{-2}	40.7458	gal/ft^2
Stabilization ponds and lagoons				
Organic loads				
kb BOD_5/ha surface area · d	kg/ha · d	0.8922	1.1209	lb/acre · d
Volumetric load				
kg BOD_5/m^3 basin volume · d	$kg/m^3 \cdot d$	62.4280	0.0160	$lb/10^3\ ft^3 \cdot d$
Chlorination				
Feed rate				
kg chlorine/d	kg/d	2.2046	0.4536	lb/d
Sludge thickening				
Sludge loading				
kg dry solids fed/m^2 surface area · d	$kg/m^2 \cdot d$	0.2048	4.8824	$lb/ft^2 \cdot d$
Surface overflow rate				
m^3 wastewater/m^2 surface area · d	$m^3/m^2 \cdot d$	24.5424	0.0407	$gal/ft^2 \cdot d$
m^3 wastewater/m^2 surface area · d	$m^3/m^2 \cdot d$	0.0170	58.6740	$gal/ft^2 \cdot min$

Table A-3 (continued)

Parameter (in SI units)	SI units	→	←	U.S. units
	To convert, multiply in direction shown by arrows			
Sludge digestion				
Gas production				
m^3 gas/kg volatile solids fed	m^3/kg	16.0185	0.0624	ft^3/lb
m^3 gas/capita	m^3/capita	35.3147	0.0283	ft^3/capita
Loading rate				
kg BOD_5/m^3 digester volume · d	kg/m^3 · d	62.4280	0.0160	lb/10^3 ft^3 · d
Sludge heating				
W/m^2 surface area · °C	W/m^2 · °C	0.1763	5.6735	Btu/ft^2 · °F · h
Volatile-solids loading				
kg volatile solids/m^3 digester volume · d	kg/m^3 · d	62.4280	0.0160	lb/10^3 ft^3 · d
Sludge drying beds				
Dry-solids loading				
kg dry solids/m^2 area · yr	kg/m^2 · yr	0.2048	4.8824	lb/ft^2 · yr
m^2 area/capita · yr	m^2/capita · yr	10.7639	0.0929	ft^2/capita · yr
Vacuum filtration				
Dry solids				
kg dry solids/m^2 surface area · h	kg/m^2 · h	0.2048	4.8824	lb/ft^2 · h
Pressure applied				
kPa (kN/m^2) pressure	kPa	0.1450	6.8948	lb_f/in^2 (gage)
Sludge feed				
m^3 wet sludge/m^2 surface area · h	m^3/m^2 · h	3.2808	0.3048	ft^3/ft^2 · h
Vacuum applied				
kPa (kN/m^2) vacuum	kPa (kN/m^2)	0.2961	3.3768	in Hg (60°F)
Heat drying				
kJ heat energy required/kg water evaporated (sludge cake)	kJ/kg	0.4303	2.3241	Btu/lb
kg water evaporated/h	kg/h	2.2046	0.4536	lb/h
kg wet sludge/m^2 heating surface · h	kg/m^2 · h	0.2048	4.8824	lb/ft^2 · h

Incineration				
kJ heat energy/kg moisutre evaporated	kJ/kg	0.4303	2.3241	Btu/lb
kg sludge/m² heating surface area	kg/m²	0.2048	4.8824	lb/ft² · h
kg sludge/m³ combustion chamber volume · h	kg/m³ · h	0.0624	16.0185	lb/ft³ · h
Land disposal				
kg mass/ha field area	kg/ha	0.8922	1.1208	lb/acre
bu yield/ha field area · yr	bu/ha · yr	0.4047	2.4711	bu/acre · yr
Mg loading/ha field area	Mg/ha	0.4461	2.2417	tons/acre
m³ wastewater/ha field area · d	m³/ha · d	106.9064	0.0094	gal/acre · d
Surface or in-depth filters				
L wastewater (backwash)/m² surface area · min	L/m² · min	0.0245	40.7458	gal/ft² · min

Table A-4 Values of useful constants

Acceleration due to gravity, g = 9.807 m/s^2 (32.174 ft/s^2)
Standard atmosphere = 101.325 kN/m^2 (14.696 lb_f/in^2)
= 101.325 kPA (1.013 bar)
1 bar = 10^5 N/m^2 (14.504 lb_f/in^2)
Standard atmosphere = 10.333 m (33.899 ft) of water
1 metre head of water (20°C) = 9.790 M/m^2 (1.420 lb_f/in^2)
= 0.00979 N/mm^2 (1.420 lb_f/in^2)
= 9.790 kN/m^2 (1.420 lb_f/in^2)

APPENDIX

B

PHYSICAL PROPERTIES OF WATER

The principal physical properties of water are summarized in Table B-1 in SI units and in Table B-2 in U.S. customary units. They are described briefly below [1].

B-1 SPECIFIC WEIGHT

The specific weight γ of a fluid is its weight per unit volume. In the SI system, it is expressed in kilonewtons per cubic meter. The relationship between γ, ρ, and the acceleration due to gravity g is $\gamma = \rho g$. At normal temperatures γ is 9.81 kN/m^3 or 62.4 lb$_f$/ft^3.

Table B-1 Physical properties of water (SI units)[a]

Temperatue, °C	Specific weight, γ, kN/m^3	Density, ρ, kg/m^3	Modulus of elasticity,[b] $E/10^6$, kN/m^2	Dynamic viscosity $\mu \times 10^3$, N · s/m^2	Kinematic viscosity $\nu \times 10^6$, m^2/s	Surface tension,[c] σ, N/m	Vapor pressure, p_v, kN/m^2
0	9.805	999.8	1.98	1.781	1.785	0.0765	0.61
5	9.807	1000.0	2.05	1.518	1.519	0.0749	0.87
10	9.804	999.7	2.10	1.307	1.306	0.0742	1.23
15	9.798	999.1	2.15	1.139	1.139	0.0735	1.70
20	9.789	998.2	2.17	1.002	1.003	0.0728	2.34
25	9.777	997.0	2.22	0.890	0.893	0.0720	3.17
30	9.764	995.7	2.25	0.798	0.800	0.0712	4.24
40	9.730	992.2	2.28	0.653	0.658	0.0696	7.38
50	9.689	988.0	2.29	0.547	0.553	0.0679	12.33
60	9.642	983.2	2.28	0.466	0.474	0.0662	19.92
70	9.589	977.8	2.25	0.404	0.413	0.0644	31.16
80	9.530	971.8	2.20	0.354	0.364	0.0626	47.34
90	9.466	965.3	2.14	0.315	0.326	0.0608	70.10
100	9.399	958.4	2.07	0.282	0.294	0.0589	101.33

[a] Adapted from Ref. 2.
[b] At atmospheric pressure.
[c] In contact with air.

Table B-2 Physical properties of water (U.S. customary units)[a]

Temperature, °F	Specific weight, γ, lb/ft³	Density,[b] ρ, slug/ft³	Modulus of elasticity,[b] $E/10^3$, lbf/in²	Dynamic viscosity, $\mu \times 10^5$, lb · s/ft²	Kinematic viscosity, $\nu \times 10^5$, ft²/s	Surface tension,[c] σ, lb/ft	Vapor pressure, p_v, lbf/in²
32	62.42	1.940	287	3.746	1.931	0.00518	0.09
40	62.43	1.940	296	3.229	1.664	0.00614	0.12
50	62.41	1.940	305	2.735	1.410	0.00509	0.18
60	62.37	1.938	313	2.359	1.217	0.00504	0.26
70	62.30	1.936	319	2.050	1.059	0.00498	0.36
80	62.22	1.934	324	1.799	0.930	0.00492	0.51
90	62.11	1.931	328	1.595	0.826	0.00486	0.70
100	62.00	1.927	331	1.424	0.739	0.00480	0.95
110	61.86	1.923	332	1.284	0.667	0.00473	1.27
120	61.71	1.918	332	1.168	0.609	0.00467	1.69
130	61.55	1.913	331	1.069	0.558	0.00460	2.22
140	61.38	1.908	330	0.981	0.514	0.00454	2.89
150	61.20	1.902	328	0.905	0.476	0.00447	3.72
160	61.00	1.896	326	0.838	0.442	0.00441	4.74
170	60.80	1.890	322	0.780	0.413	0.00434	5.99
180	60.58	1.883	318	0.726	0.385	0.00427	7.51
190	60.36	1.876	313	0.678	0.362	0.00420	9.34
200	60.12	1.868	308	0.637	0.341	0.00413	11.52
212	59.83	1.860	300	0.593	0.319	0.00404	14.70

[a] Adapted from Ref. 2.
[b] At atmospheric pressure.
[c] In contact with the air.

B-2 DENSITY

The density ρ of a fluid is its mass per unit volume. In the SI system, it is expressed in kilograms per cubic meter. For water, ρ is 1000 kg/m³ at 4°C. There is a slight decrease in density with increasing temperature.

B-3 MODULUS OF ELASTICITY

For most practical purposes, liquids may be regarded as incompressible. The bulk modulus of elasticity K is given by

$$\mathbf{K} = \frac{\Delta p}{\Delta V / V}$$

where Δp is the increase in pressure, which when applied to a volume V, results in a decrease in volume ΔV. For water, K is approximately 2.150 kN/m² at normal temperatures and pressures.

B-4 DYNAMIC VISCOSITY

The viscosity of a fluid μ is a measure of its resistance to tangential or shear stress. Viscosity is expressed in newton seconds per square meter in the SI system.

B-5 KINEMATIC VISCOSITY

In many problems concerning fluid motion, the viscosity appears with the density in the form μ/ρ, and it is convenient to use a single term ν, known as the kinematic viscosity and expressed in square meters per second or strokes in the SI system. The kinematic viscosity of a liquid diminishes with increasing temperature.

B-6 SURFACE TENSION

Surface tension is the physical property that enables a drop of water to be held in suspension at a tap, a glass to be filled with liquid slightly above the brim and yet not spill, or a needle to float on the surface of a liquid. The surface-tension force across any imaginary line at a free surface is proportional to the length of the line and acts in a direction perpendicular to it. The surface tension per unit length σ is expressed in newtons per meter. There is a slight decrease in surface tension with increasing temperature.

B-8 VAPOR PRESSURE

Liquid molecules that possess sufficient kinetic energy are projected out of the main body of a liquid at its free surface and pass into the vapor. The pressure exerted by this vapor is known as the vapor pressure p_v. The vapor pressure of water at 15°C is 1.72 kN/m^2.

REFERENCES

1. Webber, N. B.: *Fluid Mechanics for Civil Engineers,* SI ed., Chapman and Hall, London, 1971.
2. Vennard, J. K., and R. L. Street: *Elementary Fluid Mechanics,* 5th ed., Wiley, New York, 1975.

APPENDIX

C

MINOR LOSSES IN CLOSED CONDUITS

The purpose of this appendix is to present the necessary information for calculating minor losses for the most common pipeline appurtenances. The information presented in this appendix is for appurtenances equal to or larger than 100 mm (4 in) and not for plumbing fixtures. The information is presented in tabular form with an illustration for each entry. Unless a different formula is given, all head losses are evaluated by using the relationship $h_L = K(V^2/2g)$, and the value given in the table is for the coefficient K. The velocity V is the average fluid velocity in the pipe adjacent to the fitting. Where there might be some question, the correct velocity for use in the calculation is shown in the illustration. Where appropriate, references are cited for each entry in the table. For additional information, the reader should consult the references listed at the end of this appendix or any other text or handbook on hydraulics.

Table C-1 Information for calculating minor losses for common pipeline appurtenances

Cause of minor loss	K value or loss expression	Reference
Gate valve Fig. C-1 (d, D)	see gate valve table below. For larger sizes, values for 300-mm valve may be used.	2
Butterfly valve Fig. C-2 (θ)	see butterfly valve table below. Values of K depend on size of valve and design of disk. At wide-open position, K may vary from 0.3 to 1.3. Values of K at part-open positions vary widely. Consult manufacturer's catalog for values for a specific valve. The K values given are representative.	3, 5

Gate valve (d/D):

D, mm	in	1/8	1/4	3/8	1/2	3/4	1
50	2	140	20	6.5	3.0	0.68	0.16
100	4	91	16	5.6	2.6	0.55	0.14
150	6	74	14	5.3	2.4	0.49	0.12
200	8	66	13	5.2	2.3	0.47	0.10
300	12	56	12	5.1	2.2	0.47	0.07

For larger sizes, values for 300-mm valve may be used.

Butterfly valve:

ϕ	5°	10°	20°	30°	40°	50°	60°	70°
K	0.24	0.52	1.54	3.91	10.8	32.6	118	751

Values of K depend on size of valve and design of disk. At wide-open position, K may vary from 0.3 to 1.3. Values of K at part-open positions vary widely. Consult manufacturer's catalog for values for a specific valve. The K values given are representative.

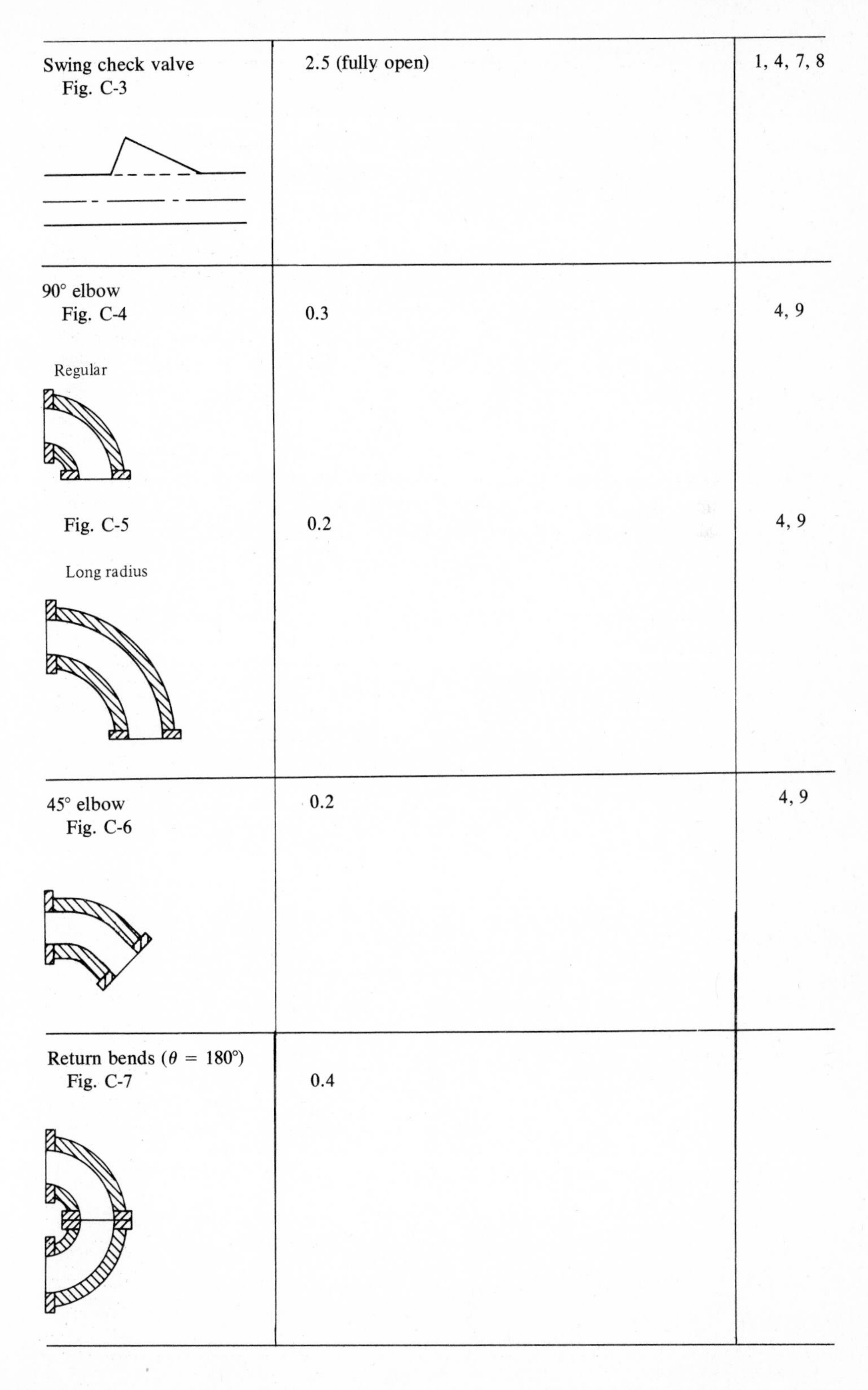

Swing check valve Fig. C-3	2.5 (fully open)	1, 4, 7, 8
90° elbow Fig. C-4 Regular	0.3	4, 9
Fig. C-5 Long radius	0.2	4, 9
45° elbow Fig. C-6	0.2	4, 9
Return bends (θ = 180°) Fig. C-7	0.4	

Table C-1 (continued)

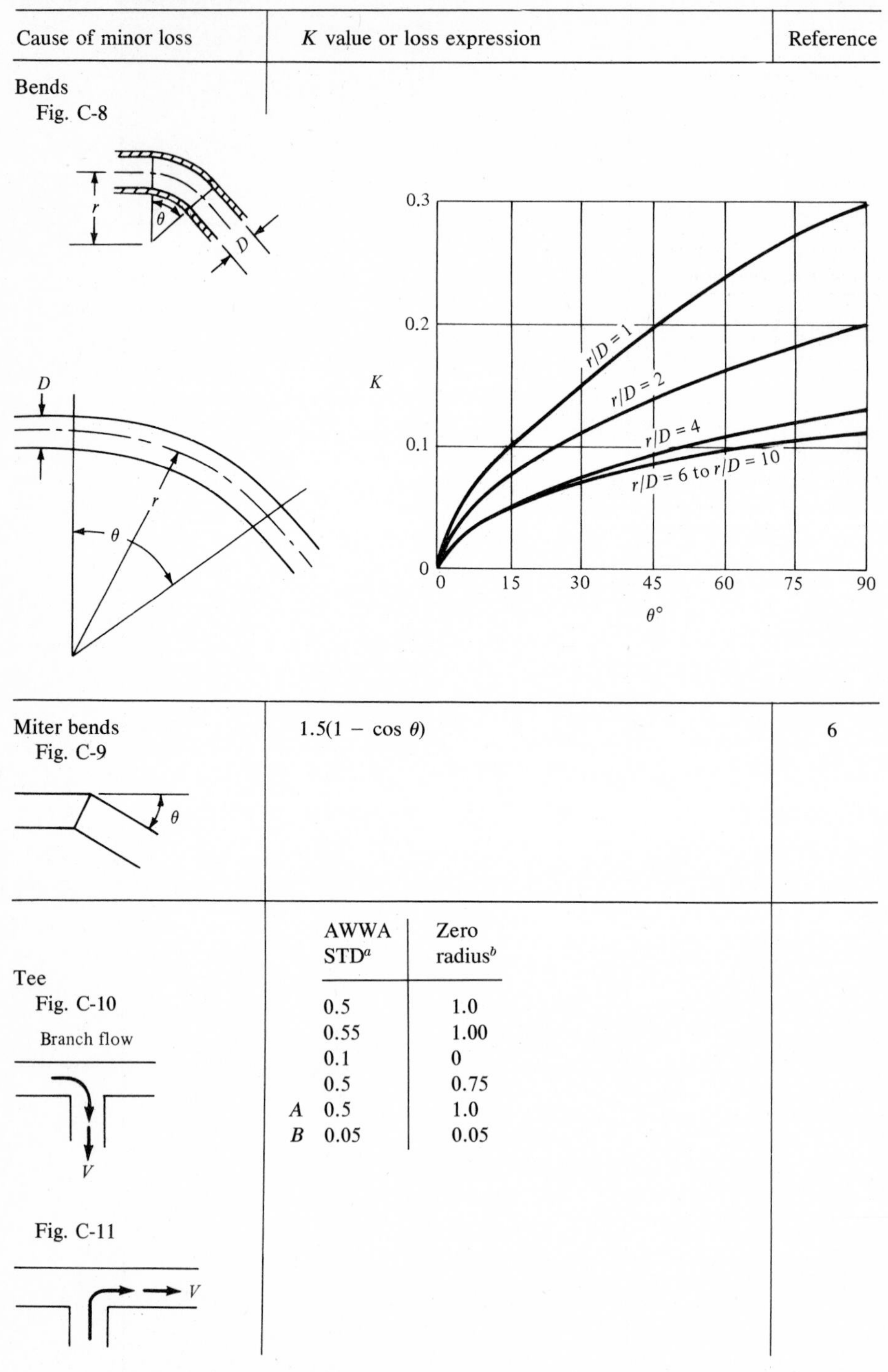

Cause of minor loss	K value or loss expression			Reference
Bends Fig. C-8				
Miter bends Fig. C-9	$1.5(1 - \cos \theta)$			6
		AWWA STD[a]	Zero radius[b]	
Tee Fig. C-10 Branch flow		0.5	1.0	
		0.55	1.00	
		0.1	0	
		0.5	0.75	
	A	0.5	1.0	
	B	0.05	0.05	
Fig. C-11				

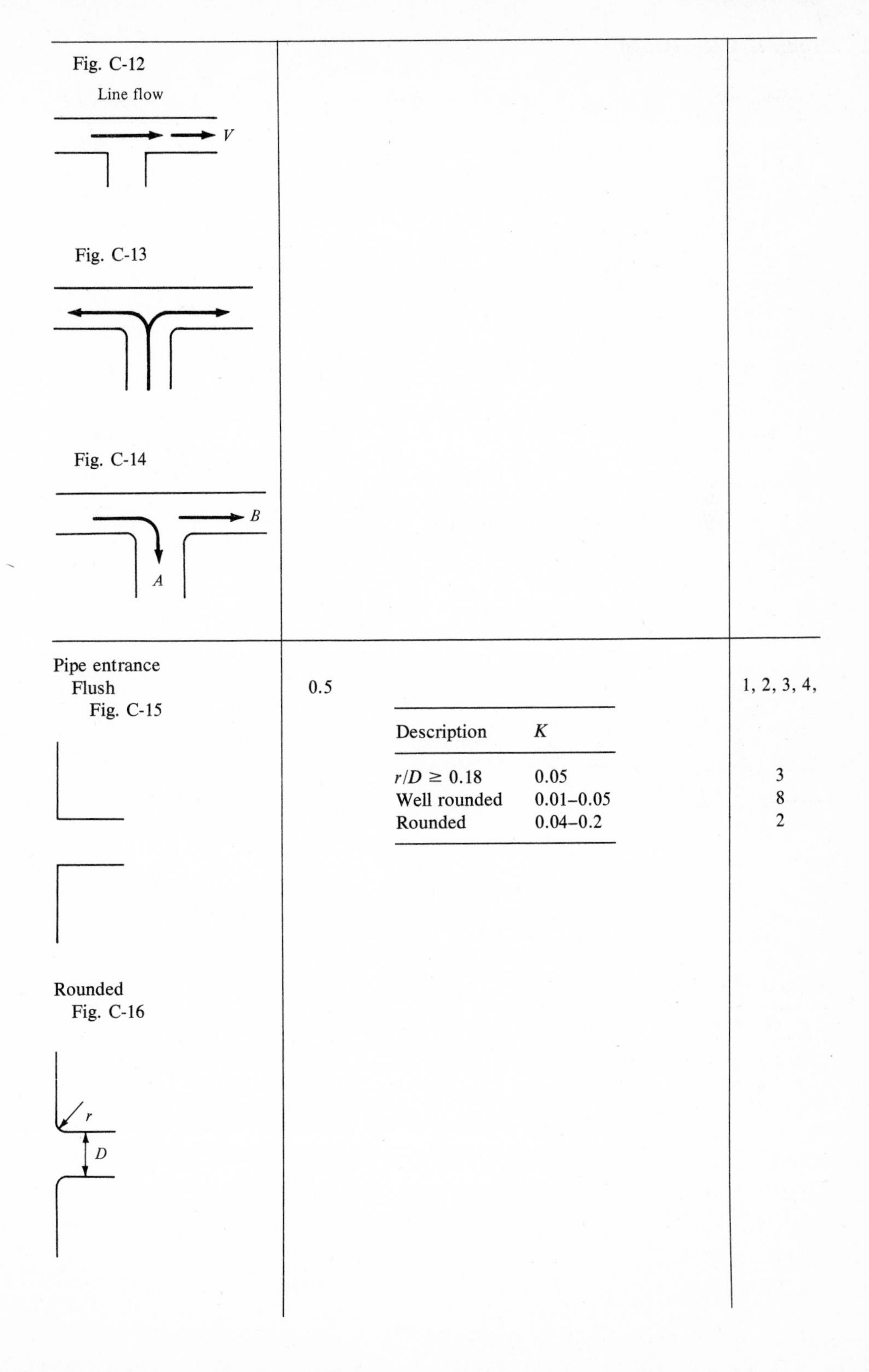

Fig. C-12 Line flow		
Fig. C-13		
Fig. C-14		
Pipe entrance Flush Fig. C-15	0.5	1, 2, 3, 4,
Rounded Fig. C-16		

Description	K	
$r/D \geq 0.18$	0.05	3
Well rounded	0.01–0.05	8
Rounded	0.04–0.2	2

Table C-1 (continued)

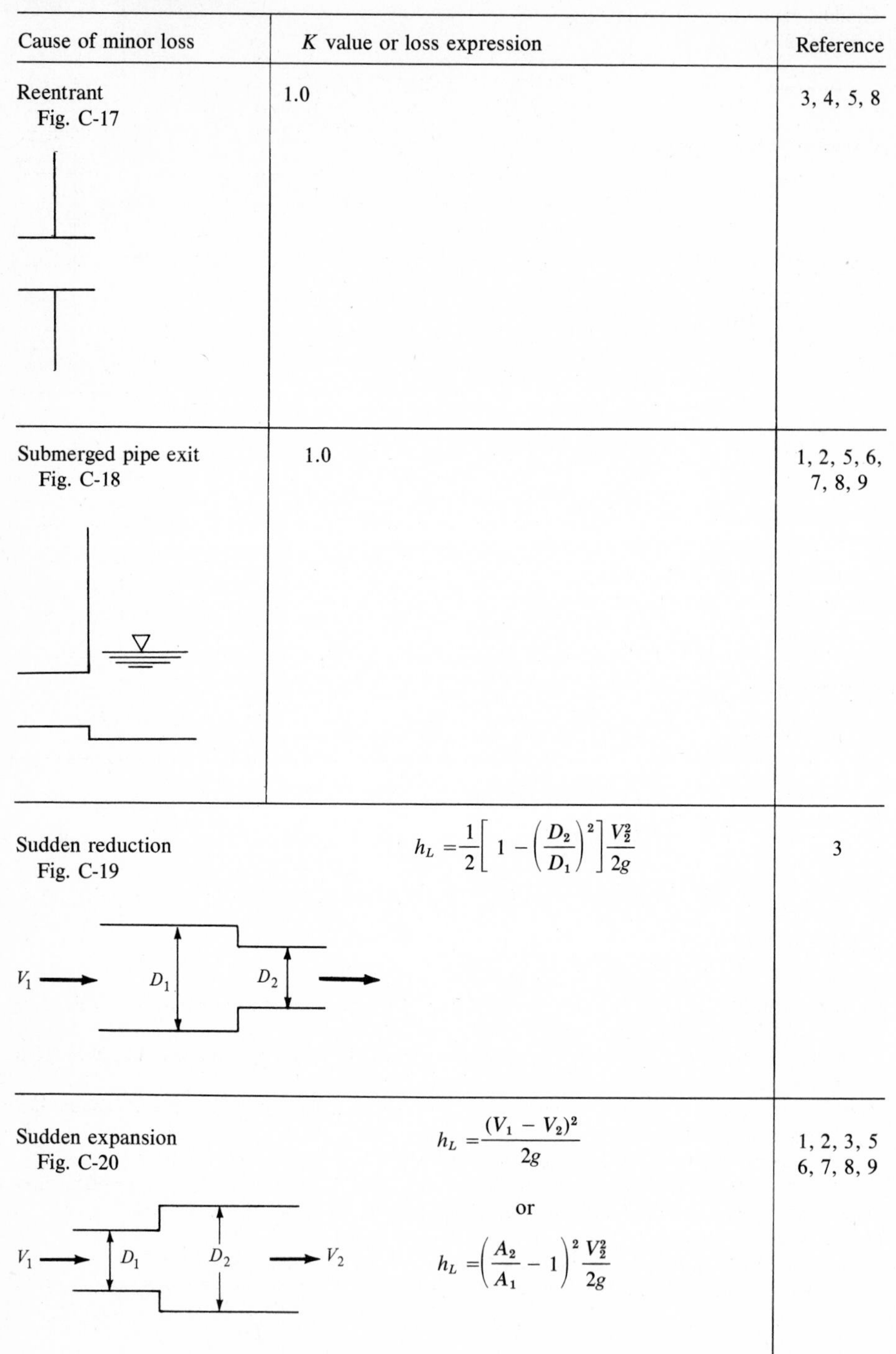

Cause of minor loss	K value or loss expression	Reference
Reentrant Fig. C-17	1.0	3, 4, 5, 8
Submerged pipe exit Fig. C-18	1.0	1, 2, 5, 6, 7, 8, 9
Sudden reduction Fig. C-19	$h_L = \frac{1}{2}\left[1 - \left(\frac{D_2}{D_1}\right)^2\right]\frac{V_2^2}{2g}$	3
Sudden expansion Fig. C-20	$h_L = \frac{(V_1 - V_2)^2}{2g}$ or $h_L = \left(\frac{A_2}{A_1} - 1\right)^2 \frac{V_2^2}{2g}$	1, 2, 3, 5 6, 7, 8, 9

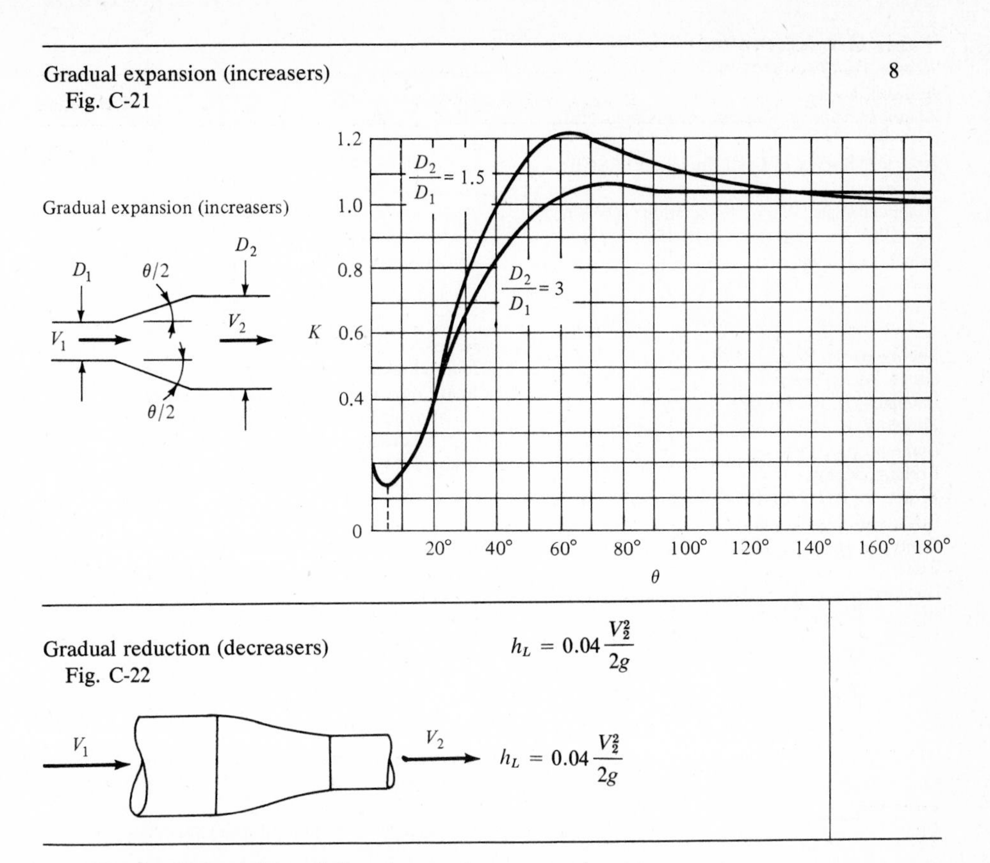

[a]Use ductile-iron pipe, reinforced-concrete pressure pipe, etc.
[b]Use for welded-steel pipe.

REFERENCES

1. Albertson, M. L., J. R. Barton, and D. B. Simons: *Fluid Mechanics for Engineers,* Prentice-Hall, Englewood Cliffs, N. J., 1964.
2. Brater, E. F., and H. W. King: *Handbook of Hydraulics,* 6th ed., McGraw-Hill, New York, 1975.
3. Degremont: *Water Treatment Handbook,* Stephen Austin and Sons, Ltd., Caxton Hill, Hertford, England, 1973.
4. FMC Corporation: *Hydraulics and Useful Information,* 1973.
5. Gibson, A. H.: *Hydraulics and Its Applications,* Constable and Company, Ltd., London, 1961.
6. Jaeger, C.: *Engineering Fluid Mechanics,* Blackie and Son, Ltd., London, 1961.
7. Simon, A.: *Practical Hydraulics,* Wiley, New York, 1976.
8. Streeter, V. L., and E. B. Wylie: *Fluid Mechanics,* McGraw-Hill, New York, 1967.
9. Vennard, J. K., and R. L. Street: *Elementary Fluid Mechanics,* 5th ed., Wiley, New York, 1975.

NAME INDEX

SUBJECT INDEX

Table 1 Base units in the international system of units (SI)

Quantity	Name	Symbol
Length	meter	m
Mass	kilogram	kg
Time	second	s
Electric current	ampere	A
Thermodynamic temperature	kelvin	K
Amount of substance	mole	mol
Luminous intensity	candela	cd
Plane angle*	radian	rad
Solid angle*	steradian	sr

*Supplementary units

Table 2 Derived SI units with special names

Quantity	SI unit symbol	Name	Units
Frequency	Hz	hertz	1/s
Force	N	newton	$kg \cdot m/s^2$
Pressure, stress	Pa	pascal	$kg/m \cdot s^2$ or N/m^2
Energy or work	J	joule	$kg \cdot m^2/s^2$ or $N \cdot m$
A quantity of heat	J	joule	$kg \cdot m^2/s^2$ or $N \cdot m$
Power, radiant flux	W	watt	$kg \cdot m^2/s^3$ or J/s
Electric charge	C	coulomb	$A \cdot s$
Electric potential	V	volt	$kg \cdot m^2/s^3 \cdot A$ or W/A
Potential difference	V	volt	$kg \cdot m^2/s^3 \cdot A$ or W/A
Electromotive force	V	volt	$kg \cdot m^2/s^3 \cdot A$ or W/A
Capacitance	F	farad	$A^2 \cdot s^4/kg \cdot m^2$ or C/V
Electric resistance	Ω	ohm	$kg \cdot m^2/s^3 \cdot A^2$ or V/A
Conductance	S	siemens	$s^3 \cdot A^2/kg \cdot m^2$ or A/V
Magnetic flux	Wb	weber	$kg \cdot m/s^2 \cdot A$ or $V \cdot s$
Magnetic flux density	T	tesla	$kg/s^2 \cdot A$ or Wb/m^2
Inductance	H	henry	$kg \cdot m^2/s^2 \cdot A^2$ or Wb/A
Luminous flux	lm	lumen	$cd \cdot sr$
Illuminance	lx	lux	$cd \cdot sr/m^2$ or lm/m^2
Activity (radionuclides)	Bq	becquerel	l/s
Absorbed dose	Gy	gray	m^2/s^2 or J/kg